History of Women in the Sciences

READING S FROM *ISIS*

History of Women in the Sciences

READINGS FROM *ISIS*

EDITED BY

SALLY GREGORY KOHLSTEDT

The University of Chicago Press

Chicago and London

The essays in this volume originally appeared in *Isis*. Acknowledgment of the original publication date can be found on the first page of each article.

The University of Chicago Press, Chicago, 60637
The University of Chicago Press, Ltd., London
© 1976, 1978, 1980, 1982, 1984, 1986, 1987, 1988, 1991, 1993, 1994, 1997 by the History of Science Society
© 1999 The University of Chicago
All rights reserved. Published in 1999
Printed in the United States of America
ISBN: (cl)   0-226-45069-4
ISBN: (pa)   0-226-45070-8

00 99 98 97 96   5 4 3 2 1

Library of Congress Cataloging in Publication Data

History of women in the sciences : readings from Isis / edited by Sally Gregory Kohlstedt.
    p.   cm.
  Includes bibliographical references and index.
  ISBN 0-226-45069-4.—ISBN 0-226-45070-8 (pbk.)
  1. Women in science.   I. Kohlstedt, Sally Gregory, 1943–   .
II. Isis.
Q130.H58   1999
508.2—dc21   99-14693
                                                          CIP

CONTENTS

Cover image courtesy of Harvard College Observatory.

Introduction

Sally Gregory Kohlstedt

WOMEN HAVE ALWAYS investigated their world, exploring, analyzing, and using what they discover about the living and nonliving elements around them. They have shared their knowledge and have inevitably been part of the enterprise that became western science, however obscure their participation has become in the historical record. This volume presents a few glimpses of women's involvement in the scientific world and reveals the ways in which historians have skillfully sought and recovered the stories of women's involvement in science. Fundamental shifts within the history and philosophy of science have influenced some dimensions of the history of women in science, while feminist studies underscore the importance of rethinking the history of science, technology, and medicine. There is much more to be done, and these essays only begin to demonstrate how analyses of language, metaphor, and meaning may be combined with concern about class, race, gender, and ethnicity to open new historical arenas of scientific activity for closer scrutiny.

Women's history and research on gender have arguably been in the vanguard of the effort to understand and to teach ways in which sustainable science and its practices rely heavily on the surrounding culture. Class privilege has typically been identified with scientific work that, in turn, involves the support and enthusiasm of others who hold resources of time, money, and cultural imprimatur; but the boundaries prove permeable. Studying women in science also has meant acknowledging the ways in which domestic life fosters or inhibits scientific inquiry and the importance of collaboration and complementarity in framing and implementing experimental investigations. The stories of women in science pull us back into frameworks where the conditions nearly defy our imagination and where patterns of accommodation and resistance can seem peculiar and foreign. At the same time these stories capture moments that resonate with a spike of recognition or constitute a point of continuity between past and present.

This volume contains essays written during the past 15 or so years that highlight most of the important themes that characterize the study of women in science. The journal *Isis* has played an important role in fostering this scholarship, as has the *British Journal for the History of Science,* which has published pioneering essays on aspects of gender and science. The early scholarship of the 1970s, which emphasized the discovery and recovery of individual women, has yet to accomplish its aspirations, but the rediscovery persists within the changing historiographical and theoretical frameworks of feminist studies and within the history of science. Margaret Rossiter has investigated accounts of women in American science at multiple layers of the scientific enterprise and in every discipline. As a result, her comprehensive books provide comparative lenses through which to view the territorial and hierarchical segregation of women in American science and elsewhere. Biographical essays in this volume are richly documented and present detailed accounts that underscore the gender gap in science, the role of outsider frequently ascribed to women, the arbitrary authority that often denied qualified women access to facilities, and the personal pain of such exclusions. At the same time, the accounts represent the capacity for multitasking that served women in science well, the ways in which knowledge offered from a patronizing stance could be turned to income and inspiration, and the multiple measures to assess women's accomplishments within science. Primarily organized chronologically, these essays pursue basic themes in the history of science as they intersect with women's participation. In some cases, the linkages reinforce familiar patterns, such as the growing access to formal scientific education in the nineteenth century and the importance of formal scientific institutions. In other cases, the authors offer distinctive and unpredicted accounts through their feminist reappraisals of older texts in botany or by recounting the curiously circumvented career paths of women in astronomy and physiology.

In many ways, Carolyn Merchant's opening essay constitutes a call to action. She deliberately juxtaposes the positivist ideals of the widely acknowledged founder of the discipline of the history of science, George W. Sarton, with the symbolic patron he chose for the journal—*Isis*. Cognizant of the irony involved, Merchant demonstrates how *Isis* now serves its mistress as a disseminator of interdisciplinary research of a quite different mode than its founder envisioned. Merchant's call for a feminist history of science begins with a challenge to simplistic assumptions about objectivity and an appeal to look closely at language and symbol within scientific work. Building on that account, the essay proposes a feminist agenda that stresses the history of women in science that not only celebrates their achievements but also documents their often formidable obstacles. The agenda of this essay resonated among contemporaries when at first appeared in the early 1980s and is often cited by a generation of scholars who, whether they agree or disagree in detail, recognize the foresight of Merchant's own important revisionist account of the scientific revolution in *The Death of Nature*.

A recently written, prize-winning essay takes us back to Elizabethan England, where natural philosophy was literally embedded in domesticity. Deborah E. Harkness traces the dynamics in Jane Fromond Dee and John Dee's household, the latter perhaps the most renowned philosopher of his generation. Science was conducted in the negotiated space between the public and private sites of a complex household near London that contained family, servants, scientific assistants, and important visitors. While the domestic setting worked remarkably well in many ways, the efficiencies were offset by tensions surrounding the ambiguity of authority in overlapping domains between husband and wife. The subsequent movement of science to the more familiar sites of the laboratory and academy, Harkness argues, may well be related to the generation of natural philosophers following Dee who sought greater autonomy for their scientific practices and conversations.

Another kind of direct and learned partnership took place in a quasi-domestic setting a century later, one in which domestic tasks were clearly segregated and managed apart from the deliberately scientific facilities within the Academy of Sciences in Berlin. Londa Schiebinger's account of astronomer Maria Winkelmann demonstrates how a determined young woman might excel in astronomy in the seventeenth century if she had the right family connections—in Winkelmann's case, marriage to Germany's leading astronomer, Gottfried Kirch. Their subsequent collaboration and her own independent observations and calculations brought Winkelmann satisfaction, recognition, and remuneration. Finally, however, other leading scientists, concerned about the precedent of appointing a woman to fill the Academy post vacated by her husband's death, refused Winkelmann's request to continue the work in which she and her husband collaborated. Schiebinger argues that this and similar decisions in other relatively young, learned societies by the early eighteenth century marked a strong turning point away from science in domestic settings. Challenges to women's participation and to public acknowledgment of their contributions took their cue from the increasingly insistent polarity of public and private spheres.

As this collection of essays continuously reminds us, however, historical context is critical. Location in time as well as in geographical and cultural settings must always be taken into account when studying the lives of women in science. Italy, with its long tradition of intellectual women within the aristocracy, provides a fascinating exception to the rest of Europe with the appointment of Laura Bassi as a lecturer at the University of Bologna during the Enlightenment era. Even in Italy, however, Bassi evidently was a "curiosity" rather than the norm, a well-positioned woman who skillfully took advantage of the ambiguities surrounding her special status and who consciously, if subtly, pushed against barriers. It was her talent and female presence that served to bolster Bologna's academic aspirations, and she never freed herself from the identifying markers of gender. The accolades that accumulated bore a curious relationship to her still somewhat obscure contributions, according to essayist Paula Findlen. Despite being pulled into a genuine public life of science, Bassi's individual written achievements were few. Thus the very strategies that marked her achievement may have subtly undermined her capacity to publish work that might have recorded her intellect for posterity.

Unlike most sciences, botany began to be framed for and by women in the late eighteenth century. It was, of course, a domain that women had long understood through their work with food preparation and herbal medicine, but the scientific niche gained prestige with the new systematics. Carl Linnaeus's taxonomic system made botany and zoology an increasingly popular pastime to be pursued by individuals and groups with the help of field guides and handbooks. Contemporaries, however, found that the "sexual system" that relied on identification of stamen and pistils was problematic as it raised questions about whether polite female society should be exposed to such discussions. In her probing account of Erasmus Darwin, Janet Browne finds that his epic poem, *The Love of Plants,* provides important commentary on contemporary gender; in significant ways, Browne's essay reconfigures Merchant's discussion of gendered language embedded in the sciences and its impact on women engaging in scientific study. *The Love of Plants,* with its dramatic metaphors about the sex life of flowers, extends a didactic analogy that is intended for women readers. Tracing out the plant "characters" in the poem reveals a preoccupation with morality, particularly in managing the numbers of pistils and stamen from representative orders and dealing with the inevitable promiscuity involved. Browne argues that this account directed at "lady readers" was more than a teaching tool for botany; it was a way of reminding female readers of a dominant masculine view of female behavior. The scientific metaphor gained a weighty significance at the opening of an era sometimes designated the "century of science." Still, the participation of women in botanical science, as suggested by Darwin's assumptions that ladies would be the primary audience, indicates an important and expanding dimension of popular science.

In response to the writings of Erasmus Darwin, the prominent European man of letters, Johann von Goethe, penned a volume of verse, *The Metamorphosis of Plants,* with a somewhat different, although still paternal and occasionally sexual, outlook. As Lisbet Koerner points out, Goethe perceived himself as a ladies' man and a science teacher, who had clear, but limited, notions about women's capacity for doing intellectual work. In his view, women were in some essential way linked to nature, but female familiarity and empathy were never scientific. Nowhere was this more evident than in his acceptance of the gardening skills of his lover and later wife, Christiane Vulpius, who nonetheless remained in "carefully managed female ignorance" of science. Koerner's highly interpretative account demonstrates the extensive popular attraction of the study of botany in the early nineteenth century while indicating the limits of this highly socialized enterprise for those women attracted to it.

Picking up the theme of popular science, M. Susan Lindee moves beyond the tendency of early nineteenth century male naturalists to treat women as an audience in order to reveal the ways in which women produced texts that were popular among men, women, and children. Textbook writing became, in some subjects, a highly profitable way for middle-class women to earn a living. For example, Jane Marcet's *Conversations on Chemistry* was among the earliest and arguably one of the most successful textbooks. Testing the demands of audience, Lindee examines Marcet's continuously updated textbook to show that it kept up-to-date in terms of theory as it went through twenty-three editions and provided workable experiments through which students acquired skill and textbook knowledge. The fact that *Conversations on Chemistry* was widely used by young women in North America attests to the importance of that science among those leading the women's academies and collegiate education. Science held value for a variety of reasons, including religious and domestic justifications, but the actual instruction in Marcet's best-selling chemistry text was quite definitely on theory and experiment.

American historians have argued that one important reason why girls and women had access to popular scientific topics was because mothers were responsible for the early education of future citizens. As a result, whether at home, in small classrooms, or in public lecture halls, women gained access to science. For the first half of the nineteenth century, science was a popular entertainment and a badge of intellectual acuity that found its place in a range of public and private spaces accessible to women. Thus, amateur science served, among other things, as a cradle for the early training of a middle-class generation who would make their living in science and move American science ever higher in international rankings. Sally Gregory Kohlstedt argues that parlors, primers, and public schooling provided multiple levels of direct experience with scientific ideas and apparatus that could amuse and advance those girls and women with aptitude and serve as foundation for further study.

In a classic early essay that suggested that girls' academies in the antebellum period may have even surpassed some boys' academies in contemporary science and language study, Deborah Jean Warner investigates the study of curriculum, texts, and scientific equipment. Her analysis of leading academies and colleges for young women concludes that their holdings and instruction in science were substantive and extensive. This concentration of scientific books and instruments suggests that a more modern curriculum was possible for young women who would never need the classics for gaining access to prestigious men's colleges. Astronomy, chemistry, natural philosophy, as well as natural history were all part of the curriculum in a system that prepared at least some women to teach science, write textbooks, and engage in scientific investigations. This tradition continued in the women's colleges founded during and after the Civil War.

Even as educational opportunities expanded in the United States, women touched by the waves of revolution and themes of exploitation across the Atlantic also sought higher education, some hoping that science would afford more sure answers to important questions. Ann Hibner Koblitz traces the careers of several radical young women, members of the Russian intelligencia, who left during the tsarist repression of the 1860s and found opportunities in Western Europe—especially at the University of Zurich. Not only did they gain access to higher education but a few also gained doctorates in medicine and science during the exciting, open decade between 1863 and 1873. Some of these women went on to realize exceptional achievements and established a useful precedent for other European and American women in later decades. In the unstable political era in Russia, however, these ambitious women graduates found few employment opportunities.

While the emphasis on formal education was important, it should not obscure the fact that much science continued to be learned and practiced outside academic settings. Peggy Aldrich Kidwell's account of British women astronomers between 1880 and 1930 points out their obstacles as well as their opportunities. Some women attended classes informally, did routine work in observatories, performed some night-time observations, attended professional meetings, and participated in observation expeditions. None of this access was ever guaranteed, however, and opportunities could erode. Kidwell observes that a period of increasing opportunities in astronomy for women into the first decade of the twentieth century was subsequently followed by a decline in later decades in terms of actual numbers and proportionate representation of women. Brief sketches of the careers of individual women, noting struggles and achievements, give some clues, but much remains to be studied about individual disciplines in relationship to women members.

The fact that British women, particularly those of the upper classes, continued to participate in science is demonstrated by the opportunities at the Balfour Biological Laboratory for Women at Cambridge University from 1884 to 1914. Marsha L. Richmond traces the complex web of interactions among women privileged to work in such independent laboratory subcultures. Cambridge was the undisputed center for life science research, and the Balfour Laboratory, which had access to certain classes taught by leading male faculty in the natural sciences, essentially provided an unprecedented opportunity for young women who had the means to attend the Cambridge women's colleges. The founders, moreover, hired newly qualified women scientists to direct students, thus providing employment opportunities, role models, and research facilities in which the instructors could pursue their own investigations. Given the emphasis on team projects, the women enjoyed a nurturing scientific subculture that fostered their individual capabilities. Still, the opportunities were never equivalent to those of the men at Cambridge who attended classes, had well-established male scientists as mentors, and gained access to more advanced laboratories. The early years of advancement were followed by some setbacks before the 1914 decision to admit the women of Newnham and Girton Colleges to

all-University classes. The issue of sex segregation, with its advantages of camara-
derie and mutual respect and its limitations of isolation and second-class environ-
ments, is a theme that runs through this and other essays.

The intense debate about women's access to higher education was linked to ideas
about sex and race—topics that were rooted in scientific theory in the nineteenth
century. Nancy Leys Stepan argues that race and sex served as polarities in evolutionary thinking in late nineteenth-century natural sciences. A race-sex analogy codified eighteenth-century ideas about human differences as well as similarities in the
reported connections made between white women and Africans. While the differences had been "obvious" to Victorians, the analogy linking so-called lower races
and women also gained credence in anthropomorphic, medical, and biological discourse. Stepan's early and relatively short "critique and contention" paper is a milestone in subsequent discussions of gender and science that has moved in multiple
directions among feminist philosophers, historians, and biologists.

As women gained skills and challenged stereotypes about their capacity for
science, they entered the employment marketplace with optimism, intending to challenge barriers in employment just as they had those in higher education. The segregation that persisted in parts of higher education were reflected in a distinctive set
of employment options as well. Margaret W. Rossiter discusses "women's work" in
science at the turn of the century in both well-established fields like astronomy as
well as in the newly created field of home economics. In the former example, the
need for staff to do analytical work and calculations based on photographic plates
generated by the new generation of astrophysicists created employment for women
with talent and ambition. Many came from women's colleges prepared to do this as
entry-level "computing" work temporarily only to find that, unlike young men who
would become observers, assistants, and finally even directors of observatories, they
were unlikely to move up any career ladder in astronomy. Labeling the sex-typing
of certain research tasks as hierarchical segregation, Rossiter points to the innovative
new field of home economics as territorial segregation. Founded by Ellen Swallow
Richards, who had a chemistry degree from Massachusetts Institute of Technology,
the new field initially emphasized research on pure foods and nutrition and gradually
took on more responsibility for educational programs that would be useful both in
institutional settings and in [the] homes. The two strategies—working lower in the
hierarchy or creating entire new fields of expertise—gave women opportunities and,
thus, had immediate advantages. Both, however, became familiar, indeed limiting,
patterns in the twentieth century.

By the late nineteenth century, women had joined the scientific ranks and were
educated within or parallel to the growing academic scientific disciplines. Given the
importance of women's health and interest in women's bodies, it is not surprising
that women were influential in physiological teaching and research at the turn of the
century. Toby A. Appel's "Physiology in American Women's Colleges" perceptively
analyzes the women physiologists at five women's colleges, while simultaneously
following the transformation of the field within this context. Here again, the environment of sex education in women's higher education allowed women to create their
own subculture, defining physiology in terms of personal hygiene and advocating
for women's health in the 1860s and 1870s. While debates about the suitability of
women's minds and bodies for education attracted public attention and debate, the
physiology and hygiene courses offered a place for discussion and critique by

women themselves. By the end of the century, as biomedicine took over physiology, the programs at women's colleges also became more experimental. The faculty skillfully integrated research and teaching, but they and their students were little recognized, although a few women proved exceptional. The influence persisted, and, as late as 1950, a majority of women in physiology had graduated from one of the country's liberal arts colleges for women.

The lack of credit for work accomplished, whether masked by joint publications, explicit exclusion, or simple neglect, has been a theme among those who first investigated pioneering women scientists. Stephen G. Brush's note on Nettie Stevens's work on chromosomes covers one tantalizing case in point where the oversight of peers played some role. More recently, Margaret R. Wright has analyzed Stevens's contemporary, Marcella O'Grady Boveri, citing her as an example where collaboration, family responsibilities, and commitment to education framed a woman's choices over a life-long course. Like other recent work on women in science, Wright's account deals explicitly with the multiple threads women often weave into their lives, incorporating collaboration and scientific work with family life and educational service. O'Grady moved along the fast track after a degree from the Massachusetts Institute of Technology, graduate work at Bryn Mawr College (under direction from faculty at Johns Hopkins University), and research at the Marine Biological Laboratory at Woods Hole, Massachusetts. A position at Vassar College in 1890 gave her intellectual colleagues and eager students, but she took a leave in 1896 to pursue more graduate work at the University of Würzburg. There she married Theodor Boveri, the head of the Zootomical Institut—a man previously skeptical about the capacity of women for science—and became his partner in research. Even O'Grady Boveri's final, or third-stage career as a faculty member at Albertus Magnus College included research along with other teaching responsibilities. The ebb and flow of individual women's scientific activities becomes another theme in the literature, one that belies any assumption that only younger scholars produce significant results.

The essays in this volume represent significant and pioneering research on women in science, written during the past two decades while women have been entering science and engineering careers in unprecedented numbers. Collectively, these essays spotlight significant, albeit intermittent, views of the history of women in western science since the seventeenth century that are yet to be captured in traditional textbooks. The authors mark the complex patterns of gender on the scientific landscape and document the persisting quest of women to gain access and to contribute to the scientific enterprise—although it should be noted that women also have been well represented among the skeptics of science and technology. Readers will find the repetition and continuities of such themes as women's exclusion, lack of recognition, and efforts to balance scientific work with family life that persist throughout modern western science. Simultaneously here are themes about the creativity and resilience of women whose strategies included confrontation, assimilation, and reconstituting their domestic and scientific aspirations in ways appropriate to their historical location. History of science is viewed differently from a subaltern position.

The author thanks several people, particularly Olivia Walling, Margaret Rossiter, and Donald Opitz. She also is grateful to the contributors for their original work and their readiness to participate.

RELATED BOOKS ON WOMEN AND SCIENCE

Abir-Am, Pnina, and Dorinda Outram. 1987. *Uneasy Careers and Intimate Lives: 1789–1979.* New Brunswick, NJ: Rutgers University Press.

Ainley, Marianne G. 1990. *Despite the Odds: Essays on Canadian Women and Science.* Montreal: Vehicule Press.

Bleier, Ruth. 1984. *Science and Gender: A Critique of Biology and Its Theories on Women.* New York: Pergamon Press.

Bleier, Ruth, ed. 1986. *Feminist Approaches to Science.* New York: Pergamon Press.

Cadden, Joan. 1993. *The Meanings of Sex Difference in the Middle Ages: Medicine, Natural Philosophy, and Culture.* Cambridge: Cambridge University Press.

Cowan, Ruth Schwartz. 1983. *More Work for Mother: The Ironies of Household Technology from the Open Hearth to the Microwave.* New York: Basic Books.

Easlea, Brian. 1981. *Science and Sexual Oppression: Patriarchy's Confrontation with Woman and Nature.* London: Weidenfeld and Nicholson.

Fausto-Sterling, Anne. 1985. *Myths of Gender: Biological Theories about Men and Women.* New York: Basic Books.

Haraway, Donna. 1991. *Primate Visions: Race, Gender, and Nature in the World of Modern Science.* London: Routledge.

———. 1991. *Simians, Cyborgs and Women.* New York: Routledge.

Harding, Sandra. 1986. *The Science Question in Feminism.* Ithaca, N.Y.: Cornell University Press.

———. 1991. *Whose Science? Whose Knowledge?* Ithaca, NY: Cornell University Press.

Hubbard, Ruth, M. S. Henifin, and B. Fried, eds. 1979. *Women Look at Biology Looking at Women.* Cambridge: Schenkman.

Jordanova, Ludmilla. 1998. *Sexual Visions: Images of Gender in Science and Medicine between the Eighteenth and Twentieth Centuries.* Madison: University of Wisconsin Press.

Kass-Simon, G., and Patricia Farnes. 1990. *Women of Science, Righting the Record.* Bloomington: Indiana University Press.

Keller, Evelyn Fox. 1983. *A Feeling for the Organism: The Life and Work of Barbara McClintock.* New York: W. E. Freeman.

———. 1985. *Reflections on Gender and Science.* New Haven, CT.: Yale University Press.

———. 1992. *Secrets of Life, Secrets of Death.* New York: Routledge.

———. 1995. *Refiguring Life.* New York: Columbia University Press.

Koblitz, Ann Hibner. 1993. *A Convergence of Lives: Sofia Kovalevskaia.* New Brunswick, NJ: Rutgers University Press, rev. ed.

Kohlstedt, Sally Gregory, and Helen Longino, eds. 1997. *Women, Gender, and Science: New Directions.* Chicago: University of Chicago Press.

Laslett, Barbara, Sally Gregory Kohlstedt, Helen Longino, and Evelynn Hammonds, eds. 1996. *Gender and Scientific Authority.* Chicago: University of Chicago Press.

Longino, Helen. 1990. *Science as Social Knowledge.* Princeton, N.J.: Princeton University Press.

Merchant, Carolyn. 1980. *The Death of Nature: Women, Ecology, and the Scientific Revolution.* San Francisco: Harper and Row.

———. 1989. *Ecological Revolution, Gender, Nature, and Science in New England.* Chapel Hill: University of North Carolina Press.

———. 1995. *Earthcare: Women and the Environment.* New York: Routledge.

Mozans, H. J. 1974 [1913]. *Woman in Science.* Cambridge: MIT Press.

Newman, Louise Newman. 1985. *Men's Ideas / Women's Realities: Popular Science, 1870–1915.* New York: Pergamon Press.

Norwood, Vera. 1993. *Made from This Earth: American Women and Nature.* Chapel Hill: University of North Carolina Press.

Ogilvie, Marilyn Bailey. 1986. *Women in Science: Antiquity through the Nineteenth Century.* Cambridge, Mass.: MIT Press.

Pycior, Helena, Nancy Slack, and Pnina Abir-Am, eds. 1996. *Creative Couples in the Sciences.* New Brunswick, N.J.: Rutgers University Press.

Rose, Hilary. 1994. *Love, Power, and Knowledge: Towards a Feminist Transformation of the Sciences*. Bloomington: Indiana University Press.

Rosser, Sue. 1993. *Biology and Feminism: A Dynamic Interaction*. New York: Twayne.

Rossiter, Margaret. 1982. *Women Scientists in America: Struggles and Strategies to 1940*. Baltimore: Johns Hopkins University Press.

———. 1995. *Women Scientists in America: Before Affirmative Action, 1940–1972*. Baltimore: Johns Hopkins University Press.

Russett, Cynthia Eagle. 1989. *Sexual Science: Victorian Construction of Womanhood*. Cambridge, Mass.: Harvard University Press.

Schiebinger, Londa. 1989. *The Mind Has No Sex? Women in the Origins of Modern Science*. Cambridge, Mass.: Harvard University Press.

———. 1993. *Nature's Body: Gender in the Making of Modern Science*. Boston: Beacon Press.

Shteir, Ann B. 1996. *Cultivating Women, Cultivating Science: Flora's Daughters and Botany in England, 1760–1860*. Baltimore: Johns Hopkins University Press.

Traweek, Sharon. 1988. *Beamtimes and Lifetimes: The World of High Energy Physics*. Cambridge, Mass.: Harvard University Press.

Tuana, Nancy, ed. 1989. *Feminism and Science*. Bloomington: Indiana University Press.

Weinbard, Phyllis Holman and Rima Apple, eds. 1993. *The History of Women and Science, Health, and Technology: A Bibliographic Guide to the Professions and the Disciplines*. Madison: University of Wisconsin System Women's Studies Librarian. (http://www.library.wisc.edu/librarian/womensStudies/homemore.htm#biblibgraphies

Isis' Consciousness Raised

By Carolyn Merchant

W HEN GEORGE SARTON first published in 1913 what became the journal of the History of Science Society, he named it *Isis* after the Egyptian mother goddess associated with the annual flooding of the Nile. Isis, according to Sarton in his *History of Science*, "began her foreign conquests in the seventh century, if not before. Herodotus says that . . . the women of Cyrene worshipped her. . . . Temples and inscriptions to Isis and other Egyptian gods can be found in many of the Islands, even in the sacred Delos. . . ." In Greece, Sarton continued, she was celebrated at the mysteries of Eleusis as "Demeter, the glorification of motherly love (cf. Isis)." She "says of herself 'I am everything which existed, which is now and will ever be, no mortal has ever disclosed my robe.'"[1] For Sarton, as for the Greeks, Isis is symbolic of nature, and her robe conceals nature's secrets. Only those initiated through the mysteries (later through science) could glimpse the reality "which is now and will ever be."

What led from the Egyptian worship of the mother goddess, Isis, to Sarton's approach to the history of science as the "acquisition and systematization of positive knowledge," a process that can be symbolized by the disclosure of Isis' secrets? Does such imagery still pervade the writings of scientists and historians of science? Can feminist history of science contribute to a new perspective on our discipline and the symbolism associated with our patron goddess, Isis?

Feminist history of science involves a female perspective on science, nature, and society, the study of female challenges to traditional scientific roles, and a female consciousness concerning the origins of women's lower position and consequent exclusion from historical literature. A feminist approach to science and history can reveal hidden biases in a field that in recent years has considered itself free of the cultural assumptions of the present when treating the science of the past. Beyond this it can offer alternative interpretations of the rise of science, scientific professionalization, and the scientific world view, and it can create new syntheses in our field.

I am grateful to the following for references: John Sinton, Michael Reardon, Adrianne Mohr, Mary Dee Bowers, Susan Feierabend, John Lesch, Spencer Weart, Charles Muscatine, and Brookes Spencer. An earlier version was read before the History of Science Society, Toronto, 17 Oct. 1980.

[1]George Sarton, *A History of Science*, 2 vols. (Cambridge: Harvard Univ. Press, 1959), Vol. I, pp. 125, 152; Sarton cites Herodotus 4.186 and quotes Plutarch, *Isis and Osiris* 354c. Sarton's own image of Isis seems to be derived from an Egyptian wall painting showing her leading Queen Nefretere to her tomb. He refers the reader to the painting, with which "the author [Sarton] is very familiar," as reproduced by Nina de Garis Davies in *Ancient Egyptian Paintings, Selected, Copied, and Described*, 3 vols. (Chicago: Univ. Chicago Press, 1936), Vol. II, Plate XCI (see Fig. 1). Davies describes the goddess Isis as "clad in a sheath-like red dress with a network of beads." In contrast to Isis, Queen Nefretere "wears a flowing robe, the transparency of which is well indicated" (Vol. III, p. 177).

*Sarton's Isis (see note 1): Isis conducting Queen Nefretere to her tomb. Repro-
duced with permission from Nina de Garis Davies,* Ancient Egyptian Paintings,
Selected, Copied and Described, *3 vols. (Chicago: University of Chicago Press,
1936), Volume II, Plate XCI.*

Here I discuss three areas in which a feminist perspective can provide a critique
of science and its history and suggest new questions for investigation. The first
centers on the Western scientific world view, its historical origins, and the way
this perception has permeated the history of science; the second on the role played
by language, image, and metaphor in science and the writing of its history; the
third on the way women, women's roles, and women scientists are portrayed by

historians. This division is consistent with a theory of historical explanation based on the interaction between superstructure (e.g., world view, ideology, conceptual schemes) and substructure (e.g., social roles, behavior, production, reproduction) mediated through symbolic normative structures (e.g., image, metaphor, description, prescription, representation)—the kind of explanation I have tried to develop in my book *The Death of Nature*.[2]

AN IDEOLOGY OF OBJECTIVITY

At the level of superstructure, or ideology, the philosophy of nature that has guided the work of many modern scientists has been logical positivism. Positivism assumes that valid, verifiable, hence positive knowledge of the world derives ultimately from experience obtained through the senses or experiment and interpreted via the conventions and rules of mathematical language and logic. Scientific knowledge is rule-governed, context-free, and empirically verifiable and as such claims to be objective, that is, independent of the influence of particular historical times and places. Yet the positivist approach itself to nature and history relies conceptually on an interlocking structure of dualities that is context-bound and rooted in history: the dualities of subject and object, activity and passivity, male and female, and culture and nature.

The basic dichotomy is that between subject and object, and indeed objectivity, the hallmark of logical positivism, depends upon it. The objectification of nature is rooted in Aristotle's locus of reality in the objects of the natural world and made explicit in Descartes's separation of mind from matter, that is, of thinking subject from external object. The dualism between activity and passivity hypothesizes an active subject—man—who receives, interprets, and reacts to sense data supplied by a passive object—nature. Nature as object, whether conceived as things (in the Aristotelian framework) or as corpuscles (in the Cartesian) is composed of dead passive matter set in motion by efficient or final causes (Aristotle) or the transfer of motion (Descartes). Stemming from the same Aristotelian roots as the ideology of objectivity is the association of passivity with femaleness and activity with maleness. As Aristotle put it, "the female, as female, is passive and the male, as male, is active, and the principle of movement comes from him."[3] The male semen contributes power and motion to the embryo; the female supplies the matter, or passive principle. Finally, culture is identified with the active subject and thus with the male, as a passage from the philosopher Georg Simmel makes clear:

> The requirements of . . . correctness in practical judgments and objectivity in theoretical knowledge . . . belong as it were in their form and their claims to humanity in general, but in their actual historical configuration they are masculine throughout. Supposing that we describe these things, viewed as absolute ideas, by the single word "objective," we find that in the history of our race the equation objective = masculine is a valid one.[4]

[2]Carolyn Merchant, *The Death of Nature: Women, Ecology, and the Scientific Revolution* (San Francisco: Harper & Row, 1980).

[3]Aristotle, *De generatione animalium*, trans. Arthur Platt (Oxford: Clarendon Press, 1910), 1.19, 729b13.

[4]See Georg Simmel (1858-1918), *Philosophische Kultur*, as quoted in Karen Horney, "The Flight from Womanhood," in *Women and Analysis*, ed. Jean Strouse (New York: Grossman, 1974), p. 172. See also Evelyn Fox Keller, "Gender and Science," *Psychoanalysis and Contemporary Thought*, 1978, *1*:409–433, on p. 409.

The Aristotelian identification of the female principle with passivity and the further association of passivity with object and the natural world have furnished the basic philosophical framework of Western culture.

Sociologists of the Frankfurt school have pointed out how the subject-object and attendant dualities of mainstream Western thought entail a philosophy of domination. Because an active controlling subject is separate from and dominant over a passive controlled object, the scientific rationale of objectivity can legitimate control over whatever has been assigned by culture to a lower place in the "natural" order of things.[5] It thus maintains a hierarchical domination of subject over object, male over female, and culture over nature. In particular, this conceptual system can justify the subordination of women when compounded by the separation of productive (public, male), and reproductive (female, private) spheres in modern industrialized society. Historically nature and the female have been identified, and cultural ideology has legitimated the domination of both. This identification appears in the science of such men as Francis Bacon, William Harvey, Thomas Hobbes, Joseph Glanvill, and Robert Boyle, to mention only a few of the "fathers" of modern science, and it permeates the history of science.

How has this ideology of objectivity and its associated sexual bias manifested itself in the work of historians of science? One place to look for these connections is in the writings of those who argue that the history of science ought to portray the progress of objective knowledge. A few quotations from Charles Gillispie's *Edge of Objectivity* will exemplify the implicit bias against women and nature as object.

> After Galileo science could no longer be human in the deep internal sense of its forerunner in classical antiquity. Bacon makes science what it has become in part and what the public tends to wish it were in its entirety: an innocuous instrument of human betterment which requires of him who would master it not difficult abstract thought but only patience and right method. . . . Scientific thought itself is bound to be far more abstract, elegant, and intellectually aristocratic than Bacon foresaw or would have approved. But scientists are likely to be humane men who wish to do good and like to be told that they do. . . . It is the materialistic commitment of a Bacon, at once tough-minded and humanitarian, rather than the delicacy of a mind of a Pascal, which has shaped the technical tradition.[6]

This passage makes clear that for Gillispie masculine qualities of toughness and mastery are attributes of the scientific mind and technical tradition, that science is an aristocratic elite pursuit, that scientists are clearly male, and that a "feminine" quality such as delicacy will not lead to human progress and betterment. Lest

[5]See, e.g., Vincent di Norcia, "From Critical Theory to Critical Ecology." *Telos*, 1974/75, No. 22, pp. 85–95, on pp. 88–89; Jurgen Habermas, "Technology and Science as Ideology," in Habermas, *Toward a Rational Society* (London: Heineman, 1971), pp. 81–122; William Leiss, *The Domination of Nature* (New York: Braziller, 1972); Theodor W. Adorno *et al.*, *The Positivist Dispute in German Sociology*, trans. Glyn Adey and David Frisby (New York: Harper & Row, 1976). For a feminist extension of the Frankfurt critique of objectivity see Marcia Westkott, "Feminist Criticism of the Social Sciences," *Harvard Educational Review*, Nov. 1979, 49:422–430. See also Dorothy Smith, "Women's Perspective as a Radical Critique of Sociology," *Sociological Inquiry*, 1974, 44:7–13; Sandra Harding, "Objectivity in Social Science Revisited: Gaps in the 'Text' of the Dispute," and Nancy Hartsock, "The Natural Science Model in Social Science: Shifting the Boundary Between Nature and Culture," both papers presented to the Philosophy of Science Association, Toronto, Ontario, Canada, 17 Oct. 1980; and Judith Long Laws, "Patriarchy as Paradigm: The Challenge from Feminist Scholarship," paper presented at the American Sociological Association Meeting, New York, Aug. 1976.

[6]Charles C. Gillispie, *The Edge of Objectivity* (Princeton, N.J.: Princeton Univ. Press, 1960), pp. 81–82.

Gillispie's association of male qualities with science seem problematic in the passage just quoted, it is abundantly clear in the following: "Indeed, Diderot was the Spinoza of biology before ever the science had its name—or its Newton. His was no feminine dislike of precision, no soulful sense of God in nature, but a philosophy of necessitarian organism." Gillispie accepts without question the association of nature and the female with object and berates and ridicules Goethe for his failure to dissect, mathematize, and objectify nature as had Newton:

> Nor though he looked through a prism, did Goethe believe in experiment. On the contrary, Newton's errors were the price he paid for his methods, mathematicizing nature into abstractions, torturing her with instruments, with telescopes, prisms, and mirrors, until she expires like a butterfly on a pin. . . . It is impossible to read the *Farbenlehre* without an acute sense of embarrassment at the painful spectacle of the author, a great man, making a fool of himself. . . . The historian is bound to represent this Goethean intrusion as profoundly hostile to science, hostile to physical science, and misleading even if stimulating to biology.

Gillispie evaluates his great men of science according to his standard of tough-mindedness and precision, eschewing Goethean romanticism as weak and permeated by emotion. Descartes, a great man with a "subtle" mind, was led astray by the assumption that "what is simple is nature herself whereas every neat-handed physicist knows that nature is very complex."[7] This historical association of objectivity with masculinity not only reinforces the tendency for scientists to be predominantly male, but also supports the identification of nature as object with femaleness, emotion, soulfulness, and sentience.

THE ROLE OF SYMBOL

The second area in which a feminist perspective offers an opportunity to reveal sexual bias and to reformulate traditional interpretations of the history of science is the realm of symbolic structure. Male and female symbolism and metaphor mediate between a society's ideological superstructure and its daily activities, providing insights into the origins of the linguistic patterns that still permeate modern scientific and historical writing. Images, metaphors, myths, and modes of description can either legitimate dominant conceptual systems or present alternatives to the mainstream view. Such image systems have a normative function, mediating between a society and its conceptual ideology and reinforcing its behavior. The role played by language in structuring the perception of subject-object relations, the interaction theory of metaphor, and the politics of metaphor, topics currently addressed by philosophers of science, should interest feminist historians of science. The most powerful such image is the identification of nature with the female, especially a female harboring secrets.[8]

Female imagery and myths or beliefs featuring female figures reflect a culture's changing values. The symbolism associated with Nature deified that began with

---

[7]*Ibid.*, pp. 192, 195–197, 93; see also p. 201.

[8]On the philosophy of metaphor see Philip Wheelwright, *Metaphor and Reality* (Bloomington: Indiana Univ. Press, 1962); Max Black, *Models and Metaphor* (Ithaca, N.Y.: Cornell Univ. Press, 1962); Mary Hesse, *Models and Analogies in Science* (Notre Dame, Ind.: Univ. Notre Dame Press, 1966); Richard Olson, ed., *Science as Metaphor* (Belmont, Calif.: Wadsworth, 1971); George Lakoff and Mark Johnson, *Metaphors We Live By* (Chicago: Univ. Chicago Press, 1980); Stanley Brandes, *Metaphors of Masculinity* (Philadelphia: Univ. Pennsylvania Press, 1980); Robin Lakoff, *Language and Woman's Place* (New York: Harper & Row, 1975).

Isis' refusal to disclose her robe undergoes significant change in the Middle Ages and after. In the twelfth century the Neoplatonic cathedral school of Chartres depicts her as the goddess *Natura*, the lower form of the Platonic world soul. In Alain de Lille's allegory *Nature's Complaint*, Natura laments her exposure to the view of the vulgar as her garments of modesty are torn by the wrongful assaults of men aggressively penetrating the secrets of heaven. "A damask tunic . . . pictured with embroidered work, concealed the maiden's body. . . . In its principal part man laid aside the idleness of sensuality, and by the direct guidance of reason penetrated the secrets of the heavens." Natura, whose "features are bedewed with a shower of weeping" is questioned about her torn robe:

> "I marvel," then I said, "wherefore certain parts of thy tunic, which should be like the connection of marriage, suffer division in that part of their texture where the fancies of art give the image of man." "Now from what we have touched on previously," she answered, "thou canst deduce what the figured gap and rent mystically show. For since, as we have said before, many men have taken arms against their mother in evil and violence, they thereupon, in fixing between them and her a vast gulf of dissension, lay on me the hands of outrage, and themselves tear apart my garments piece by piece, and, as far as in them lies, force me, stripped of dress, whom they ought to clothe with reverential honor, to come to shame like a harlot. This tunic then is made by this rent, since by the unlawful assaults of man alone the garments of my modesty suffer disgrace and division."[9]

Such symbolism suggests the rape or sexual conquest of both women and nature. Just as nature aggressively investigated is depicted as a woman molested, so femininity is symbolized as an enclosure, often one associated with nature's bounty, that can be breached. Thus medieval artists depict the goddess Venus or the Virgin Mary in enclosed gardens or stone circles symbolic of the female womb and of love, fruitfulness, and pleasure. Chaucer sets comic stories in enclosed gardens in which the lover in gaining access to the garden symbolically penetrates the womb. In *The Merchant's Tale*, based on the biblical "Song of Solomon," Damyan fashions a key to unlock the circular garden and subsequently makes love to a maiden situated in a fruit-bearing tree.[10]

In the seventeenth century the disclosure of Isis is carried beyond her robe into the interior of her body as Francis Bacon advises his new man of science to wrest from nature the secrets harbored in her womb, to search into the bowels of nature for "the truth that lies hid in deep mines and caves" and "to shape her on the

[9]Alain de Lille, *The Complaint of Nature*, trans. Douglas Moffat (New York: Henry Holt, 1908), pp. 15, 33, 41. For original see Alanus de Insulis, *De Planctu Naturae*, in Thomas Wright, ed., *The Anglo-Latin Satirical Poets and Epigrammatists of the Twelfth Century* (London: Longman & Trubner, 1872), Vol. II, pp. 429–522; esp. pp. 441, 467. On the proper role of nature as teacher in unveiling her truths to mankind see p. 457 (Alain, *Complaint*, p. 31). For a commentary see George D. Economou, *The Goddess Natura in Medieval Literature* (Cambridge, Mass.: Harvard Univ. Press, 1972), esp. pp. 72–80; Merchant, *Death of Nature*, pp. 10–20, 31–33.

[10]Yvonne Noble, "What Became of the Image of the Virgin as *Hortus conclusus* in the Augustan Age," paper presented at the 11th Annual Meeting of the American Society of Eighteenth Century Studies, San Francisco, Apr. 1980; Merchant, *Death of Nature*, pp. 8, 10–11. See also the French painting "St. Genevieve with Her Flock" (16th cent.), depicting the virgin with her flock of sheep within a protective stone circle on a hillside of trees and blooming flowers; reproduced in John Michell, *The Earth Spirit* (New York: Avon, 1975). The imagery of the enclosed garden (*hortus conclusus*) as a scene of love stems from the love of the maiden for the shepherd in the Song of Solomon, 4:12. The biblical images appear in Geoffrey Chaucer, "The Merchant's Tale," see *Works*, ed. F. N. Robinson (Boston: Houghton Mifflin, 1957), lines 2044–2046; 2143–2146. For more on the garden symbolism see Stanley Stewart, *The Enclosed Garden: The Tradition and Image in Seventeenth Century Poetry* (Madison: Univ. Wisconsin Press, 1966).

anvil." "Nature must be taken by the forelock, being bald behind," he asserted. "Nor ought a man to make scruple of entering and penetrating into these holes and corners, when the inquisition of truth is his whole object." For Bacon's apologist Joseph Glanvill, who said that nature must be "mastered" and "managed" by "searching out the depths . . . and intrigues of remoter nature," nothing was more helpful than the microscope, for "the secrets of nature are not in the greater masses, but in those little threads and springs which are too subtle for the grossness of our unhelped senses." In the *Vanity of Dogmatizing*, Glanvill pointed out that "Nature's coarser wares" are "exposed to the transient view of every common eye; her choicer riches are locked up only for the sight of them that will buy at the expense of sweat and oil." In achieving such insights, however, true understanding is often misled by the emotions, for "the woman in us, still prosecutes a deceit, like that begun in the Garden: and our understandings are wedded to an Eve, as fatal as the mother of our miseries."[11]

By the nineteenth century nature is removing her own veil and voluntarily exposing her secrets. A sculpture by Louis Ernest Barrias, *La Nature se devoilant devant la science* ("Nature revealing herself to science"), is appropriately located in the entry to the School of Medicine at the Sorbonne in Paris. A naked woman (based on the nymph or nature goddess in a sixteenth-century engraving) picnics on the grass with two fully clothed gentlemen in Edouard Manet's *Le Dejeuner sur l'herbe* (1863). *Mother Earth Laid Bare* by Alexander Hogue (1936) portrays the female shape taken on by eroded earthen mounds in drought-ridden Oklahoma in the 1930s.[12]

In the twentieth century we find scientists fervently hoping that the veil of nature can be lifted from matter itself (traditionally feminine) so that all may view the hidden secrets of the atom. One may hope "to be able to lift a corner of the veil that conceals creation. . . . Each of us hopes that . . . a sensational application of radium will completely tear away the veil and that truth will appear before everyone's eyes," announced the inaugural editorial from *Le Radium* in 1904. "The notion of impenetrable mysteries has been dismissed," wrote Sir William Crookes in 1903. "A mystery is a thing to be solved—and 'man alone can master the impossible.'" Hans Reichenbach in 1933 charged nuclear physicists with the task of the "unveiling of the secrets surrounding the inner structure of matter," and the editors of *Harper's* (1924) applauded the "laying bare" of its structure.[13]

---

[11]Francis Bacon, "De dignitate et augmentis scientarum," *Works*, ed. James Spedding, Robert Ellis, and Douglas Heath, 14 vols. (London: Longmans Green, 1857–1874), Vol. IV, pp. 343, 287; Bacon, "The Refutation of Philosophies," in Benjamin Farrington, ed. and trans., *The Philosophy of Francis Bacon* (Liverpool: Liverpool Univ. Press, 1964), p. 130; Bacon, "De dignitate et augmentis scientarum," *Works*, Vol. IV, pp. 294, 296; Joseph Glanvill, *Plus Ultra* (1668; Gainesville, Fla.: Scholar's Facsimile Reprints, 1958), pp. 87, 10, 56; Glanvill, *The Vanity of Dogmatizing* (1661; New York: Columbia Univ. Press, 1931), pp. 247, 118.

[12]George Mauner, *Manet: Peintre-philosophe* (University Park, Pa.: Pennsylvania State Univ. Press, 1975), pp. 7–45, esp. pp. 10–12, 32–33, 40–43. Corky Bush, "Cultural Images of Women and Technology," in *Women and Technology: Deciding What's Appropriate* (Missoula, Mont.: Women's Resource Center, 1979), pp.11–17, 34.

[13]Inaugural editorial, *Le Radium*, Jan. 1904, *1*:2, trans. Spencer Weart (I thank Dr. Weart for this and the three following quotations); Sir William Crookes, "Modern Views on Matter," *Scientific American Supplement*, July 1903, *56*:23014; Hans Reichenbach, *Atom and Cosmos: The World of Modern Physics*, trans. and rev. Edward S. Allen (New York: Macmillan, 1933), p. 222; editor's note in *Harper's*, July 1924, *149*:251, as quoted in Daniel Kevles, *The Physicists* (New York: Knopf, 1978), p. 174.

Such language has by no means vanished from current science. Newscasters describing the May 1980 erruption of Mount St. Helens in Washington interviewed a geologist who had become an instant vulcanist:

> Question: What is going to happen next to Mount St. Helens?
> Geologist: We can't tell what she's going to do. Her flanks are shuddering . . . . We don't know her intentions. Scientists haven't been able to probe her deeply enough with their instruments.[14]

While such quotations may be suggestive of sexual assaults on nature, and can be so viewed when placed in the perspective of the historical evolution of the language and metaphor of science, further work is needed. The meaning of scientific metaphor changes over time and is integrally tied to its historical context. Only context can shed light on the meaning of language for a given society at a given historical moment.

Historians of science have appropriated these same culturally derived sexual metaphors in their presumedly objective histories of scientific development. Reproductive metaphor betraying a masculine bias abounds in the writings of both male and female historians and philosophers of science who eulogize the rationality of the scientific world view. A woman historian of science writes:

> Whereas the physicists believed themselves to be approaching the position of Laplace's omniscient intelligence, the philosophers came to abandon the hope that scientific methods can lead to certainty or even penetrate the veil of appearances. . . . Even Laplace could not penetrate into the "secret springs and principles" producing these phenomena. . . . For Hume a science based on appearances can never penetrate into the real essences of things and yield necessary laws of nature.[15]

Like scientists, historians of science accept without hesitation or critical comment the linguistic identification of nature with the female. When translating from languages with a feminine gender, they retain feminine forms rather than translate feminine articles as "it." Thus John Heilbron quotes Fontenelle: "Or so it was for those who held with Fontenelle that 'nature is never so admired as when she is understood.'" Heilbron may be preserving Fontenelle's intent, but philosopher and systems theorist C. West Churchman appropriates Baconian sexual language in his description of the classical laboratory: "Finally we should note the basic assumption of the classical laboratory—namely, that nature is neither capricious nor secretive. If nature were capricious, she would tell one observer one thing and another observer a quite different thing. . . . Also nature is not secretive, in the sense that she will not forever hide certain aspects of her being. . . ."[16]

Judged within the context of these examples, the sexual overtones of the following quotation from George Sarton become increasingly clear:

---

[14]Geologist Marvin Beeson, Portland State University, paraphrased excerpts from two television interviews, May 1980. I thank John Sinton and Michael Reardon for this information. Ironically the name St. Helens, which lends credence to the identification of the mountain as female, is not named for a (nonexistent) female saint, but for Alleyne Fitzherbert, Baron St. Helens (1753–1839).

[15]Margaret J. Osler, "Certainty, Scepticism, and Scientific Optimism: The Roots of Eighteenth-Century Attitudes toward Scientific Knowledge," in Paula Backscheider, ed., *Probability, Time, and Space in Eighteenth-Century Literature* (New York: AMS Press, 1979), pp. 3–28; on pp. 3, 21.

[16]John Heilbron, *Electricity in the 17th & 18th Centuries* (Berkeley: Univ. California Press, 1979), p. 43; C. West Churchman, *The Systems Approach and Its Enemies* (New York: Basic Books, 1979), p. 57.

I read this morning in the paper that a man called John O'Brien died suddenly in Boston while he was watching a wrestling match. . . . His was probably a heart case, and the wrestling excited him overmuch. I have no trouble in understanding that, and my sympathy goes out to him, for I have been deeply moved time after time while I was contemplating my fellow men wrestling not with other men but with nature herself, trying to solve her mysteries, to decode her message.[17]

The final goal of positivist science is to wrest from Isis the secrets she harbors within. The real meaning Sarton attached to Isis as patron of the history of science, then, is to be found within the linguistic tradition of the sexual conquest of nature. Through such examples historians of science can become aware of the ways in which sexual biases have permeated their own work and help to liberate Isis from culturally derived sexual values.

WOMEN AND SCIENCE

The third area in which a feminist perspective can generate new interpretations in the history of science is on the level of substructure: the influence of social roles on scientific theory, the role of women in science, and the sciences historically associated with women. Much history of science has followed George Sarton in associating the progress of science, the rise of human culture, and the fulfillment of human destiny with men:

> We have some degree of interest in every *man and woman* whom we approach near enough. Should we not be even more interested in those *men* who accomplish more fully the destiny of the race? . . . The same instinct which causes sport-lovers to be insatiably curious about their heroes causes the scientific humanist to ask one question after another about the great men to whom he owes his heritage of knowledge and culture. In order to satisfy that sound instinct it will be necessary to prepare detailed and reliable biographies of the men who distinguished themselves in the search for truth.[18]

By contrast, female roles in science and female scientists are now being resurrected from obscurity and reinterpreted, not according to a great woman theory paralleling the great man approach of Sarton, but from the perspective of women as a sociological group challenging cultural norms that militate against their participation in science. Women with feminist perspectives on science, such as Margaret Cavendish, Mary Astell, and Mary Wollstonecraft, are being studied along with women's scientific work in such fields as astronomy, mathematics, chemistry, and geology. How some scientific fields became professionalized along gender lines that functioned to exclude women, while others such as nutrition, home economics, and teaching became feminized, is also being addressed.[19]

[17]George Sarton, *The Study of the History of Science* (New York: Dover, 1936), pp. 41–42.
[18]*Ibid.* Italics added.
[19]Bibliographies on women in science include Audrey B. Davis, *Bibliography on Women: With Special Emphasis on Their Roles in Science and Society* (New York: Science History Publications, 1974); Phyllis Zweig Chinn, *Women in Science and Mathematics Bibliography* (Washington, D.C.: American Association for the Advancement of Science, 1979); Michele Aldrich, "Review Essay: Women in Science," *Signs*, Autumn 1978, 4(1):126–135; John Ernest, *Mathematics and Sex* (Santa Barbara, Calif.: Mathematics Dept., Univ. California, n.d.). On Margaret Cavendish and Mary Astell see Hilda Smith, "Feminism and the Methodology of Women's History," in Berenice A. Carroll, ed., *Liberating Women's History* (Urbana: Univ. Illinois Press, 1976), pp. 368–384, on pp. 378–380; Merchant, *Death of Nature*, pp. 268–274; Gerald Dennis Meyer, *The Scientific Lady in England* (Berkeley: Univ. California Press, 1980). On Mary Wollstonecraft see Lois Magner, "Women and

One example of a field traditionally associated with women is botany. Women healers, women's knowledge of herbal properties, the cultivation of herb gardens by women, and the study of Elizabethan herbal handbooks written by women offer rich areas for further investigation. The role of sexual stereotyping in plant study is not completely straightforward. In the eighteenth century, as Ann Shteir has shown, the Linnaean sexual system of classification led to the use of language derived from human behavior as categories of plant taxonomy—virgins, concubines, eunuchs, polygamists. This sexualization of the field then prompted men, as Richard Polwhele did in 1798, to condemn female botanizing as a lascivious form of vicarious sex. But botany provided many women with an intellectual pursuit in which they enthusiastically participated.[20]

Indeed so culturally ingrained had been the association of women with plants that botany had some difficulty establishing itself as a profession appropriate for men. In 1887 J. F. A. Adams felt compelled to write an article entitled "Is Botany a Suitable Study for Young Men?" "An idea seems to exist in the minds of some young men," he began, "that botany is not a manly study; that it is merely one of the ornamental branches, suitable for young ladies and effeminate youths but not adapted for able-bodied and vigorous-brained young men who wish to make the best use of their powers." His defense of botany has a familiar ring. Botany offered men thorough mental training and opportunities to "harden their muscles" and "amid the solitudes of nature, to penetrate her wondrous mysteries." Not only have botanists had difficulty demonstrating that botany was conducive to "a vigorous mind and body and a robust character," but historians of botany seem to have

---

the Scientific Idiom," *Signs*, Autumn 1978, *4*(1):61–80. Works on women scientists include H. J. Mozans, *Woman in Science* (1913; Cambridge, Mass.: MIT Press, 1977) [see *Isis*, 1977, *68*:111—eds.]; Lynn S. Osen, *Women in Mathematics* (Cambridge, Mass.: MIT Press, 1974); Marie Louis Dubreil-Jacotin, "Women Mathematicians," in *Great Currents of Mathematical Thought*, ed. Le Lionnais (New York: Dover, 1970); Helen Wright, *Sweeper of the Sky: The Life of Maria Mitchell, First Woman Astronomer* (New York: Macmillan, 1949); Eve Merriam, "Maria Mitchell," in *Growing Up Female in America: Ten Lives* (New York: Doubleday, 1971); Herman S. Davies, "Women Astronomers, 400 A.D.–1750," *Popular Astronomy*, May 1898, *6*:128–229; Deborah Warner, "Women Astronomers," *Natural History*, May 1979, *88*(5):12–26; P. V. Rizzo, "Early Daughters of Urania," *Sky and Telescope*, 1954, *14*:7–10; Lois Barber Arnold, "American Women in Geology: A Historical Perspective," *Geology*, 1977, *5*:493–494; Susan Schacher, ed., *Hypatia's Sisters: Biographies of Women Scientists—Past and Present* (Seattle: Feminists Northwest, 1976); George Basalla, "Mary Somerville: A Neglected Popularizer of Science," *New Scientist*, 1963, *17*:531–533; Elizabeth C. Patterson, "Mary Somerville," *British Journal for the History of Science*, 1969, *4*:311–339; Sherida Houlihan and John H. Wotiz, "Women in Chemistry Before 1900," *Journal of Chemical Education*, 1975, *52*:362–364; Joan Hoff Wilson, "Dancing Dogs of the Colonial Period: Women Scientists," *Early American Literature*, 1973, *7*:225–235; Joan N. Burstyn, "Women in American Science," *Actes du XIᵉ Congress International d'Histoire des Sciences*, 1965 (pub. 1968), 2:316–319; Deborah Warner, "Science Education for Women in Antebellum America, " *Isis*, 1978, *69*:58–67; Carolyn Merchant Iltis, "Madame du Châtelet's Metaphysics and Mechanics," *Studies in History and Philosophy of Science*, 1977, *8*:29–48; Anne Sayre, *Rosalind Franklin and DNA* (New York: Norton, 1975); Eve Curie, *Madame Curie*, trans. Vincent Sheean (Garden City, N.Y.: Doubleday, 1937); Joan Dash, *A Life of One's Own* (New York: Harper & Row, 1973), ch. on Maria Goeppert-Mayer. On professionalization see Margaret Rossiter, "Women's Work in Science, 1889–1910," *Isis*, 1980, *71*:381–398; Sally G. Kohlstedt, "In from the Periphery: American Women in Science, 1830–1880," *Signs*, Autumn 1978, *4*(1):81–96; Margaret Rossiter, "Women Scientists in the United States Before 1920," *American Scientist*, 1974, *62*:312–323.

[20]Ann Shteir, "With Bliss Botanic: Women and Plant Sexuality," and Susan Groag Bell, "Women Create Gardens in Male Landscapes," papers read at the 11th Annual Meeting of the American Society for Eighteenth Century Studies, San Francisco, Apr. 1980; Richard Pohlwhele, *The Unsexed Females* (1798; New York: Garland, 1974), pp. 8–9. On the early history of botanic drugs see Edith Grey Wheelwright, *The Physick Garden: Medicinal Plants and Their History* (Boston: Houghton Mifflin, 1935); Joseph E. Meyer, *Nature's Remedies* (Hammond: Indiana Botanic Gardens, 1934).

had equal trouble dissociating it from "the markedly feminine connotation that has been botany's doubtful fate."[21]

Equally rewarding would be an investigation of zoology, including the roles women have played in the development of the field and its professionalization along lines that may have helped exclude them, but especially the influence of social assumptions about women on zoological studies and theory. Historically women have been associated with animality, particularly in regard to sexual desire. Renaissance writers associated the supposed sexual lust of women with a greater preponderance of animal passions, the Christian church saw female sexual desire as the downfall of the male, and the Protestant John Knox (1505–1572) equated the "inordinate appetites" of untamed women with those of beasts. More recently Donna Haraway has investigated the influence of social structure on the history of primate investigation. She examines scientific assumptions and values concerning male-female dominance hierarchies, the infusion of economic and sex-biased language into scientific description, and differences in interpretations made by male and female scientists.[22]

The evolving controversial field of sociobiology, which explicitly seeks the roots of social behavior in animal behavior, offers especially egregious examples of culture-laden language. Sociobiologists' use of such terms as adultery, rape, divorce, monogamy, polygamy, infanticide, and prostitution either legitimates human sexual activities as "natural" or condemns them as "unnatural." Violent attacks by male bluebirds on females who have committed "adultery," "rape" by male mallard ducks when a surplus of males exists, "homosexuality" in acanthocephalan worms, "divorce" in kittiwakes, and "lesbianism" among California gulls are more than just catchy words used to popularize the new field. One sociobiologist, David Barash, asks: "Is it mere coincidence that when a woman is raped, her husband or lover often responds either by rejecting her (mountain bluebird style), or by being remarkably 'turned on' (like the mallard duck)?"[23]

The history of scientific theories of sex differences, reproduction, and childbirth offers further examples of the influence of culture on science as it affects women. Historians have examined ancients—Plato, Aristotle, and Galen—and moderns—William Harvey, Charles Darwin and their followers—who wrote on male-female differences in generation, sexuality, and intellectual activity. The biology of Aristotle, which assumed the activity of the male and passivity of the

---

[21]Quoting J. F. A. Adams, M.D., "Is Botany a Suitable Study for Young Men?" *Science*, 1887, 9:117–118; David Allen, *The Naturalist in Britain* (London: A. Lane, 1976), p. 28. See also Emmanuel D. Rudolph, "How it Developed that Botany was the Science Thought Most Suitable for Victorian Young Ladies," *Children's Literature*, 1973, 2:92–97; and (on botany, plant pathology, and professionalization) Rossiter, "Women's Work," pp. 387–388.

[22]Donna Haraway, "Animal Sociology and a Natural Economy of the Body Politic," *Signs*, Autumn 1978, 4(1):21–36, 37–60; Haraway, "The Biological Enterprise: Sex, Mind, and Profit from Human Engineering to Sociobiology," *Radical History Review*, 1979, 20:206–237. On women, the animal passions, and violence see Merchant, *Death of Nature*, pp. 132–140; Natalie Zemon Davis, "Men, Women, and Violence: Some Reflections on Equality," *Smith Alumnae Quarterly*, Apr. 1977; John Knox, *The First Blast of the Trumpet Against the Monstrous Regiment of Women*, in Edward Arber, ed., *The English Scholars Library* (London, 1878), Vol. II, p. 30; Vern Bullough, *The Subordinate Sex* (Baltimore: Penguin, 1974), p. 98.

[23]For examples of sexual metaphor in sociobiology see David Barash, "Sexual Selection in Birdland," *Psychology Today*, March 1978, pp. 82–86, quoting p. 86; Barash, "Sociobiology of Rape in Mallards (*Anas platyrhynchos*): Responses of the Mated Male," *Science*, 1977, 197:788–789; Barash, "Male Response to Apparent Female Adultery in the Mountain Bluebird (*Sralia currucoides*): An Evolutionary Interpretation," *American Naturalist*, 1976, 110:1097–1101; Lawrence G. Abele and Sandra Gilchrist, "Homosexual Rape and Sexual Selection in Acanthocephalan Worms," *Science*, 1977, 197:81–83. For a feminist response see Marian Lowe, "Sociobiology and Sex Differences," *Signs*, Autumn 1978, 4(1):118–125.

female to be "natural," was reinforced by William Harvey in the seventeenth century; the assumption continued to hold implications for women's roles and social position. After the Darwinian revolution variability, the basis for evolutionary progress, was used to explain women's intellectual inferiority, since more of women's energy was thought to be directed toward pregnancy, lactation, and nurture with less available for learning and reasoning. Historians have focused on the contemporary challenges to these theories made by both women and men and on alternative theories.[24]

The study of women's actual scientific achievements is insufficient unless the social factors that have excluded women from the scientific professions are also considered. Yet not all claims to have examined these factors are reliable. The recent book *Fair Science* by Jonathan Cole, which presents a sociological analysis of women in science, attempts to demonstrate that science is fair and that "the measureable amount of sex-based discrimination against women is small." To feminist critics, however, his analysis is grossly inadequate because of the nature of his assumptions, the biased interpretation of data, the lack of qualitative sources, and the skewing that results from his drawing data from fields with higher female entry—biology, chemistry, psychology, and sociology. Cole's conclusions might well have been altered had he included mathematics, physics, engineering, and computer science, or considered female isolation, women's lower access to the means of scientific production, and the importance of tenure and higher academic status in determining visibility and access to honorifics.[25]

CONCLUSION

Having delineated what seem to me to be three useful approaches to criticizing and recasting the history of science, I would like to conclude with a plea for new historical syntheses in all fields and periods of the history of science. A feminist perspective can help to redefine the broad periods of scientific change. Those in which scientific or technological advance may seem most marked from a male point of view may appear retrograde when women's issues are included. Feminist history of science offers the potential for syntheses with traditional approaches that could lead to major new interpretations in our discipline as a whole.

[24]On the ancient world see Maryanne Cline Horowitz, "Aristotle and Woman," *Journal of the History of Biology*, 1976, 9:183–213; Caroline Whitbeck, "Theories of Sex Difference, " *Philosophical Forum*, 1973/74, 5:54–80; Anne Dickason, "Anatomy and Destiny: The Role of Biology in Plato's Views of Women," *ibid.*, pp. 45–53. On early modern Europe see Merchant, *Death of Nature*, pp. 149–163; Hilda Smith, "Gynecology and Ideology in Seventeenth-Century England," in Carroll, ed., *Liberating Women's History* (cit. n. 19), pp. 97–114; Jean Donnison, *Midwives and Medical Men* (New York: Schocken, 1977). On the modern period see Stephanie A. Shields, "Functionalism, Darwinism, and the Psychology of Women," *American Psychologist*, 1975, 30:739–754; Rosalind Rosenberg, "In Search of Women's Nature, 1850–1920," *Feminist Studies*, 1975, 3(1–2):141–154; Diana Long Hall, "Biology, Sex Hormones, and Sexism in the 1920s," *Phil. Forum*, 1973/74, 5:81–96; Elizabeth Fee, "The Sexual Politics of Victorian Social Anthropology," in Mary Hartman and Lois Banner, eds., *Clio's Consciousness Raised* (New York: Harper Colophon, 1974), pp. 86–102; Estelle Ramey, "Sex Hormones and Executive Ability," *Annals of the New York Academy of Sciences*, 1973, 208:237–245; Joan N. Burstyn, "Brain and Intellect: Science Applied to a Social Issue, 1860–1875," *Actes XIIe Cong. Int. Hist. Sci.*, 1968 (pub. 1971), 9:13–16; J. Burstyn, "Education and Sex: The Medical Case Against Higher Education for Women," *Proceedings of the American Philosophical Society*, 1973, 117:79–89; Eliza Gamble, *The Sexes in Science and History: An Inquiry into the Dogma of Women's Inferiority to Men* (New York: Knickerbocker, 1916).

[25]Jonathan R. Cole, *Fair Science: Women in the Scientific Community* (New York: Free Press, 1979), quoting p. 86. For critical reviews see Karen Oppenheim Mason, "Sex and Status in Science," *Science*, 1980, 208:277–278; and Margaret Rossiter, "Fair Enough?" *Isis*, 1981, 72:99–103.

Managing an Experimental Household

The Dees of Mortlake and the Practice of Natural Philosophy

By Deborah E. Harkness

ABSTRACT

Jane Dee was married to Elizabethan England's most eminent natural philosopher, John Dee. As John Dee's wife Jane was expected to perform a variety of traditional housewifely duties—managing servants, supervising domestic arrangements, and serving as a model of virtue for her family and the community. As the wife of a natural philosopher, however, Jane Dee was expected also to be her husband's helpmeet and partner—helping to make sure that the "business" of natural philosophy was profitable, overseeing the work of his philosophical assistants, and protecting her husband's privacy. This essay examines how Jane Dee negotiated her two roles within a busy and complicated household—how she interacted with her husband and his natural philosophy—and argues that the domestic sphere is an important site of knowledge production in the early modern period.

FROM THE OUTSIDE, little would have distinguished the house of Elizabethan England's most eminent natural philosopher, John Dee, from those of his neighbors in the town of Mortlake—except perhaps its size. Close to both the village church and the river Thames, Dee's residence comprised a house, a garden, an open courtyard, and a number of smaller structures.[1] From a middle distance it would have looked like the home of an Englishman of some means but no great title or position—which it was. Were we to draw closer, however, we might have picked up clues that this was no ordinary gentry residence. For one thing, several of the smaller, independent structures contained smelly

Versions of this essay were presented at the History of Science Society annual meeting in 1995, the History and Philosophy of Science Colloquium at the University of California, Davis, and as the spring 1996 Lax Lecture of Mount Holyoke College. I am grateful to all those who offered their suggestions and criticisms, especially Paula Findlen, Margaret Jacob, Karen Halttunen, Mary Terrall, Margaret Rossiter, Kathleen Whalen, Michael Dietrich, and two anonymous *Isis* readers.

[1] The best description of the exterior of the Dee household and how it was situated in relationship to the town of Mortlake comes from Elias Ashmole's report on Dee contained in Oxford, Bodleian Library, Ashmole MS 1788, fols. 147r–149r.

and noisy alchemical stills. And, though generous hospitality was the norm in the period, we surely would have noted with surprise the high volume of visitors approaching the main door: clients seeking astrological advice, messengers, potential patrons from the court of Elizabeth, and students of navigation and natural philosophy, to identify just a few.

Peeps through other, less conspicuous doors and windows would have yielded further insights into the complex workings of the house. We might, for instance, have seen Walter Hooper, the Chiswick gardener who kept the Dees' hedges trimmed and in good order; or the melancholy Roger Cook as he shuttled between the alchemical stills where he worked long hours and the main house where he ate his meals. Dee's manservant George might have been up on the ladder from which he was known to take occasional falls. Were we able to climb George's ladder, we might peer into the private second-story study where another of Dee's assistants, the scryer Edward Kelly, carried out his task of talking to angelic "scholemasters" about natural philosophy and the apocalypse with the aid of a crystal stone.[2] In the courtyard where Queen Elizabeth once stopped on her way to Richmond Palace we might have seen Mary Goodwyn, governess to Dee's daughters, or Mr. Lee, the Mortlake schoolmaster, who stopped by regularly to collect the stipend Dee provided. We probably would not have seen John Dee himself; he would have been closeted in his study, library, or alchemical houses. But we would certainly have seen the mistress of the house, Jane Fromond Dee, as she moved about the property and managed the hundred details that were her responsibility as John Dee's wife.

Though the Dee household has received no scholarly attention from historians of science, its activities can reveal much about the practice of natural philosophy during the early modern period, at a time when sites of knowledge production were in transition.[3] By the late sixteenth century, an increasing number of natural philosophers were, through marriage and Reformation, excluded from the two chief medieval sites of knowledge—the monasteries and the universities—but not yet absorbed by the independent laboratories and academies that would soon take center stage in scientific culture. Recently, a number of scholars have helped to clarify our understanding of the diversity and problematic status of the sites where natural philosophical knowledge was produced in the early modern

---

[2] Scryers were divinators of the early modern period who used shiny objects (pools of water, mirrors, stones, and even polished fingernails) to find lost items and see into the future. For more information on scrying see Keith Thomas, *Religion and the Decline of Magic* (New York: Scribner's, 1971); and Christopher Whitby's introduction to John Dee, *John Dee's Actions with Spirits*, ed. Whitby, 2 vols. (New York: Garland, 1988) (hereafter cited as **Dee, Actions with Spirits, ed. Whitby**), Vol. 1, pp. 75–93. For a full discussion of the angel conversations, their relevance to Dee's natural philosophy, and their cultural and intellectual contexts see Deborah E. Harkness, "The Scientific Reformation: John Dee and the Restitution of Nature" (Ph.D. diss., Univ. California, Davis, 1994); Harkness, "Shows in the Showstone: A Theater of Alchemy and Apocalypse in the Angel Conversations of John Dee (1527–1608/9)," *Renaissance Quarterly*, 1996, *49:*707–737; and Harkness, *Talking with Angels: John Dee and the End of Nature* (Cambridge: Cambridge Univ. Press, forthcoming).

[3] No coherent picture of the Dee household now exists, but hints can be drawn from a number of sources, including John Dee, "Compendious Rehearsal," in *Autobiographical Tracts of Dr. John Dee, Warden of the College of Manchester*, ed. James Crossley (London: Chetham Society, 1851), pp. 1–45 (hereafter cited as **Dee, "Compendious Rehearsal"**); Dee, *Diary, for the Years 1595 to 1608*, ed. Jon Eglinton Bailey (Privately printed, 1880); Dee, *Actions with Spirits*, ed. Whitby; Dee, *A Letter, Containing a Most Briefe Discourse Apologeticall (1599)* (English Experience, 502) (Amsterdam: Theatrum Orbis Terrarum, 1973); Dee, *Private Diary of Dr. John Dee*, ed. J. O. Halliwell (Camden Society, 19) (London, 1842) (hereafter cited as **Dee, Private Diary, ed. Halliwell**); Dee, *A True and Faithful Relation . . .* , ed. Meric Casaubon (London, 1659) (hereafter cited as **Dee, True and Faithful Relation, ed. Casaubon**); and C. H. Josten, "An Unknown Chapter in the Life of John Dee," *Journal of the Warburg and Courtauld Institutes*, 1965, *28:*223–257. Despite the scattered evidence, the Dee household and its activities have provided fuel for modern novelists. See Peter Ackroyd, *The House of Doctor Dee* (London: Penguin, 1994); and Claude Postel, *John Dee, la mage de la Ruelle d'Or* (Paris: Belles Lettres, 1995).

period. For a relatively brief time in the sixteenth and seventeenth centuries, the household bridged the gap between monastery and laboratory as a site for the practice of natural philosophy. During that period, natural philosophy was very different from the "world without women" recently described by David Noble.[4] Instead, the world of sixteenth-century natural philosophy was the domestic sphere, where women presided as wives, partners, and managers.

The relocation of natural philosophy to the domestic sphere occurred during a period when the internal operations of the household were themselves undergoing a major transition. In the sixteenth and seventeenth centuries, self-sufficient medieval households were giving way to more modern households structured around the husband's wage earning and the wife's shopping in the marketplace for the necessities of daily life. Household size and organization—always an index of social and economic status—had become more important during the course of the sixteenth century, as increasing numbers of professionals and merchants attempted to make the difficult leap to the lower rungs of the gentry.[5] Social and economic historians debate the precise nature and timing of these changes, but few would dispute that they were taking place. Some of the most important changes affected the women who managed these homes, for they were expected to abandon a significant, productive role in the household economy for a nonproductive role. Increasingly, women were encouraged to enjoy time-consuming busywork that would display a husband's prosperity—keeping fresh flowers on the table, candying small violets for the delight of guests, and embroidering needlework that they immediately unpicked to evade the charge of having idle hands.

The movement of natural philosophical practice to the household, during a period when the household itself was undergoing a major transformation, created a problematic and sometimes volatile situation for both Jane Dee and John Dee. According to the sixteenth-century ideal, husbands and wives were to form a managerial partnership (see Figure 1). The Dees' partnership came under stress because their roles were not clearly defined during this transitional period. An additional complication was that Dee's social status as a professional natural philosopher was unclear. Still, both Dees clung to the partnership ideal with a self-conscious determination that allows us to examine closely these overlap-

---

[4] For recent studies on the sites of natural philosophical knowledge in the early modern period see Paula Findlen, *Possessing Nature: Museums, Collecting, and Scientific Culture in Early Modern Italy* (Berkeley: Univ. California Press, 1994), esp. pp. 97–150; Findlen, "Masculine Prerogatives: Gender, Space, and Knowledge in the Early Modern Museum," in *The Architecture of Science*, ed. Emily Thompson and Peter Galison (Cambridge, Mass.: MIT Press, forthcoming); Owen Hannaway, "Laboratory Design and the Aim of Science," *Isis,* 1986, *77:*585–610; C. R. Hill, "The Iconography of the Laboratory," *Ambix,* 1975, *22:*102–110; Adi Ophir, "A Place of Knowledge Re-Created: The Library of Michel de Montaigne," *Science in Context,* 1991, *4:*163–189; Steven Shapin, "The House of Experiment in Seventeenth-Century England," *Isis,* 1988, *79:*373–404; Shapin, " 'The Mind Is Its Own Place': Science and Solitude in Seventeenth-Century England," *Sci. Context,* 1990, *4:*191–218; and William H. Sherman, *John Dee: The Politics of Reading and Writing in the English Renaissance* (Amherst/ Boston: Univ. Massachusetts Press, 1995). Contrast David Noble, *A World without Women: The Christian Clerical Culture of Western Science* (New York: Knopf, 1992).

[5] The literature on early modern households and their organization is sizable and growing. Some of the most provocative and insightful studies include Alice T. Friedman, *House and Household in Elizabethan England* (Chicago: Univ. Chicago Press, 1989); Susan Cahn, *Industry of Devotion: The Transformation of Women's Work in England, 1500–1660* (New York: Columbia Univ. Press, 1987); and Anne Laurence, *Women in England, 1500–1760: A Social History* (New York: St. Martin's, 1994), pp. 108–124. For an anthology of primary texts and images of the household in the period, along with a superb introduction and commentary, see Lena Cowen Orlin, *Elizabethan Households: An Anthology* (Washington, D.C.: Folger Shakespeare Library, 1995). For a discussion of the English and Welsh gentry during the early modern period see Felicity Heal and Clive Holmes, *The Gentry in England and Wales, 1500–1700* (Stanford, Calif.: Stanford Univ. Press, 1994). Heal and Holmes specifically address the inclusion of professional men like Dee in the gentry on p. 7.

Figure 1. *The orderly Elizabethan family is portrayed in this illustration, which accompanied* The Whole Booke of Psalmes *(London, 1563). The wife sits on a stool, hands folded obediently, with her children clustered around her. Opposite, her husband sits in a grand chair, ticking off the points that he is making. (By permission of the Folger Shakespeare Library, Washington, D.C.)*

ping domestic and natural philosophical changes. Jane Dee's life reveals the strain that natural philosophy put on traditional notions of the household and the duties of the house-wife. At the same time, John Dee's experience reveals the tension between his role as husband and father and his role as natural philosopher. Where the two perspectives meet, we gain a new appreciation for the dilemmas Elizabethans faced as they tried to find a place for natural philosophy within existing social and cultural systems.

We are fortunate, in the case of Jane and John Dee, to be able to open the front doors of their home and step inside.[6] Today psychologists would be tempted to label John Dee a clinical narcissist or anal retentive for his exhaustive documentation of everything from his astrological consultations to his wife's menstrual cycles and the injuries his children received while playing in the neighborhood. That these notes are widely scattered through-out manuscripts now in the possession of over a dozen libraries and archives in the United States and Great Britain has, I think, hindered scholarly attention to them. But Dee's notes

---

[6] It has been difficult for historians to enter the rooms and chambers of the past and witness the pressures and tensions that I have mentioned—something that Barbara Hanawalt noted in her study of the intricate workings of medieval households and Steven Shapin has observed as well in his studies of the practice of seventeenth-century science. See Barbara Hanawalt, *The Ties that Bound* (Oxford: Oxford Univ. Press, 1986); and Shapin, "House of Experiment" (cit. n. 4).

about his household practice of natural philosophy are unmatched in extent and detail for the sixteenth century. They invite us to go behind the closed doors of the natural philosopher's study, past the curtains of the bed where he spoke to his wife about his day, and into the library where he discussed natural philosophy and conversed with angels.

* * *

To begin, we turn to Jane Dee. As I noted earlier, natural philosophy in the sixteenth century was not a world without women but a world *among* women, for natural philosophy was the guest of the household during this period. Even in the seventeenth century it was common for natural philosophers to work in the home, as did Robert Boyle (who lived with his sister Katherine, Lady Ranelagh, in her house in Pall Mall) and Robert Hooke (whose "Bohemian" domestic quarters at Gresham College were shared with his niece Grace and supervised by several housekeepers).[7] If Jane Dee's case is representative, the women who supervised natural philosophical households did not lead easy lives. Their central challenge came from the tension between natural philosophy and two early modern ideals of feminine conduct—the supportive "helpmeet" or partner and the virtuous woman or wife. As the wife of a natural philosopher, Jane Dee often had to deny or challenge contemporary notions of feminine virtue in order to support her husband's professional aspirations. In addition, John Dee's natural philosophy interfered with Jane's management of the household, intruded into the intimate relationship between husband and wife, and damaged the household's image in the eyes of the local community.

Jane Fromond Dee would have been trained from an early age to understand the importance of domestic management and of her own impeccable virtue. Though the details of her youth are obscure—we know only that her father's name was Bartholomew, that she was from East Cheam in Surrey, and that she was associated with the entourage of Lady Clinton at the court of Elizabeth—it is likely that Jane Fromond was the daughter of a relatively prosperous member of the gentry. Her education, like that of all girls of the period, would have included a thorough grounding in virtuous housewifery. Even the future Queen Elizabeth had been taught to cook and sew as a girl, and Jane would have learned those skills as well as the importance of money management and thrift, how to manage a staff of domestic servants, and how to offer the hospitality that was an important feature of any Christian household. Conduct literature written to help housewives cope with these responsibilities flooded Europe during Jane's lifetime. Although it is difficult to gauge contemporary adherence to the ideals of female perfection presented by the authors of conduct books, the virtuous woman who emerges from their pages is so complex that every aspect of the household required discussion, from cooking and cosmetic recipes to the management of servants, clothing, children, money, and wagging tongues.[8] The over-

---

[7] Shapin, "House of Experiment," pp. 380–382. Shapin questions whether Hooke's unorthodox living arrangements made up a "home" in the seventeenth-century sense of the word. For our purposes, what is important is that women were present and played a key role in ordering the household.

[8] For insights into the lives of early modern women as household managers see Felicity Heal, *Hospitality in Early Modern England* (Oxford: Clarendon, 1990), esp. pp. 300–388; Susan Dwyer Amussen, *An Ordered Society: Gender and Class in Early Modern England* (Oxford: Blackwell, 1988); Mary Abbott, *Family Ties: English Families, 1540–1920* (London: Routledge, 1993), esp. pp. 96–131; Laurence, *Women in England* (cit. n. 5); Friedman, *House and Household* (cit. n. 5); and Cahn, *Industry of Devotion* (cit. n. 5). Secondary studies of conduct literature include Ruth Kelso, *Doctrine for the Lady of the Renaissance* (Urbana: Univ. Illinois Press, 1956); Katherine Usher Henderson and Barbara F. McManus, *Half Humankind: Contexts and Texts of the Controversy about Women in England, 1540–1640* (Urbana/Chicago: Univ. Illinois Press, 1985); and Suzanne W. Hull, *Chaste, Silent, and Obedient: English Books for Women, 1475–1640* (San Marino, Calif.: Huntington Library, 1982).

arching message to women was clear: feminine virtue was fragile and difficult to maintain, but it was a desirable commodity on the marriage market, the cornerstone of an orderly household, and the pillar of a godly community.

After her marriage to John Dee at the age of twenty-three, however, Jane was expected to manage a unique and problematic household that the authors of conduct literature had never imagined. Heinrich Bullinger's confident assertion that the husband worked outside the home and the "wyves workyng place is within her house, there to oversee and set al thynges inn good ordre," was made without consideration for the complications of life in a natural philosophical household.[9] John Dee's house was not just a home, it was a work-shop; and it was no ordinary workshop, it was a natural philosophical workshop. Though the union of domestic and professional activities under one roof was common in early modern England—many trades and professions were based in the home—natural philos-ophy had only recently moved into the domestic sphere. Thus Jane had no models of virtuous natural philosophical household management to follow but was forced to master her responsibilities *in media res,* as both her husband and the community at large watched attentively for signs of disorder.

Thrift and hospitality were two of the most visible signs of managerial success in the period. For Jane Dee, these aspects of housewifery were especially problematic because of the gap between her husband's social status as the "Queen's philosopher" and his unstable income. A man as highly regarded as Dee attracted visitors from every step in the social hierarchy—the queen, noblemen, other natural philosophers, and the humble residents of Mortlake. It was Jane's responsibility to welcome them all, find them food and drink if required, and send them on their way praising the household's virtue and hospitality. Jane must have been successful in this regard, though John Dee received the credit for her actions; nearly seventy years after his death, a resident of the town of Mort-lake commented upon the Dees' "plentifull Table." Such lavish and memorable hospitality required money. Despite Dee's constant cries of poverty, the household did have an income sufficient to meet its inhabitants' needs as members of the gentry. But Dee's income was insufficient to fund the expenses of natural philosophy: employing additional servants and buying expensive books and equipment.[10] The only way to acquire such funds was to attract rich and royal patrons. So Jane was caught in a double bind: she had to sustain the appearance of a prosperous, genteel, hospitable, and orderly household so that patrons would be encouraged to sponsor her husband, but she could not spend too much of the family's limited assets in the process.

While thrift and hospitality were tangible signs of her managerial abilities, Jane's biggest challenge was maintaining household order and efficiency so she could present a virtuous front to the outside world. In any household that could afford them, the most significant impediment to domestic order was servants—a mobile labor force that worked for room and board, clothing, and sometimes wages. The Dees employed at least twenty-seven female servants and twenty-nine male servants during the course of their marriage. At times the number of individuals living within the household was quite high; in 1592 the Dee household numbered seventeen, nine members of the immediate family and eight

---

[9] Heinrich Bullinger, *The Christian State of Matrymonye,* trans. Miles Coverdale (London, 1575), fol. 73r.

[10] The remark about the Dees' hospitality appears in a report of Elias Ashmole's trip to Mortlake to gather information about John Dee. Ashmole spoke to a "Goodwife Faldo" who had lived near the Dees and was born around 1591. See Bodleian Library MS Ashmole 1788, fol. 149r. The rising price of distillation equipment had already made it difficult for housewives to continue to engage in home brewing, for example, and Dee's equip-ment would have been even more complicated and extensive. See Cahn, *Industry of Devotion* (cit. n. 5), pp. 45–56.

servants.[11] Most of the latter were domestic servants, including nurses, governesses, and wet nurses for the children, housemaids and cooks, gardeners, and manservants. In addition, John Dee employed an assortment of men to assist him in his practice of natural philosophy. We have already met the alchemist Roger Cook, who served for over fifteen years in Dee's stills. Dee employed at least three other men to see angelic visions in his crystal stone, and several boys were put into the Dee household as students of natural philosophy.[12] Jane Dee's management of her experimental household thus involved not only governing the normal range of domestic servants but also watching over her husband's assistants and apprentices.

The usual run of personnel problems among domestic servants seems to have been multiplied in the Dee household. Two maids, Jane Gaele and Mary Cunstable, set their room on fire twice in a single year. Nurse Ann Frank confessed to being "tempted by a wycked spirit." John Dee, his natural philosophical curiosity stirred, may have made the situation more difficult by examining the woman and pronouncing her possessed. Despite Dee's repeated ministrations and several applications of holy oil, nurse Frank tried to throw herself down the well and was later found dead in a room upstairs. Their manservant George, he who tumbled off ladders, was finally dismissed after he stayed out all night drinking and returned home to heap verbal abuse on the rest of the household.[13]

And that was only the domestic staff—John's assistants were even more difficult to manage. Problems with the natural philosophical assistants were virtually guaranteed, since John sought out men with the melancholic temperament that increased their susceptibility to divine and astral influences.[14] From Jane's perspective, however, melancholia just disturbed the household by filling it with exceptionally erratic, moody, and difficult residents. Roger Cook, for instance, kept "pycking and devising occasions . . . to depart on the sudden," removals that John Dee apologetically explained away with references to his "melancholik nature." Though educated and of a higher social status than the domestic servants, John's assistants also tended to have disreputable pasts. Edward Kelly, for example, was accused of counterfeiting money and sported at least one alias. Three assistants—Kelly, Richard Walkdyne, and Robert Web—had legal difficulties while living with

[11] See Dee, "Compendious Rehearsal," p. 38. The size of the Dee household at this time—which can serve as an index of prosperity and prestige—would have placed it among the highest levels of the gentry in a comparable London suburb of the period. See Marjorie K. McIntosh, "Servants and the Household Unit in an Elizabethan Community," *Journal of Family History,* 1984, 9:3–23, esp. p. 8.

[12] A total of twenty-nine servants were described as "maids" or "manservants," without specific reference to their duties. Nine servants were described as nurses, and eight of them were wet nurses (both residential and nonresidential) hired at different times for the children. The practice of hiring wet nurses was common in the period, and one recent study has concluded that merchant and professional families like the Dees were as likely to have their children nursed as aristocratic families and more likely to send children away to be nursed. See Gillian Clark, "A Study of Nurse Children, 1550–1750," *Local Population Studies,* 1987, 39:8–23. Roger Cook and Robert Gardener both served Dee as alchemists; Cook entered the household when he was fourteen, a common age for apprentices to go into service. In addition, Morrice Kyffin served Dee in some unspecified philosophical capacity, and Barnabas Saul, Edward Kelly, and Bartholomew Hickman all were employed as scryers. Three young boys—two sons of Goodman Hilton and Benjamin Lock, the son of the merchant adventurer Michael Lok—resided temporarily in the Dee household; they may have received part of their education there.

[13] Dee, *Private Diary,* ed. Halliwell, pp. 7, 35–36, 18.

[14] As Robert Burton, the author of the most famous treatise on melancholy, argued, "melancholy men of all others are most witty, [and their melancholy] causeth many times divine ravishment, and a kind of enthusiasmus . . . which stirreth them up to be excellent Philosophers, Poets, Prophets, &c." See Robert Burton, *The Anatomy of Melancholy,* ed. Thomas Faulkner, Nicolas Kiessling, and Rhonda Blair, 3 vols. (Oxford: Clarendon, 1989), Vol. 1, p. 400. See also E. Patricia Vicari, *The View from Minerva's Tower* (Toronto: Univ. Toronto Press, 1989); and Lawrence Babb, *The Elizabethan Malady* (East Lansing: Michigan State College Press, 1951), esp. pp. 58–67.

the Dees. Magistrates came to the house to take them into custody; legal inquiries were launched; property was confiscated.[15]

The staff's bad behavior reflected poorly upon Jane's managerial skills, which probably explains why she sometimes flew into what John described as her "mervaylous rages" at the servants' behavior, especially that of his assistants. While the wife was traditionally responsible for governing all servants under her roof, it was not clear who should exercise authority over apprentices living in the master's household. This ambiguity rendered Jane's ability to curb her husband's assistants problematic, as can be seen in the case of Edward Kelly. Jane developed a dislike for Kelly from the moment he appeared in March 1582 under the name "Edward Talbot." In the following weeks Edward forced another philosophical assistant out of the house, was discovered to have been traveling under an alias, falsely claimed that he had never been formally educated, demonstrated abundant evidence of a melancholic temperament, and monopolized Dee's time and attention with his unique access to angelic visions.[16]

John Dee's intense interest in angels has been one of the most puzzling features of his intellectual life. Dee's conversations with angels stemmed from a period of dissatisfaction with his study of the natural world and the authoritative texts that attempted to disclose its secrets. By the 1570s and 1580s, as Dee explained in a memorandum included at the beginning of his surviving angel diaries, he concluded that he "could fynde no other way, to . . . true wisdome" about the natural world except through the "extraordinary gift" of divine revelation.[17] In an effort to be worthy of God's great gift and to facilitate the entertainment of angelic visitations, Dee set out to lead a godly life and employed scryers to see visions in a collection of crystal stones. Together, Dee and his scryer would set a crystal stone near a window where it could catch the natural light entering the house; after much prayer and humble supplication, the scryer was usually able to detect angels within the stone. Once this level of success had been achieved, the angels revealed to Dee information about the end of nature, the divine language shared by God and the angels that would help to reorder the chaotic world, the heavily alchemical natural philosophy that would be practiced in the world to come, and the role that Dee himself would play in the upcoming apocalypse.[18]

With Kelly and the angels in her house, Jane could not schedule regular meals, was forced to rearrange her rooms to afford her husband and Kelly a more private spot to talk to the angels, feared admitting guests lest they interrupt activities that would further the

---

[15] Dee, *Private Diary,* ed. Halliwell, pp. 12, 47–48, 60.

[16] Dee's annotations in Bodleian Library MS Ashmole 487, entry for 5 May 1582 (rages); and Dee, *Private Diary,* ed. Halliwell, pp. 14–15. The published version of Dee's *Private Diary* edited by Halliwell was drawn from Bodleian Library MSS Ashmole 487 and 488, but it must be used with caution, as Halliwell omitted and mistranscribed a number of entries.

[17] Dee, *Actions with Spirits,* ed. Whitby, Vol. 2, pp. 8–12. The two most influential studies of Dee's life and intellect are Peter French, *John Dee: The World of an Elizabethan Magus* (London: Routledge, 1972); and Nicholas Clulee, *John Dee's Natural Philosophy: Between Science and Religion* (London: Routledge, 1988). Biographical studies of Dee have emphasized the mathematical, alchemical, and navigational aspects of his intellectual career while understating the importance of his sustained intellectual interest in the angels. For two studies that place Dee's angel conversations at the center of analysis see Whitby's introduction to Dee, *Actions with Spirits,* ed. Whitby; and Harkness, *Talking with Angels* (cit. n. 2). The manuscript angel diaries that survive are scattered in libraries throughout England, but a good (though not complete) sense of their contents can be ascertained from Dee, *Actions with Spirits,* ed. Whitby; Dee, *True and Faithful Relation,* ed. Casaubon; and Josten, "Unknown Chapter in the Life of John Dee" (cit. n. 3).

[18] For a description of these scryers and the preparations required for angel conversation see Harkness, "Scientific Reformation" (cit. n. 2), pp. 325–392. On the information revealed by the angels see *ibid.,* pp. 394–550; and Harkness, "Shows in the Showstone" (cit. n. 2).

public belief that her husband was a conjurer, worried about the family finances (angels were not nearly as profitable as astrology), and suffered through Edward's tantrums, threatened departures, and public scenes. Kelly's presence placed Jane Dee in a difficult position. Her husband needed Kelly for his natural philosophy, yet Kelly damaged the order and virtue of *her* household. Jane had two options: she could maintain silence and be the supportive helpmeet, as most conduct books would have advised; or she could speak out, which those same books would have condemned. Jane preferred to speak out: John's diaries are filled with anxious references to her anger and displeasure with Kelly. John was always grateful when Jane's anger could be soothed, as on 26 April 1583, when Kelly returned to their house after a falling-out and John noted that Jane was "very willing" to welcome him back and was "quietted in mynde, and very frendely to E[dward] K[elly] in Word, and cowntenance."[19] Matrimonial love and affection, his own anxiety about Kelly's conduct, and the chaos ensuing from the disruption of household routine help to explain John's meticulous accounts of these volatile events.

When Jane's anger spilled over into meddling with Dee and Kelly's shared intellectual projects, however, she was quickly reminded of her inherent female inferiority and her properly subservient role within the home. In March 1585 Jane decided to approach the angelic advisers, through the intermediaries of her husband and Kelly, to remind them that her household was in dire financial straits because of natural philosophy. Jane argued that she and her husband required "sufficient and needful provision, for meat and drink for us and our Family." She made it clear that while natural philosophy might be the family business, God surely did not want her to pawn what she described as "the ornaments of our House, and the coverings of our bodies" in order to pursue it. An angel reproached Jane, reminding her that, as a woman, she was forbidden to "come unto the Synagogue"— the room where her husband communicated with angels. This passage makes it clear that Dee's natural philosophy had not only entered the house—it had taken over. The house was no longer Jane's; it was a sacred space, a place where angels discussed natural philosophy with her husband. After vague promises that God would look after her children and further reminders to be "faithful and obedient . . . as thou art yoked," the angel exhorted Jane to stop "murmurring . . . and sweep your Houses."[20] Jane's response to the angel's suggestions is unfortunately not recorded.

The angel's comments placed Jane in another double bind: it was her responsibility to maintain household order and efficiency, but it was not her right to question her husband's professional activities in that household. The angel, speaking through Jane's nemesis Edward Kelly, clearly wanted to remind her that the silent and obedient fulfillment of housewifely duties was the seat of female virtue.[21] This argument was familiar, and though authors of conduct books disagreed on precisely how virtue could be manifested, the commonplace "chaste, silent, and obedient" captures its essence. Actual behavior, as I have pointed out, may not have coincided with the ideal. Jane Dee certainly did not fit into this neat model of female virtue. The problem was not simply a matter of Jane's moral limitations; rather, the model failed to accommodate the peculiar conditions arising from

---

[19] On problems with Kelly see, e.g., Dee, *Actions with Spirits,* ed. Whitby, Vol. 2, pp. 137, 351 (quotation), 407; and Dee, *True and Faithful Relation,* ed. Casaubon, p. 23.

[20] Dee, *True and Faithful Relation,* ed. Casaubon, pp. 389–390. John Dee acknowledges that he did, in fact, pawn Jane's jewelry, although he is not specific as to when this took place. See Dee, "Compendious Rehearsal," p. 35.

[21] The sentiment was common. See, e.g., Barnaby Rich, *The Excellencie of Good Women* (London, 1613), p. 24, quoted in Kelso, *Doctrine for the Lady of the Renaissance* (cit. n. 8), p. 116.

the presence of her husband's natural philosophy in the most important showplace for female virtue: the household.

Another major aspect of the feminine ideal made problematic for Jane Dee by her husband's practice of natural philosophy was chastity. Chaste wives were faithful to their husbands within marriage and mindful of God's instructions to be fruitful and atone for the sins of Eve through the pains of childbirth. Chastity did not denote sexual abstinence but, instead, was an ingredient in maintaining a happy, and fertile, family. Both men and women needed to manage their sexual impulses, however, and guard against carnal lust. Sexuality was thus an important issue in the period. But sexuality was of special concern in the Dee household because sex, unlike melancholy, could undermine the success of the natural philosophy practiced there. Alchemy, astrology, and communicating with angels all demanded spiritual and sexual purity on the part of their practitioners if positive results were to be obtained.[22]

Abstinence may have made John Dee a better natural philosopher, but it made Jane Dee a poor wife because it limited her ability to be fruitful. John Dee responded to this contradiction between his natural philosophical success and his wife's reproductive responsibilities by scrutinizing and orchestrating their sexual relations. He kept secretive but exacting records of Jane's menstrual patterns in his personal diaries, using the astrological sign for Aquarius with comments such as "Jane had a small show" encoded in Greek characters on each day of her menstrual periods. References to sexual intercourse appear in John's use of the astrological sign for Mars—the planet of lust and desire as well as war. Sometimes, intriguing notes accompany the mark of Mars in Dee's manuscript diaries (though not in the printed edition)—as on 17 January 1601, when John Dee indicated that Jane initiated sex. A rough translation of the note reads "[Sex with] Jane around 10 at night. The woman voluntary"—literally, "femina voluntaria."[23] Dee may have included the remark to absolve himself of any natural philosophical implications arising from the union, since it was his wife's idea.

These detailed and copious records indicate that managing the couple's sexual and reproductive life became yet another feature of John Dee's natural philosophy. Moral implications alone cannot account for Dee's intense and sustained intellectual interest in his wife's body. Still, Dee's brief comments, provocative as they are, do not clearly reveal his reasons for keeping such odd records. One possibility is that Dee was studying his wife's fertility. In addition, he may have been collecting data for contemporary astrological debates as to whether it was better to calculate nativities according to the date of birth or the date of conception. Tracking sexual activity, menstrual cycles, and his children's dates of birth may have given Dee insights into the question. Often his references to sexual activity were accompanied by astrological annotations on planetary positions and relationships, suggesting that astrological issues were involved. And when Jane miscarried in the beginning of their marriage, Dee examined the fetus and puzzled over its shapelessness, wondering in his manuscript annotations, "Where are the separate limbs?"[24] Natural phi-

---

[22] One of the clearest statements of this belief appears in a work attributed to Agrippa, which advocated that practitioners should be confessed of their sins, contrite of heart, chaste, abstinent, and fasting. See Henry Cornelius Agrippa, *His Fourth Book of Occult Philosophy,* trans. Robert Turner (London, 1655), pp. 59–60.

[23] Bodleian Library MS Ashmole 488, entry for Jan. 1601. The note that accompanies the symbol for Mars reads "Jane hora circiter 10 nocte. [sign for Venus] voluntaria." Dee's notes concerning Jane's menstruation appear at one-month intervals throughout his personal diary, with the exception of nine-month periods when she was pregnant. See, e.g., Dee's annotations in Bodleian Library MSS Ashmole 487 and 488.

[24] Bodleian Library MS Ashmole 487, entry for 3 July 1580.

losophy, which had invaded Jane's domestic space, went on to invade her body as well. Her body became another of her husband's laboratories, a site of knowledge, like Dee's crystal show stone, whose purity was important.

It is all the more surprising, therefore, that Dee's natural philosophy was responsible for one of the most notorious episodes in their marriage. In April 1587 John Dee's intellectual life was dominated by two activities: conversations with angels and alchemical experiments. Assisting him in both areas was the difficult and disreputable Edward Kelly, who had introduced a wife, Joan, into the Dee household, now located in the Central European city of Trebona. Dee's angelic teachers had instructed the pair to make the move in order to hasten the world's progress toward apocalypse and to secure like-minded patrons such as King Stephen of Poland and the Holy Roman Emperor, Rudolf II. At this time, the partnership between Dee and Kelly was showing signs of strain. Jane's dislike of Edward Kelly had grown intensely; Kelly was beginning to strike out on his own as an alchemist at Rudolf's court; and John was frustrated by the enormous task of trying to decipher God's new and improved natural philosophy, which came to him through the mediation of the angels, a crystal stone, and his increasingly distracted scryer.

The angels offered a solution: Dee and Kelly should share all things between them—alchemical secrets, angelic revelations, and their wives, thus demonstrating their faith in their philosophical partnership. God, in turn, would bless them with additional insights into the practice of natural philosophy and the workings of the natural world. John Dee was horrified but, over dinner, dutifully informed his wife and Joan Kelly of the plan for "the common and indifferent using of Matrimonial Acts amongst any couple of us four." In what must be the understatement of a lifetime, Dee noted in his diary that the idea was "strange to the women." Despite the unease Dee and the women shared, further discussions with the angels failed to yield a compromise between sixteenth-century virtue and God's newest commandment. John went to bed and found his wife awake and anxious for news. "Jane," Dee told her, "I see that there is no other remedy, but as hath been said . . . so it must needs be done." Jane "fell a weeping and trembling for a quarter of an hour," and John "pacified her as well as I could." Finally, after entreaties and explanations, John Dee convinced his wife to go along with the plan, and Jane "shewed her self to be . . . resolved to be content for God his sake and his secret Purposes." Contracts and agreements for the "pactum sacrum"—the sacred covenant between the Dees and the Kellys—were drawn up and signed. A terse note—"Pactum factum"—in Dee's personal diary on 11 May 1587 suggests that the covenant was soon consummated.[25] Once the two couples had physically consummated the relationship, the angelic revelations increased, including important information about the upcoming apocalypse and the new heaven and earth that would result.

This was the most severe blow that natural philosophy struck against Jane Dee, virtuous Elizabethan housewife. Her house had been invaded, her body examined, and her marriage irreparably subverted—all in the service of natural philosophy. The infusion of natural philosophy into every aspect of Jane Dee's life suggests how further study of natural philosophers' wives might help us understand the complex, transitional world of early modern science. The case of Jane Dee suggests that natural philosophy's move from the domestic sphere was not only about issues of privacy, prestige, and solitude—it was about who should control the physical and social settings where natural philosophy was practiced. Jane's activities remind us that as long as natural philosophy was practiced in the home women were not only present, but influential. From calls to dinner and ultimatums against

---

[25] Dee, *True and Faithful Relation,* ed. Casaubon, pp. *11–12, *13; and Bodleian Library MS Ashmole 488, entry for 11 May 1587.

philosophical assistants to angelic petitions and a trip to Central Europe to further patronage opportunities, Jane Dee's life was inextricably bound up with her husband's professional interests. Sometimes as a helpmeet and sometimes as a hindrance, Jane negotiated the boundaries of natural philosophy and challenged them whenever they came into conflict with her own duties and responsibilities.

* * *

Having viewed the "experimental household" through Jane's eyes, I would like to look at the same situation from John's perspective. Specifically, I would like to examine more closely the tension between Dee's need for privacy—to practice the occult features of his natural philosophy and to read and write in his library, for example—and his need to open his doors to a larger public so that he could gain patrons and sponsors. This tension exacerbated the heavy burdens on Jane and confused the traditional division of labor within the household. How was Jane to provide open hospitality if her husband needed privacy? How was she to protect his privacy if she also needed to be his helpmeet and attract patrons?

From John's perspective, however, the issue was female authority in his workspace. Traditionally, the workplace was defined as male space (even if women were present, as in merchant and craft households), and male authority was sacrosanct there. Yet the management of the household was considered to be his wife's primary responsibility, and John had to negotiate continually for the space in which he practiced his natural philosophy. Because John had to construct a private space within a public household, it must at times have seemed as though Jane was actually in control of the parameters of his professional life.[26] Throughout Elizabethan England there was widespread concern over women gaining the upper hand through their preeminence in the domestic sphere, and it is difficult to pick up a conduct book or see a play from the period that does not include a representation of the henpecked husband and the overbearing wife.

Along with concerns about negotiating with the woman in control of his workplace, a more fundamental issue troubled John Dee: how to negotiate the tricky boundary separating public and private in the practice of natural philosophy. John was caught between a socially accepted desire for a place where he could practice natural philosophy in peace and quiet and his family's financial needs. Natural philosophers articulated their desire for privacy in a code of professional conduct that advocated withdrawal from the public gaze. The reasons for this behavior were twofold: theoretically, the natural philosopher was then able to devote himself to God and to the contemplation of God's mysteries in the natural world; practically, the natural philosopher's own mysteries were concealed and thus less susceptible to misconstrual by the ignorant. No matter the reason, such a strategy was difficult to enact for anyone who depended on revenue from patrons to survive. While Steven Shapin has viewed such a combination of private and public impulses with suspicion, arguing that an early modern natural philosopher's plea for solitude "rarely meant absolute aloneness" but instead was often "an intensely public pose," he does not take into account how much was at stake when a natural philosopher maintained a truly open, public household. Dee and his wife worried, for instance, about their neighbors' reactions to the odd things that went on in the house. And with good cause: the house was ransacked twice during the course of their lives, a later account reports that the children of Mortlake

---

[26] For the difficulties associated with this juxtaposition of public and private from the distinct vantage point of the gentry see Heal and Holmes, *Gentry in England and Wales* (cit. n. 5), pp. 278–280.

"avoided [Dee] because he was counted a conjurer," and on one occasion some of his scientific instruments were destroyed.[27]

This situation was made more difficult by John's interest in the occult sciences, and Jane must have agonized repeatedly over who should—and should not—be admitted to the house given what was often going on behind closed doors. On more than one occasion, for example, John's angel conversations had been interrupted by unexpected visitors. Since he was using a crystal ball, a table painted with strange letters and sigils, and other occult trappings, the visitors must have been justifiably curious about what the good doctor was doing. To help protect his conversations from inquiring eyes, John secluded them in a "private study" set off from the household by a set of double doors. Sometimes he closed both doors as a signal that strict privacy was desired; if only one door was closed, he could be interrupted. Only Jane ignored the signal—she called him to dinner regardless. John made the situation more difficult by the frequency with which he forgot to close both doors, enabling a patron's messenger to burst into the study in the midst of an angel conversation. John, "fearing his rash opinion afterward of such things, as he could not perceive perfectly what my Companion and I were doing," asked the angels for advice on how to rearrange things to ensure more privacy. Given the limited domestic space, the angels resorted to gentle reminders when John forgot to shut the door rather than recommending redecoration.[28]

These intrusions into John's private studies underscore the impossibility of reaching a compromise between the contemplative life of the scholar and the active life of the courtier and natural philosopher within a domestic space. John's last hope for a solution was to find a bigger house. While the Dees' home in Mortlake was probably larger than that of the average citizen in the town, it was not nearly so impressive or spacious as John felt the house of someone who aspired to be the Queen's official natural philosopher should be. He made this clear in a petition to Elizabeth in 1592. Once again in England following his fall from influence at the court of Rudolf II, Dee requested that Elizabeth give him the hospital of St. Cross, near Winchester, specifically because it was larger than their current residence. First, he noted, St. Cross possessed more rooms for "learned men to be entertained and lodged in, in far better manner" than Dee and his wife were able to provide in what he disparagingly called "the Mortlake Hospital for Wandering Philosophers." This

[27] Shapin, " 'The Mind Is Its Own Place' " (cit. n. 4), p. 195. See Ashmole's account of John Aubrey's visit to Mortlake to inquire about Dee and his household in Bodleian Library MS Ashmole 1788, fol. 147r. For Dee's remarks about the damage done to his library see Dee, "Compendious Rehearsal," pp. 25–32. Julian Roberts and Andrew Watson have questioned the tale that historians have repeated concerning a "mob" that marched on the Mortlake house in Dee's absence and destroyed part of the library. See Julian Roberts and Andrew Watson, *John Dee's Library Catalogue* (London: Bibliographical Society, 1990), pp. 49–52. Diane Shaw has explained how important—and contentious—privacy issues could be in the period, and from her examination of the cases brought before the Assize of Buildings and the Assize of Nuisance in London (cases that involved complaints about smells, noises, and heavy traffic), we can surmise that the Dees would not have been considered ideal neighbors in sixteenth-century England. See Diane Shaw, "The Construction of the Private in Medieval London," *Journal of Medieval and Early Modern Studies,* 1996, 26:447–466.

[28] See Dee, *True and Faithful Relation,* ed. Casaubon, pp. 252, 23 (quotation); and Dee, *Actions with Spirits,* ed. Whitby, Vol. 2, p. 63. In the domestic sphere, interruptions were the greatest threat to natural philosophical productivity. In Italy, Alberti, Cortesi, and others recommended that the philosopher's study or library be placed in the innermost reaches of the house; see Findlen, *Possessing Nature* (cit. n. 4), pp. 110–112. The French philosopher Montaigne handled potential interruptions within his household by retreating to the upper levels of a tower library; see Ophir, "Place of Knowledge Re-Created" (cit. n. 4), pp. 168–169. Dee's Mortlake home did not offer him such a remote site for his studies, but while in a rented house in Central Europe he was able to use a tower for private study and angel conversations; see Josten, "Unknown Chapter in the Life of John Dee" (cit. n. 3), p. 240.

document shows that John, too, was affected by the logistical difficulties Jane faced in the day-to-day operation of their home. Moreover, though the Mortlake house did not have adequate kitchens, St. Cross had ample storage space for food to feed the "true and earnest students" who would be permanent residents.[29] These remarks make it clear that Dee wanted to expand, rather than contract, his coterie of apprentices and assistants—a plan that Jane would hardly have endorsed.

John's petition for St. Cross foregrounds once again the fact that he and his wife were at cross-purposes when it came to what was best for the prosperity of the household and the furtherance of his natural philosophy. While John wanted more room to ensure the privacy of his natural philosophical activities, Jane was already frustrated by the problems of staffing such a large household given the lack of patrons—and money—her husband managed to attract. John was unclear about what should be done to resolve the situation. While he agreed that the benefits of a house on the high road to court could be substantial in terms of patronage, in the petition John expressed a desire to "retyre . . . from the multitude and haunt of my common friends, and other[s], who visit me. Which thing without offense, and loss, or breach of some folkes friendship, cannot be conveniently performed, while I continually am at my house at Mortlake; the passage and way to my house there is so easy, neere, and of light cost from London or court." Retirement would not have sat well with Mrs. Dee, who often found herself at court asking for financial assistance and support from her former employers. So, for the sake of his family, Dee made references to the hotbed of intellectual life St. Cross could become. He looked forward to welcoming scholars who would flock there "from all parts of Christendome . . . when . . . with me . . . in such a solitary and commodious place, they may dwell in freedome, security, and quietnes." Dee surveyed maps of the area and discovered that St. Cross was particularly accessible to foreign visitors; he noted that its proximity to the coast of England would permit "the secret arrival of special men . . . unto me there . . . some [who] . . . would be loath to be seene or heard of publickly in court or city."[30]

This is hardly the image of solitary study that we would associate with the contemplative life of a scholar. Foreign philosophers scurrying to and fro—many of them arriving surreptitiously to escape detection—assistants, and students would have occupied an enormous amount of Dee's time and drained the household's purse. The petition for St. Cross—which ultimately failed—highlights, therefore, many of the problems that faced the Dees. First, Jane had to have adequate space to entertain patrons and scholars. Second, the house should ideally be accessible to the court—but not so much so that the curious could drop in night and day. Third, the house must provide enough room for John to perform his experiments without his wife having to worry about what visitors might see. Fourth, Jane had to provide Dee's experimental assistants with room and board—and therefore John had to have a regular source of income from patrons.

Caught, as even this brief description demonstrates, between the library and the laboratory, the court and the city, the active and the contemplative life, John Dee practiced

---

[29] Dee, "Compendious Rehearsal," p. 40. For an interpretation of these conflicting ideals in a slightly later period see Shapin, " 'The Mind Is Its Own Place' " (cit. n. 4).

[30] Dee, "Compendious Rehearsal," pp. 40–41. Jane's connections to the court and the influence she may have wielded there on behalf of her husband and family are worthy of further study for the light they may shed on the actions of other women of the gentry. John Dee's diaries specifically mention one trip to the court and two occasions on which she appeared before Queen Elizabeth. See Dee, *Private Diary,* ed. Halliwell, pp. 5, 49, 51. On 7 Dec. 1594 Jane appears to have been in the intimate royal Privy Garden at Somerset House and delivered what John described as a "supplication to the Quene's Majestie" to one of Elizabeth's courtiers, Sir Thomas Heneage.

Figure 2. In this illustration from Thomas Heywood, A Curtaine Lecture *(London, 1637), the vulnerability of husbands to their wives' opinions is made clear. The wife looms over her husband as she lectures, "I proclaim the truth." The husband responds defensively, "Don't believe women." The detailed rendering of the domestic space and the husband's entrapment within the confines of the bed and its curtains attest to the idea that the husband is weakest in the domestic sphere where his wife is most powerful. (By permission of the Folger Shakespeare Library, Washington, D.C.)*

and produced natural philosophical knowledge at a time when both the locations and the boundaries of science and scientific discourse were continually being reconfigured and redefined. At his side was his wife, and though Jane Dee comes to us through the mediating hand of her husband, she emerges as a person very much concerned with her husband's work and with the effects of that work on their family, one with a shrewd ability to judge how the wide variety of activities that took place in the household might damage their reputation. In addition, there is evidence that Jane Dee had the power to shape and influence—though not entirely control—her husband's practice of natural philosophy through her interactions with servants, by interrupting him when natural philosophy upset carefully managed domestic schedules and arrangements, and occasionally by intervening in the natural philosophy itself.

* * *

The Dee household proved to be experimental in two significant ways. First, scientific experiments took place there. The resituation of natural philosophy not only transferred alchemy and astrology from the monastery to the household; it also generated new scientific practices, as mundane domestic events were subjected to scientific observation and documentation. Second, the household arrangements surrounding these invasive new activities necessarily became experimental in the family's effort to contain natural philosophical practices. As experiments go, the latter was not a clear success: eventually other venues were tested and found more acceptable. Academies, scientific societies, and universities proved more compatible with the increasingly experimental culture of science than had the domestic sphere. In addition, natural philosophers could reign supreme in these spaces, while in the household they were forever at risk of being dominated by their wives. As Thomas Heywood made clear in his collection of stories and anecdotes about women, *A Curtaine Lecture* (1637), it was not possible to shut out the voices of women when in the household (see Figure 2). Though it might be argued that women such as Jane Dee were merely on the margins of natural philosophy during the early modern period, we should be wary of dismissing them too quickly. If we ever hope to understand how science moved from the monastery and university to the laboratory and academy, we must examine not only these end points of historical transition but also the liminal space between them: the early modern household. And, when examining the household as a site of knowledge, we must not forget that the early modern household was, first and last, domestic and feminine space.

Maria Winkelmann
at the Berlin Academy

A Turning Point for Women in Science

By Londa Schiebinger

> If one considers the reputations of Madame Kirch [Maria Winkelmann] and Mlle Cunitz, one must admit that there is no branch of science . . . in which women are not capable of achievement, and that in astronomy, in particular, Germany takes the prize above all other states in Europe.
> —ALPHONSE DES VIGNOLES, vice president of the Berlin Academy, 1721

EUROPEAN SCIENCE WAS in many ways a new enterprise in the seventeenth and eighteenth centuries, an enterprise that (at least ideologically) welcomed a broad participation. The regulations of the newly founded Berlin Academy stressed that modern science could flourish only with contributions from men of all social classes, nationalities, and religions.[1] Was this ideological largess to be extended to women as well? After centuries of proscribing women from active participation, were centers of European intellectual life now to open their doors to them?

The major European academies of science were founded in the seventeenth century—the Royal Society of London in 1662, the French Académie Royale des Sciences in 1666, and the Berlin Akademie der Wissenschaften in 1700.[2] Women were not, however, to become regular members of these academies for three

Support for this research and for my larger project on women and the origins of modern science was provided by the Deutscher Akademischer Austauschdienst, the Rockefeller Foundation, and the National Endowment for the Humanities. My thanks to those friends and critics who read and commented on this essay: Robert Proctor, Richard Kremer, Margaret Rossiter, Roger Hahn, Merry Wiesner, Robert Westman, and Lyndal Roper.

[1] "General-Instructions für die Societät der Wissenschaften vom 11. Juli 1700," in Adolf von Harnack, *Geschichte der Königlich Preussischen Akademie der Wissenschaften zu Berlin*, 3 vols. (1900; Hildesheim: Georg Olms, 1970), Vol. II, p. 106. Thomas Sprat made a similar point in *The History of the Royal Society of London* (London, 1667), pp. 62–63, 72. Harnack, who was the father of German feminist Agnes von Zahn-Harnack, gave the highlights of Winkelmann's story in his three-volume work. Such a sensitivity to the plight of women is rarely found in official (and essentially laudatory) histories of national institutions.

[2] The Berlin Academy first bore a Latin name, the Societas Regia Scientiarum. From its founding, it was also commonly known as the Brandenburgische or Berlin Societät der Wissenschaften. In the 1740s it took a French name, Académie Royale des Sciences et Belles-Lettres. In the 1780s it became the Königlich Preussische Akademie der Wissenschaften, which it remained until its reorganization after World War II, when it took its present name, the Akademie der Wissenschaften der Deutschen Demokratischen Republik. For simplicity I refer to the Societät der Wissenschaften as the Berlin Academy or the Academy of Sciences.

39

centuries. At the Royal Society in London, the first women members—Marjory
Stephenson and Kathleen Lonsdale—were not elected until 1945. The prestigious
Académie des Sciences in Paris did not admit a woman as a full member until
1979 (Yvonne Choquet-Bruhat). Even Marie Curie, the first person ever to win
two Nobel prizes, was denied membership. No woman scientist was awarded
membership at Berlin until 1949, when Lise Meitner was made a corresponding
member.[3]

Why were women denied membership in the scientific academies of Europe? It
would be a mistake to think there were no qualified women scientists when the
academies first opened their doors. There were, in fact, a significant number of
women trained in the sciences. The exclusion of women was not a foregone
conclusion but resulted from a process of extended negotiation between these
women and the academy officials.

The case of Maria Winkelmann at the Academy of Sciences in Berlin was a
decisive and pivotal one.[4] In 1712 she lost her year-long battle to become Academy astronomer. Already a seasoned astronomer when her husband, the Academy astronomer Gottfried Kirch, died in 1710, Winkelmann asked to be appointed in her husband's stead. But although Leibniz was among her backers,
her request was denied, setting an important precedent for women's participation
in the scientific work of the Academy.

The story of Maria Winkelmann's rejection by the Academy is a compelling
one. But more important, it illustrates patterns in women's participation in early
modern science. When Winkelmann petitioned the Academy in the early years of
the eighteenth century, she was caught in conflicting social trends. Craft traditions, on the one hand, fostered women's participation in science. Through apprenticeships, women gained access to the secrets and tools of a trade, whether
illustrating manuscripts or using telescopes. As we will see, Winkelmann's petition to become Academy astronomer drew legitimacy from guild traditions that
recognized the right of a widow to carry on the family business. Craft traditions,
however, were counterbalanced by other trends, both old and new. For centuries, women had been excluded from universities, as they were to be excluded
from the new scientific academies. In many ways, the new trend of professionalization served to reaffirm the traditional exclusion of women from intellectual
culture. Indeed, there were those at the Berlin Academy who judged it improper

---

[3] See *Notes and Records of the Royal Society of London*, 1946, 4:39–40; and Lise Meitner "The
Status of Women in the Professions," *Physics Today*, 1960, *13*(8):16–21. See also Kathleen Lonsdale's thoughtful "Women in Science: Reminiscences and Reflections," *Impact of Science on Society*, 1970, *20*(1):45–59. The Berlin Academy, unlike its fraternal counterparts in London and Paris,
awarded honorary membership to a few women of high social standing—Catherine the Great (elected
1767), Duchess Juliane Giovane (1794), and Maria Wentzel (1900) (see Section VIII below).

[4] When writing the history of women, we immediately face problems even in such small matters as
what name to use for our main character. To employ the married name reveals a nineteenth- and
twentieth-century bias. I have elected to use Maria Winkelmann's maiden name throughout since this
is the name she used for her publications. (Her name as it appears here has been modernized from the
original Winckelmannin.) The use of the maiden name is also consistent with the practice of the day.
Astronomer Maria Cunitz, for example, published under her maiden name; midwives listed in eighteenth-century Prussian address lexicons also used their maiden names (their married names were
given in parentheses). The use of Winkelmann's maiden name also makes it easier to distinguish her
from her husband without falling into the somewhat degrading habit of using her first name. Winkelmann did also refer to herself as "Kirchin" (the feminine form of Kirch, her husband's name) in
correspondence with Leibniz and Academy officials, playing (I assume) on her husband's name in her
quest for employment. Academy officials referred to her as "Kirchin" or the "widow Kirch."

for a woman to practice the art of astronomy and suggested that Winkelmann return to her "distaff" and "spindle."[5]

I. WOMEN AND CRAFT TRADITIONS IN EARLY MODERN ASTRONOMY

Edgar Zilsel was among the first to point to the importance of craft skills for the development of modern science in the West. Zilsel located the origin of modern science in the fusion of three traditions: the tradition of letters exemplified by the literary humanists; the tradition of logic and mathematics exemplified by the Aristotelian scholastics; and the tradition of practical experiment and application exemplified by the empirical artist-engineers. Astronomy drew from each of these traditions. It was, however, the craft aspects of astronomy that were especially important in the sixteenth and seventeenth centuries. The astronomer was both theoretician and technician; he or she was versed not just in Copernican theory and mathematics, but also in the arts of glass grinding, copperplate engraving, and instrument making. These were skills of the artisan, not of the scholar. Thus the astronomer of the seventeenth century bore a close resemblance to the guild master or apprentice.[6]

Astronomers were never, of course, officially organized into guilds. Yet craft traditions that molded all aspects of working life in early modern Europe were very much alive in astronomical practices. In many ways the legal and political structures of the guild merely codified these wider practices and traditions. This was especially true in Germany, where stirrings of industrialization came late. Whereas in England and Holland guilds declined from the mid-seventeenth century, in Germany they remained an important economic and cultural force well into the nineteenth century.[7]

The new value attached to the traditional skills of the artisan allowed for broader participation in the sciences. Of the various institutional homes of astronomy, only the artisanal workshop welcomed women. Women were not newcomers to the workshop: it was in craft traditions that the fifteenth century writer

---

[5] Alphonse des Vignoles, "Eloge de Madame Kirch à l'occasion de laquelle on parle de quelques autres Femmes & d'un Paison Astronomes," *Bibliothèque germanique*, 1721, *3*:115–183, on p. 181.

[6] See Edgar Zilsel, "The Sociological Roots of Modern Science," *American Journal of Sociology*, 1942, *47*:544–562, on pp. 545–546; see also Arthur Clegg, "Craftsmen and the Origin of Science," *Science and Society*, 1979, *43*:186–201; Paolo Rossi, *Philosophy, Technology, and the Arts of the Early Modern Era*, trans. Salvator Attansasio (New York: Harper & Row, 1970); and Rupert Hall, "The Scholar and the Craftsman in the Scientific Revolution," *Critical Problems in the History of Science*, ed. Marshall Clagett (Madison: Univ. Wisconsin Press, 1959), pp. 3–23. Astronomers also held chairs as mathematicians in universities and served powerful patrons at royal courts. Noble patrons themselves became avid amateurs of astronomy. The astronomer has also been likened to a feudal lord: see Robert Westman, "The Astronomer's Role in the Sixteenth Century: A Preliminary Study," *History of Science*, 1980, *18*:105–147, on pp. 124–125; and Ernst Zinner, *Die Geschichte der Sternkunde von den ersten Anfängen bis zu Gegenwart* (Berlin: Julius Springer, 1931), pp. 587–590.

[7] Jean Quataert has warned against conflating important distinctions between guilds and households; see Quataert, "The Shaping of Women's Work in Manufacturing: Guilds, Households, and the State in Central Europe, 1648–1870," *American Historical Review*, 1985, *90*(5):1122–1148, on p. 1134. For the case of astronomy, however, the larger danger has been to ignore almost entirely both these forms of production. Here I use the term *craft* to refer to household production, and *guild* to refer to regulated crafts. Guilds emerged throughout Europe between the twelfth and fifteenth centuries, when they dominated most urban economies. See Anthony Black, *Guilds and Civil Society in European Political Thought from the Twelfth Century to the Present* (Ithaca: Cornell Univ. Press, 1984), p. 123; and Quataert, "Women's Work in Manufacturing," p. 1125.

Christine de Pizan had located women's greatest innovations in the arts and sciences—the spinning of wool, silk, and linen, and "creating the general means of civilized existence."[8] In the workshop, women's (like men's) contributions depended less on book learning and more on practical innovations in illustrating, calculating, or observing. It was the intersection of these traditions—craft traditions in astronomy and the tradition of women in the crafts—that fostered women's participation in astronomy.

The craft aspects of astronomy, then, are crucial for understanding the prominence of women in the field. It is important to keep in mind that Winkelmann was not the lone woman astronomer. Between 1650 and 1720, women constituted a little over fourteen percent of German astronomers.[9] The strength of the artisan in Germany may also explain the observation by Alphonse des Vignoles, quoted in the epigraph, that for women's accomplishments in astronomy, Germany took the prize. There were more women astronomers in Germany at the turn of the century than in any other European country.

Though guild traditions gave women access to the practice of science, it is important not to see this in romantic terms. Women's position in astronomy was similar to their position in the guilds—important, but subordinate. Only a few women, such as Maria Cunitz or Maria Winkelmann, directed and published their own work. More often a woman served in various support positions, editing her husband's writings or performing astronomical calculations. While guild traditions gave women an initial toehold in the sciences, the limitations built into those traditions allowed women a creative role only in exceptional cases.

II. WINKELMANN'S EDUCATION

The apprentice system provided the key to women's training in astronomy. Maria Margaretha Winkelmann was born in 1670 at Panitzsch (near Leipzig), the daughter of a Lutheran minister. She was educated privately by her father and, after his death, by her uncle. The young Winkelmann made great progress in the arts and letters; from an early age, she took a special interest in astronomy. She received advanced training in astronomy from the self-taught Christoph Arnold, who lived in the neighboring town of Sommerfeld. Like Winkelmann, Arnold benefited from the openness of astronomy. It was characteristic of this period that the farmer and self-taught Arnold was recognized for his contributions.[10] At

---

[8] Christine de Pizan, *The Book of the City of Ladies*, trans. Earl Jeffrey Richards (1405; New York: Persea Books, 1982), pp. 70–80.

[9] In his work on artists in Nuremberg (among whom he included astronomers and the like), Joachim von Sandrart recorded the names of three women among some fifty entries: *Teutsche Academie der Edlen Bau-, Bild-und Mahlerey-Künste* (Frankfurt, 1675). In his section on mathematicians and astronomers in Silesia, Friedrich Luce recorded one woman's name along with the names of about ten men; see Luce, *Fürsten Kron oder eigentliche wahrhaffte Beschreibung ober und nieder Schlesiens* (Frankfurt am Main, 1685). In his *Historia astronomiae* (Wittenberg, 1741), Friedrich Weidler listed 3 women along with 22 men astronomers working in Germany. Weidler's count is based on published work and includes men only marginally involved in astronomy (such as Leibniz and Christian Wolff). It is significant that a number of the lexicons made a point of including women. The titles of both Georg Will's and Christian Jöcher's works announce that the achievements of "both sexes" are to be included; see Will, *Nürnbergisches Gelehrten-Lexicon oder Beschreibung aller Nürnbergischen Gelehrten beyderley Geschlechtes . . .* (Nürnberg, 1755–1758); and Jöcher, *Allgemeines Gelehrten-Lexicon, darinne die Gelehrten aller Stände sowohl männ- als weiblichen Geschlechts* (Leipzig, 1760).

[10] Arnold discovered the comets of 1683 and 1686, and observed the transit of Mercury across the

Arnold's house, Winkelmann served as an unofficial apprentice, learning the art of astronomy through hands-on experience in observation and calculation.

Winkelmann's education followed a pattern commonly found in the trades, and a common one for women. In guild families, household and economic concerns were closely associated, if not identical. The importance of the household as a social and economic unit conferred on daughters and wives important responsibilities and privileges. Women's position in the guilds was stronger than has generally been appreciated. Before 1600 in Nuremberg, for example, women were active in nearly all areas of production. Craftswomen in fifteenth-century Cologne also held strong positions. Of the nearly forty guilds that Margret Wensky described in her study of working women in Cologne, between twenty and twenty-four had women members.[11] Women entered these guilds either as apprentices or through marriage.

There were, however, also differences in male and female apprenticeships. Women trained as apprentices but did not have the journeymen years. Journeymen might travel from master to master. Young women, in contrast, took what training was available in their homes. If a young woman did not have good training available at her doorstep, it is unlikely that she could have traveled to the master of her choice. At least within the sciences, there is no example of a woman apprentice traveling from master to master. In the seventeenth and early eighteenth centuries, the most important factor determining a woman's future in science was her father. Winkelmann trained with Arnold, outside her home; but her case was extraordinary because she was an orphan.

Astronomy in late seventeenth-century Germany was not, however, organized entirely along guild lines, and women's exclusion from university education created additional differences between women's and men's preparation. Had Maria Winkelmann been male, she would probably have continued her studies at the nearby universities of Leipzig or Jena. Leading male astronomers—Johannes Hevelius, Georg Eimmart, Gottfried Kirch—held university degrees, though not degrees in astronomy. Hevelius, for example, was by profession a brewer and was educated in jurisprudence. This was not uncommon; mathematics and astronomy in this period were not autonomous disciplines. Consequently, most astronomers studied law, theology, or medicine.[12]

sun. For his efforts he was awarded a pension by the Senate of Leipzig: Vignoles, "Eloge de Madame Kirch" (cit. n. 5), pp. 169–171. Although many scientific papers of the Kirch family have survived, few of their personal papers have. On the whereabouts of the Kirch papers see Diedrich Wattenberg, "Zur Geschichte der Astronomie in Berlin im 16. bis 18. Jahrhundert, II," *Die Sterne*, 1972, 49(2):104–116. Most of what we know of Winkelmann's life comes from the eulogy (cit. n. 5) by Alphonse des Vignoles, vice president of the Berlin Academy and a family friend. Most biographical notes follow Vignoles. Parts of this eulogy were reprinted in the *Neue Zeitungen von gelehrten Sachen*, August 1722, pp. 642–647. See also Gottfried Kirch and Maria Margaretha Winkelmann, *Das älteste Berliner Wetter-Buch 1700–1701*, ed. G. Hellmann (Berlin, 1893); and P. Aufgebauer, "Die Astronomenfamilie Kirch," *Die Sterne*, 1971, 47(6):241–247.
[11] Margret Wensky, *Die Stellung der Frau in der stadtkölnischen Wirtschaft im Spätmittelalter: Quellen und Darstellungen zur hansischen Geschichte* (Cologne: Böhlau, 1981), pp. 318–319. For women's role in guilds in Germany see Ute Gerhard, *Verhältnisse und Verhinderungen: Frauenarbeit, Familie und Rechte der Frauen im 19. Jahrhundert* (Frankfurt am Main: Suhrkamp, 1978); Carl Bücher, *Die Frauenfrage im Mittelalter* (Tübingen, 1882); Peter Ketsch, *Frauen im Mittelalter: Quellen und Materialien*, 2 vols. (Düsseldorf: Schwann, 1983) Vol. I, Ch. 6.1; and Merry E. Wiesner, *Working Women in Renaissance Germany* (New Brunswick, N.J.: Rutgers Univ. Press, 1986), esp. Ch. 5.
[12] See Johann Westphal, *Leben, Studien und Schriften des Astronomen Johann Hevelius* (Königsberg, 1820). Similarly, Copernicus studied law and medicine; see Westman, "Astronomer's Role"

Though women's exclusion from universities set limits to their participation in astronomy, it did not exclude them entirely.[13] Debates over the nature of the universe filled university halls, yet the practice of astronomy—the actual work of observing the heavens—took place largely outside the universities. In the seventeenth century, the art of observation was learned under the watchful eye of a master. Gottfried Kirch, for example, studied at Hevelius's private observatory in Danzig; this was as important for his astronomical career as his study of mathematics with Erhard Weigel at the University of Jena. An astronomical apprenticeship was also important for Kirch's son. When Christfried Kirch applied for the position of astronomer at the Berlin Academy, his training with his father (and mother) was a more important credential than his year at university.[14]

After her apprenticeship, a scientifically minded woman often married a scientist in order to continue practicing her trade. It was at the astronomer Christoph Arnold's house, where Maria Winkelmann served her unofficial apprenticeship, that she met Gottfried Kirch, Germany's leading astronomer. Though Winkelmann's uncle wanted her to marry a young Lutheran minister, he consented to her marriage to Kirch.[15] By marrying Kirch—a man some thirty years her senior —Winkelmann secured her place in astronomy. Knowing she would have no opportunity to practice astronomy as an independent woman, she moved, in typical guild fashion, from being an assistant to Arnold to becoming an assistant to Kirch. Kirch also benefited from this marriage. In Winkelmann, he found a second wife who could care for his domestic affairs, and also a much-needed astronomical assistant who could help with calculations, observations, and the making of calendars.[16]

In 1700 Kirch and Winkelmann took up residence in Berlin, the newly expanding cultural center of Brandenburg. This move represented an advance in social standing for both husband and wife. Yet in the late seventeenth century, the route to Berlin was very different for men and for women. A university education at Jena and apprenticeship to the well-known astronomer Hevelius afforded Kirch the opportunity to move from the household of a tailor in the small town of Guben to the position of astronomer at the Royal Academy of Sciences. Maria Winkelmann's mobility, on the other hand, came not through education, but through marriage. Though coming via different routes, both served at the Berlin Academy: Gottfried as Academy astronomer, Maria as an unofficial but recognized assistant to her husband.[17]

(cit. n. 7), p. 117. See also Diedrich Wattenberg, "Zur Geschichte der Astronomie in Berlin im 16. bis 18. Jahrhundert, I," *Die Sterne*, 1972, *48*(3):161–172, on p. 161.

[13] Zinner has shown that a university education was not absolutely required for the practice of astronomy: *Die Geschichte der Sternkunde* (cit. n. 7), p. 590.

[14] F. Herbert Weiss, "Quellenbeiträge zur Geschichte der Preussischen Akademie der Wissenschaften," *Jahrbuch der Preussischen Akademie der Wissenschaften*, 1939, pp. 214–224, on pp. 221–222.

[15] Vignoles, "Eloge de Madame Kirch" (cit. n. 5), p. 173.

[16] *Ibid.*, p. 172; and "Lebens Umstände und Schicksale des ehemahles berühmten Gottfried Kirchs, "Königl. Preuß. Astronomi der Societät der Wissenschafften zu Berlin," *Dresdenische gelehrte Anzeigen*, 1761, No. 49, pp. 769–777, on p. 775.

[17] See Erik Amburger, *Die Mitglieder der deutschen Akademie der Wissenschaft zu Berlin 1700–1950* (Berlin: Akademie-Verlag, 1950), p. 173: "1700–1710 Kirch, Gottfried, Astronom, unterstützt von seiner Frau, Kirch, Maria Margaretha geb. Winkelmann, geb. Panitzsch bei Leipzig 25.2.1670, gest. Berlin 29.12.1720. 1716–1720 Kirch, Christfried, Astronom, unterstützt von seiner Mutter."

III. COMETS AND CALENDARS: WINKELMANN'S SCIENTIFIC ACHIEVEMENT

In 1710 Winkelmann petitioned the Academy of Sciences for a position as calendar maker. Was she merely a wifely assistant engaged at the periphery in what has been defined as "women's work"? Or was she a qualified astronomer capable of setting and carrying out her own researches?

Though Maria Winkelmann is little known today, she was well regarded in her time.[18] Her scientific accomplishments during her first decade at the Berlin Academy were many and varied. Every evening, as was her habit, she observed the heavens beginning at nine o'clock.[19] During the course of an evening's observations in 1702, she discovered a previously unknown comet—a discovery that should have secured her position in the astronomical community. (Her husband's position at the Academy rested partly on his discovery of the comet of 1680.) There is no question about Winkelmann's priority in the discovery. In the 1930s F. H. Weiss published her original report of the sighting of the comet (see Figure 1).[20] In his notes from that night, Kirch also recorded that his wife found the comet while he slept:

> Early in the morning (about 2:00 A.M.) the sky was clear and starry. Some nights before, I had observed a variable star, and my wife (as I slept) wanted to find and see it for herself. In so doing, she found a comet in the sky. At which time she woke me, and I found that it was indeed a comet. . . . I was surprised that I had not seen it the night before.[21]

---

[18] In early histories of astronomy, Winkelmann received at least a mention. During her lifetime, she was cited in the German editions of Christian Wolff's *Mathematisches Lexicon* (1716, in Wolff, *Gesammelte Werke,* ed. J. E. Hofmann [Hildesheim: Georg Olms, 1965]), Pt. I, Vol. XI, p. 972. Wolff reported that "of special glory to the German nation is Kirch's widow who is well studied in astronomical observation and calculation"; this tribute was dropped, however, in the 1741 Latin edition of the work. Friedrich Weidler picked up the review of her publications from the *Acta eruditorum* in his *Historia astronomiae* (cit. n. 9) of 1741, p. 556, as did Joseph Jérôme Le Français de Lalande in his *Bibliographie astronomique; avec L'histoire de l'astronomie* (Paris, 1803), p. 359. Jérôme de Lalande included Winkelmann in the short history of women astronomers that introduced his popular astronomy textbook for women, *Astronomie des dames* (Paris, 1786), "Préface historique". J. E. Bode, astronomer of the Berlin Academy, gave her an entry in his "Chronologisches Verzeichniss der berühmtesten Astronomen, seit dem dreizehnten Jahrhundert, ihrer Verdienste, Schriften und Entdeckungen," *Astronomisches Jahrbuch für das Jahr 1816* (Berlin, 1813), p. 113. In the nineteenth and twentieth centuries, Winkelmann appeared most often in popular histories of astronomy or in articles about women astronomers. In his semipopular *Geschichte der Astronomie,* Rudolf Wolf gave much attention to women's achievements, beginning with those of Hypatia (Munich, 1877, p. 458). See also E. Lagrange, "Les femmes-astronomes," *Ciel et terre,* 1885, 5:513–527, on pp. 515–516; Alphonse Rebière reprinted Lagrange's account in his *Les femmes dans la science* (2d ed., Paris, 1897), pp. 153–154; H. J. Mozans (pseud. of John Augustine Zahm) also repeated this account in his *Woman in Science,* (1913; rpt. Cambridge, Mass.: MIT Press, 1974), pp. 173–174. For more recent accounts, see R. V. Rizzo, "Early Daughters of Urania," *Sky and Telescope,* 1954, *14*(1):7–9, on p. 8; and Diedrich Wattenberg, "Frauen in der Astronomie," *Vorträge und Schriften,* 1963, *14*:1–8. Lettie S. Multhauf included highlights from Winkelmann's life in an article on the Kirch family in the *Dictionary of Scientific Biography,* ed. Charles Gillispie, 16 vols. (New York: Scribner's, 1973), Vol. VII, pp. 373–374. Winkelmann also received a note in her husband's entry in the *Allgemeine deutsche Biographie* (Berlin, 1882)', Vol. XV, p. 788, and in the more recent *Neue deutsche Biographie* (Berlin, 1972), Vol. XI, pp. 634–635. Though we have official Academy portraits of her husband and son, we have no portraits of her or her daughters.

[19] Winkelmann to G. W. Leibniz, Leibniz Archive, Niedersächsische Landesbibliothek, Hannover, Kirch, No. 472, p. 11. I thank Gerda Utermöhlen and staff of the Leibniz Archive for their assistance.

[20] Weiss, "Quellenbeiträge zur Geschichte der Preussischen Akademie" (cit. n. 14), pp. 223–224, from his private collection. A copy of Winkelmann's report can be found in the Kirch papers, Paris Observatory, MS A.B. 3.7, No. 83, 41, B. I thank the staff of the Paris Observatory for their assistance.

[21] "Früh um 2 Uhr war es hell gestirnet. Meine Ehefrau hat (nach dem ich etliche Nächte zuvor

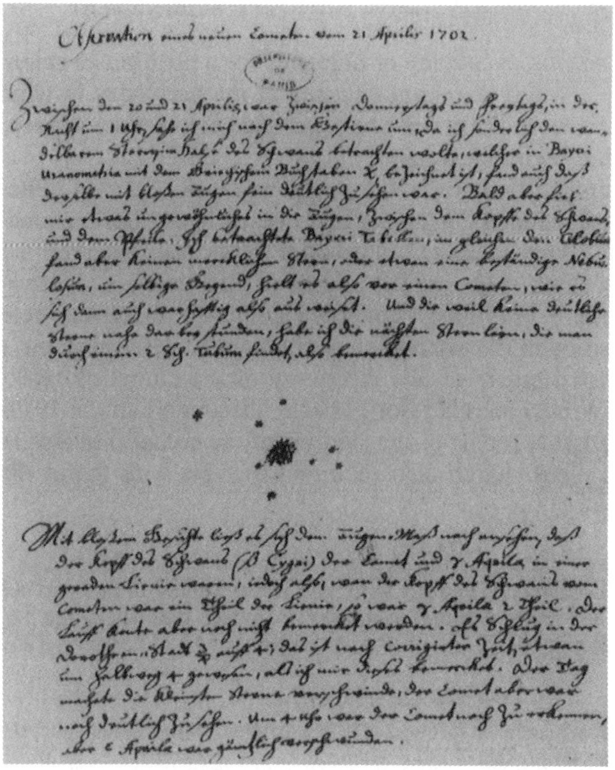

Figure 1. Winkelmann's report of her discovery of the comet of 1702. Reproduced with kind permission of the Observatoire de Paris.

As the first "scientific" achievement of the young Academy, a report of the comet was sent immediately to the king. The report, however, bore Kirch's, not Winkelmann's name.[22] Published accounts of the comet also bore Kirch's name, which unfortunately led many historians to attribute the discovery to him alone.[23]

Why did Winkelmann let this happen? Surely, she knew that recognition for her achievements could be important to her future career. Nor was she hesitant about publishing; she was to publish three tracts under her own name between 1709 and 1711. Her inability to claim recognition for her discovery hinged, in part, on her lack of training in Latin—the shared scientific language in Germany at the time—which made it difficult for her to publish her discovery in the *Acta eruditorum,* then Germany's only scientific journal. Her own publications were all in German.

More important to the problem of credit for the initial sighting of the comet, however, was the fact that Maria and Gottfried worked closely together. The labor of husband and wife did not divide along modern lines: he was not fully

---

dem wandelbaren Stern am Halse das Schwans observiert, und sie ihn, als ich noch schlieff, auch gerne sehen, und selbst finden wolte) einen Comentan am Himmel gefunden. Worauff sie mich auffwachetet, da ich denn fand, dass es warhaftig ein Comet war. . . . Es wondert mich doch, dass ich den Cometan die vorigen Nächte nicht gesehen habe." Kirch papers, Paris Observatory, MS A.B. 3.5, No. 81 B, p. 33.

[22] Adolf von Harnack, "Berichte des Secretars der brandenburgischen Societät der Wissenschaften J. Th. Jablonski an der Präsidenten G. W. Leibniz (1700–1715) nebst einigen Antworten von Leibniz," *Philos.-histor. Abhandlungen der königlichen Akademie der Wissenschaften zu Berlin,* 1897, *3,* Letter No. 22.

[23] See Wattenberg, "Zur Geschichte der Astronomie in Berlin II," (cit. n. 10), p. 107.

professional, working in an observatory outside the home; she was not fully a housewife, confined to hearth and home. Nor were they independent professionals, each holding a chair of astronomy. Instead, they worked very much as a team and on common problems. As Vignoles put it, they took turns observing so that their observations followed night after night without interruption. At other times they observed together, dividing the work (he observing to the north, she to the south) so that they could make observations that a single person could not make accurately.[24] After Winkelmann's sighting of the comet on 21 April, both Kirch and Winkelmann followed its course until 5 May.

Though Gottfried Kirch published the report under his own name and as if he alone had made the discovery, it would be too simple to fault him for "expropriating" his wife's achievement. According to Vignoles, a family friend, Kirch was timid about acknowledging his wife's contributions to their common work and so published the first report of the comet without mentioning her. Later, however, someone (we do not know who) told him "that he could feel free to acknowledge her contributions." Thus when the report of the comet was reprinted eight years later in the first volume of the journal of the Berlin Academy, *Miscellanea Berolinensia,* Kirch mentioned Winkelmann's part in the discovery. This report opened with the words: "My wife . . . beheld an unexpected comet."[25]

In addition to their scientific work, Kirch and Winkelmann took an active interest in the development of astronomical facilities at the Academy. The Academy of Sciences in Berlin was founded primarily to promote astronomy. In 1696 Sophie Charlotte, electress of Brandenburg, later queen of Prussia, had directed her minister Johann Theodor Jablonski to build an observatory, a project that took a decade to complete.[26] The Kirch family struggled long and hard, squeezing money from Academy and royal purses, to create the conditions necessary for good astronomical observations. Winkelmann took an active part in these efforts. On 4 November 1707 she wrote to Leibniz (adviser to Sophie Charlotte and president of the Academy), describing her sighting of the northern lights. In her letter, Winkelmann enticed Leibniz with reports of northern lights "the likes of which my husband has never seen," yet her real motive in writing was to secure housing for the astronomers more convenient to the observatory. She asked for Leibniz's intervention.[27]

During the years of their acquaintance at the Berlin Academy, Leibniz had expressed a high regard for Winkelmann's scientific abilities. Though none of his letters to her have been preserved, her letters to him reveal his interest in her scientific observations.[28] In 1709 Leibniz presented her to the Prussian court,

---

[24] Vignoles, "Eloge de Madame Kirch" (cit. n. 5), p. 174.

[25] See Gottfried Kirch, "Observationes cometae novi," *Acta eruditorum,* 21 Apr., 1702, pp. 256–258; Vignoles, "Eloge de Madame Kirch," pp. 175–176; and Gottfried Kirch, "De cometa anno 1702: Berolini observato," *Miscellanea Berolinensia,* 1710, *1*:213–214.

[26] The observatory could not be used until 1706, and the official opening did not take place until 1711: Harnack, *Geschichte der Akademie zu Berlin* (cit. n. 1), Vol. I, pp. 48–49.

[27] Winkelmann (as Kirchin) to Leibniz, 4 Nov. 1707, Leibniz Archive, Kirch, No. 472, pp. 11–12. The astronomers eventually received quarters in 1708: Harnack, *Geschichte der Akademie zu Berlin* (cit. n. 1), Vol. I, p. 152.

[28] Winkelmann often sent Leibniz special reports of her observations. She knew him well enough to drop by to announce that her book was finished and would arrive from the publisher in a few hours. See Winkelmann to Leibniz, n.d. Leibniz Archive, Kirch, No. 472, p. 10.

where Winkelmann was to explain her sighting of sunspots. In a letter of introduction Leibniz wrote:

> There is [in Berlin] a most learned woman who could pass as a rarity. Her achieve-
> ment is not in literature or rhetoric but in the most profound doctrines of astronomy.
> . . . I do not believe that this woman easily finds her equal in the science in which she
> excels. . . . She favors the Copernican system (the idea that the sun is at rest) like all
> the learned astronomers of our time. And it is a pleasure to hear her defend that
> system through the Holy Scripture in which she is also very learned. She observes
> with the best observers, she knows how to handle marvelously the quadrant and the
> telescope (*grandes lunettes d'approche*).

He added that if only she had been sent to the Cape of Good Hope instead of
Peter Kolb (Baron von Krosigk's apprentice astronomer), the Academy would
have received more reliable observations.[29]

Maria Winkelmann apparently made a good impression at the court of Frederick I. On 17 July she reported in a letter to Leibniz that the ambassador of
Denmark had visited the Royal Observatory and had praised her for the aid and
assistance she offered her husband in his astronomical work. While at court,
Winkelmann also distributed copies of her astrological pamphlet "Vorstellung
des Himmels bey der Zusammenkunfft dreyer Grossmächtigsten Könige."[30]
Leibniz, commenting on Winkelmann's tract, remarked on "an astrological note
that on the second of that month that the sun, Saturn and Venus would be in a
straight line. One supposes that there is significance in this."[31]

Maria Winkelmann's three pamphlets published between 1709 and 1711 were
all astrological. In his 1721 eulogy, Vignoles tried to explain away her interest in
astrology. "Madame Kirch," as he called her, "prepared horoscopes at the request of her friends, but always against her will and in order not to be unkind to
her patrons."[32] Perhaps Winkelmann's interest in astrology was purely financial,
as Vignoles suggested. Yet her correspondence with Leibniz reveals her belief
that nature was something more than matter in motion. In her description of the
extraordinary northern lights of 4 November 1707 she wrote to Leibniz, "I am
not sure what nature was trying to tell us."[33] Another of Winkelmann's

[29] Leibniz to Sophie Charlotte, Jan. 1709, in Gottfried Wilhelm Leibniz, *Die Werke von Leibniz,* ed.
Onno Klopp, 11 vols. (Hannover: Klindworths, 1864–1888), Vol. IX, p. 295–296. Leibniz is referring
to the attempt to get an exact measurement of the lunar parallax, which failed because Kolb was
irresponsible and only occasionally made observations; see Hans Ludendorff, "Zur Frühgeschichte
der Astronomie in Berlin," *Vorträge und Schriften der Preussischen Akademie der Wissenschaften,*
1942, 9:3–23, on p. 15. Vignoles also reported that Leibniz often tested Winkelmann's knowledge of
certain subjects; she was, he wrote, a zealous partisan of the Copernican system. See Vignoles,
"Eloge de Madame Kirch" (cit. n. 5), p. 182.
[30] Winkelmann to Leibniz, 17 July 1709, Leibniz Archive, Jablonski, No. 440, pp. 111–112; and
Maria Margaretha Winkelmann, "Vorstellung des Himmels bey der Zusammenkunfft dreyer Grossmächtigsten Könige" (Potsdam, 1709). This pamphlet was originally housed in the Preussische
Staatsbibliothek, where one still finds a card for it in the catalogue; through the vagariés of war the
only extant copy is in the Biblioteka Jagielloński in Krakow.
[31] Leibniz's note in the margin of Winkelmann's letter to him, 17 July 1709, Leibniz Archive;
reprinted in Harnack, "Berichte des Secretars Jablonski an der Präsidenten Leibniz" (cit. n. 22),
Letter No. 87.
[32] Vignoles, "Eloge de Madame Kirch" (cit. n. 5), p. 182.
[33] Winkelmann to Leibniz, 4 Nov. 1707, Leibniz Archive, Kirch, No. 472, pp. 11–12. Though
astrology had begun losing ground in Germany in the sixteenth century, it continued to exercise
considerable influence even within scientific circles; see Zinner, *Die Geschichte der Sternkunde* (cit.
n. 6), pp. 558–564. Erhard Weigel, Gottfried Kirch's teacher at the University of Jena, prepared

pamphlets, "Die Vorbereitung zur grossen Opposition," predicting the appearance of a new comet, was reviewed favorably in the *Acta eruditorum*. The reviewer praised her talents, ranking her "skill in observation and astronomical calculation" as equal to that of her husband. Even though Winkelmann made "concessions" to the art of astrology, the reviewer judged her work valuable. The review closed with a lavish tribute to this woman who "understood matters . . . that are not understood without the force of intelligence and the zeal of hard work."[34]

Several months after her pamphlet appeared in 1711, Academy Secretary Jablonski reported favorably that Winkelmann was becoming famous. Nowhere is there a hint that Academy authorities objected to her astrological work. In fact, her son, Christfried Kirch, continued to publish astrological calendars some years later during his tenure as Academy astronomer.[35]

Winkelmann mixed astrology and astronomy in calendar making, a project of both scientific and monetary interest for her and the Academy. Unlike many major European courts, the Prussian court did not yet have its own calendar. In 1700 the Reichstag at Regensburg ruled that an improved calendar similar to the Gregorian calendar was to be used in German lands.[36] Thus the production of an astronomically accurate calendar became a major project for the Academy of Sciences, founded in the same year. In addition to fixing the days and months, each calendar predicted the position of the sun, moon, and planets (calculated using the Rudolphine tables); the phases of the moon; eclipses of the sun or moon to the hour; and the rising and setting of the sun within a quarter of an hour for each day.

The monopoly on the sale of calendars was one of the two monopolies granted to the Academy by the king in 1700. Throughout the eighteenth century, the Berlin Academy of Science derived a large part of its revenues from the sale of calendars. This income (some 2,500 talers per year in the early 1700s; over 19,000 talers in the 1770s, with the added income of the Silesian calendar) made the position of astronomer particularly important.[37]

---

horoscopes and held astrological beliefs. Though he later turned against calendar makers who profited from "ill-founded" astrological predictions, Weigel continued to believe that comets were omens of good or ill fortune. See Weigel, *Unterschiedliche Beschreibung-und Bedeutung sowohl der Cometen insgemein als in Sonderheit des Wunder-Cometen* (1681); and Otto Knopf, *Die Astronomie an der Universität Jena von der Gründung der Universität im Jahre 1558 bis zur Entpflichtung des Verfassers im Jahre 1927* (Jena: Gustav Fischer, 1937), pp. 59–63.

[34] Maria Margaretha Winkelmann, *Vorbereitung, zur grossen Opposition, oder merckwürdige Himmels-Gestalt im 1712* (Cölln an der Spree, 1711). To my knowledge, the only extant copy of this pamphlet is at the Paris Observatory, acquired (I assume) along with other papers of the Kirch family by Joseph-Nicolas Delisle. For the review of her work see "Praeparatio ad Oppositionem magnam, sive notabilis Coelisacies ad Annum 1712, quam sequenti 1713 excipit oppositio triplex Saturni & Jovis, delineata a Maria Margaretha Winkelmannia, Kirchii Vidua, Astronomiae & Astrologiae Cultrice," *Acta eruditorum*, 1712, pp. 77–79.

[35] See Christfried Kirch, "Bestehende in einem Prognostico Astronomico-Astrologico Vorinnen Diejenigen merckwürdigste Vergebenheiten welcher sich an Sonne, Mond, und Sternen, in diesem 1726 Jahr . . . ," *Curieuser Astronomischer und Historischer Kalender Auf das Jahr Christi 1726 Berechnet, und Auf der Stadt Danzig und umliegender Ort Horizont mit Fleiss gerichtet* (Danzig, 1726), appendix.

[36] Since the Gregorian calendar reform of 1582, Catholics and Protestants had used calendars which differed by ten days: Wattenberg, "Zur Geschichte der Astronomie in Berlin, I" (cit. n. 12), p. 165. The "improved" Protestant calendar prepared by the Academy astronomer was similar to the Gregorian calendar except that Easter was calculated differently.

[37] Aufgebauer, "Die Astronomenfamilie Kirch" (cit. n. 10), p. 246. The other monopoly was silk, which never produced the desired revenues for the Academy.

Though calendrical reform was an important scientific project for the Academy, the sale of calendars depended on their more popular aspects. Calendars—which Leibniz called "the library of the common man"—had been issued since at least the fourteenth century and drew much of their popular appeal from astrology. Until 1768 there was little distinction between calendars and farmer's almanacs; both predicted the best times for haircutting, bloodletting, conceiving children, planting seeds, and felling timber.[38]

Weather prediction was also an important part of the Academy calendars and an important part of the duties of the Academy astronomer. Between 1697 and 1774 different members of the Kirch family kept a daily record of the weather.[39] Winkelmann's weather prediction partook of the "art of astrology."[40] In her *Wetter-Buch* of 1701 she wrote: "In God's name I have recorded the weather daily with diligent attention, and in order to see from which aspects [of the sky] the changes of weather may come." Her weather predictions also had an empirical basis. Her daily observations were made with the aid of a "weather-glass," a term used at that time for both the barometer and thermometer. Daily observation, she noted, sharpens prediction and can be very useful, especially in agriculture and navigation. It was Winkelmann's hope that "weather can be more accurately forecast, if more diligence is applied."[41]

IV. THE ATTEMPT TO BECOME ACADEMY ASTRONOMER

Gottfried Kirch died in 1710. It fell to the executive council of the Academy—President Leibniz, Secretary J. Th. Jablonski, his brother and Court Pastor D. E. Jablonski, and Librarian Cuneau—to appoint a new astronomer. The council needed to make the appointment quickly, as the Academy depended on the yearly revenues from the calendar; but apart from one in-house candidate, Jablonski could think of no one qualified for the position.[42] Ten years earlier the

---

[38] Harnack, *Geschichte der Akademie zu Berlin* (cit. n. 1), Vol. I, p. 124; Wolf, *Geschichte der Astronomie* (cit. n. 18), pp. 94–105. An excellent explanation of the calendar is given in a manual for women calendar users written in 1737 by Sidonia Hedwig Zäunemannin. See her *Curieuser und immer wahrender Astronomisch Meteorologisch-Oeconomischer-Frauenzimmer-Reise-und Hand-Kalender* (Erfurt, 1737). For more information about the Prussian calendars see Ludendorff, "Frühgeschichte der Astronomie" (cit. n. 29), pp. 19–22; and Knopf, *Astronomie an der Universität Jena* (cit. n. 33), p. 49. In 1702 there were six kinds of calendars sold by the Academy: the Improved Astronomical and Economic Calendar, the Improved Calendar of Curiosities, the Improved Conversation Calendar, the Improved Writing Calendar, and the Improved Historical and Geographical Calendar. The Academy also produced a wall calendar; see *Verbesserter Haushaltungs-Kalender* (Berlin, 1702), foreword.

[39] Kirch and Winkelmann traded off observing the weather from 1697 to 1702. Winkelmann then observed daily from 1702 to 1714 and 1716 to 1720. Her daughter Christine carried on after her mother's death, observing the weather daily from 1720 to 1751, 1755 to 1759, and 1760 to 1774: Hellmann in Kirch and Winkelmann, *Das älteste Berliner Wetter-Buch* (cit. n. 10), p. 12. The offices of astronomer and meteorologer were split after Winkelmann's death in 1720; see Alphonse des Vignoles, "Eloge de M. Kirch le Fils, Astronome de Berlin," *Journal litteraire d'Allemagne de Suisse et du Nord,* 1741, *1*:300–351, on p. 328.

[40] See Societät der Wissenschaften, *Historische-und Geographischer Calender* (Berlin, 1729), appendix. The conjunction of planets was thought to influence winds and temperatures. Thus the conjunction of "cold and dry" Saturn with "cold and moist" Venus forebodes much snow and hail in winter, while "warm and moist" Jupiter with "hot and dry" Mars brings warm and thundery weather: Zäunemannin, *Curieuser und immer wahrender Kalender* (cit. n. 38), pp. 112, 118.

[41] Kirch and Winkelmann, *Das älteste Berliner Wetter-Buch* (cit. n. 10), pp. 20, 12, 20–21.

[42] Harnack, "Berichte des Secretars Jablonski an der Präsidenten Leibniz" (cit. n. 22), No. 112. From 1700 to 1711 the Berlin Society of Sciences rarely held meetings, and no proceedings of meetings were kept or preserved. Thus these letters, which passed between Jablonski, the Academy

council had settled on Gottfried Kirch, who despite his advanced age (sixty-one) was the best in the field. Though there were few candidates, Maria Winkelmann's name did not enter the deliberations in 1710.[43] This is even more surprising when one considers that her qualifications at that time were not that different from her husband's earlier. They both had long years of experience preparing calendars (before coming to the Berlin Academy of Sciences, Kirch had earned his living by selling Christian, Jewish, and Turkish calendars); they had both discovered comets—Kirch in 1680, Winkelmann in 1702; and they both prepared ephemerides and recorded numerous other observations. What Winkelmann did not have, which nearly every member of the Academy did, was a university degree.

Kirch died in July; in August, since her name had not come up in discussions about the appointment, Winkelmann submitted it herself, along with her credentials. In a letter to Secretary Jablonski, she asked that she and her son be appointed assistant astronomers in charge of preparing the Academy calendar (see Figure 2).[44] Winkelmann made it clear that she was asking only for a position as assistant calendar maker. I would not, she wrote, be so "bold as to suggest that I take over completely the office [of astronomer]." Her argument for her candidacy was twofold. First, she argued, she was well qualified, since she was instructed in astronomical calculation and observation by her husband. Second, and more important, she had been engaged in astronomical work since her marriage and had, de facto, been working for the Academy since her husband's appointment ten years earlier. Indeed, she reported, "for some time, while my dear departed husband was weak and ill, I prepared the calendar from his calculations and published it under his name." She also reminded Jablonski that he had had occasion to remark on how she "lent a helping hand to her husband's astronomical work"—work for which she was paid a wage. In addition, she asked to be allowed to stay in the astronomer's quarters. For Winkelmann, a position at the Berlin Academy was not just an honor, it was a way to support herself and her four children. Her husband, she reported, had died leaving her with no means of support.

Secretary Jablonski was aware that the Academy's handling of the Winkelmann case would set important precedents for the role of women in it. In September 1710 he cautioned Leibniz: "You should be aware that this approaching decision could serve as a precedent. We are tentatively of the opinion that this case must be judged not only on its present merits but also as it could be judged for all time, for what we concede to her could serve as an example in the future."[45] Winkelmann's repeated requests for an official appointment at the obser-

---

secretary, and Leibniz, president *in absentia,* provide a rare written record of the Academy's early activities; see *ibid.,* pp. 5–6.

[43] The director of the Astrophysical Observatory in Potsdam in the 1940s, Hans Ludendorff, considered the only men qualified to be Academy astronomer at the turn of the century to have been E. Weigel of Jena (already dead in 1699), G. C. Eimmart of Nuremberg (already too old), and J. P. von Wurtzelbau of Nuremberg. He judged Winkelmann to be among the leading astronomers of her day. Ludendorff, "Frühgeschichte der Astronomie" (cit. n. 29), p. 12.

[44] Winkelmann (as Kirchin) to the Berlin Academy, 2 Aug. 1710; original in Kirch papers, Archives of the Akademie der Wissenschaften der DDR, I–III, 1, pp. 46–48; copy in the Leibniz Archive, Jablonski, 440, pp. 154–165. I would like to thank Christa Kirsten, Director, and the staff of the Zentrales Akademie-Archiv, der Akademie der Wissenschaften der DDR, for assistance.

[45] "Wenn E. Excell. Dero Gedanken darüber zu eröfnen belieben, so können dieselben bei bevorstehender überlegung derselben zur Richtschnur dienen. Hier ist man vorläufig der Meinung, daß

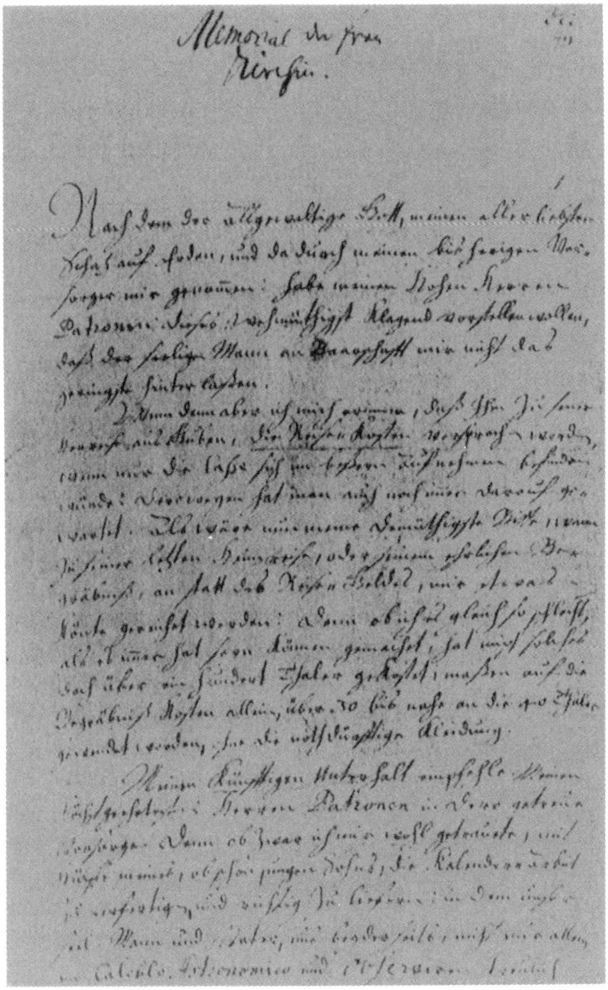

Figure 2. *The first page of Winkelmann's six-page letter to the Academy asking to be appointed assistant astronomer. Reproduced with kind permission of the Zentrales Akademie-Archiv, Akademie der Wissenschaften der DDR.*

vatory were not welcomed by the Berlin Academy because of concern about the effect on its reputation of hiring a woman. Jablonski wrote Leibniz: "That she be kept on in an official capacity to work on the calendar or to continue with observations simply will not do. Already during her husband's lifetime the society was burdened with ridicule because its calendar was prepared by a woman. If she were now to be kept on in such a capacity, mouths would gape even wider." By rejecting Winkelmann's candidacy, the Academy ensured that the social stigma attached to women would not further tarnish its already dull reputation.[46] In 1667

man die Sache so ansehen müsse, wie sie nicht nur gegenwärtig, sondern auch in Zukunft allezeit bestehen könne, immassen wass ihr eingeräumet würde, denen Künftigen zum Exempel dienen werde." Harnack, "Berichte des Secretars Jablonski an der Präsidenten Leibniz" (cit. n. 22), No. 115. Unfortunately, Leibniz's response to Jablonski has not been preserved.

[46] Jablonski to Leibniz, 1 Nov. 1710, in Harnack, "Berichte des Secretars Jablonski an der Präsidenten Leibniz" (cit. n. 22), No. 116. "Der Frau Kirchin gönnet Jedermann alles Gutes und wird ihr Niemand entgegen sein, ihr alles zuzuwenden, was möglich und anständig ist. Das sie aber bei der Calenderarbeit oder Observiren gebraucht und beibehalten werde, würde sich darum so viel weniger schicken, weil schon zu ihres Mannes Lebzeit sich spötter gefunden, so der Societät aufgebürdet,

a similar fear had prompted the members of the Royal Society of London to think long and hard before allowing Margaret Cavendish, Duchess of Newcastle, to visit a session.[47]

Leibniz was one of the few at the Academy who supported Winkelmann. In the council meeting of 18 March 1711 (one of the last meetings at which he presided before leaving Berlin), Leibniz argued that the Academy, considered as either a religious or an academic body, should provide a widow with housing and salary for six months as was customary. At Leibniz's behest, the Academy granted Winkelmann the right to stay in its housing a while longer; the proposal that she be paid a salary, however, was defeated. Instead the council paid her forty talers for her husband's observation notebooks. Later that year, the Academy showed some goodwill toward Winkelmann by presenting her with a medal.[48]

After Leibniz left Berlin, Winkelmann took her case to the king, and again her petition to be appointed Academy astronomer was placed before the council. With Leibniz gone, however, the council became more adamant in denying her requests. In 1712, after one and a half years of active petitioning, Winkelmann received a final rejection. Concerning her request for appointment as assistant astronomer, the council reported: "Frau Kirch's request is in many ways unseemly (*ungereimt*) and inadmissable (*unzulässig*). We must try and persuade her to be content and to withdraw of her own accord; otherwise we must definitely say no."[49]

The Academy never spelled out its reasons for refusing to appoint her to an official position, but Winkelmann traced her misfortunes to her sex. In a poignant passage, she recounted her husband's assurances that God would show his grace through influential patrons. This, she wrote, does not hold true for the "female sex." Her disappointment was deep: "Now I go through a severe desert, and because . . . water is scarce, . . . the taste is bitter." It was about this time that Winkelmann felt compelled to defend women's intellectual abilities in the preface to one of her scientific works. Citing Biblical authority, she argued that the "female sex as well as the male possesses talents of mind and spirit." With experience and diligent study, she wrote, a woman can become as "skilled as a man at observing and understanding the skies."[50]

Thus although Winkelmann had been involved in preparing the calendar for ten years and knew the work well, the position of Academy astronomer was awarded to Johann Heinrich Hoffmann.[51] Hoffmann had been a member of the Academy

---

dass ihre Calender durch ein Weib verfertiget werden, denen man hiermit das Maul noch weiter aufsperren würde." In 1706, the Academy was already coming under attack for its inactivity; see Harnack, *Geschichte der Akademie zu Berlin* (cit. n. 1), Vol. I, pp. 155–156.

[47] Samuel Pepys wrote in his diary that there was "much debate, *pro* and *con*, it seems many being against it, and we do believe the town will be full of ballads of it." See Samuel Mintz, "The Duchess of Newcastle's Visit to the Royal Society," *Journal of English and Germanic Philology*, 1952, *51*(2):168–176.

[48] Protokollum Concilii, Societatis Scientiarum, 15 Dec. 1710, 18 Mar. 1711, 9 Sept. 1711, DDR Academy Archives, I, IV, 6, Pt. 1, pp. 54, 65–66, 93. Unfortunately, we do not know why Winkelmann received a medal.

[49] *Ibid.*, 3 Feb. 1712, p. 106.

[50] Winkelmann to the council of the Berlin Academy, 3 Mar. 1711, DDR Academy Archives, Kirch papers, I, III, 1, p. 50; and Winkelmann, *Vorbereitung zur grossen Opposition* (cit. n. 34), pp. 3–4.

[51] Hoffmann has been almost entirely forgotten today. His name does not appear in the comprehensive, multivolumed *Allgemeine deutsche Biographie* or the *Neue deutsche Biographie*. Nor is he

since its founding in 1700 and had long hoped to be appointed Academy astronomer. Yet, his tenure was not a happy one. By December 1711 he was already behind in his work. Jablonski wrote to Leibniz complaining that Hoffmann was guilty of neglecting his work. Jablonski suggested that perhaps Hoffmann needed an assistant; ironically he suggested "Frau Kirchin, for example, who would spur him on a bit." In 1712 Jablonski again had occasion to complain to Leibniz about Hoffmann's performance. Hoffmann had not completed the yearly observations as he should have, and the calendar was still not ready. Hoffmann was officially censured by the Academy for his poor performance. While Hoffmann was being reprimanded, Winkelmann was becoming, as Jablonski reported, "rather well known" for her pamphlet on the conjunction of Saturn and Jupiter.[52]

During this period, conflict arose between Winkelmann and Hoffmann, each of whom considered the other a competitor at the observatory. Jablonski reported to Leibniz that Winkelmann had complained that "Hoffmann used her help secretly, yet denounced her publicly, and never let her use the observatory." Unemployed and unappreciated for her scientific skills, Winkelmann moved across Berlin in October 1712 to the private observatory of Baron Bernhard Friedrich von Krosigk. This did not end Hoffmann's·problems with the Academy, however. In 1715 Jablonski complained once again to Leibniz that Hoffmann was neglecting his duties.[53]

V. THE CLASH BETWEEN CRAFT TRADITIONS AND PROFESSIONAL SCIENCE

Did Winkelmann have a legitimate claim to the post of assistant astronomer? How was it possible in 1700 for a woman to hold a semiofficial position (as Winkelmann did) as assistant to her husband at the Berlin Academy? How did she imagine that her requests to continue on the calendar project would be taken seriously?

Winkelmann owed her position at the Academy to the perpetuation of guild traditions. These were as alive in the Academy as they were in Germany as a whole. Wolfram Fischer has argued that the relation of apprentice-journeyman-master provided a model for many German institutions. Fischer gave the example of the masons; W. V. Farrar has developed the example of the universities. According to Farrar, the guild character of the university system survived longer in Germany than elsewhere.[54]

But while retaining vestiges of the guild system, the Berlin Academy incorporated other traditions. We should distinguish two levels of participation in the Academy. At the top was a tier of university-educated, internationally renowned

---

mentioned in histories of astronomy such as Wolf's *Geschichte der Astronomie*. Maria Winkelmann is recognized for her work in each of these sources.

[52] Harnack, "Berichte des Secretars Jablonski an der Präsidenten Leibniz, Nos. 112, 133, 143, 144. In this last Jablonski was probably referring to Winkelmann's 1711 *Vorbereitung zur grossen Opposition*.

[53] Krosigk's observatory was built in 1705. See Jablonski to Leibniz, 29 Oct. 1712, in Harnack, "Berichte des Secretars Jablonski and der Präsidenten Leibniz," No. 143; and *ibid.*, No. 167.

[54] Wolfram Fischer, *Handwerksrecht und Handwerkswirtschaft um 1800* (Berlin: Duncker & Humblot, 1955), p. 18. W. V. Farrar found the corporation of master-tradesmen of the guild analogous to the academic body of professors, the journeyman's *Wanderjahre* similar to the student's traveling from university to university, and the masterpiece of the tradesman similar to the university M.A. See Farrar, "Science and the German University System: 1790–1850," in *The Emergence of Science in Western Europe*, ed. Maurice Crosland (London: Macmillan, 1975), p. 181.

members. This aspect of the organization had nothing in common with the guilds; rather, class standing was important for membership at this level. Like members of the Royal Society in London and the Académie des Sciences in Paris, many "gentlemen" members of the Berlin Academy were of noble standing. It was its financial structure that set the Berlin Academy apart from its counterparts in Paris or London and nearer craft traditions. Roger Hahn has shown that members of the Académie des Sciences in Paris drew pensions directly from the king's purse in order to distance themselves from traditional trades and professions, considered "mere occupations."[55] The Berlin Academy, in contrast, drew much of its revenues directly from two trades—calendar making and silk making—and hired artisans, the second tier of participants in its activities, to carry out the tasks required.

The Academy astronomer was in fact caught between the two tiers of the Academy hierarchy: as a university-educated mathematician, he was a distinguished gentleman; as calendar maker, he was an artisan who worked for his employer. The "gentlemen" of the Academy (except the president and secretary) were not paid, nor did they pay for their membership. The astronomer, however, like the other artisans of the Academy, derived his living (500 talers per year) from its coffers. It should be noted that though Maria Winkelmann asked to continue as Academy calendar maker, she never asked to become a member of the Academy (nor was she granted membership).[56]

It was as the wife of an artisan-astronomer that Winkelmann enjoyed a modest measure of respect at the Academy. When she petitioned the council to continue as assistant calendar maker, she was invoking (although not explicitly) age-old principles well established in the organized crafts and free arts. In most cases, guild regulations gave a widow the right to run the family business after the death of her husband. Guild regulations were local and varied from region to region, craft to craft; yet general patterns can be identified. In her study of thirty-eight Cologne guilds in the late Middle Ages, Margret Wensky found that eighteen of those guilds allowed a widow to continue the family business after her husband's death.[57] The rights of widows followed three general patterns. In some guilds, the widow was allowed to serve as an independent master as long as she lived. In others, she was allowed to continue the family business but only with the help of journeymen or apprentices. In still others, she filled in for one or two years to provide continuity until her oldest son came of age.[58] Within lower echelons of the Academy, widows were allowed to continue in their husband's position. Pont, widow of the keeper of the Academy mulberry trees, was allowed to complete the last four years of her husband's six-year contract.[59]

This is what Maria Winkelmann also tried to do. After the death of her husband, she tried to carry on the "family" business of calendar making as an inde-

[55] Roger Hahn, *The Anatomy of a Scientific Institution: The Paris Academy of Science 1660–1803* (Berkeley/Los Angeles: Univ. California Press, 1971), p. 39.

[56] Harnack, *Geschichte der Akademie zu Berlin* (cit. n. 1), Vol. I, p. 370.

[57] Wensky, "Die Stellung der Frau" (cit. n. 11), pp. 58–59.

[58] See, e.g., in Ketsch, *Frauen im Mittelalter* (cit. n. 11), Vol. I, p. 210: the regulations for the Lübeck dyers guild (No. 296); p. 204: the regulations of the Cologne hatmakers' guild (No. 276); and p. 29.

[59] Protokollum Concilii, Societatis Scientiarum, 23 Sept. 1716, DDR Academy Archives, I, IV, 6, Pt. 2, pp. 230–232.

pendent master. Yet, as we have seen, she found that traditions which had once secured women a (limited) role in science were not to apply in the new institutions.

Though the Academy retained vestiges of an older order, it also contained the seeds of a new. The founding of the Academy in 1700 represented a first step in the professionalization of astronomy in Germany. Earlier observatories—those of Hevelius in Danzig and Eimmart in Nuremberg—had been private. The Academy's observatory, however, was a public ornament of the Prussian state. Astronomers were no longer owners and directors of their own observatories, but employees of the Academy, selected by a patron on the basis of personal merit rather than family tradition. This shift of the character of scientific institutions from private to public had dramatic implications for the role of women in science. As astronomy moved more and more out of the private observatories and into the public world, women lost their toehold in modern science.

VI. A BRIEF RETURN TO THE ACADEMY

Although Winkelmann could not remain at the Berlin Academy, she did continue her astronomical work. In October 1712 she moved with her family to Baron von Krosigk's private observatory, where she and Gottfried Kirch had worked while the Academy observatory was under construction. She was thus able to keep in touch with astronomical work in Berlin.

At Krosigk's observatory, Winkelmann reached the height of her career. With her husband dead and her son away at university, she enjoyed the rank of "master" astronomer. She continued her daily observations and—now the master—had two students to assist her. The published reports of their joint observations bear her name. Many of Winkelmann's observations from this period—the conjunction of Saturn and Mars, several eclipses of the moon, and several sightings of sunspots—were published in her son's *Ephemeriden* of 1714 and 1715.[60] During this period, she also supported herself and her daughters by preparing calendars for Breslau and Nuremberg.[61]

When Krosigk died in 1714, Maria Winkelmann left his observatory, taking a position in Danzig as assistant to a professor of mathematics.[62] This part of her life remains sketchy. When this position fell through, Winkelmann again found a patron. The family of Hevelius (Gottfried Kirch's teacher) invited her and her son, Christfried, now a student in Leipzig, to reorganize the deceased astronomer's observatory and to use it to continue their own observations.

In 1716 the Winkelmann-Kirch family received an invitation from Peter the Great of Russia to become astronomers in Moscow.[63] The family decided instead

---

[60] Christfried Kirch, *Teutsche Ephemeris* (Nuremberg, 1715), p. 82; *ibid.* (1714), pp. 76–77, 80; and *ibid.* (1715), pp. 78–80, 82–84.

[61] Winkelmann was given permission by the Berlin Academy to prepare these calendars: Protokollum Concilii, Societatis Scientiarum, 18 Oct. 1717, DDR Academy Archives, I, IV, 6, Pt. 2, p. 280. Though Herbert Weiss has suggested that Winkelmann did not sign her name as author of these calendars because a woman astronomer was accorded little respect ("Quellenbeiträge zur Geschichte der Preussischen Akademie" [cit. n. 14], p. 216), a look at calendars of the eighteenth and nineteenth century shows that very few were signed.

[62] The only report on this is from Vignoles, "Eloge de Madame Kirch" (cit. n. 5), p. 180.

[63] *Ibid.*

to return to Berlin when Christfried was appointed one of two observers for the Academy following the death of Hoffmann.[64] The Academy had grave reservations about the abilities of their newly appointed astronomers: Christfried Kirch was not well grounded in astronomical theory and could not express himself decently in German or Latin; J. W. Wagner was weak in astronomical calculation. Academy funds, however, were insufficient to support the appointment of a "celebrated" astronomer who would require a higher salary, better housing, and assistants. Under these circumstances, a factor weighing in Kirch's favor was that, along with him, the Academy received an extra astronomical hand—Winkelmann—with skills very similar to those of Kirch and Wagner. Thus, Winkelmann returned once again to the work of observation and calendar making for the Academy, this time as assistant to her son.[65]

But all was not well. The opinion was still prevalent that women should not do astronomy, at least not in a public capacity.[66] In 1717 Winkelmann was reprimanded by the Academy council for talking too much to visitors at the observatory. The council cautioned her to "retire to the background and leave the talking to Wagner and her son." A month later, the Academy again reported that "Frau Kirch meddles too much with Society matters and is too visible at the observatory when strangers visit." Again the council warned Winkelmann "to let herself be seen at the observatory as little as possible, especially on *public* occasions."[67] As Vignoles reported, there were those who found it wrong for a woman to practice astronomy. Maria Winkelmann was forced to make a choice. She could either continue to badger the Academy for a position of her own, or, in the interest of her son's reputation, she could retire, as the Academy requested, to the background. Vignoles reported that she chose the latter option. Academy records show, however, that the choice was not hers to make. On 21 October 1717 the Academy resolved to remove Winkelmann—who apparently had paid little heed to their warnings—from Academy grounds. She was forced to leave her house and the observatory. The Academy did not, however, want her to abandon her duties as mother; officials expressed the hope that Winkelmann "could find a house nearby so that Herr Kirch could continue to eat at her table."[68]

In 1717 Winkelmann quit the Academy's observatory and continued her observations only at home, as was thought appropriate, "behind closed doors," a move which Vignoles judged detrimental to the progress she might have made in astronomy. With few scientific instruments at her disposal, she was forced to

[64] Christfried was first appointed Academy "astronomer" in 1728. For his letters of application see Weiss, "Quellenbeiträge zur Geschichte der Preussischen Akademie" (cit. n. 14), pp. 219–222. In the first letter, Kirch suggested that he do the calendar work while the society found another astronomer. Among his qualifications, he mentioned his long experience preparing calendars. In his second letter, he asked for the job of astronomer, playing very much on the good reputation of his father. The position of Academy astronomer was largely hereditary: J. E. Bode, Academy astronomer at the turn of the eighteenth century, was also related to the Kirch family.

[65] Protokollum Concilii, Societatis Scientiarum, 8 Oct. 1716, 6 Apr. 1718, DDR Academy Archives, I, IV, 6, Pt. 2, pp. 236, 318.

[66] Vignoles, "Eloge de Madame Kirch" (cit. n. 5), pp. 181.

[67] Protokollum Concilii, Societatis Scientiarum, 18 Aug. 1717, DDR Academy Archives, I, IV, 6, Pt. 2, pp. 269, 272–273 (emphases added).

[68] Vignoles, "Eloge de Madame Kirch," p. 181; Protokollum Concilii, Societatis Scientiarum, 21 Oct. 1717, DDR Academy Archives, I, IV, 6, Pt. 2, pp. 275–276.

quit astronomical science. Maria Winkelmann died of fever in 1720. In Vignoles's opinion, "she merited a fate better than the one she received."[69]

VII. WOMEN ASTRONOMERS IN GERMANY

Maria Winkelmann was not the only woman astronomer in late seventeenth-century Germany. Between 1650 and 1710 a surprisingly large number of women— Maria Cunitz (1610–1664), Elisabetha Hevelius (1647–1693), Maria Klara Eimmart (1676–1707), Maria Winkelmann (1670–1720) and her daughters Christine Kirch (1696–1782) and Margaretha (active in the 1740s)—worked in German astronomy. The group comprised, as noted above, fourteen percent of German astronomers for this period.[70] All these women worked in family observatories: Hevelius built his private observatory in 1640 and again in 1687; Eimmart built his in 1678.[71] Of this group, only Maria Cunitz was not the daughter or wife of an astronomer who, in guildlike fashion, assisted a master in his trade. As in the case of Winkelmann, the perpetuation of craft traditions allowed these women access to the secrets and tools of the astronomical trade.

It is perhaps unfair to include the example of Maria Cunitz among women working within the crafts tradition, for her father was a landowner. Nonetheless, her education too depended on training given her by her father, the learned medical doctor Heinrich Cunitz, lord of the estates of Kunzendorf and Hoch Giersdorf near Schweidnitz in Silesia. Sometimes called the "second Hypatia," Cunitz learned from her father six languages—Hebrew, Greek, Latin, Italian, French, and Polish—as well as history, medicine, mathematics, painting, poetry, and music.[72] Her principal occupation, however, was astronomy. In 1630 she married Eliae von Lowen, a medical doctor and amateur astronomer. During the Thirty Years' War her family took refuge in Poland, where she prepared her astronomical tables, published in 1650 as *Urania propitia*. The main purpose of this work was to simplify Kepler's Rudolphine Tables, used for calculating the position of

---

[69] Vignoles, "Eloge de Madame Kirch," pp. 181, 182.

[70] See note 9 above.

[71] Ernst Zinner, *Deutsche und niederländische astronomische Instrumente des 11.–18. Jahrhunderts* (Munich: C. H. Beck, 1956), pp. 221–223. Hevelius's observatory spanned the roofs of three adjoining houses; Eimmart's was built on the city wall. See also Wolf, *Geschichte der Astronomie* (cit. n. 18), p. 458.

[72] Like a number of women of her time, Cunitz provided some biographical information in the preface of her book. See Maria Cunitz, *Urania Propitia, sive Tabulae Astronomica mirè faciles, vim hypothesium physicarum à Kepplero proditarum complexae; facillimo calculandi compendio, sine ullą Logarithmorum mentione, phaenomenis satisfacientes* (Oels, 1650), esp. p. 147. See also Joanne Hallervordio, *Bibliotheca curiosa* (Frankfurt, 1676), p. 260; and Vignoles, "Eloge de Madame Kirch" (cit. n. 5), pp. 163–168. In his *Mathematisches Lexicon* (cit. n. 18), Christian Wolff reported that Cunitz simplified the Rudolphine Tables (p. 1360); he also mentioned Cunitz in his *Elementa matheseos universae*, Vol. IV, in *Gesammelte Werke*, Vol. XXXIII, p. 112. See also Weidler, *Historia astronomiae* (cit. n. 18), pp. 489–490. Cunitz received a note under Winkelmann's entry in Bode's *Astronomisches Jahrbuch* (cit. n. 18), p. 113. Since Cunitz was not the daughter or wife of an astronomer, she is one of the few women to receive her own entry in many histories of astronomy. See also Wolf, *Geschichte der Astronomie*, pp. 305–306; Lagrange, "Les femmes-astronomes" (cit. n. 18), p. 517; Mozans, *Woman in Science* (cit. n. 18), pp. 170–171, who gives the reference to a "second Hypatia"; Rizzo, "Early Daughters of Urania" (cit. n. 18), p. 8; and Ingrid Guentherodt, "Maria Cunitz and Maria Sibylla Merian: Pionierinnen der deutschen Wissenschaftssprache im 17. Jahrhundert," *Zeitschrift für germanistische Linguistik*, 1986, *14*(1):23–49. For an excellent bibliographic source on women in this period see Jean Woods and Maria Fürstenwald, *Schriftstellerinnen, Künstlerinnen und gelehrte Frauen des deutschen Barock: Ein Lexikon* (Stuttgart: J. B. Metzler, 1984).

the planets; but Maria Cunitz was not merely a calculator: her book also treated the art and theory of astronomy.[73]

Maria Klara Eimmartin-Müller, another woman practicing astronomy at the end of the seventeenth century, fits squarely into craft traditions.[74] The daughter of Georg Christoph Eimmart, astronomer and director of the Nuremberg Academy of Art from 1699 to 1704, Maria Eimmart learned French, Latin, drawing, and mathematics from her father. As a young girl she also learned the art of astronomy at her father's observatory, where she worked alongside his other students. She owed her place in astronomy largely to the strong position of women in the graphic arts. Much of Eimmart's scientific achievement depended on her ability to make exact sketches of the sun and moon. Between 1693 and 1698, she prepared 250 drawings of phases of the moon in a continuous series, thus laying the groundwork for a new lunar map. She also made two drawings of the total eclipse of 1706.[75] A few sources claim that in 1701 Eimmart published a work on ancient views of the sun, *Ichnographia nova contemplationum de Sole,* under her father's name, but there is no evidence that this was her work.[76]

After training as an apprentice to her father, the scientifically minded Eimmart secured her position in astronomy by marrying the astronomer Johann Heinrich Müller in 1706. Müller was professor of physics at a Nuremberg *Gymnasium* and since 1705 director of her father's observatory. Through this marriage Maria Eimmart ensured that she could continue her astronomical work at her father's observatory, now as wife of the director. Johann Müller also benefited from this marriage. Through the principle of daughter's rights, the Eimmart observatory became part of his daughter's inheritance, passing through the daughter to her husband.[77] Maria Müller's astronomical career was cut short when she died in childbirth in 1707.

Elisabetha Koopman (later Hevelius) of Danzig also married with care to ensure her career in astronomy. In 1663 she married a leading astronomer, Johannes Hevelius, a man thirty-six years older than herself. Hevelius, a brewer by trade, took over the lucrative family beer business in 1641. His first wife, Catherina Rebeschke, had managed the household and brewery, leaving Hevelius free to serve in city government and to pursue his avocation, astronomy. When she died in 1662, Hevelius married Elisabetha Koopman, who had been interested in

---

[73] As is common in German women's books of the period, the text was given in Latin and German. The work also presented a guide to astronomy for the layperson. Astronomy, Cunitz taught, has four parts: observation, which must be carefully recorded; mechanics or the craft of making instruments; hypotheses or theory of the heavens; and calculus or tables of predictions. See also Vignoles, "Eloge de Madame Kirch" (cit. n. 5), pp. 148–149.

[74] See the mention of Maria Klara Eimmart under her father's entry in Weidler, *Historia astronomiae* (cit. n. 18), p. 543. See also Wolf, *Geschichte der Astronomie* (cit. n. 18), p. 104; and Kurt Pilz, *600 Jahre Astronomie in Nürnberg* (Nuremberg: Carl, 1977).

[75] Johann Gabriel Doppelmayr, *Historische Nachricht von den nürnbergischen Mathematicis und Künstlern* (Nuremberg, 1730; rpt. Hildesheim: Olms, 1972), pp. 259–260; and Jöcher, *Allgemeines Gelehrten-Lexicon* (cit. n. 9), Vol. III, p. 743.

[76] See, e.g., J. C. Poggendorff, *Handwörterbuch zur Geschichte der exacten Wissenschaften* (Leipzig: Barth, 1863), Vol. I, p. 651. Eighteenth-century lexicons, however, which list her works in great detail, attribute the *Ichnographia* to her father. See Doppelmayr, *Historische Nachricht* (cit. n. 75), p. 126; and Will, *Nürnbergisches Gelehrten-Lexicon* (cit. n. 9).

[77] According to Peter Ketsch, family trades passed more often to the daughter than to the son: Ketsch, Frauen im Mittelalter (cit. n. 11), Vol. I, p. 29.

astronomy for many years.[78] In appropriate guild fashion, Elisabetha Hevelius served as chief assistant to her husband, both in the family business and in the family observatory.

Margaret Rossiter has defined and described the notion of "women's work" in nineteenth- and twentieth-century science, and especially in astronomy.[79] Women's work in science—tedious computation, support positions, and the like —is a legacy of the guild wife. Elisabetha Hevelius is perhaps the best example of a wife who served as chief assistant to her astronomer husband. The role of the guild wife, however, cannot be collapsed into that of a mere assistant. The very different structure of the workplace—in the seventeenth century the observatory was in the home, not part of a university—gave the wife a more comprehensive role. For twenty-seven years Elisabetha Hevelius collaborated with her husband, observing the heavens in the cold of night by his side (see Figure 3).[80] After his death, Elisabetha Hevelius edited and published their joint work, *Prodromus astronomiae,* a catalogue of 1,888 stars and their positions.[81]

The "astronomical wife" was not an exception, but an established tradition. When Gottfried Kirch studied with Hevelius in Danzig, he learned through the example of Elisabetha Hevelius the difference an astronomical wife could make.

VIII. "INVISIBLE ASSISTANTS": WOMEN'S PARTICIPATION IN THE BERLIN ACADEMY

Maria Winkelmann was not the only woman present at the founding of the Berlin Academy of Sciences. Sophie Charlotte, queen of Prussia, was important as an ambassador of scientific ideas at the court in Berlin. Working closely with Leibniz and her ministers, Sophie Charlotte carried forth plans and negotiations for the founding of the Berlin Academy with such vigor that Leibniz claimed it "the role of women of elevated mind more properly than men to cultivate knowledge."[82] Frederick II, her grandson, credited her with establishing the Academy. He wrote that "she founded the royal Academy and brought Leibniz and many other learned men to Berlin. She wanted always to know the first principle of things." Since she died shortly after its founding, it remains unclear whether Sophie Charlotte intended to take an active part in the Academy or to serve merely as a patron.[83]

---

[78] Eugene McPike, *Hevelius, Flamsteed, and Halley* (London: Taylor & Francis, 1937), pp. 4–5. Elisabetha Hevelius receives short entries in Rebière, Mozans, and Rizzo (see n. 18).

[79] Margaret Rossiter, "Women's Work in Science, 1880–1910," *Isis,* 1980, *71*(258):381–398; see also Rossiter, *Women Scientists in America: Struggles and Strategies to 1940* (Baltimore: Johns Hopkins Univ. Press, 1982), pp. 51–72.

[80] Elisabetha Hevelius is shown in three plates; two depict her working at the sextant with Johannes, and a third shows her using a telescope: Johannes Hevelius, *Machina Coelestis, pars prior; Organographiam, sive Instrumentorum Astronomicorum omnium, quibus Auctor hactenus Sidera rimatus* (Danzig, 1673), plates following pp. 222, 254, 450.

[81] Johannes Hevelius, *Prodromus astronomiae* (Danzig, 1690).

[82] Sophie Charlotte, princess of Hannover, was privately tutored by Leibniz from an early age, well read in Latin, well traveled, and a devotee of French culture. See Leibniz to Sophie Charlotte, Nov. 1697, in Harnack, *Geschichte der Akademie zu Berlin* (cit. n. 1), Vol. II, p. 44.

[83] Frederick II, "Mémoire de l'Académie," 1748, reprinted in Jean-Pierre Erman, *Mémoire pour servir à l'histoire de Sophie Charlotte, reine de Prusse* (Berlin, 1801), p. 382. In a nineteenth-century Academy calendar Sophie Charlotte was credited with giving the order for the founding of the Academy according to Leibniz's plan; see *Adress Calender* (Berlin, 1845), p. 113. Harnack argued that it

The founding statutes of the Berlin Academy of Sciences did not bar women from membership. In fact, Leibniz thought women should benefit from participation. In his sketch of Academy regulations of 1700, he wrote that a scientific academy would foster good taste, solid understanding, and an appreciation of God's handiwork, not only among German nobility, "but among other people of high standing (as well as among women)."[84] Yet despite his intentions, women were not admitted. Perhaps the decision to use the scientific societies of London and Paris as models for the Academy in Berlin reinforced the exclusion of women. Although neither the London nor the Paris society had regulations excluding women, neither society admitted them.

The fate of Winkelmann's daughters—Christine and Margaretha—reveals a process of privatization of women within the Academy. Trained (in guild fashion) in astronomy from the age of ten, both Kirch daughters worked for the Academy as assistants to their brother, Christfried. According to Vignoles, "Margaretha, the youngest, usually took a telescope; Christine, the oldest, most often took the pendulum in order to mark exactly the time of each individual observation."[85] Yet having witnessed the lost battles of their mother, Christine and Margaretha did not ask (as Winkelmann had) for official positions. Nor did they exude the fire of their mother, badgering the Academy for housing or greeting foreign visitors. Rather, they molded their behavior to fit Academy prescriptions, becoming "invisible helpers" to their brother. Again, Vignoles describes the sisters' situation: "They helped their brother carry out his professional duties; . . . nonetheless they remained very private and spoke with no one but their close friends. By the same modesty, they avoided going to the observatory when there was to be an eclipse or other observation that might attract strangers."[86]

When Christfried died in 1740, the Kirch sisters lost their male protector and were forced to observe more often at home. Although they watched the heavens daily, conditions made serious astronomical work almost impossible. When Christine sent their observations of the comets of 1742 and 1743 to Joseph Nicolas Delisle, director of the Paris Observatory, she complained: "We observed daily [the course of the comet] as well as we could . . . but our observations were done under very bad conditions and with inferior instruments, namely with a two foot (*zwei Schühe*) telescope. . . . We could not use a larger telescope because our house had no window large enough to accommodate it."[87]

---

was not Leibniz but Sophie Charlotte who initiated plans for the Academy: *Geschichte der Akademie zu Berlin* (cit. n. 1), Vol. I, pp. 48–49. Perhaps the decision in Maria Winkelmann's case would have gone in her favor had Sophie Charlotte been alive to intervene.

[84] "Leibnizens Denkschrift in Bezug auf die Einrichtung einer Societas Scientiarum et Artium in Berlin vom 26. März 1700, bestimmt für den Kurfürsten," in Harnack, *Geschichte der Akademie zu Berlin* (cit. n. 1), Vol. II, p. 80. I thank Gerda Utermöhlen of the Leibniz Archive, Hannover, for calling this passage to my attention.

[85] Vignoles, "Eloge de M. Kirch le Fils" (cit. n. 39), p. 349.

[86] *Ibid.*

[87] Christine Kirch to Delisle, 24 July 1744, Paris Observatory, Delisle papers, MS A. B. 1. IV, No. 12a; and 28 Apr. 1745, No. 42. There are eight letters from Christine Kirch to Delisle at the Paris Observatory. Delisle initiated the correspondence in an attempt to buy the observation notebooks of Gottfried and Christfried Kirch. It should be noted that he did not ask for her astronomical observations; she volunteered them. Hers are the only letters from a woman in his sixteen volumes of correspondence. On Delisle see Roger Jaquel, "L'astronome Français Joseph-Nicolas Delisle (1688–1768) et Christfried Kirch (1694–1740), Directeur de l'Observatoire de Berlin (1716–1740)," *Actes du 97e Congrès National des Sociétés Savantes*, 1972, pp. 407–432.

Figure 3. *Like Gottfried Kirch and Maria Winkelmann, Elisabeth and Johannes Hevelius collaborated in astronomical work. This illustration shows them working together with the sextant. From Johannes Hevelius,* Machinae coelestis *(Danzig, 1673), facing page 222. By permission of the Houghton Library, Harvard University.*

Though Christine and Margaretha Kirch had little opportunity to go to the observatory after their brother's death, Christine continued to prepare the Academy calendar—silently and behind the scenes—from at least 1720 until her death in 1782. This is not surprising. By the 1740s, calendar making was no longer on the cutting edge of astronomical science, but tedious and time-consuming work. Never married, Christine supported herself through her calendar work, for which she received a small pension of 400 talers per year.[88]

After Christine Kirch retired, no other women did scientific work for the Berlin Academy of Sciences until well into the twentieth century.[89] During the eighteenth century, the Academy did, however, grant honorary membership to some women of the noble classes. The first to be granted honorary membership at the (then) Académie Royale des Sciences et Belles-Lettres was one of the most powerful persons in Europe at the time, Catherine the Great of Russia. Rank still spoke loudly in Prussia, and the prestige of her rank outweighed the liabilities of her sex. Catherine's position in the Academy was wholly honorary.[90] After Frederick the Great's tenure as president, few women were elected. One exception was poet and writer Duchess Juliane Giovane, who was awarded honorary membership in 1794. No other woman was elected for 106 years, and even then it was for purely nonscientific reasons: in 1900 Maria Wentzel was awarded honorary membership for her gift of 1,500,000 marks.[91]

It is clear that before 1949 only women of the very highest social standing were admitted to membership in the Berlin Academy of Sciences. Though Catherine the Great and Juliane Giovane were women of intellectual stature, they were also women of social rank. Maria Winkelmann, however, was a tradeswoman who dirtied her hands in the actual work of astronomy (she was referred to by Academy officials as a "Weib," and not a "Frauenzimmer"). The election of a woman purely on scientific merit had to wait until 1949, when the physicist Lise Meitner was elected, but as only a corresponding member. Meitner was followed by the chemist Irène Joliot-Curie, daughter of Marie Curie, and then by the medical

[88] Harnack, *Geschichte der Akademie zu Berlin* (cit. n. 1), Vol. I, p. 491.

[89] I am unaware of other women doing scientific work like that of the Kirch sisters. This is not surprising since it was craft traditions that secured their positions, and by the 1780s, when Christine Kirch died, these traditions were growing feeble.

[90] It should be pointed out that Catherine was elected in 1767, when Frederick the Great, as president of the Academy, personally oversaw all appointments. The following year, Frederick decreed that her membership in the Academy should be elevated from honorary status to that of a regular foreign member: Harnack, *Geschichte der Akademie zu Berlin* (cit. n. 1), Vol. I, pp. 369, 473; and Werner Hartkopf, *Die Akademie der Wissenschaften der DDR: Ein Beitrag zu ihrer Geschichte* (Berlin: Akademie-Verlag, 1983), pp. 219–220. During Frederick's presidency several women had their work read before the Academy. On the evening of 26 Jan. 1769, e.g., the Countess of Skorzewska's "Considérations sur l'origine des Polonais" was read at a public session. Though she was present at the reading, the assembly would not seat her: Harnack, *Geschichte der Akademie zu Berlin* (cit. n. 1), Vol. I, p. 370. In 1770 another woman—the French anatomist and writer Marie Geneviève Charlotte Thiroux d'Arconville—held a discourse on "L'amour-propre" at an Academy session ("Essai sur l'amour-propre envisagé comme principe de morale," *Discours prononcé a l'assemblée ordinaire de l'académie royale des sciences et belles-lettres de Prusse*, 1770, pp. 1–32). This essay has also been attributed to Frederick the Great. As one might also expect, women of royalty attended some Academy sessions, not as members but as guests. In 1772, for example, a host of royalty including the queen of Sweden (sister of Frederick), the princess Amélie of Prussia, and the Abbess of Quedlimbourg were present at a public session (*Nouveaux Mémoires de l'Académie Royale des Sciences et Belles-Lettres*, 1772, p. 5).

[91] Maria Wentzel's endowment of 1894 was in honor of her architect husband and factory-owning father: Harnack, *Geschichte der Akademie zu Berlin* (cit. n. 1), Vol. I, p. 1019.

doctor Cécilie Vogt in 1950. The first woman to be awarded full membership was the historian Liselotte Welskopf, in 1964.

Since the founding of the Academy of Sciences in Berlin in 1700, only fourteen of its twenty-nine hundred members have been women. Of those fourteen, only four have enjoyed full membership.[92] As of 1983, no woman had ever served in any leadership role as president, vice president, general secretary, or head of any of the various scientific sections.

IX. THE CONSEQUENCES FOR WOMEN'S PARTICIPATION IN SCIENCE

As the case of Maria Winkelmann illustrates, the poor representation of women in the Berlin Academy cannot be traced simply to an absence of qualified women.[93] Instead, the exclusion of women resulted from policies consciously implemented at an early period in the Academy's history. These decisions, made in the early eighteenth century, held serious consequences for women's later participation.

The Academy did not, however, make its decisions in a vacuum. Larger developments in both science and society set parameters within which the Academy maneuvered. The professionalization of the sciences (a gradual process which took place over the span of two centuries) weakened craft traditions within the sciences and weakened, in turn, women's position in science. With the gradual professionalization of astronomy, astronomers ceased working in family attics doubling as observatories as they had done in the days of the Kirch family. As late as 1704, Gottfried Kirch recorded in his diary: "July 4, [The sky was] light early. But I was unable to use the floor [to make observations through windows in the ceiling] since the washing from two households was hanging there. It was a pity, because I missed the conjunction of Jupiter and Venus."[94] Kirch's complaint reveals a striking juxtaposition of science and private life that began to disappear in the course of the eighteenth century. With the increasing polarization of public and private life, the family moved into the private sphere of hearth and foyer, while science migrated to the public sphere of the university and industry.[95] This polarization held important consequences for women's participation in science.

With the privatization of the family, husbands and wives ceased to be partners in the family business, and women were increasingly confined to the domestic role of wives and mothers. A wife like Maria Winkelmann-Kirch could no longer

---

[92] For information on Academy membership see Hartkopf, *Die Akademie der Wissenschaften der DDR* (cit. n. 90).

[93] In addition to the fourteen women who became Academy members over the past three and a half centuries, fifteen other women won Academy prizes: *ibid.*

[94] Quoted in Zinner, *Die Geschichte der Sternkunde* (cit. n. 6), p. 583, and Wattenberg, "Zur Geschichte der Astronomie in Berlin, I" (cit. n. 12), p. 166.

[95] See Jean-Louis Flandrin, *Families in Former Times: Kinship, Household, and Sexuality,* trans. Richard Southern (1975; Cambridge: Cambridge Univ. Press, 1979); Edward Shorter, *Making of the Modern Family* (New York: Basic Books, 1975); Werner Conze, ed., *Sozialgeschichte der Familie in der Neuzeit Europas* (Stuttgart: Klett, 1976); Michael Mitterauer and Reinhard Sieder, *The European Family: Patriarchy to Partnership from Middle Ages to Present,* trans. Karla Oosterveen and Manfred Hoerzinger (1977; Chicago: Univ. Chicago Press, 1982); Richard Evans and W. R. Lee, *The German Family: Essays on the Social History of the Family in Nineteenth- and Twentieth-Century Germany* (London: Croom Helm, 1981); and Heidi Rosenbaum, *Formen der Familie* (Frankfurt am Main: Suhrkamp Verlag, 1982).

become assistant astronomer to a scientific academy through marriage. Such positions became reserved for those with public certification of their qualifications.

With the changes in the social structure of science, women's participation changed. On the one hand, women attempted to follow the course of public instruction and certification through the universities, like their male counterparts. These attempts, however, were not successful until nearly two centuries later, at the turn of the twentieth century.[96] A second option open to women was to continue to participate within the (now private) family sphere as increasingly "invisible" assistants to a scientific husband or brother. These invisible assistants are difficult to distinguish from the unpaid artisan wife and represent a legacy of that tradition.

Changes in both the structure of science and the structure of the family served to distance wifely assistants from the professional world of science. Whereas Gottfried Kirch acknowledged his wife's work in scientific publications in 1710, Hermann von Helmholtz in the mid-nineteenth century praised his wife for her help in his experiments only privately and never publicly acknowledged her help either in his books or in his papers.[97]

In eighteenth-century Germany, modern science was a new enterprise forging new institutions and norms. With respect to the problem of women, science (and society) at this time may be seen as standing at a fork in the road. Science could either affirm and broaden practices inherited from craft traditions and welcome women as full participants, or it could reaffirm academic traditions and continue to exclude them. As the case of Maria Winkelmann demonstrates, the Berlin Academy of Science chose to follow the latter path.

[96] Women were not formally admitted to European universities until the 1860s in Switzerland, 1870s in England, 1880s in France, and 1900s in Germany. See Rita McWilliams-Tullberg, "Women and Degrees at Cambridge University 1862–1897," in *A Widening Sphere: Changing Roles of Victorian Women,* ed. Martha Vincinus (Bloomington: Indiana Univ. Press, 1977), pp. 117–146; and Laetitia Böhm, "Von dem Anfängen des akademischen Frauenstudiums in Deutschland," *Historisches Jahrbuch,* 1958, 77:2298–2327.

[97] I thank Richard Kremer, Department of History, Dartmouth College, for information on Helmholtz.

Laura Bassi as a Petrarchan muse, 1732. From Biblioteca Comunale dell'Archiginnasio, Bologna, Gabinetto Disegni e Stampe, Collezione dei ritratti, A/5, cart. 30, n. 3.

Science as a Career in Enlightenment Italy

The Strategies of Laura Bassi

By Paula Findlen

When Man wears dresses
And waits to fall in love
Then Woman should take a degree.[1]

IN 1732 LAURA BASSI (1711–1778) became the second woman to receive a university degree and the first to be offered an official teaching position at any university in Europe. While many other women were known for their erudition, none received the institutional legitimation accorded Bassi, a graduate of and lecturer at the University of Bologna and a member of the Academy of the Institute for Sciences (Istituto delle Scienze), where she held the chair in experimental physics from 1776 until her death in 1778. In Germany she was held up as a model to encourage other learned women to receive formal recognition for their studies. In France she earned the praise of contributors to the article on "Woman" that appeared in the *Encyclopédie;* the *Journal des Dames* devoted an article to her accomplishments in March 1775. Bassi left behind little of her scientific work, but her correspondence amply testifies to her accomplishments in and centrality to the learned world. She evoked the admiration of Voltaire and Francesco Algarotti, corresponded with natural philosophers such as Roger Boscovich, Charles Bonnet, Jean Antoine Nollet, Giambattista Beccaria, Paolo Frisi, and Alessandro Volta, and numbered her cousin Lazzaro Spallanzani among her pupils. As Voltaire wrote to her in 1744, "There is no

The research for this essay was funded by a Humanities Graduate Research Grant, University of California, Berkeley, and a Faculty Research Grant, University of California, Davis. Thanks to Marta Cavazza, Beate Ceranski, Alberto Elena, Giuliano Pancaldi, and the anonymous *Isis* readers for providing me with ideas, bibliography, and sound editorial advice; to Massimo Zini for giving me access to the archive of the Istituto delle Scienze; and to Richard Lombardo for unearthing a much-needed article. Londa Schiebinger, Dorinda Outram, and the members of the Cross-Cultural Women's History Group at UC Davis, particularly Betty Jo Dobbs and Susan Mann, offered their encouragement and suggestions.

[1] In Giancarlo Roversi, "Conquistavano il sapere ma per gli uomini erano sempre . . . le dottoresse ignoranti: Un divertente esempio di polemica antifemminista nella Bologna del '700," *Donne di Garbo,* 1984, 2:27–29, on p. 29. The poems were originally written in 1728, but published in 1732.

Bassi in London, and I would be much happier to be added to your Academy of Bologna than to that of the English, even though it has produced a Newton."[2]

Bassi is one of the most interesting women natural philosophers of the early modern period.[3] During her long tenure as a professor at the University of Bologna and the most prominent female member of Italy's leading scientific society, she played a central role in the introduction of new forms of learning into the university science curriculum and in the constitution of a network of experimenters that connected Italy to the scientific culture of France and England. Other women natural philosophers published more than she did—for example, Margaret Cavendish, Maria Sybilla Merian, and Voltaire's celebrated companion, Émilie du Châtelet—but Bassi was best at inserting herself within the academic world of science. This essay explores the conditions that made her success possible. While considering the limitations that the social and institutional framework of science placed upon Bassi as a female practitioner in light of the cultural expectations of learned women in Enlightenment Italy, I also wish to indicate how Bassi used the patronage system and her unique role within it as both patron and emblem of the new science to carve a niche for herself in the scientific community of the eighteenth century.

Bassi's activities began in the period that ushered in early discussions about Newton, in the form of poems about the *Opticks* and cautious explorations of the *Principia;* by the time she died the general principles of Newtonianism had become a basis for explorations of all facets of the natural world. When Bassi first became associated with the University of Bologna, Newtonianism had only just begun to enter Italian academic discourse. Bassi studied primarily Aristotelian and Cartesian philosophy prior to receiving her degree, and she did not explore Newton's thought until the mid 1730s, under the tutelage of the mathematician Gabriele Manfredi. She became one of the first scholars in Italy to teach Newtonian natural philosophy, beginning with her lectures on the less controversial *Opticks* in the late 1730s and continuing with the course in experimental physics that she conducted in her home and ultimately at the Institute. In 1749 she presented a dissertation on the problem of gravity and in 1763 one on refrangibility before her colleagues at the Institute; in 1757 she published a paper on hydraulics in the *Commentaries of the Bolognese Institute and Academy for Sciences and Arts* that worked out certain theorems posed by Newton. From the 1760s onward, in collaboration with her husband Giuseppe Veratti, she made Bologna a center for experimental research in electricity, attracting the interest of well-known scholars of this subject such as Abbé Nollet.[4] Right until

[2] Quoted in Ernesto Masi, "Laura Bassi ed il Voltaire," in *Studi e ritratti* (Bologna: Zanichelli, 1881), p. 167 (here and elsewhere, translations are mine unless otherwise indicated). On praise of Bassi in contemporary publications see Londa Schiebinger, *The Mind Has No Sex? Women in the Origins of Modern Science* (Cambridge, Mass.: Harvard Univ. Press, 1990), p. 252; Boucher d'Argis, "Femme" (*Jurisp.*), *Encyclopédie ou dictionnaire raisonné des sciences, des arts et du métiers* (Paris, 1751), Vol. 6, p. 475; and Nina Gelbart, *Feminine and Opposition Journalism in Old Regime France: Le Journal des Dames* (Berkeley/Los Angeles: Univ. California Press, 1987), p. 187.

[3] For a more detailed biographical treatment see Elio Melli, "Laura Bassi Veratti: Ridiscussioni e nuovi spunti," in *Alma mater studiorum: La presenza femminile dal XVIII al XX secolo* (Bologna: CLUEB, 1988), pp. 71–79; and Alberto Elena, " 'In lode della filosofessa di Bologna': An Introduction to Laura Bassi," *Isis*, 1991, *82*:510–518. For contemporary accounts see, in particular, Biblioteca Comunale dell'Archiginnasio, Bologna (BCAB), *Mss. Bassi, Laura*, Box I, fasc. 1 (*Notizie riguardanti Laura Bassi*); and Giovanni Fantuzzi, *Elogio della Dottoressa Laura Maria Caterina Bassi Veratti scritto da Giovanni Fantuzzi: Aggiungesi un'orazione del Dott. Matteo Bazzani* (Bologna, 1778).

[4] Unfortunately, the standard study of electricity mentions the work of Bassi and Veratti only in passing: John L. Heilbron, *Electricity in the Seventeenth and Eighteenth Centuries: A Study of Early Modern*

the end of her career, she made the dissemination of Newtonian ideas one of her principal goals. As late as 1774 she lectured on the work of Stephen Hales, well known for his applications of Newtonian ideas to explorations of fluids in works such as his *Vegetable Staticks* (1727).

Despite Bassi's importance to the scientific culture of Enlightenment Italy, we know very little about her intellectual activities because so few of her dissertations and lectures have survived. In addition to the forty-nine theses printed for her doctoral defense and various poems, Bassi published only four works in her lifetime: *De acqua corpore naturali elemento aliorum corporum parte universi* (1732), *De aeris compressione* (1745), *De problemate quodam hydrometrico* (1757), and *De problemate quodam mechanico* (1757). Another was published posthumously: *De immixto fluidis aere* (1792). These represent only a fraction of the dissertations that she prepared and defended annually at the Institute academy. And, as Alberto Elena observes, they tell us very little about Bassi's experimental and pedagogical activities.[5] Unlike Émilie du Châtelet, Bassi did not derive her fame from her publications. In addition to the normal obstacles that faced women writers, she had to divide her time between teaching, experimenting, and raising a family of eight children (five of whom survived to adulthood).[6] Equally important, Bassi did not need to publish to stake her claim within the community of natural philosophers. Instead, the unique opportunities that early modern Italian academic culture afforded Bassi made it possible for her to achieve recognition for her work in physics and mathematics through her actions rather than her pen. Her correspondence and contemporary reports of her activities allow us to follow her career as a natural philosopher in Enlightenment Italy.

By the early eighteenth century science increasingly was a legitimate pursuit for exceptional women of unquestionable virtue. Women had become an audience for philosophical speculations, and their role as patrons and consumers of natural philosophy was increasingly acknowledged by contemporaries. The popularity of scientific activities in the Baroque Italian courts gave noblewomen the opportunity to observe and moderate the culture of experimentation and debate. In this context, women such as the grand duchess of Tuscany, Christina of Lorraine, and Queen Christina of Sweden emerged as patrons of science.[7] The visibility of female patrons at court and the expansion of roles for women in the literary, artistic, and scientific academies of Italy set the background for the increased participation of socially prominent and intellectually gifted women in natural philosophy. Equally important were the possibilities that Cartesian philosophy offered. Descartes's famous dedi-

---

Physics (Berkeley: Univ. California Press, 1979), p. 354. For a complete list of Bassi's Institute lectures see Melli, "Laura Bassi Veratti," p. 79, n. 40. On Newtonianism in Italy see Paolo Casini, *Newton e la conscienza europea* (Bologna: Mulino, 1983), pp. 173–227; and Vincenzo Ferrone, *Scienze, natura, religione: Mondo newtoniano e cultura italiana nel primo Settecento* (Naples: Jovene, 1982).

[5] Elena, "Introduction to Laura Bassi" (cit. n. 3), p. 515. With the exception of the 1732 lecture, all of Bassi's publications appeared in *De Bononiensi Scientiarum et Artium Instituto atque Accademia Commentarii*. The citations are, respectively: 1745, 2(1):347–353; 1757, 4:61–73; 1757, 4:74–79; and 1792, 7:44–47.

[6] Beate Ceranski is completing a thesis at the University of Hamburg that deals with Bassi and Veratti's marriage in greater detail.

[7] Galileo Galilei, *Letter to the Grand Duchess Christina* (1615), in *Discoveries and Opinions of Galileo*, ed. and trans. Stillman Drake (New York: Anchor, 1957), pp. 145–216; and Susanna Åkerman, *Queen Christina of Sweden and Her Circle: The Transformation of a Seventeenth-Century Philosophical Libertine* (Leiden: Brill, 1991).

cation of his *Principles of Philosophy* (1644) to Elizabeth of Bohemia was seen by
a number of scholars, male and female, as clear evidence that women were capable
of philosophizing. Taking Descartes's praise of Elizabeth of Bohemia literally, Giu-
seppa-Eleonora Barbapiccola translated the work into Italian in 1722 in order "to
make it accessible to many others, particularly women, who, as the same René says
in one of his letters, are more apt at philosophy than men."[8] Bassi, in her studies
of Cartesian philosophy in preparation for her thesis defense, may even have read
Barbapiccola's translation. Certainly her family and mentors acted in accordance
with these principles when they chose to make natural philosophy an important part
of her education. Learned women had long been known for their skill in ancient
languages and, in a city such as Bologna, law. Mastery of natural philosophy, particularly the tenets of the new scientific learning, added a novel dimension to this topos.

During the same period, Italian intellectuals explored the desirability of women's
education. Most famously, the Accademia de' Ricovrati, known for its admission
of women, conducted a series of debates in 1723 about whether "women should be
allowed to study Sciences and the Fine Arts" that expressed well the range of opinions of this issue.[9] While many natural philosophers continued to oppose the entry
of women into academic discourse, others, such as the president of the Accademia
de' Ricovrati, the anatomist Antonio Vallisneri, encouraged their participation. Thus
the Academy of the Institute for Sciences had important precedents to draw upon
when it chose to add several women to its ranks, first among them Bassi. While
claiming to imitate the new scientific societies such as the Royal Society, the Paris
Academy of Sciences, and the Accademia del Cimento, none of which included
women as members, the Institute academy nonetheless acknowledged its debt to the
flourishing tradition of Italian academies that made the presence of women in their
assemblies a necessary part of their composition.

The presence of women within academic institutions in Italy stood in marked contrast to the situation elsewhere. In countries like France women's participation occurred primarily in the salons. Excluded from the universities, women were allowed
no role in organizations such as the Paris Academy of Sciences. As François Poulain
de la Barre wrote in 1673, "They [men] founded Academies to which women were
not invited; and in this way [women] were excluded from learning as they were from
everything else." In contrast, the Italian academies welcomed French women of
learning, capitalizing on the dearth of possibilities within French academic culture.
Upon hearing of Émilie du Châtelet's admission to the Institute academy in 1746,
one philosophe lamented:

> When Bologna proudly displays, in Italy,
> Its register adorned with the fair name of Émilie

---

[8] "La traduttrice a' lettori," *I principi della filosofia di Renato Des-cartes tradotti . . . da Giuseppa-
Eleonora Barbapiccola tra gli Arcadi Mirista* (Turin, 1722), n.p. On Cartesian philosophy and the role
and image of women see Michel Delon, "Cartésianisme(s) et féminisme(s)," *Europe,* 1978, 56:73–86;
Erica Harth, *Cartesian Women: Versions and Subversions of Rational Discourse in the Old Regime*
(Ithaca, N.Y.: Cornell Univ. Press, 1992); and Schiebinger, *The Mind Has No Sex?* (cit. n. 2), pp. 23,
171–172, 175–178.

[9] *Discorsi accademici di vari autori viventi intorno agli studi delle donne; la maggior parte recitati
nell'Accademia de' Ricovrati di Padova* (Padua, 1729). For the broader context see Luciano Guerci, *La
discussione sulla donna nel Settecento* (Turin: Tirrenia, 1987).

> Why is the fair sex, so greatly loved by us,
> Excluded, in France, from the Academy?[10]

While other regions discarded the Renaissance model of the academy as a place in which learned men and women of high social standing could interact, Italian practice enhanced this image. Rather than creating a salon culture that served to formalize the separation of the world of the academies from the society of learned women, scholars in numerous Italian cities formed academies that linked the university, the salon, and the leisure activities of the urban patriciate. In Bologna, Bassi soon became the centerpiece of such a network. Her presence strengthened preexisting ties between different sectors of the political and cultural elite and contributed to the enhancement of the city's position within the Republic of Letters.

While the academies and salons provided a setting in which men and women intermingled, enacting versions of the conversations fictionalized in works such as Bernard Le Bovier de Fontenelle's *Conversations on the Plurality of Worlds* (1686) and Algarotti's *Newtonianism for Ladies* (1737), few women had access to the universities in any capacity. Despite the fact that the University of Bologna celebrated a handful of women who had either attended or given lectures in the Studium during the Middle Ages, no woman in any part of Europe had been offered a degree before 1678 or an official teaching position before 1732. Most universities implicitly excluded women from any form of participation and would continue to do so until the late nineteenth and twentieth centuries. Women's learning was represented as a separate enterprise from the education of men. In the early seventeenth century, when the theologian Gisbert Voet allowed Anna Maria van Schurman to attend his lectures at the University of Utrecht, he maintained the tradition of having her listen behind a curtain in order to separate her from the male scholars.[11] Bassi's integration into the university culture of Bologna upon the receipt of her degree marked a departure from these practices. Benefiting from earlier traditions, which permitted women occasional access to the classroom, Bassi nonetheless was distinguished by the permanency of her position and by the fact that she performed many of the same functions as her male colleagues, as a salaried professor, lecturer, and experimenter.

The attention showered upon Bassi was the culmination of several abortive attempts to install a woman within the Italian university system. In 1678 over twenty thousand spectators crowded into Padua to see the Venetian noblewoman Elena Cornaro Piscopia receive a degree in philosophy. Immediately afterward, the rectors of the University of Padua agreed to admit no more women; when Piscopia died in 1684, she was celebrated as the first and, for the time being, the last female graduate of that university. By the 1720s the leading citizens of Bologna had begun to con-

---

[10] François Poulain de la Barre, *De l'egalité des deux sexes* (Paris, 1673), p. 28, quoted in Harth, *Cartesian Women* (cit. n. 8), p. 135; and P.-R. de Cideville, quoted in Esther Ehrman, *Mme. du Châtelet: Scientist, Philosopher, and Feminist of the Enlightenment* (Leamington Spa: Berg, 1986), p. 39. On salon culture see Carolyn Lougee, *Le Paradis des Femmes: Women, Salons, and Social Stratification in Seventeenth-Century France* (Princeton, N.J.: Princeton Univ. Press, 1976); Dena Goodman, "Enlightenment Salons: The Convergence of Female and Philosophic Ambitions," *Eighteenth-Century Studies*, 1989, 22:329–350; and Dorinda Outram, "Before Objectivity," in *Uneasy Careers and Intimate Lives*, ed. Pnina Abir-Am and Dorinda Outram (New Brunswick, N.J.: Rutgers Univ. Press, 1987), pp. 22–25.

[11] Joyce L. Irwin, "Anna Maria van Schurman: The Star of Utrecht," in *Female Scholars: A Tradition of Learned Women before 1800*, ed. J. R. Brink (Montreal: Eden, 1980), pp. 86–100; and Una Birch, *Anna van Schurman: Artist, Scholar, Saint* (London: Longmans, Green, 1909).

ceive of the idea of having their own woman graduate. The city abounded with learned women such as Teresa and Maddelena Manfredi, sisters of Eustachio and Gabriele and their assistants in matters astronomical, or the *bella cartesiana,* Laura Bentivoglio Davia, who so disparaged "the noisy conferral of the doctorate" on Bassi.[12] However, the first attempt occurred in the domain of law rather than natural philosophy. In 1722, ten years before Bassi embarked on her career as a public intellectual figure, the Bolognese noblewoman Maria Vittoria Delfini Dosi defended several legal theses at the Spanish College. While subsequent attempts on the part of her father to convince the University of Bologna to grant her a degree in jurisprudence failed, Delfini Dosi's public display of learning reopened the debate about the place of women in academic culture in Bologna. By the 1750s two women had received degrees—Bassi and Cristina Roccati—and three—Bassi, the mathematician Maria Gaetana Agnesi, and the anatomist and wax modelist Anna Morandi Manzolini—had been offered teaching positions. Bologna, as one critic complained, now had "a platoon of women teachers."[13]

Despite this evaluation, Bassi was the only woman in the eighteenth century whose circumstances allowed her the opportunity to engage fully in the activities that her male counterparts took for granted. Situated in a climate in which the idea of a "career" as an experimental philosopher was only gradually beginning to emerge, she took advantage of the ambiguous parameters of the scientific community and of the veneration accorded learned women to extend the range of her responsibilities.[14] In less than a century, the patronage that women offered to the sciences and their growing presence as spectators and members of various learned academies had opened the way to their participation. And Italy, more than any other region in Enlightenment Europe, offered learned women diverse circumstances in which to exercise and display their erudition.

BASSI TAKES HER DEGREE

When Laura Bassi accepted the invitation to defend forty-nine theses in front of the notables of Bologna on 17 April 1732, and the subsequent offer of a position as a university lecturer, she entered a complex social world. The scientific culture of early modern Europe was constituted within a dense network of patrons, brokers, and clients whose shifting relationships forged the boundaries of the learned commu-

[12] Laura Bentivoglio Davia to Giovanni Bianchi, 14 June 1732, Bologna, in Gian Ludovico Masetti Zannini, "Laura Bassi (1711–1778): Testimonianze e carteggi inediti," *Strenna Storica Bolognese,* 1979, 29:219–241, on p. 222. On the Manfredi sisters see Ilaria Magnani Campanacci, "La cultura extra-accademica: Le Manfredi e le Zanotti," in *Alma mater studiorum* (cit. n. 3), pp. 39–67.

[13] Bienvenuto Robbio, *Disgrazia di Donna Urania ovvero degli studi femminili* (Florence, 1798), p. 122. On Piscopia see Francesco Ludovico Maschietto, *Elena Lucrezia Cornaro Piscopia (1646–1684) prima donna laureata nel mondo* (Padua: Antenore, 1978); on Delfini Dosi see Emilio Oriolo, "Una cultrice di diritto a Bologna nel secolo XVIII," *L'Archiginnasio,* 1911, 6:25–31; on Agnesi see Giovanna Tilche, *Maria Gaetana Agnesi: La scienziata santa del '700* (Milan: Rizzoli, 1984); and Carla Vettori Sandor, "L'opera scientifica ed umanitaria di Maria Gaetana Agnesi," in *Alma mater studiorum,* pp. 105–118; on Morandi Manzolini see Vittoria Ottani Gabriella Giuliani-Piccari, "L'opera di Anna Morandi Manzolini nella ceroplastica anatomica bolognese," *ibid.,* pp. 81–103; on Roccati see Paola Savaris, "Cristina Roccati: Una rodigina del '700 tra scienza e poesia" (thesis, Facoltà di Magistero, Univ. degli Studi di Ferrara, 1990–1991) (I would like to thank Clelia Pighetti—an advisor, with Marco Mondadori—for bringing this work to my attention).

[14] See Brendan Dooley, "Science Teaching as a Career at Padua in the Early Eighteenth Century: The Case of Giovanni Poleni," *History of Universities,* 1984, 4:115–151.

nity.[15] Aspiring natural philosophers could no more make their way in the world without patrons than they could lay claim to the title of "philosopher" without having read Aristotle. Social connections in conjunction with learning made acceptance into the scientific community possible; they mediated the awarding of university positions and promotions and paved one's entry into the courts, salons, and academies, where knowledge was put on display in front of a largely patrician audience.

Within this system, women were perceived primarily as *facilitators*. Like Baldassar Castiglione's ideal female courtiers, their presence provided a necessary impetus to conversation, but they spoke rarely, if at all. While early modern culture recognized, indeed encouraged, female patrons of science, it accommodated women who sought rather than dispensed patronage less easily. Bassi was precisely this sort of individual. Genuinely committed to teaching and research, she attempted to build a career out of her position by using the same patrons and institutions as her male counterparts. While always acknowledging her special status, she nonetheless refused to accept the limitations placed upon her. Like Madame du Châtelet, she continued to profess astonishment at the fact that while women were accepted as rulers in several countries, "there is none in which we are elevated to think," and she devoted her life to making Bologna an environment in which at least one woman contradicted this general rule.[16]

The scientific climate of Bologna in 1732 was particularly conducive to public and institutional recognition of a learned woman, especially one versed in the latest mathematical and experimental philosophies. The final decades of the seventeenth century had witnessed a major decline in the international reputation of the University of Bologna, and even the efforts of committed reformers like Anton Felice Marsili to revitalize the curriculum had yielded disappointingly few results and little acknowledgment from the outside world. The founding of the Institute for Sciences by Luigi Ferdinando Marsili in 1714 institutionalized these reforms, establishing a teaching and research facility specifically designed to supplement the traditional curriculum, particularly in the experimental sciences.[17] Where late seventeenth-century academies had failed, the Institute succceeded, bringing Bologna again to the attention of the learned world through its importation of the best ideas of the new philosophy to Italy. In 1728 the young Algarotti first replicated Newton's optical experiments be-

---

[15] Scientific patronage has been the subject of a growing body of secondary literature. I will indicate only some of the most relevant studies: Mario Biagioli, "Galileo's System of Patronage," *History of Science,* 1990, *28*:1–62; David S. Lux, *Patronage and Royal Science in Seventeenth-Century France: The Académie de Physique in Caen* (Ithaca, N.Y.: Cornell Univ. Press, 1989); Bruce Moran, ed., *Patronage and Institutions: Science, Technology, and Medicine at the European Courts, 1500–1750* (Woodbridge, Suffolk: Boydell, 1991); Lisa T. Sarasohn, "Nicolas-Claude Fabri de Peiresc and the Patronage of the New Science in the Seventeenth Century," *Isis,* 1993, *84*:70–90; Alice Stroup, *A Company of Scientists: Botany, Patronage, and Community at the Seventeenth-Century Parisian Royal Academy of Sciences* (Berkeley/Los Angeles: Univ. California Press, 1990); and Richard Westfall, "Scientific Patronage: Galileo and the Telescope," *Isis,* 1985, *76*:11–30. For a later period see also Dorinda Outram, "Before Objectivity" (cit. n. 10), pp. 19–30; and Outram, *Georges Cuvier: Vocation, Science, and Authority in Post-Revolutionary France* (Manchester: Manchester Univ. Press, 1984).

[16] Baldassar Castiglione, *The Book of the Courtier,* trans. Charles S. Singleton (New York: Anchor, 1959), esp. pp. 15–26, 201–282; and Émilie du Châtelet, preface to her translation of Mandeville's *Fable of the Bees,* in Elisabeth Badinter, *Émilie, Émilie: L'ambition féminine au XVIIIème siècle* (Paris: Flammarion, 1983), p. 448.

[17] For a comparative perspective see Dooley, "Science Teaching as a Career at Padua" (cit. n. 14). On the history of the Istituto and its predecessors see Marta Cavazza's excellent *Settecento inquieto: Alle origini dell'Istituto delle Scienze di Bologna* (Bologna: Mulino, 1990); see also *I materiali dell'Istituto delle Scienze* (Bologna: CLUEB, 1979); and Richard Rosen, "The Academy of Sciences of the Institute of Bologna, 1690–1804" (Ph.D. diss., Case Western Reserve Univ., 1971).

fore the Institute members, the beginning of his efforts to introduce Newtonianism into Italy. Best known for his *Newtonianism for Ladies,* completed at Cirey in the company of Voltaire and Châtelet, Algarotti was an active participant in the diffusion of Newtonianism in Italy. While we most often think of this work in relation to Châtelet's translation of Newton's *Principia,* we should not neglect Algarotti's activities preceding the trip to France. Algarotti included a portrait of Châtelet in the engraved frontispiece, but he made several references to Bassi in the text. Completed only four years after Bassi had defended several Newtonian theses in public, and five years after Algarotti had witnessed her degree ceremonies (about which he wrote several poems), *Newtonianism for Ladies* was as much a tribute to Bassi and the Institute as it was to the activities of the French philosophes.[18]

While the Institute commenced with a great flourish, the death of its principal patron Marsili in 1730 and the lack of tangible results had diminished its public splendor. This intellectual torpitude was further exacerbated by the flagging fortunes of the Bolognese patriciate. By the 1720s the ranks of the major families—the Aldrovandi, Bentivoglio, Cospi, Paleotti, and Ranuzzi, to list a few—that had dominated the political culture of the city had thinned greatly.[19] Their replacement by members of the minor nobility only furthered the perception that Bologna was no longer the splendid center it once had been. Thus the university, the academy, and the city needed Bassi as much as she needed them. Publicizing Bassi's accomplishments, and enhancing them beyond anything achieved by earlier learned women, would add luster to the reputation of Bologna.

The daughter of a lawyer, Bassi received her early tutelage at home from Gaetano Tacconi, the family physician, a professor at the university, and a member of the Institute academy. As Bassi's accomplishments grew, pressure for her to appear in public mounted. Bassi was reputed to be a "monster in philosophy"; her fluency in Latin as well as Cartesian and Newtonian philosophy was noted by many contemporaries, among them the poet Giampietro Zanotti, who proclaimed her Latin to be better than his Bolognese. For Bassi and most educated women in Italy, science was not an alternative to classical learning, as it often was advertised in France and England, but its supplement; this provided the necessary element of continuity between Renaissance and Enlightenment views of learned women.[20] In the early months of 1732 Tacconi finally allowed a select group of professors and gentlemen, among them the secretary of the Institute, Francesco Maria Zanotti, and the new archbishop of Bologna, Prospero Lambertini, to hear Bassi dispute privately on various subjects.

---

[18] On Algarotti see Franco Arato, *Il secolo delle cose: Scienza e storia in Francesco Algarotti* (Genoa: Marietti, 1991); and Ida Frances Treat, *Un cosmopolite italien du XVIIIe siècle: Francesco Algarotti* (Trévoux: Jules Jeannin, 1913). See also Mauro De Zan, "Voltaire e Madame du Châtelet: Membri e corrispondenti dell'Accademia delle Scienze di Bologna," *Studi e Memorie dell'Istituto per la Storia dell'Università di Bologna,* 1987, 6:141–158, esp. p. 142. On Bassi and Algarotti see Marta Cavazza, "L''aurata luce settemplice': Algarotti, Laura Bassi e Newton," in *Settecento inquieto,* pp. 237–256; and the introduction to Francesco Algarotti, *Dialoghi sopra l'ottica neutoniana,* ed. Ettore Bonora (1969; Turin: Einaudi, 1977).

[19] Alfeo Giacomelli, "La dinamica della nobiltà bolognese nel XVIII secolo," in *Famiglie senatorie e istituzioni cittadine a Bologna nel Settecento* (Bologna: Istituto per la Storia di Bologna, 1980), pp. 55–112, esp. pp. 76–80.

[20] BCAB, *Ms. Hercolani,* fasc. 382, letters 32, 34, in Masi, "Laura Bassi ed il Voltaire" (cit. n. 2), p. 162. For a contrasting view on the role of ancient learning in England and France see Patricia Philips, *The Scientific Lady: A Social History of Women's Scientific Interests, 1520–1918* (London: Weidenfeld & Nicolson, 1990), pp. ix, 27–28; and Harth, *Cartesian Women* (cit. n. 8), p. 25.

As news of her remarkable erudition spread, she "found herself constrained to make almost a continuous spectacle of herself in the City." Cardinal Lambertini soon persuaded her to appear in public, and Tacconi and Zanotti proposed her election to the Institute academy, to which she was aggregated on 20 March 1732. Less than a month later, on 17 April, she engaged in a public dispute with five university professors, among them the physicist-chemist Jacopo Bartolomeo Beccari and Gabriele Manfredi.[21]

From the moment that Bassi agreed to participate in the public debates, her social position shifted. No longer simply a woman whose learning made her an object of curiosity and a participant in the civil discourses of the urban patriciate, she had become the symbol of the scientific and cultural regeneration of the city. The conferral of a degree on 12 May and the Senate's decision to award her a university chair on 29 October 1732 formalized the terms of the new relationship. Detailed reports of the defense, the degree ceremony, and her first lecture as a university professor illuminate the emergence of Bassi as a public figure. All these events were attended not only by the university faculty and students, but also by the principal political and religious figures of the city—the papal legate and vice-legate, the archbishop of Bologna, the gonfaloniere, the elders (Anziani), senators, and magistrates. Additionally, "all the Ladies of Bologna and all the Nobility," along with foreigners and curious onlookers, filled the rooms to hear her speak.[22] The composition of the audience testified to the close ties between the academic, civic, and patrician life of the city and underscored Bassi's importance to all three domains. In her passage from private citizen to public ornament, Bassi had become an emblem of Bologna's efforts to regain its foothold in the learned world.

Contemporary descriptions and illustrations give us a fairly precise idea of the performative nature of these events. The defense and the degree ceremony occurred not in the Archiginnasio, home to the Studium, but in the Palazzo Pubblico, seat of the local and papal government.[23] There, accompanied for decorum's sake by Countess Maria Bergonzi Ranuzzi and Marchesa Elisabetta Ercolani Ratta, two prominent noblewomen, Bassi engaged in her first public disputes. Arriving at the Palazzo Pubblico, Bassi went first to the quarters of the current gonfaloniere, Filippo Aldrovandi, where she was joined by the archbishop, the legate Girolamo Grimaldi, Cardinal Melchiorre de Polignac—a dignitary visiting Bologna on his way from Rome to Paris—and other "eminences." From these rooms she made her way to the Sala d'Ercole, where the degree was awarded by the chancellor of the Studium, Alessandro Formagliari, and the prior of the college, Matteo Bazzani. Both ceremonies ended with the retreat of the young "philosophess" (filosofessa) to the gonfaloniere's quarters within the Palazzo Pubblico, where she was privately feted, accompanied by "all the Ladies." In between her arrival and departure, Bassi traversed the space

---

[21] BCAB, Mss. Bassi, Laura, Box I, fasc. 1, cc. 1v–2r; and G. B. Comelli, "Laura Bassi e il suo primo trionfo," Stud. Mem. Stor. Univ. Bologna, 1912, Ser. 1, 3:197–256, on p. 205.

[22] Biblioteca Universitaria, Bologna (BUB), Codex 212 (116), no. 23, c. 94r. I am aware of three contemporary reports of the degree ceremonies: BUB, Codex 212 (116), no. 23, cc. 94–95; BCAB, Gozzadini 140, cc. 12–13; and Elisabetta Ratta to Francesco Algarotti, 19 Apr. 1732, in Comelli, "Laura Bassi e il suo primo trionfo," p. 213. My discussion of the ceremonies draws on these as well as Fantuzzi, Elogio (cit. n. 3), pp. 6–10. On Bassi's importance to the city see Cavazza, Settecento inquieto (cit. n. 17), pp. 249–256, passim.

[23] On the structure of the Bolognese government see Paolo Colliva, "Bologna dal XIV al XVIII secolo: 'Governo misto' o signoria senatoria?" in Storia dell'Emilia Romagna, ed. A. Berselli (Imola: Edizioni Santerno, 1977), Vol. 2, pp. 13–34.

separating her podium from the canopied throne (*baldacchino*) on which Archbishop Lambertini and Cardinal de Polignac were seated in order to pay her respects and receive their compliments.[24]

In an oration written especially for the degree ceremony, Matteo Bazzani praised Bassi for displaying her talents "in the most prosperous civic theater." Drawing attention to the long heritage of learned women associated with the University of Bologna, he presented Bassi as the culmination of a tradition that stretched back to the early days of the university in the later Middle Ages. As Bassi's colleague at the Institute, Beccari, was later to write to Maria Gaetana Agnesi when she was offered an honorary position at the university in 1750, "From the most ancient times Bologna has had people of your sex [lecture] from the public chairs."[25] Thus Bassi was invited to enhance and reconstitute this tradition in 1732. Her skills in ancient languages as well as modern sciences made her eminently qualified to fulfill her role.

Like Bazzani, Bassi was extremely conscious of her audience and its expectations. In the opening of her thesis defense she acknowledged frequently the presence of all the principal parties who had made the dispute possible. More pointedly, in the course of her first public lecture after the conferral of a university chair, on 18 December 1732, Bassi made particular reference to the governing body as patron of her efforts: "the magnanimity of the Senate raised me, beyond what I asked for and dreamed of, to the highest dignity of speaking in public," she proclaimed. The active role of the Senate in facilitating Bassi's position was confirmed by her biographer Giovanni Fantuzzi, who described it as "a spectator up until this [moment]," intervening with the offer of a position to prevent her from lapsing into "leisure and mediocrity" only after witnessing the success of her defense.[26] While her tutor Tacconi and his circle of friends, among them the influential Lambertini and Zanotti, had first brought this remarkable woman to the attention of Bologna, the Senate, as the body that governed all university appointments, gave her an official position.

MISPLACED EXPECTATIONS?

While offering Bassi a permanent role in the academic culture of the city, the Senate initially defined her duties as a lecturer restrictively. As one 1778 eulogy to Bassi described it, they "wished to inscribe her in the roll of Professors of Philosophy without the burden of exercising this commitment." During the deliberations about her position, beginning on 25 August and concluding on 29 October 1732, the Senate voted to award Bassi an annual stipend of 100 *scudi* "on the condition, however, that she should not read in the public schools except on those occasions when her Superiors commanded her, because of [her] Sex." The dates for these lectures were to be determined jointly by the papal legate and the gonfaloniere. Less than a year later, the governing board of the university (*Assunti di Studio*) recommended to the

---

[24] Bassi's thesis defense, *laurea*, and first lesson, along with her appearance at the 1734 carnival anatomy, are illustrated as part of a famous series of pictures detailing the civic life of the city: Archivio di Stato, Bologna (ASB), *Insignia degli Anziani*, Vol. 13, cc. 94, 95, 98, and 105, respectively.

[25] Matteo Bazzani, *Oratio ad egregiam virginem D. Lauram Mariam Catharinam Bassi*, in Fantuzzi, *Elogio* (cit. n. 3), p. 32; and Jacopo Bartolomeo Beccari to Maria Gaetana Agnesi, 8 July 1750, Bologna, Biblioteca Ambrosiana, Milan, ms. 0.201 sup., c. 12.

[26] Laura Bassi, *Praefectio primae conclusionis in aula palatii magistratus*, in Comelli, "Laura Bassi e il suo primo trionfo" (cit. n. 21), pp. 224–225; Bassi, *Praelectio*, in Luisa Caterina Cavazzuti, "Nuovi testi sull'attività scientifica el filosofica di Laura Bassi" (thesis, Facoltà di Magistero, Univ. Bologna, 1964–1965), p. 75; and Fantuzzi, *Elogio*, p. 10.

city government (*Reggimento*) that Bassi give one lecture "every Trimester."[27] Quite likely they imagined that she would speak at fall convocation, at the public anatomy (usually held in early February), and at the conclusion of the academic year in June. Thus the appearance of the local celebrity marked the rhythm of the academic year.

The nature of the events, both public and private, at which Bassi was invited to appear illustrates well the expectations her patrons had of her. Within days of the degree ceremony, Bassi entertained the visiting Cardinal de Polignac by disputing with four doctors from the university. On 15 June 1733 she participated in a public debate with Giuseppe Azzoguidi on "Poisons and Their Antidotes." From 1734 until 1778 she appeared annually as one of the disputants at the famed carnival anatomy and participated in 105 public disputes formally mandated by the university. From 1746 until 1777 she presented one formal dissertation yearly at the Institute academy as part of her responsibilities as a member of its elite core, the Benedictines (*Benedettini*). Many of these coincided with "the occasion of the Public Academies," held periodically by the Institute, to which nonmembers could come.[28] This pattern of activity, dictated by the institutional calendar, was occasionally punctuated by singular events, command performances at the request of patrons. For example, the arrival of a new legate in 1734 and the election of Carlo Grassi in 1745 were noteworthy enough to demand Bassi's presence. Similarly, the awarding of a degree to Cristina Roccati in 1751 and the fortieth anniversary of the Institute in 1754 required her participation. Had Maria Gaetana Agnesi chosen to come to Bologna to accept her honorary chair in mathematics in 1750, she certainly would have engaged in a public dispute with Bassi; this was surely one of the reasons why Benedict XIV and the members of the Institute were so keen on affiliating Agnesi with the scientific institutions of Bologna. Undoubtedly they were disappointed when the opportunity to pit the two most famous women natural philosophers in Italy against each other did not come to pass.[29]

In addition to her participation in various ceremonies, Bassi attended important social gatherings for the nobility. "All the gentlemen of Bologna make a great display of this girl, and depict her everywhere as the miracle of our age," wrote Giovanni Bianchi from Rimini in 1732. Friends such as the noblewoman Elisabetta Ratta had informal literary gatherings at their homes, to which Bassi was invited. Her close association with local poets like the Zanotti brothers made her sought after for her eloquence as well as her learning. Patrons like Senator Filippo Aldrovandi, gonfaloniere when Bassi received her degree, hosted well-publicized "Sunday evenings" for the Bolognese patriciate. As the *Avvisi di Bologna* reported in 1736, on one such evening in November Bassi debated both Matteo Bazzani and Francesco Maria Zanotti, alternately in Italian and Latin, in front of "all the Nobility in formal attire,"

[27] BCAB, B.2727, cc. 14v–15r (eulogy); and BCAB, *Gozzadini* 337, c. 89 (as reported in the *Avvisi di Bologna*, no. 45, 4 Nov. 1732). For the original deliberations see ASB, *Assunteria di Studio: Partitorum*, Vol. 49, fol. 49v, in Comelli, "Laura Bassi e il suo primo trionfo," p. 241; and ASB, *Assunteria di Studio: Atti* (1730–1735), Vol. 22 (9 Oct. 1733).

[28] BUB, Codex 212 (116), no. 23, c. 95r; ASB, *Assunteria di Studio: Atti* (1730–1735), Vol. 22 (13 June 1733); and BCAB, *Mss. Bassi, Laura*, Box I, fasc.1(i). The information about Bassi's Institute lectures is culled from Rosen, "Academy of Sciences" (cit. n. 17), pp. 224–264; and BCAB, *Mss. Bassi, Laura*, Box I, fasc. 1(a), c. 3v. All the *Benedettini* were assigned times by drawing lots, with the exception of Bassi, who could choose when she would lecture.

[29] BCAB, *Gozzadini* 337, c. 105; *Mss. Bassi, Laura*, Box I, fasc. 1(i), c. 4v. The possibility of Agnesi debating Bassi is suggested in Cornelia Benazzoli, *Maria Gaetana Agnesi* (Milan: Fratelli Bocca, 1939), p. 104.

including "over 120 Ladies." Appearing at the conclusion of the "noble Symphonies in diverse parts of the Palace," Bassi received the "universal applause" of the Bolognese elite as well as the "many Foreign Nobles" present.[30] Despite Bassi's initial reticence about appearing in public, after receiving her degree she was quickly integrated into the social circuit of the city.

Most revealing, however, was the frequency of her appearance at the carnival anatomy. The anatomy was a central feature of the public life of the university; like Bassi's degree ceremony, it enhanced the ritual life of the city. Unlike Bassi's degree ceremony, it was performed annually—the one dissection during the year that was open to anyone able to buy tickets. While other professors rotated in and out of the lectureships assigned to the public dissection, Bassi's presence was essential to the success of this popular event. Lasting ten to fifteen days, the annual dissection, held before Lent, entailed an elaborate ceremonial in which the leading professors, senators, and dignitaries of the town participated. Formal invitations were sent to prominent members of the community and important foreigners.[31] In addition, a riotous crowd of carnival revelers filled the anatomy theater with shouts and jeers as they watched the local intellectuals try to acquit themselves on any subject deemed worthy of conversation, as well as the proposed topic of the anatomy. The fortunes of the carnival anatomy curiously paralleled Bassi's own. Revived just when the university's reputation had reached its nadir, the anatomy was perceived as both an "honor to the Studium" and an "attraction to foreign scholars."[32] Given its significance, Bassi's presence was imperative in the eyes of the Senate, once her position made such public performances possible.

Her first appearance in this setting occurred on 23 February 1734 (see Figure 1). Domenico Gusmano Galeazzi presented a dissertation on sight, *De visu*. The new legate, Giovanni Battista Spinola, had arrived in the city shortly after New Year's. Accompanied by the gonfaloniere and the *Anziani,* Spinola attended the first and last lectures and the one at which Bassi debated Galeazzi: "With most subtle and learned arguments and rare expositions of experiments regarding the Sense of Sight, she demonstrated a high and profound understanding of this material, rousing widespread applause from the large Audience, most especially from the most Excellent signor Cardinal Legate, who understood for the first time the great Virtue and Knowledge of the lauded *Dottoressa*." What the official acts of the governing body of the university did not mention, however, were the "many Foreigners and a great many Maskers present."[33] While the content of her disputes may have appealed chiefly to a small learned constituency able to follow the intricacies of academic Latin, among them the legate and the archbishop, her presence at the anatomy only heightened the

[30] Bianchi to Antonio Leprotti, 18 May 1732, Rimini, in Zannini, "Laura Bassi" (cit. n. 12), p. 230; and BCAB, *Mss. Bassi, Laura,* Box I, fasc. 1(i), c. 2v (*Avvisi di Bologna,* no. 45, 6 Nov. 1736).

[31] On 19 Dec. 1745 Bassi invited Canonico Pier Francesco Poggi to her "public lesson at the anatomy theater"; see Elio Melli, "Epistolario di Laura Bassi Veratti" (hereafter cited as **Melli, "Epistolario"**), in *Studi e inediti per il primo centenario dell'Istituto Magistrale Laura Bassi* (Bologna: N.p., 1960), p. 118. See also the deliberations and expenditures of the *Assunti di Studio*: ASB, *Assunteria di Studio: Atti* (1749–1755), Vol. 24 (7 Feb. and 10 Mar. 1749).

[32] Giovanna Ferrari, "Public Anatomy and the Carnival: The Anatomy Theater of Bologna," *Past and Present,* 1987, *117*:50–106, on pp. 76, 94. The remarks are drawn from the "Memoriale del dottor Laghi," ASB, *Assunteria di Studio: Anatomia publica,* file 6, as reproduced by Ferrari.

[33] ASB, *Assunteria di Studio: Atti* (1730–1735), in G. Martinotti, "L'insegnamento dell'anatomia in Bologna prima del secolo XIX," *Stud. Mem. Stor. Univ. Bologna,* 1911, 2:132; and BCAB, *Gozzadini* 337, c. 105.

Figure 1. *Laura Bassi at the carnival anatomy, 1734. From Archivio di Stato, Bologna,* Insignia degli Anziani, *Vol. 13, c. 105.*

perception of the event as the epitome of the world turned upside down. For the "infinity of Foreigners and People" who crowded into the theater on the second floor of the Archiginnasio, only in the carnival setting of the public anatomy could a woman truly shake off the weight of custom and become learned. Just as members of the audience, in the spirit of carnival, cloaked their true identities by wearing masks, Bassi too "disguised" herself with her knowledge. The organizers of the public anatomy manipulated this tradition to their advantage, linking the "virtue" of Bassi's presence in the Studium with the "honor" of the carnival festivities.[34] Since this was the one time during the year when the general public could see an anatomical dissection *and* Europe's only female professor, it was a rare occasion indeed.

The Senate's decision to restrict Bassi's public appearances to ceremonial occasions, *ratione sexu,* matched well the cultural image of Bassi developed by her admirers in the first years of her lectureship. Returning to the degree ceremony itself, we can see the crafting of an image at work. Despite reports that Bazzani "awarded her a degree according to the usual forms," the ceremony diverged in several important ways. Bassi received her degree gratis, without any of the presents, payments, and banquets that graduates customarily gave their professors and patrons. Rather than giving gifts, as tradition dictated, she was herself the recipient of lavish presents: the silver, jewel-encrusted crown of laurels—a gift of the Countess Ranuzzi—that replaced the traditional beret of male graduates, the medal struck for the occasion (see Figure 2), and the poems written in her honor. Bazzani, putting the

---

[34] Apropos of this imagery, see the stanza of a poem circulated in 1732 that is reproduced at the beginning of this essay.

Figure 2. *Laura Bassi depicted as Minerva. From Giovanni Fantuzzi,* Elogio della Dottoressa Laura Maria Caterina Bassi Veratti *(Bologna, 1778). (By permission of the Biblioteca Universitaria, Bologna.)*

silver laurels upon her head, compared her transfiguration to the metamorphosis of Ovid's chaste Daphne, transformed into a laurel tree to flee the unwanted attentions of her lover Phoebus.[35]

Other gifts, while part of the customary degree ceremony, took on added significance. Along with the crown of laurels, Bassi also received a ring and an ermine cape. Placed on her "connubial finger" by the prior Bazzani, the ring signified not simply her membership in the academic community but her virtual "marriage" to the city and the Studium. Writing to Algarotti, Elisabetta Ratta described "the most vigorous praise of our Beccari towards the young scholar, who never finished without saying: *egregie, virgo sapientissima, egregie*" during the thesis defense.[36] Recalling both the religious tradition of women in orders as brides of Christ and the civic tradition of virgins whose chastity cemented the foundations of republican government, Bassi, as a "most learned virgin," found herself fulfilling these ancient topoi.

[35] BUB, Codex 212 (116), no. 23, c. 94v (quotation); Fantuzzi, *Elogio* (cit. n. 3), p. 16; and Bianchi to Leprotti, 18 May 1732 (cit. n. 30). For the comparison to Daphne see Fantuzzi, *Elogio,* p. 36; the reference is to Ovid, *Metamorphoses,* trans. Mary M. Innes (New York: Penguin, 1955), pp. 41–44.

[36] Fantuzzi, *Elogio,* pp. 32, 36 (on the ring); and Ratta to Algarotti, 19 Apr. 1732 (cit. n. 22). For the context of this imagery see Stephanie H. Jed, *Chaste Thinking: The Rape of Lucretia and the Birth of Humanism* (Bloomington/Indianapolis: Indiana Univ. Press, 1989); and Edward Muir, *Civic Ritual in Renaissance Venice* (Princeton, N.J.: Princeton Univ. Press, 1981), pp. 119–134.

Given the continued strength of such bonds, even in the eighteenth century, the murmurings against Bassi's decision to marry Giovanni Giuseppe Veratti in February 1738 were quite understandable. "Thus you would blemish your glory? [*Dedisti maculam in gloria tua?*]," accused one anonymous pamphleteer.[37]

The association between female learning, virtue, and sanctity was still strong in the eighteenth century, though not as pervasive as it had been in the fourteenth through the seventeenth centuries. But learning and virginity did not only suggest sanctity. They also imbued a woman with the qualities of Minerva. Numerous elogies presented Piscopia as the "Venetian Minerva," and Voltaire described Châtelet in similar terms. Under the portrait commissioned by the Accademia degli Infecondi of Rome on the death of Piscopia in 1686, the inscription read, "Surely you believe the image which you see to be Minerva? [*Quam cernis pictam ne credas esse Minervam?*]."[38] Bassi, the "marvel of her sex and an ornament of our *Patria*," fulfilled a similar role in the eyes of her admirers. As the inscription on the obverse of the medal commemorating her degree declared: "*Soli cui fas est videsse Minervam*" (see Figure 2). Contemporaries must have associated this vision of Bassi with the frequent appearance of Minerva in the allegorical imagery of the Studium and the Institute.[39] Casting her light across the city, Bassi illuminated the path to wisdom, so recently obscured, permitting the citizens of Bologna once again to "see Minerva."

Described variously by contemporaries as the "new light of philosophy," the "luminous mirror of Science," and the "alumna of the Muses," Bassi became the emblem of *scientia* in all its forms. Compared to the sixteenth-century poetess Vittoria Colonna and most frequently to Petrarch's Laura, whom she surpassed by combining a learning and eloquence equal to Petrarch's with womanly grace, Bassi was the quintessential Enlightenment Muse (see Frontispiece).[40] With this designation came a variety of social burdens. Not only the subject of numerous poems, among them Algarotti's famous "Non la lesboa," Bassi also was expected to write poetry for important public functions. In 1737 she contributed to a volume of poems, collected by Giampietro Zanotti, for the wedding of Carlo Emanuele, King of Sardinia, and Elizabeth of Lorraine; the same year she repaid Maria Ranuzzi's gift of the silver

[37] Quoted in Comelli, "Laura Bassi e il suo primo trionfo" (cit. n. 21), pp. 220–221. For more on Bassi's decision to marry see Beate Ceranski, "Il carteggio tra Laura Bassi e Giovanni Bianchi 1733–1745," *Nuncius* (forthcoming).

[38] For a broader discussion of these issues see Margaret L. King, "'Book-lined Cells': Women and Humanism in the Early Renaissance," in *Beyond Their Sex: Learned Women of the European Past,* ed. Patricia H. Labalme (New York: New York Univ. Press, 1980), pp. 66–90; and King and Albert Rabil, eds., *Her Immaculate Hand: Selected Works by and about the Woman Humanists of Quattrocento Italy* (Binghamton, N.Y.: Center for Medieval and Early Renaissance Studies, 1983). On Châtelet as Minerva see Badinter, *Émilie, Émilie* (cit. n. 16), pp. 257, 280–281. On Piscopia see Giovan Nicolò Bandiera, *Trattato degli studi delle donne* (1740), Vol. 1, p. 339; Massimiliano Deza, *Vita di Helena Lucretia Cornara Piscopia* (Venice, 1686), p. 11; *Helenae Lucretia (Quae & Scholastica) Corneliae Piscopiae Virginis Pietate, & Eruditione admirabilis; Ordini D. Benedicti Privatis voris adscriptae Opera quae quidem haberi potuerunt* (Parma, 1688), p. 163; and *Le Pompe Funebri celebrate da' Signori Accademici Infecondi di Roma per la morte dell'Illustrissima Signora Elena Cornara Piscopia* (Padua, 1686), n.p.

[39] BCAB, *Mss. Bassi, Laura,* Box II, no. 6, p. 69. The original medal is reproduced in Fantuzzi, *Elogio* (cit. n. 3), p. 2; see also Marta Cavazza, "Scienziati in Arcadia," in *La Colonia Renia: Profilo documentario e critico dell'Arcadia bolognese,* ed. Mario Saccenti (Modena: Mucchi, 1988), Vol. 2, p. 432, n. 35. For the context of this sort of imagery see Londa Schiebinger, "Feminine Icons: The Face of Early Modern Science," *Critical Inquiry,* 1988, *14*:661–691.

[40] BCAB, B.2727, c. 11v; BCAB, *Mss. Bassi, Laura,* Box II, no. 8 (*Rime in lode della Signora Laura Maria Cattarina Bassi . . . prendendo la laurea dottorale in filosofia* [Bologna, 1732]); Antonio Magnani, *Elogio di Laura Bassi* (Venice, 1806), p. 11; and Roseann Runte, "Women as Muse," in *French Women and the Age of Enlightenment,* ed. Samia I. Spencer (Bloomington/Indianapolis: Indiana Univ. Press, 1984), pp. 143–154.

crown of laurels (*laurea d'argento*) by writing poems for the wedding of the count-ess's niece. Even as late as 1744, when Gian Lodovico Bianconi wrote to Giuseppe Veratti in gathering information on Bassi for his book of contemporary authors, he was more interested in the number of "poetic Academies" to which she had been admitted than in her scientific activities.[41]

While Bassi welcomed the opportunity to display her talents in a public forum, she was increasingly uncomfortable with her literary apotheosis. Letters such as the one written by the learned Francesca Manzoni in 1737—"you fly so high, where I do not dare address my thoughts," effused the Milanese poetess—must have only reinforced her desire to carve an image for herself removed from these laudations. By February 1737 she was complaining to Zanotti that she had no desire "to compose poetry ever again," and she resumed with vigor her attempts to get the Senate to expand her pedagogical duties.[42] By a somewhat different path than Châtelet, whose interest in mathematics was nurtured in the salon culture of Paris and through her association and correspondence with various philosophes, Bassi too had discovered natural philosophy as her vocation.

The late 1730s were, in many respects, crucial years for the formation of Bassi's career. Recently married and firmly established as an arbiter of learning in the city, Bassi, with the help of her husband, friends, and patrons, began to test the limits of her authority. In 1738, frustrated at the restrictions placed upon her teaching, she initiated a series of private lessons at home in the tradition of many university pro-fessors. Soon she found herself managing a lively scientific salon. Arriving in Bo-logna only a year later, Charles de Brosses described the balance of Bassi's private activities and public duties: "Indeed she wears the gown and ermine cape when she gives public lectures; this happens rarely and only on certain festival days, because it was not considered decent for a woman to show the hidden things of nature to all-comers daily. In recompense, philosophical conferences are held at her house from time to time." Initially intended for university students, the private lessons soon were attended exclusively by "nobles as well as scholars," and her fame as a teacher and orator grew.[43]

One year later, following the procedure used by other professors, Bassi submitted a request for a salary increase to the university; at the same time she petitioned for a reconsideration of the parameters of her duties. As the letter presented to the *As-sunti di Studio* on her behalf explained: "She has sustained many disputes in the presence of Cardinals, Princes and other noteworthy subjects and continually re-ceives Foreigners in her house with the obligation to respond to the Questions put

[41] *Lettere inedite alla celebre Laura Bassa scritte da illustri italiani e stranieri con biografia* (Bologna: G. Cenerelli, 1885) (hereafter cited as *Lettere inedite*), p. 166; Melli, "Epistolario," p. 65; and Gian Lodovico Bianconi to Giuseppe Veratti, 26 Nov. 1744, Dillinga, in *Lettere inedite*, p. 201. Marta Ca-vazza has discussed the use of poetry about Bassi as a means of introducing Newtonian subjects into Italy in *Settecento inquieto* (cit. n. 17), pp. 237–256, *passim*. For other poetry written in celebration of Bassi see Maria Elisabetta Machiavelli, *De rebus praeclaris gestis a clarissima philosopho doctore collegiata Laura Maria Catherina Bassi cive bononiensi* (17 May 1732); and, more generally, *Rime per la famosa laureazione ed acclamatissima aggregazione al Collegio Filosofico della Illustrissima ed ec-cellentissima Signora Laura Maria Caterina Bassi Accademica nell'Istituto delle Scienze e Cittadina Bolognese* (Bologna, 1732).

[42] Francesca Manzoni to Laura Bassi, 22 May 1737, Milan, in *Lettere inedite*, p. 86; and Bassi to Giampietro Zanotti, 9 Feb. 1737, Bologna, in Melli, "Epistolario," p. 82.

[43] *Lettres d'Italie du Président de Brosses*, ed. Fréderic d'Agay (Paris: Mercure de France, 1986), Vol. 1, p. 268; and ASB, *Assunteria di Studio: Requisiti dei Lettori*, Box II, n. 1 (Laura Bassi, 1748).

to her and to hold literary discourses at their pleasure, often entailing formal disputes in her House on such occasions with the intervention of many Gentlemen and Scholars." This was in addition to her mathematical studies with Gabriele Manfredi and the teaching of a "Course in the Experiments of Newton regarding light and color." In two separate meetings held that December, Bassi was granted an increase of 160 *lire* and the restrictions placed upon her public teaching were substantially reduced.[44]

Possibly the modifications in the Senate's policy toward Bassi in 1739 were a reflection of the 1737 reform of the university that paved the way for the introduction of the experimental sciences in the curriculum. Bassi's colleague Beccari accepted the first university chair in chemistry as a result of this reorganization.[45] More likely, however, they testified to the growing strength of her connections in the Senate, where supporters like Aldrovandi rallied to her cause. Less than a decade after her installation as a "celebrated woman," Bassi's presence had achieved what her patrons hoped it would do: more foreigners were coming to Bologna, and the visibility, if not exactly the reputation, of the Studium had risen accordingly. As one of her biographers put it, "no scholar would pass through Bologna without being eager for her learned conversation." Contemporaries delighted in knowing about the "visitors of consequence" like Joseph II or the son of the Polish king who graced her philosophical conferences.[46] By acceding to the demands placed upon her by various patrons, Bassi eventually found herself in a position of strength, for she could expect some degree of reciprocation. This took the form of concessions to her desire to participate more actively in the culture of the Studium and the allocation of money to support the costly equipment and materials for the physical and electrical experiments that she and Veratti began to do at home. Still convinced that her situation was not all that it could be, Bassi gradually began to accelerate attempts to increase her responsibilities at the university and the Institute, as her circle of patrons grew.

ENTER THE PATRON: BENEDICT XIV AND EDUCATIONAL REFORM

While Bassi enjoyed the patronage of many contemporaries, Prospero Lambertini (1678–1758) shaped the early and most significant stages of her career. "Among these [admirers], Benedict XIV was one of the most insatiable in praising her," wrote the author of Bassi's obituary in the *Avvisi di Bologna* in 1778.[47] Cardinal Lambertini had only recently returned to Bologna as its archbishop when Bassi took her degree. The timing of these two events was hardly coincidental. Lambertini, passionately devoted to his birthplace and keenly interested in the sciences, was precisely the sort of patron that Bassi needed to catapult her from relative obscurity to international fame. As archbishop of Bologna (1731–1740), he encouraged the university and the Senate to recognize Bassi's accomplishments. As pope (1740–1758), he show-

---

[44] ASB, *Assunteria di Studio: Requisiti dei Lettori*, Box II, n. 21 (Laura Bassi, 1739) (quotations); *Assunteria di Studio: Atti* (1735–1743), Vol. 23 (5 Dec. 1739); and Melli, "Laura Bassi Veratti" (cit. n. 3), p. 74.

[45] Cavazza, "Scienzati in Arcadia" (cit. n. 39), p. 433; and Cavazza, *Settecento inquieto* (cit. n. 17), p. 77. For more on university reform see F. Baldelli, "Tentativi di regolamentazione e riforme dello Studio bolognese nel '700," *Carrobbio*, 1984, *10*:9–23. Beccari had already held that position at the Institute since 1734.

[46] BCAB, B.2727, cc. 18v–19r; *Lettere inedite*, p. 201; and Comelli, "Laura Bassi e il suo primo trionfo" (cit. n. 21), pp. 222–223.

[47] *Avvisi di Bologna*, 25 Feb. 1778, in Cavazzuti, "Nuovi testi sull'attività scientifica e filosofica di Laura Bassi" (cit. n. 26), p. 66.

ered further gifts upon the Institute, of which Bassi was a member, and intervened when other academicians attempted to exclude Bassi from the activities of the Institute or limit her participation. Without the support of Lambertini, Bassi undoubtedly would have been a woman of note, but not someone who excited the admiration of princes and philosophes and earned a position in the Republic of Letters.

Despite his importance to the culture of Enlightenment Italy, Lambertini has not received the sort of scholarly attention that he merits. Holder of a degree in theology from La Sapienza, the University of Rome, Lambertini rose through the ecclesiastic ranks to become custodian of the Vatican Library (1712–1726). It was at the end of his tenure there that Monsignor Lambertini was called upon to arbitrate the acrimonious disputes between Luigi Ferdinando Marsili and the Senate of Bologna over the proposed reform of the Studium in 1726; he decided in favor of Marsili's more "modern" program. Five years later he returned to his native city as its archbishop. Once installed in Bologna, Lambertini renewed his contacts with the Institute and the Studium and strove to complete the work that Marsili had left unfinished.[48] Nurtured in the Italian Republic of Letters shaped by Lodovico Antonio Muratori and Scipione Maffei, and the heterodox scientific culture of eighteenth-century Rome embodied by his friend, the Jesuit Boscovich, Lambertini's plans for educational and religious reform reflected the optimism of the early decades of the eighteenth century, when much seemed possible. As pope, he would loosen the restrictions censuring Copernicus and Galileo and initiate a reorganization of La Sapienza that included the introduction of a chair in experimental physics.[49] For the moment, however, he confined himself to improving the state of learning in Bologna.

As a friend of Aldrovandi, Beccari, Eustachio Manfredi, the anatomist Ercole Lelli, the publisher Lelio della Volpe, and the Zanotti brothers, Lambertini quickly associated himself with the social and intellectual avant-garde. Through them, he must first have heard of the female prodigy of learning in their midst. Despite his reputed "abhorrence of women," Lambertini was so taken with Bassi that, according to one contemporary chronicler, he "encouraged her to take a degree and ordered the College of Doctors of Philosophy to award her a degree without considering to what extent the Imperial faculties of the same Collegians extended the awarding of degrees to women." Here we should recall Lambertini's presence at all of the ceremonies surrounding Bassi's intellectual ascension in the Studium. The archbishop was not simply a mute witness to her philosophical and oratorical skills. As two different witnesses tell us, "Signor Cardinal Prospero Lambertini wished to participate in the exam as a doctor in philosophy." On 1 June 1732 he visited Bassi at home to congratulate her on her recent success.[50]

[48] Cavazza, *Settecento inquieto* (cit. n. 17), p. 235. The most comprehensive treatment of Lambertini can be found in the articles collected in *Benedetto XIV (Prospero Lambertini): Convegno internazionale di studi storici*, ed. Marco Cecchelli, 2 vols. (Cento: Centro Studi "Girolamo Baruffaldi," 1981); Renée Haynes, *Philosopher King: The Humanist Pope Benedict XIV* (London: Weidenfeld & Nicolson, 1970), p. 37; and Rosen, "Academy of Sciences" (cit. n. 17), p. 56.

[49] Giuseppe Cenacchi, "Benedetto XIV e l'Illuminismo," in *Benedetto XIV*, ed. Cecchelli, Vol. 2, pp. 1079–1102, on p. 1094; and Heilbron, *Electricity in the Seventeenth and Eighteenth Centuries* (cit. n. 4), pp. 145–146. For more on the intellectual climate in Enlightenment Italy see Brendan Dooley, *Science, Politics, and Society in Eighteenth-Century Italy: The "Giornale de' letterati d'Italia" and Its World* (New York: Garland, 1991).

[50] BCAB, B.517, c. 4r (Giacomo Amadei, *Libro delle cose che vanno accadendo in Bologna . . . dal 1732 al 1743*), in *Benedetto XIV*, ed. Cecchelli, Vol. 1, p. 177 (urging the degree); BCAB, *Gozzadini*

Although never a member of the Institute academy, Lambertini was its most important benefactor after its founder Marsili. As he wrote to Marchese Paolo Magnani from Rome in 1744, "the Institute is capable of rendering famous our *patria,* as the university did in other times." While showering the Institute and its academicians and professors with gifts and preferments, he nonetheless was very much an absent patron. The records of the Institute indicate that he appeared only twice in any formal capacity: to pose two paradoxes for the academicians to solve in 1736, and to introduce the *motu proprio* for the formation of the *Benedettini* in 1745. Perhaps his most enduring legacy was the enormous library he bequeathed to the Institute upon his death. Benedict XIV, as Bassi's husband Giuseppe Veratti noted, remade the Institute through his patronage. "Now by the highest and incomparable beneficence of Our Great and Good Pontifex Benedict XIV, the Academy of this Institute for Sciences no less than the Institute itself recently has received a new form, indeed one could say that it was refounded," wrote Veratti in his *Physico-Medical Observations on Electricity* (1748).[51]

While Lambertini began with the regeneration of intellectual life in Bologna, as pope he strove to expand his reforms to encompass all of Italy. Women continued to have a place in his endeavors, since their accomplishments reflected the success of the new scientific learning as the core of an educational program that, he imagined, would eventually revitalize the Catholic world. Bassi was the first woman to whom he offered his patronage, but certainly not the last. After browsing through a copy of Agnesi's *Analytical Institutions* (1749), he encouraged the Milanese mathematician to become a professor for the "glory of Italy." A day later, he wrote again to urge her to accept the offer of membership in the academy of the Institute. By October 1750 the Senate had awarded Agnesi an honorary chair in mathematics, despite her lack of a degree, thanks to "the sovereign attention of His Beatitude, excited by your merits."[52] Thus Agnesi, like Bassi, was drawn into the orbit of the Institute and the Studium as a result of Lambertini's patronage.

Coming only four years after Châtelet had been admitted to the Institute, and within a year of the French *newtonienne's* death, the embracing of Agnesi by the academicians in Bologna looked suspiciously like the act of a society in search of external luster. Francesco Maria Zanotti, for example, described the decision to admit Madame du Châtelet as motivated by "a sort of ambition . . . that you, who are

---

140, c. 12r; and BUB, Codex 212 (116), no. 23, c. 94r; see also Haynes, *Philosopher King* (cit. n. 48), p. 67. On Lambertini's circle in Bologna see Mario Fanti, "Prospero Lambertini, Arcivescovo di Bologna (1731–1740)," in *Benedetto XIV,* ed. Cecchelli, Vol. 1, pp. 165–210, on p. 173; and Haynes, *Philosopher King,* pp. 47–48.

[51] In Paolo Prodi, "Carità e Galateo: La figura di Papa Lambertini nelle lettere al marchese Paolo Magnani (1743–1748)," in *Benedetto XIV,* ed. Cecchelli, Vol. 1, pp. 447–471, on p. 463; and Giovanni Giuseppe Veratti, *Osservazioni fisico-mediche intorno alla elettricità* (Bologna, 1748), sig. a.7r. See also Rosen, "Academy of Sciences" (cit. n. 17), pp. 216, 233. On the phenomenon of the "absent patron" see Mario Biagioli, "Scientific Revolution, Social Bricolage, and Absolutism," in *The Scientific Revolution in National Context,* ed. Roy Porter and M. Teich (Cambridge: Cambridge Univ. Press, 1992), pp. 26–32; Biagioli, "Galileo's System of Patronage" (cit. n. 15), pp. 36–38; and Prodi, "Carità e Galateo," p. 465, n. 34.

[52] Benedict XIV to Agnesi, 21 June 1749, Biblioteca Ambrosiana, ms. O.202 sup., c. 2; Benedict XIV to Agnesi, 22 June 1749, Biblioteca dell'Accademia delle Scienze dell'Istituto di Bologna, *Antica Accademia: Lettere ricevute,* fasc. 2 (1741–1750), A–B; and Camillo Grossi, ed., *Maria Gaetana Agnesi da Milano Professoressa Onoraria di Matematiche all'Università di Bologna l'Anno MDCCL: Documenti e note* (Bologna, 1843), letter 6 (8 Oct. 1750).

such an illustrious ornament of France, would begin in this way to be an ornament of Italy, particularly Bologna." Châtelet's untimely death in 1749 significantly reduced the ranks of illustrious women associated with the Institute. With the aggregation of Agnesi, corresponding members of the Institute academy like Giovanni Poleni in Padua could write that "Italy can no longer envy France, which boasts Madame du Châtelet." Benedict XIV, "content to see that the beautiful sex is devoted to the progress of the sciences," perceived the association between learned women and the scientific culture of Bologna to be a happy, albeit calculated, reflection of the *renovatio* of the sciences in his native land. And he expected Bassi, through her continued presence in the institutions of learning that he held so dear, to complete this circle by communicating her congratulations to the other women that he brought to the Institute's attention.[53]

CHANNELS OF COMMUNICATION

Benedict XIV may have been Bassi's principal patron, but her communication with him was by no means direct. Intermediaries played a central role in her ability to accumulate privileges.[54] Her most important contact in this regard was Flaminio Scarselli. A member of the Institute since 1722, Scarselli was secretary to the Bolognese ambassador at the papal court. The primary intermediary between Benedict XIV and the people of Bologna, Scarselli returned to his native city in 1760; he served as secretary (*segretario maggiore*) to the Senate until his death in 1776.

Bassi's education in the vicissitudes of patronage began soon after she obtained her degree. Her contact with some of the most important men of the city—Aldrovandi, Lambertini, and Zanotti, to name a few—prior to her public appearance gave her instant cachet. With their backing, no palazzo was closed to her. No doubt at the encouragement of some of these patrons, she began to introduce herself to the wider learned community in Italy. As Ruth Perry notes, epistolary communication was as much a female as a male prerogative, and an appropriate means of expanding one's circle of acquaintances. In a letter to Giambattista Morgagni in 1733, for example, Bassi wrote: "Since I consider it to my greatest advantage to hear my insufficiency accredited by the greatest *letterato* that Italy has, I will never cease to profess to you my most obsequious and sincere obligation for many signal favors. May this gratify your Most Illustrious Signor, since I beg you dearly to continue [giving] me the honor of your most esteemed grace and patronage."[55] Through her adept use of the language of patronage, Bassi established a foothold in the world of learning by obliging scholars like Morgagni to protect and encourage her as a sign of their munificence. In a community bound by concepts such as honor and trust, to refuse the praise of a learned woman, a Minerva on the horizon, would be to

[53] Zanotti to Émilie du Châtelet, 7 Apr. 1746, Bologna, in De Zan, "Voltaire e Madame du Châtelet" (cit. n. 18), p. 156; Giovanni Poleni to Agnesi, 5 July 1749, Padua, Biblioteca Ambrosiana, ms. O.201 sup., c. 22v; Gino Evangelisti, *Arguzia petroniana nei motti di spirito di Papa Lambertini* (Bologna: Ponte Nuovo Editrice, 1990), p. 36 (quoting Benedict); and Bassi to Agnesi, 18 June 1749, Bologna, Biblioteca Ambrosiana, ms. O.201 sup., c. 10.

[54] On the brokerage system in Italy see Biagioli, "Galileo's System of Patronage" (cit. n. 15), esp. pp. 6–13; and Paula Findlen, "The Economy of Scientific Exchange in Early Modern Italy," in *Patronage and Institutions*, ed. Moran (cit. n. 15), pp. 5–24.

[55] Ruth Perry, "Radical Doubt and the Liberation of Women," *Eighteenth-Cent. Stud.*, 1985, *18*:472–493, on p. 476; and Bassi to Giambattista Morgagni, 22 Aug. 1733, Villa di Secerno, in Melli, "Epistolario," p. 80.

ignore a scientific etiquette that depended on such relationships. Bassi, luminescent in her efforts to repay the honor of her patrons, was a client that no virtuoso could possibly refuse.

While mentors in Bologna encouraged her to cultivate the arbiters of the Italian Republic of Letters, like Morgagni, others introduced her to well-positioned individuals like Scarselli, who offered more concrete rewards to clients through his proximity to the pope. In 1742 Bassi's cousin, the canon Giambattista Bassi, praised the Bolognese ambassador, his secretary Scarselli, Countess Bolognetti, Cardinal Pompeo Aldrovandi, and "so many others working for the Pope," as "all most able to obtain for you more than you could seek." With the ascension of Lambertini to the papacy in 1740, Scarselli became an invaluable source of information and negotiator for the Bolognese *letterati*. Very little occurred in the academic and political world of the city that did not reach Benedict XIV's attention via Scarselli. Bassi was not the only person to communicate her frustrations and desires to the papal secretary. Scarselli was the main intermediary between the Institute, as a whole, and Lambertini.[56] When his cousin Beccari became president of the Institute in 1749, the web of relations between the pope, the secretary, and the Institute was completed.

The relationship between Scarselli and Bassi was one of friendship as well as patronage. They exchanged poetry and local gossip, and the secretary offered Bassi advice on the placement and preferment of various relatives within the ecclesiastical patronage network. In turn, Bassi informed him about the local scientific and political scene—a situation from which he was increasingly removed, the longer he stayed in Rome. While Scarselli advanced Bassi's position in Bologna through his encouragement of papal intervention, she commiserated with him about his wife's fragile health and had her husband—increasingly known for his electrical therapy—suggest various cures.[57]

Scarselli figured prominently in the later advancement of Bassi's career. Through his ministrations, Bassi was included as the twenty-fifth member of the *Benedettini*, and her role in the experimental culture of the academy was thus assured. The machinations surrounding this particular appointment tell us much about the importance of the brokers who helped Bassi expand the parameters of her position. On 22 June 1745 Benedict XIV issued a *motu proprio* establishing a new category of membership in the Institute, which he presented in person on 25 August. In response to waning attendance at the academy meetings and a noticeable decrease in the number of scientific papers presented, Lambertini decided to create new incentives for the academicians to produce original research. The twenty-four individuals designated *Benedettini* would receive a 50-*lire* stipend on the condition that they present one dissertation a year, describing the results of new work. Modeled in part on the practice of the Paris Academy of Sciences, which paid a select number of members to do research, such initiatives were designed to ensure a certain degree of stability in the educational reforms recently enacted by encouraging experimental studies. As Lambertini wrote to Paolo Magnani in January 1746, shortly after the *Benedettini* had been established, "our common *patria* will resume the true title of Mother of studies in the opinion of polite men, if not in law at least in the physical sciences."[58]

[56] *Lettere inedite*, p. 186 (quotation); and Rosen, "Academy of Sciences" (cit. n. 17), p. 7.

[57] The correspondence between Bassi and Scarselli is found in Melli, "Epistolario," pp. 89–157, *passim;* and *Lettere inedite*, pp. 108–124.

[58] In Prodi, "Carità e Galateo" (cit. n. 51), p. 463.

Benedict XIV's designs did not go unnoticed by the academicians. By April 1745 rumors about his proposal and the composition of the list of twenty-four were circulating. As a general rule, fourteen—the president and secretary, all the professors, and their assistants—were to be appointed by the Senate; the remaining ten were to be chosen by the *Benedettini*. To initiate the process, however, Lambertini handpicked all twenty-four. Bassi's name did not appear anywhere on that list. Made aware of this fact, she wrote immediately to Scarselli to pose her dilemma. While couching the letter in terms of her acceptance of the wisdom of the pope's decision, she nonetheless underscored her interest in belonging to the elite core of the Institute:

> I know that I am not among the nominated, and am glad not to be when placing me there would have resulted in the exclusion of some of those who are there, and deserve such an honor more than I do. Yet it would be arbitrary of His Holiness to place me in the series as I was placed in the University, *per straordinaria,* that is, as an extra. Here is the deception that I beg of you, in all secrecy, with the confidence that Our Holiness wishes to maintain, it is worth saying, the good opinion that he has had of me, [which exists] only by his bounty. On this occasion, you can help me, and no one better than you can, insinuate to His Holiness to give me some reply, as a *motu proprio.*

As she assured Scarselli, she had many dissertations ready to present to the Institute but was hesitant to do so, unless she was added to the *Benedettini,* "for fear that perhaps some would conceive [of the idea] that I would aspire to one of the first vacancies of these posts."[59]

With Bassi obliquely reminding Scarselli of the unusual nature of her position at the Institute and her high level of productivity relative to many of those selected as *Benedettini,* the papal secretary mustered all available resources to bring about an alteration in the original plan. A week later he responded. Urging Bassi to exercise discretion, he informed her that the matter had been brought to the attention of the pope and was now in the hands of the papal legate:

> Now nothing should be easier than to move the spirit of His Eminence, who knows you well and esteems you, to propose your addition to the Pensioners as a supernumerary, in the same fashion that you were assigned the Lectureship in universal Philosophy above the number of seventy-two. Since it is unsuitable for you to do this part, an honest and efficacious friend could do it, and should, if he loves the merit and honor of the Academy and *Patria.*

June produced a flurry of correspondence between Bologna and Rome as Bassi and Scarselli made arrangements to add her name to the list. The proposal was presented not as Bassi's own, but as coming from Scarselli and his associates; as she later observed, "I would not have been able to procure this honor for myself without a trace of presumption." Bassi scrupulously followed the secretary's advice in drafting her supplication to Benedict XIV, once it was requested, and thanked Scarselli for the "shrewdness and caution which you exercised in proportion to your love and kindness towards me." By 19 June, only a few days before the public announcement

---

[59] Bassi to Flaminio Scarselli, 21 Apr. 1745, Bologna, in Melli, "Epistolario," pp. 105–106. This correspondence is also reproduced in *Lettere di quattro gentildonne bolognesi: Bassi–Tambroni–Dalle Donne–Martinetti* (Bologna: Monti, 1883), pp. 24–26.

of the composition of the first *Benedettini,* Bassi could write to Scarselli that the "excellency of the means adopted make me certain of the success of the affair."[60]

Scarselli's intervention on Bassi's behalf did not end here, however. By October he was actively encouraging Bassi and Veratti to submit printed dissertations, dedicated to the pope, for Benedict XIV's approval. Having observed the success that greeted the arrival of Pietro Paolo Molinelli's dissertation on aneurysms, and the attempts of Monsignor Antonio Leprotti to motivate other academicians to submit their dedicated work, he wrote encouragingly, "Now why couldn't you be numbered among these others? And why couldn't you have some dissertations printed separately, dedicating them to His Holiness? . . . Then whoever presents the book would have all the opportunity to present a petition shortly thereafter, without need of more powerful help." Less than a month later, following a pattern exercised to great advantage by earlier natural philosophers such as Galileo, Scarselli reminded Bassi of "the necessity of exploring first the spirit of His Holiness, if he is content with the dedication." Bassi responded with equal caution, assuring him that she did not wish to send the pope an "imperfect thing."[61]

In the midst of these rather ordinary negotiations, designed to conclude a successful patronage transaction, disaster nearly intervened. At a November 1745 meeting of the *Benedettini* the question of Bassi's vote had been raised; several members voiced the opinion that, given the special nature of her position, she should "remain segregated from that body in such cases." Neither Bassi nor her husband was there to defend her. Upon getting wind of this dissent, Bassi immediately requested that Scarselli have Benedict XIV arbitrate the affair. While leaving the final word with the pope, she nonetheless expressed an opinion about the "legitimate significance" of her place as the sole *Benedetta.* Reminding Scarselli that both the Institute academy and the Studium accorded her voting privileges, the latter "on the occasion of the Doctorates," she saw no obstacle to exercising such rights in her new situation. "Nor do I know how to persuade myself enough that the singular clemency of Our Signor intended to tacitly deprive me of the best prerogatives of the Academy, that is, to take part in the election of new subjects when it occurs, after having deigned to desire my participation in everything else pertaining to the above-mentioned Academy." However, she awaited the pope's decision about whether her position encompassed these privileges or was simply a "quid tertium, that is, neither yes nor no." As with the earlier frustration about her initial exclusion from the *Benedettini,* Bassi's efforts were soon repaid, and Benedict XIV handed down a decision in her favor. Two weeks later she wrote to Scarselli to thank him for quieting "the doubting minds of our Academicians."[62]

Bassi's correspondence with Scarselli throughout the next decade was filled with the mundane details of an extended patronage relationship. While she never managed to publish her own dissertations separately—the only one to appear in 1745 was on the compression of air and was published in the *Commentaries* of the Institute—she

[60] Scarselli to Bassi, 28 Apr. 1745, Rome, in *Lettere inedite,* p. 108; and Bassi to Scarselli, 5 and 19 June 1745, Bologna, in Melli, "Epistolario," pp. 107–109.

[61] Scarselli to Bassi, 23 Oct. and 16 Nov. 1745, Rome, in *Lettere inedite,* pp. 110, 112; and Bassi to Scarselli, 30 Oct. 1745, Bologna, in Melli, "Epistolario," p. 113. For a comparative view on these tactics see Mario Biagioli, "Galileo the Emblem Maker," *Isis,* 1990, *81*:230–258.

[62] Bassi to Scarselli, 27 Nov. and 11 Dec. 1745, Bologna, in Melli, "Epistolario," pp. 115–117.

aided her husband in the publication of his *Physico-Medical Observations on Electricity* and followed Scarselli's advice about the etiquette of presenting copies to key members of the papal court. In 1749 Scarselli could report that the treatise had been well received by the "right people."[63] Despite the papal secretary's encouragement, perhaps Bassi, like many women during this period, concluded that the difficulties of getting her work published outweighed the potential benefits. Already ensconced in the university and the Institute, there was little room for her further advancement.

During the same period, Bassi expanded her teaching and research activities, "passing from Theory to experiments," as one biographer put it. Beginning in 1749, she offered private lessons in experimental physics at home and began to collaborate further with Veratti in his work on electricity. With the death of Matteo Bazzani, she took over "the exercises in Physics." All of these activities were costly, and Bassi and Veratti spared no expense in acquiring the most up-to-date equipment for their domestic laboratory, visited by luminaries such as Nollet—who was singularly unimpressed by Veratti's theories about the medical uses of electricity—and Beccaria. By 1755, frustrated in her attempts to get the Senate to increase her stipend, Bassi again turned to Scarselli for help, asking him "for your wisest counsel in this in order to understand what means I could employ to procure the help of Our Signor in this affair." Scarselli's response, in this instance, was more guarded:

> But . . . from whom do you request that recompense? Whoever speaks of it to Our Signor easily will hear the response that, having often spent a great deal to provide machines and equipment, the Institute must not think of providing them still for private Houses. And as for the people [involved], he perhaps will believe that he has done enough, assigning and increasing the honors to the Professors and Academicians in that place.[64]

Ironically, Bassi had been right when she observed in 1745 that those who benefited from the creation of the *Benedettini* would receive no further favors from the pope. Her own situation now bore out this observation.

In a second letter to Scarselli, Bassi mustered all the ammunition that she could find to persuade the papal secretary that this was a battle worth fighting. Reminding him of the importance of experimental physics and of Bologna's primacy in the introduction of this field of study to Italy, she bemoaned its current state: "Now we must blush to see the damage to our University. Everywhere else but here one teaches with that method . . . , giving entire courses annually." Whether Bassi's pleas fell on deaf ears is hard to say. In March 1759, a year after the death of Lambertini, she received a raise of 140 *lire*. By 1760 she earned 1,200 *lire* annually, a salary higher than that of any of the other professors and members of the Institute, including the president Beccari and the secretary Zanotti.[65] This was undoubtedly the most

---

[63] Scarselli to Bassi, 18 Jan. and 15 Feb. 1749, Rome, in *Lettere inedite*, pp. 117–119; see also Bassi to Scarselli, 8 June 1749, Bologna, in Melli, "Epistolario," p. 129. On Bassi's dissertations see note 5.

[64] BCAB, B.2727, c. 17v; ASB, *Assunteria di Studio: Atti* (1743–1755), Vol. 24 (10 Apr. 1750); *Assunteria di Studio: Requisiti dei Lettori* (Laura Bassi, 1750); Bassi to Scarselli, 14 June 1755, Bologna, in Melli, "Epistolario," p. 149; and Scarselli to Bassi, 21 June 1755, Rome, in *Lettere inedite*, p. 123. For more on Nollet's reaction see Simon Schaffer, "Self Evidence," *Crit. Inq.*, 1992, *18*:327–362, esp. pp. 339–349.

[65] Bassi to Scarselli, 16 July 1755, Bologna, in Melli, "Epistolario," p. 151. On Bassi's raise and for her salary see ASB, *Assunteria di Studio: Atti* (1756–1777), Vol. 25, fol. 16 (3 Mar. 1759); and Elena, "Introduction to Laura Bassi" (cit. n. 3), p. 514. Beccari and Zanotti both earned 800 *lire*.

tangible result of her continued efforts to have her activities rewarded. Thanks to the strong support of patrons such as Lambertini and intermediaries such as Scarselli, whatever qualms members of the Institute and the Studium may have had about fully integrating her into their activities slowly eroded.

While cultivating influential patrons, by mid career Bassi also offered similar services to younger scholars and foreigners seeking admittance to the academic world of Bologna. Her sex undoubtedly made it easier for her largely male clients to see her in this role, since Enlightenment salon culture privileged women as arbiters of intellectual debate. As early as 1745, Francesco Maria Zanotti declared that Bassi was the "arbiter of all his sentences." Visiting scholars like Nollet saw the Institute under her guidance and met the other academicians in her home. Many were directed to Bassi by Beccaria, professor of physics in Turin since 1749, who wrote letters of introduction for English and French visitors touring Italy. Writing to Veratti in 1763, he asked to be remembered to "Donna Laura, always my esteemed patron, to thank her for the Commemoration that she wished to make of my law of refraction in Rock Crystal." Like Zanotti, he consulted with her frequently, particularly before publishing his findings on electrical fluid.[66] She was a colleague with whom to share his discoveries—but also a patron whose visibility increased his own importance.

Bassi's other noteworthy clients were Voltaire and Spallanzani. While Voltaire was a one-time client, beseeching Bassi to submit his name as a candidate for admission to the Institute academy in 1744, Spallanzani developed a relationship with his cousin that extended well beyond any obligations he might have felt toward her as a tutor and a relative. Spallanzani's correspondence with Bassi spanned roughly the last decade of her life. Shortly after his admission to the Institute in 1768, he began to share his experiments on snails and salamanders with her, sending particularly fecund samples by courier so that she could observe their generative powers on her own and confirm his results. Spallanzani constantly repaid "the honor of her Patronage" by sending promising scholars in her direction. In 1770 he introduced Charles Bonnet; a year later he encouraged Alessandro Volta to send his "youthful productions" to the learned *dottoressa*. Like Giuseppe Testa, who sent his son to Bologna to study with Bassi, Spallanzani perceived Bassi to be one of the most inspiring teachers of experimental philosophy that he had encountered and an indefatigable patron of the sciences. Writing to Veratti in 1782, four years after Bassi died, he remarked: "I am truly pleased to hear that you are teaching Experimental Physics to studious youth in your House, continuing the work of the *Signora Dottoressa,* your dearest Wife and my venerable Mistress, whom I will always remember as long as I live. I can say truthfully that what little I know, I owe in origin to her wise teachings."[67]

Negotiating her way within the scientific networks of Enlightenment Italy, Bassi was able to collect as well as dispense patronage. Through her contacts in Rome, she maintained her ties to Lambertini as he moved from the episcopate of Bologna to the papacy. Well situated in her position as the Minerva of the Institute, she

---

[66] Zanotti to Bassi, 27 Nov. 1745, Di Casa, in *Lettere inedite,* p. 164. See also Goodman, "Enlightenment Salons" (cit. n. 10); Giambattista Beccaria to Veratti, 10 Oct. 1763 and 19 Jan. 1765, BCAB, *Coll. Ant.* VI, 1745–1746; BCAB, B.2727, c. 18r; and *Lettere inedite,* pp. 45, 50–51.

[67] Masi, "Laura Bassi ed il Voltaire" (cit. n. 2), pp. 166–167, 170. The material on Spallanzani's relationship with Bassi is culled from *Lettere inedite,* pp. 125–141, 147, 157, 218; and Melli, "Epistolario," pp. 162–167.

facilitated interactions between natural philosophers. Truly, when one contemporary described Italy as "now, perhaps more than Britain, the home of true physics" in 1784, this praise indirectly acknowledged the efforts of Bassi, Lambertini, and Scarselli, whose negotiations had resulted in the institutionalization of experimental physics in that region.[68]

BASSI AT THE ISTITUTO DELLE SCIENZE

By the end of the 1750s, Bassi could take great pride in her accomplishments. Despite the ambivalence with which female ambition was regarded in the eighteenth century, her success testified to the possibility of a woman, albeit an exceptional one, taking a position in public scientific culture.[69] Her petitions to the Senate had partially eroded the restrictions placed upon her teaching and increased her stipend immensely. While still participating in the annual lectures given during the carnival anatomy, she regularly produced papers for the meetings of the Institute and managed a famous school for experimental physics. Thanks to the efforts of Scarselli, and behind him Benedict XIV, she had been awarded additional recognition as a *Benedettina*, one of the Institute elite, with all the voting rights of her male counterparts. Yet Bassi was still not completely satisfied. In 1772 the death of Paolo Balbi, professor of experimental physics at the Institute since 1770, provided her with one last battle to wage.

The decision to award Bassi the chair in experimental physics in 1776 was the product of numerous debates and discussions on the part of the other voting members of the Institute. They arrived at that decision neither quickly nor easily. The death of Balbi had left the chair vacant. Veratti had been his assistant and was therefore a logical successor. The Institute, however, deliberated slowly about what to do. By May 1776, undoubtedly at the insistence of Bassi, Veratti, and their supporters, discussion was opened about the future of the course in experimental physics (*corso di fisica sperimentale*). After considering proposals to separate investigations of electricity from the study of experimental physics, mirrored by the idea of creating a separate room in the Institute museum for electrical apparati, the academicians, led by Senator Aldrovandi, finally came to the issue of the vacant chair. Those present considered three options: to promote Veratti; to promote Lorenzo Bonacorsi, academician since 1743, so that he could become a *Benedettino;* and to consider the request of Bassi that she be made a candidate. Undoubtedly the secretary who transcribed the proceedings expressed the exasperation of many participants in the ensuing debate when he recorded the following conversation:

> finally to satisfy, if one ever can, the demands of Signora Laura Bassi who, although she has no right to be admitted among the Professors of the Institute, nevertheless has asked for this well over three Years, having nurtured some hope of this more than once. Given that she is a celebrated Woman known to the entire Republic of Letters, who truly brings great honor to her *Patria,* thus it seems that [her request] merits benign attention.[70]

---

[68] Quoted in Heilbron, *Electricity in the Seventeenth and Eighteenth Centuries* (cit. n. 4), p. 153.

[69] For more on the problem of female ambition see Goodman, "Enlightenment Salons" (cit. n. 10), p. 332; and Badinter, "Les limites de l'ambition féminine," in *Émilie, Émilie* (cit. n. 16), pp. 417–464.

[70] ASB, *Assunteria dell'Istituto: Corsi, Laboratori e Professori: Diversorum,* Vol. 15, no. 42 (6 May 1776).

In the end, Bassi was awarded the chair and maintained Veratti as her assistant. He succeeded to the position in 1778 and was in turn succeeded by their son Paolo, who held the chair until 1796. Despite the attempts of some members to give her request only "benign attention," Bassi clearly commanded the loyalty of enough of a majority for her wish to be fulfilled. She became the Institute Professor of Experimental Physics at the age of sixty-five.

During his trip to Bologna in 1739 Charles de Brosses, like many other foreigners, had availed himself of the opportunity to see the fabled Institute. "I have made the acquaintance of the best [professors]," he wrote, "who know more than their profession because they are people of good society and *galants* in the service of women."[71] While Bassi never held office in the Institute—Veratti was vice-president of the academy six times and president twice—she nonetheless achieved a position of considerable significance. Even when intercessors such as Lambertini and Scarselli were no longer there to support her cause, she could count upon other allies from within the Institute to achieve the desired result. No doubt many of the members of the Institute learned from their experiences with this formidable woman. While they came to accept Bassi, they were increasingly less certain that the admission of women was beneficial to the Institute. Upon Bassi's death, the twenty-fifth position among the *Benedettini* was retired, to be reopened only when the obstetrician Maria Dalle Donne took up a professorship in 1800. The woman who accompanied Dalle Donne to the anatomical theater to defend her theses in 1799 was the classicist Clotilde Tambroni, professor of Greek at the University of Bologna since 1790; she was subsequently admitted to the Institute as an ordinary member in 1802. Having discovered how far an ambitious and persistent woman like Bassi could insinuate herself within the structure of the Institute, the male members of its academy were noticeably reluctant to allow another woman such latitude. Yet not until the end of the eighteenth century did they consider the possibility that an ideal scientific society should be a world without women, an image that their French and English counterparts had institutionalized a century earlier.[72]

* * *

Bassi's death, like so many aspects of her life, did not go unmarked. Aside from the numerous eulogies given in her honor by members of the learned community, in Bologna and elsewhere, written notices of her accomplishments appeared in the journals serving the Republic of Letters. In Bologna, the Institute established a commission to judge a competition for a monument to their most famous female member. By 1781, the deliberations complete, work on a marble statue designed by Senator Antonio Bovio Silvestri was under way. After much discussion, it was decided that the image of Bassi should be placed above the door to the Nautical Room in the Institute, where many of Marsili's beloved model ships were housed.[73] During her

---

[71] *Lettres d'Italie du Président de Brosses,* ed. d'Agay (cit. n. 43), Vol. 1, p. 267.

[72] Renzo Tosi, "Clotilde Tambroni e il Classicismo tra Parma e Bologna alla fine del XVIII secolo," in *Alma mater studiorum* (cit. n. 3), pp. 119–134; and Olimpia Sanlorenzo, "Maria Dalle Donne e la Scuola di Ostetricia nel secolo XIX," *ibid.,* pp. 147–156. On all-male scientific societies as the ideal see David Noble, *A World without Women: The Christian Clerical Culture of Western Science* (New York: Knopf, 1992).

[73] ASB, *Assunteria dell'Istituto: Diversorum,* Vol. B6, no. 2 (*Concorso al monumento per Laura Bassi,* 1778–1781).

lifetime Bassi had constantly reminded her colleagues that she was not simply a figurehead, but a practicing experimental philosopher who wished to teach as well as perform research. She had fought long and hard to expand her role beyond its ceremonial functions, earning the respect, admiration, and, most important, support of many contemporaries in the process. Death, however, relegated her once again to a ceremonial position. Her likeness preserved in marble, positioned high above the doors through which the members of the Institute passed, she reclaimed her role as scientific muse. Smiling benignly down upon her former colleagues in effigy, and buried wearing her silver crown of laurels, Bassi reentered the realm of mythology from which she had emerged.

Despite this final apotheosis, the historical Bassi (rather than the mythical *filosofessa*) continues to elude easy categorization. Among her contemporaries, few, with the exception of Newton and perhaps Voltaire, enjoyed an equivalent degree of institutional recognition and public acclaim. Like her more illustrious predecessor Galileo, she was as keenly interested in her social advancement as in the image she would leave to posterity. While unable to create her own patronage "myths," as Galileo did when he positioned himself as the giver of the Medicean stars to Cosimo II in 1610, she nonetheless deployed the symbols placed at her disposal to fashion an identity as a female natural philosopher; the strong classicizing elements that informed the Italian academic tradition allowed her to manipulate her image as Minerva and as Muse.[74] Her sharp appreciation of the social and political complexities of the scientific world allowed her, like her most successful male colleagues, to anticipate many of the obstacles that were thrown in her way and to draw upon the resources of the local and international learned community to resolve whatever problems arose. Most important, the unusual circumstances that first made Bassi a public figure offered her the opportunity to define the nature of her role. Unlike Galileo, searching for recognition and legitimation at the most powerful courts, Bassi *began* in the world of the scientific and literary salons and *moved into,* rather than *out of,* the institutional world of science. Having begun her career as the holder of a unique position, the first authorized female professor of any discipline at any university, she strove instead to "regularize" her situation, using the resources placed at her disposal to achieve this end. Which was the more successful is hard to say, depending on how we evaluate the long- and short-term consequences. While Galileo failed in his efforts to carve a stable niche for himself in the social world of seventeenth-century science, he left a lasting imprint on a discipline and on the contours of Italian science. Bassi, immensely successful in her own lifetime, receded into obscurity outside the parameters of her native city. Though an important figure in the introduction, teaching, and dissemination of Newtonianism and experimental philosophy in eighteenth-century Italy, Bassi was best remembered, even in her own day, for her exceptional circumstances rather than her ideas.

As the trajectory of Bassi's career indicates, it is often easier to talk about the social and cultural roles of women in early modern science than about their intellectual presence. We know a great deal about how and where Bassi achieved her success, but very little about what she actually did in the classroom and the laboratory and how her students and colleagues responded. For different reasons, in the

---

[74] See particularly Mario Biagioli, "Galileo the Emblem Maker" (cit. n. 61); and Biagioli, "Galileo's System of Patronage" (cit. n. 15).

case of Châtelet, we often know more about her relationship with Voltaire and its impact on her work than about the content of her publications. Despite the paucity of sources allowing us to connect social and intellectual issues, it is nonetheless evident that Bassi's success was seen by many contemporaries as indicative of the changing philosophical climate in Enlightenment Italy. In a less dramatic way, perhaps, than in France, the Italian community of philosophers was also in the process of making the transition from Cartesianism to Newtonianism, and Bassi became a symbol of that shift. While Bassi deftly persuaded patrons and colleagues to support her advancement within the academic hierarchy, her presence in the Studium and the Institute smoothed the way for the introduction of new and controversial philosophies of science and accelerated the institutionalization of experimental philosophy. She fought her battles not because she believed that women, in some abstract sense, deserved intellectual recognition—such a concept would have been inconceivable to her and to most of her contemporaries—but because she felt that her work as a teacher and experimental philosopher merited recognition. By the end of her career Bassi had, in essence, become the patron of the "new" science in Bologna, and it was in this capacity that she rightfully claimed the role of Minerva in the *Alma mater studiorum*.

KEY of the SEXUAL SYSTEM.

MARRIAGES of PLANTS.
Florefcence.

PUBLIC MARRIAGES.
Flowers vifible to every one.

IN ONE BED.
Hufband and wife have the fame bed.
All the flowers hermaphrodite: ftamens and piftils in the fame flower.

WITHOUT AFFINITY.
Hufbands not related to each other.
Stamens not joined together in any part.

WITH EQUALITY.
All the males of equal rank.
Stamens have no determinate proportion of length.

1. ONE MALE.	7. SEVEN MALES.
2. TWO MALES.	8. EIGHT MALES.
3. THREE MALES.	9. NINE MALES.
4 FOUR MALES.	10. TEN MALES.
5. FIVE MALES.	11. TWELVE MALES.
6. SIX MALES.	12. TWENTY MALES
	13. MANY MALES.

WITH SUBORDINATION.
Some males above others.
Two ftamens are always lower than the others.
14. TWO POWERS. | 15. FOUR POWERS.

WITH AFFINITY.
Hufbands related to each other.
Stamens cohere with each other, or with the piftil.

16. ONE BROTHERHOOD.	19. CONFEDE-
17. TWO RROTHERHOODS.	RATE MALES.
18. MANY BROTHERHOODS.	20. FEMININE
	MALES.

IN TWO BEDS.
Hufband and wife have feparate beds.
Male flowers and female flowers in the fame fpecies.

21. ONE HOUSE.	23. POLYGAMIES.
22. TWO HOUSES.	

CLANDESTINE MARRIAGES.
Flowers fcarce vifible to the naked eye.
24. CLANDESTINE MARRIAGES.

Figure 1. Erasmus Darwin's translation of a synopsis of Linnaeus's classification scheme for plants, as given in The Families of Plants (Botanical Society of Lichfield, 1787), page lxxvii. Reproduced by permission of the Syndics of the Cambridge University Library.

Botany for Gentlemen

Erasmus Darwin and *The Loves of the Plants*

By Janet Browne

HISTORIANS OF SCIENCE have long been interested in the different ways in which natural phenomena have been classified and arranged into taxonomic schemes of one kind or another. The search for "essential" characters that would yield a definition of a natural group or form, the tension between the demands of logic and the intuitive recognition of affinities, the debate over the relative merits of artificial and natural schemes, have all been topics of close attention. In recent years, however, these time-honored problems have given way to a new range of questions that focus on the social processes at work during the construction and reception of individual classification schemes; and the classification schemes themselves are being shown to represent or mirror in various ways the society that brought them into existence. Taxonomic systems of the past—particularly those found in natural history, biology, and geology—are now seen to be one of the most important resources for understanding the interconnections of science and culture.

Following work along these lines by John Dean, Adrian Desmond, Dorinda Outram, Barry Barnes, James Secord, and others, I intend to draw out some of the social commitments that underpin Erasmus Darwin's taxonomic poem, *The Loves of the Plants* (1789).[1] Although poetry may not at first seem the most appropriate place to search for such links, literature, natural philosophy, art, and social theory were so closely integrated during the eighteenth century that natural philosophy was frequently presented in a stylized literary form, as in Oliver Goldsmith's translation of Buffon's *Histoire naturelle* in 1774, and the visual and literary arts were often grounded in a sophisticated awareness of the natural

I thank colleagues in the Unit for the History of Medicine, the Wellcome Institute, and the Darwin Letters Project for help with this article, as well as Sally Bragg, Heather Edwards, Joy Harvey, Desmond King-Hele, Michael Neve, Yvonne Noble, Anne Secord, James A. Secord, and the *Isis* referees.

[1] John Dean, "Controversy over Classification: A Case Study from the History of Botany," in *Natural Order: Historical Studies of Scientific Culture,* ed. Barry Barnes and Steven Shapin (Beverly Hills, Calif./London: Sage, 1979), pp. 211–230; Adrian Desmond, *Archetypes and Ancestors: Palaeontology in Victorian London, 1850–1875* (London: Blond & Briggs, 1982); Desmond, "The Making of Institutional Zoology in London, 1822–1836," *History of Science,* 1985, 23:153–185, 223–250; Dorinda Outram, "Uncertain Legislator: Georges Cuvier's Laws of Nature in Their Intellectual Context," *Journal of the History of Biology,* 1986, 19:323–368; Barry Barnes, *Interests and the Growth of Knowledge* (London: Routledge & Kegan Paul, 1977); and James A. Secord, *Controversy in Victorian Geology: The Cambrian-Silurian Dispute* (Princeton, N.J.: Princeton Univ. Press, 1986).

97

sciences of the day. Poets were philosophers, and philosophers poets. Erasmus Darwin was both of these and a physician too. *The Loves of the Plants* expressed its author's comprehensive interests to a marked degree, for Darwin intended it to be a vindication and explanation, both amusing and instructive, of Linnaeus's classification scheme for plants. In this work Darwin dramatized Linnaeus's system by portraying the stamens and pistils (the male and female organs) as men and women. As its title suggests, the text is about the relations between the sexes—ostensibly those pertaining to plants, as described by Linnaeus and on which he based his systematic arrangement, but in actual terms translated by Darwin into an extended account of human sexual behavior. The metaphor of personification served several functions in Darwin's poetry, and there was a vigorous interplay between Darwin's defense of Linnaeus, his commitment to evolutionary transformism, his thoughts about plants, and wide-ranging views about society and progress. Verse, in his opinion, was an appropriate and effective medium for conveying these diverse ideas.

Looking beyond the purely didactic or frivolous roles usually ascribed to the poem, I will argue that Darwin's botanical taxonomy was firmly located in his eighteenth-century world: that the metaphors he chose to explain Linnaeus's system reflected not so much his personal views on human nature as, more significantly, the views of his contemporaries as expressed through the conventional images and literary stereotypes of his time. The poetic imagery, in its turn, influenced the ways in which Darwin and his readers subsequently thought about the vital activities of plants and plant reproduction. Moreover, whereas other classification schemes might have primarily reflected contemporary culture, social class, or intellectual preoccupations, Darwin's poem is of particular interest because it included in addition ideas about the social position, behavior, and functions of women. His version of Linnaeus's system therefore offers an opportunity to study the ways in which gender and views about gender relations were manifested in scientific practice. Through his verses we can follow the expression of connections between the ordering of nature and human society and examine how Darwin's explanation of Linnaeus's scheme may have embodied, maintained, or otherwise served the conventions and objectives of an extended community of natural philosophers and intellectuals in late eighteenth-century England.

ERASMUS DARWIN

Erasmus Darwin's important position in eighteenth-century English culture was first documented some twenty years ago when Desmond King-Hele brought the "essential writings" of Darwin (1731–1802) to our attention; and through King-Hele's continued interest and publications Darwin's work is now well known for its vivid pictures of evolution and cultural progress interlaced with stirring accounts of science, technology, and society during the English Industrial Revolution.[2] Expansive views like these are found most obviously in Darwin's two

[2] See *The Essential Writings of Erasmus Darwin*, ed. Desmond King-Hele (London: MacGibbon & Kee, 1968); King-Hele's earlier biography, *Erasmus Darwin* (London: Macmillan, 1963); King-Hele, *Doctor of Revolution: The Life and Genius of Erasmus Darwin* (London: Faber & Faber, 1977); *The Letters of Erasmus Darwin*, ed. King-Hele (Cambridge: Cambridge Univ. Press, 1981) (hereafter **Letters of Erasmus Darwin**); and, most recently, King-Hele, *Erasmus Darwin and the Romantic Poets*

lengthy poems *The Economy of Vegetation* (Part 1 of *The Botanic Garden,* 1791) and the posthumously published *Temple of Nature* (1803). Recent works by Maureen McNeil and others have situated Darwin more specifically in the social landscape of the entrepreneurial provincial science of the English midlands.[3] Prominent in the Lunar Society of Birmingham, Darwin and his colleagues, among them James Keir, Josiah Wedgwood, Matthew Boulton, and Richard Lovell Edgeworth, were wealthy professional figures who had received a university education: cosmopolitan, well read, advocates for many of the views set out by the French *philosophes,* these were men who were by nature liberal reformers, deeply committed in one way or another to the idea of improvement in all spheres of existence through the exercise and application of natural philosophy, and who separated themselves from the political views of the Tory hierarchy and the established church without jeopardizing their respectability or status as "gentlemen." Erasmus Darwin was reputed to be an atheist and known, in later life, to be an evolutionist, but he was also a prosperous, respectable physician. To be liberal in such circles did not mean that one was a radical firebrand.

The Loves of the Plants was first issued anonymously in 1789, having been printed in Darwin's hometown of Lichfield in Staffordshire—although he himself had moved to Derby in 1781—and was from the start meant primarily to test the water for a second, more heavyweight account of the development of the earth and society. Together, the poems would constitute a two-part set entitled *The Botanic Garden.* Darwin claimed he wrote *The Loves of the Plants* solely for the money that might come his way, hoping only to make the topic of botany agreeable to "ladies and other unemploy'd scholars,"[4] and there seems no reason to deny him this practical explanation of his own motives. The intended second poem, called *The Economy of Vegetation,* was not published for another two years, and Darwin seems to have intended to stifle it if the first had not been successful.[5] The enthusiastic reception for *The Loves of the Plants* appears to have surprised even its author, who recounted his profits with great satisfaction to friends in letters and set out to complete the next part confident that his poetry was liked by most members of the polite society in which he lived and worked.[6]

(London: Macmillan, 1986). Before King-Hele's work on Darwin few full-length studies were available.

[3] Maureen McNeil, *Under the Banner of Science: Erasmus Darwin and His Age* (Manchester: Manchester Univ. Press, 1987); McNeil, "The Scientific Muse: The Poetry of Erasmus Darwin," in *Languages of Nature: Critical Essays on Science and Literature,* ed. L. J. Jordanova (London: Free Association Books, 1986), pp. 159–203; and R. E. Schofield, *The Lunar Society of Birmingham* (Oxford: Oxford Univ. Press, 1963).

[4] King-Hele, *Letters of Erasmus Darwin,* pp. 116–117.

[5] *Ibid.,* pp. 139–140. In order of publication the poems are *The Botanic Garden, Part II: Containing The Loves of the Plants, a Poem: With Philosophical Notes* (Lichfield, 1789) (hereafter **Darwin, Loves of the Plants**) and *The Botanic Garden: A Poem in Two Parts, Part I: Containing The Economy of Vegetation* (London, 1791). Both poems were issued anonymously, and Erasmus Darwin did not officially confirm his authorship of the works until 1794, when, on the title page of his medical tract *Zoonomia; or the Laws of Organic Life,* 2 vols. (London, 1794–1796), he referred to himself as "Author of The Botanic Garden."

[6] See, e.g., James Keir, *An Account of the Life and Writings of Thomas Day* (London, 1791), pp. 112–113: "that exquisite poem, the *Botanic Garden,* in which the graces themselves seem to decorate the temple of science with their choicest wreaths and sweetest blossoms." King-Hele, *Doctor of Revolution,* pp. 197–198, summarizes the favorable reception of the *Loves of the Plants;* it is also discussed at length in King-Hele, *Darwin and the Romantic Poets* (both cit. n. 2). For Darwin's response see *Letters of Erasmus Darwin,* pp. 193, 196, 197.

THE SEXES OF PLANTS

Darwin's hesitation and subsequent surprise no doubt stem from the way in which he chose to describe the sex life of flowers, for the poem was unabashedly about sex and sexual relations, about the all-pervading drive to find a mate and to reproduce. Such a focus was decidedly controversial. Darwin based *The Loves of the Plants* on the supposition that there are indeed male and female plants, that there are two sexes that join together for the purposes of reproduction. This idea was still, in Darwin's time, the subject of heated debate, being only partially confirmed by miscellaneous observations of plant fertilization. It was a matter of some importance in natural philosophy, because analogy with animal processes, so much a part of eighteenth-century thought, demanded some kind of corresponding sexuality in plants.

Throughout the eighteenth century naturalists had puzzled over the differing roles played by male and female parents in inheritance, in generation, and in fertilization and had attempted to understand plant reproduction through analogies with what was known about animals.[7] Unlike animals, however, plants rarely convey a clear picture either of male and femaleness—most flowers possess both sets of organs—or of sexual reproduction at all, since plants are quite capable of propagating their kind by purely vegetative means, and it is hard to know which part of the organism might count as a sexual individual.[8] Did a flower perhaps mate with itself, as hermaphroditic animals like snails and earthworms were popularly supposed to do, or with another flower on the same tree or bush, or with flowers of another plant altogether, thereby incurring the mechanical problem of conveying pollen from one point to another? Such differing possibilities led many naturalists to doubt the fact of sexes in plants.[9] Nor was there any clear parallel to animal spermatozoa in plants, and indeed the debate between epigeneticists and preformationists had foundered on exactly that issue, one side seeing sperm as fully formed seeds scattered in a nutrifying womb, the other as dust, or pollen, merely bringing some needed animus to a receptive ovum in which the seed already resided, somewhat similar to the way in which aphids and other parthenogenetic animals duplicated themselves without much male intervention.[10]

By 1759 the question of plant sexuality was thought sufficiently perplexing for the Imperial Academy of Saint Petersburg to offer a prize for an essay illuminating the process of fecundation and the perfection of fruit by semen, a prize widely reported to have been proposed to draw forth the views of Linnaeus, then at the height of his considerable powers.[11] And it was indeed won by Linnaeus

[7] Jacques Roger, *Les sciences de la vie dans la pensée français du XVIIIe siècle: La génération des animaux de Descartes à l'Encyclopédie* (Paris: Armand Colin, 1963); Philip C. Ritterbush, *Overtures to Biology: The Speculations of Eighteenth-Century Naturalists* (New Haven, Conn.: Yale Univ. Press, 1964); and François Delaporte, *Nature's Second Kingdom: Explorations of Vegetality in the Eighteenth Century*, trans. Arthur Goldhammer (Cambridge, Mass.: MIT Press, 1982).

[8] John Farley, *Gametes and Spores: Ideas about Sexual Reproduction, 1750–1914* (Baltimore/London: Johns Hopkins Univ. Press, 1982); see also Duncan S. Johnson, "The Evolution of a Botanical Problem: The History of the Discovery of Sexuality in Plants," *Science*, N.S., 1914, *39*:299–319.

[9] See Delaporte, *Nature's Second Kingdom* (cit. n. 7), pp. 129–130.

[10] F. J. Cole, *Early Theories of Sexual Generation* (Oxford: Clarendon Press, 1930); Shirley Roe, *Matter, Life, and Generation: Eighteenth-Century Embryology and the Haller-Wolff Debate* (Cambridge: Cambridge Univ. Press, 1981); and Farley, *Gametes and Spores* (cit. n. 8).

[11] James Edward Smith, trans., introduction to *A Dissertation on the Sexes of Plants: Translated from the Latin of Linnaeus* (London, 1786), pp. vii–viii.

with his dissertation on the sexes of plants, published in Saint Petersburg the following year.[12] In this work Linnaeus cited several examples of experiments in plant fertilization, carried out in Uppsala, confirming his previously expressed view that flowers were expressly organs of reproduction, present only to enable the perpetuation of species.[13] He also put forward the argument that plant hybrids owed their existence to a promiscuous mixing of males and females, which might also account for the origin of many vegetable species. A genus, he claimed, is nothing else than a number of plants sprung from the same mother by different fathers.[14] As for inheritance, he proposed that the male partner gave to its offspring the form of the leaves and the external parts, while the female transmitted the inside, medullary parts and the organ of fructification.[15] Each sex consequently played a material role in the process of making a new individual, either of the existing specific type or some kind of hybrid novelty.

Clearly, this prizewinning essay was closely bound up with Linnaeus's often complex views on the origin and natural hierarchy of plants as expressed in his systematic writings,[16] and his taxonomic schemes were soundly based on well-developed theories about the function and purpose of sex. Even Linnaeus's system of classifying plants solely by the number of stamens and pistils—a quantitative procedure bearing no relation to the affinities and characteristics of groups of plants found in nature—emphasized the universal necessity of sexual reproduction. Linnaeus gave a primacy to plant sexuality that no naturalist had attempted before, and thus the fate of his classification scheme was seen to hinge on the fate of ideas about plant sexes. In short, to be a Linnaean taxonomist was to believe in the sex life of flowers.

The point did not pass unnoticed among Linnaeus's critics, and anti-Linnaeans jostled to demonstrate that the sexuality of plants was nonsense. Lazzaro Spallanzani took the lead and was quick to attack Linnaeus's observations on fecundation, claiming that productive seeds were born in gourds, spinach, and hemp without any pollination—a claim going right to the heart of the doctrine of sexuality and casting doubt on the universality of Linnaeus's scheme.[17] From France other naturalists such as Charles Bonnet and John Turbeville Needham disputed Linnaeus's claims about the way pollen acted in fertilization. Further attacks came from Michel Adanson, who wrote his *Famille des plantes* (1763–1764) to counter Linnaeus's exclusive emphasis on sex and the number of sexual organs. Adanson considered that classification schemes should be based only on natural groupings of plants and animals as discerned by anatomical resemblances. Linnaeus, he felt, had sacrificed such aims and beliefs for the sake of expediency:

[12] Carolus Linnaeus, *Disquisitio de quaestione ab Academia Imperiali Scientiarum Petropol . . . Sexum plantarum argumentis et experimentis novis, &c.* (St. Petersburg, 1760).

[13] Carolus Linnaeus, *Classes plantarum, seu systemata plantarum omnia a fructificatione desumpta . . .* (Leiden, 1738), p. 441: "A flower is nothing but an act of the generation of plants" (my translation).

[14] Linnaeus, *Sexes of plants* (cit. n. 11), p. 56: "A genus is nothing else than a number of plants sprung from the same mother by different fathers."

[15] *Ibid.*, pp. 13–28.

[16] J. L. Larson, *Reason and Experience: The Representation of Natural Order in the Work of Carl von Linné* (Berkeley/Los Angeles: Univ. California Press, 1971). See also Larson, "Linnaeus and the Natural Method," *Isis*, 1967, *58*:304–320.

[17] Lazzaro Spallanzani, *Dissertazione di fisica animale e vegetabile*, 2 vols. (Modena, 1780), Vol. II. See also Delaporte, *Nature's Second Kingdom* (cit. n. 7), pp. 118–119; J. E. Smith, in his trans. of *Sexes of Plants* (cit. n. 11), criticized Spallanzani's attack on Linnaeus's doctrine of sexuality, pp. x, 43–44n.

his numerical system, though quick and easy, frequently brought together dis-
parate plants and separated similar ones, and was therefore thought to be "artificial" rather than "natural." In yet another arena, the most daring and dangerous
of the *philosophes,* Julien Offray de La Mettrie, poked fun at the terminology of
the sexual system in a little book circulated in fashionable Parisian society which
depicted the flower parts in graphic humanized parody, a single stamen representing the penis, and so forth. Meanwhile Buffon volleyed from the philosophical corner, arguing that there was no need for—and no merit in—Linnaeus's
pronounced artificiality. In Britain the francophiles Charles Alston and William
Smellie also found effective anti-Linnaean propaganda in the sexual innuendos
that could so easily be drawn out of the numerical system, with Alston taking a
high moral tone and spluttering in outraged propriety, and Smellie asserting that
Linnaeus had pushed analogy beyond all decent limits, so that it became truly
ridiculous.[18]

Stung into action, a strong coterie of English botanists struck back with a
flurry of translations and catalogues heralding Linnaeus as the prince of flowers.
Although Basil Soulsby and Frans Stafleu have admirably described the work of
these first disciples of Linnaeus, particularly in Britain, there are several names
that bear repetition here. John Berkenhout, though not the earliest translator of
Linnaeus by any means, reached a wide audience with his little *Clavis anglica
linguae botanicae* (1764). James Lee was perhaps even more widely known
through the many editions of his *Introduction to Botany* (1760). Hugh Rose
translated Linnaeus's *Philosophia botanica* in 1775, and William Curtis presented "Linnaeus's system of botany" to English readers in 1777. James Edward
Smith, later to become the most famous Linnaean of them all through his purchase of Linnaeus's collections and manuscripts and the foundation, in 1788, of
the Linnean Society of London, translated Linnaeus's *Reflections on the Study
of Nature* and then the controversial prizewinning essay on the sexes of plants.
British readers had never been subjected to so many expository texts before;
British botanists never so anxious to defend the work of their master.[19]

[18] Michel Adanson, *Familles des Plantes,* 2 vols. (Paris, 1763), discussed at length in Jean-Paul
Nicolas et al., *Adanson: The Bicentennial of Michel Adanson's "Familles des plantes,"* 2 vols. (Hunt
Monograph Series, 1) (Pittsburgh: The Hunt Botanical Library, 1963); Julien Offray de La Mettrie,
L'homme plante (Paris, 1748), ed. Francis Rougier (New York: Columbia Univ. Press, 1936); John
Lyon and Phillip Sloan, eds., *From Natural History to the History of Nature: Readings from Buffon
and His Critics* (Notre Dame, Ind.: Univ. Notre Dame Press, 1981), pp. 97–128; Charles Alston, *A
Dissertation on Botany* (London, 1754), pp. 42–71 (see also Ritterbush, *Overtures to Biology* [cit. n.
7], p. 119); and William Smellie, *The Philosophy of Natural History,* 2 vols. (Edinburgh, 1790–1799),
Vol. I, p. 248.
[19] Basil H. Soulsby, *A Catalogue of the Works of Linnaeus, and Publications More Immediately
Relating Thereto . . .,* 2nd ed. (London: British Museum [Natural History], 1933); Frans A. Stafleu,
Linnaeus and the Linnaeans: The Spreading of Their Ideas in Systematic Botany, 1735–1789 (Reg-
num Vegetabile, 79) (Utrecht: International Association for Plant Taxonomy, 1971); John Berken-
hout, *Clavis anglica linguae botanicae; or, a Botanical Lexicon; in Which the Terms of Botany,
Particularly Those Occurring in the Works of Linnaeus, and Other Modern Writers, are Applied,
Derived, Explained, Contrasted, and Exemplified* (London, 1764), a work undoubtedly owned by
Erasmus Darwin, since there is a copy in the Cambridge Univ. Library once in the possession of his
son Robert Waring Darwin and thence passed on to his grandson Charles Robert Darwin; James Lee,
*An Introduction to Botany: Containing an Explanation of the Theory of that Science, and an Inter-
pretation of Its Technical Terms* (London, 1760); *The Elements of Botany . . . Being a Translation of
the "Philosophia botanica," and Other Treatises of the Celebrated Linnaeus,* trans. Hugh Rose
(London, 1775); William Curtis, *Linnaeus's System of Botany, So Far as Relates to His Classes and
Orders of Plants* (London, 1777); Andrew Thomas Gage, *A History of the Linnean Society of London*
(London: Linnean Society of London, 1938); James Edward Smith, *Memoir and Correspondence . . . ,*

Among them was Erasmus Darwin, busy in the affairs of a new botanical society in Lichfield and tending his own botanic garden a mile or so outside the city limits. In the company of two other friends, Darwin established the grandly titled Botanical Society of Lichfield, a society that never had more than the three original members and that was created entirely in order to translate into English Linnaeus's *Species plantarum* and the slightly later volume, the *Genera plantarum*.[20] Both books were seen as essential weapons in the defense of Linnaean taxonomy, as indicated by the enthusiasm with which eminent London figures like Sir Joseph Banks and conspicuous Linnaeans like Jonas Dryander encouraged Darwin, writing on the Botanical Society's behalf, to persevere with his efforts.[21] Banks was especially gracious, helping Darwin and his botanical colleagues acquire copies of the best taxonomic catalogues and discreetly criticizing other systematic works. To Banks, the translations filled a gap in the anglicization of Linnaeus, which contemporary "introductions" and "explanations" had left embarrassingly open. Banks, who had taken Daniel Solander, Linnaeus's star pupil, round the world to Australia and back, had invested much of his scientific reputation in the Linnaean arrangement of his own outstanding herbarium in London.[22] His encouragement and endorsement of Darwin's work, while probably serving his own purposes, were nevertheless gratefully acknowledged by Darwin, who subsequently dedicated the first translation to the great man of Soho Square.

DARWIN'S LANGUAGE OF FLOWERS

Darwin's translations of Linnaeus's catalogues were capable and plain-spoken, taking their tone from the decidedly blunt originals. Linnaeus had minced no words when he described male and female forms in plants, nor did Darwin—unlike his contemporary William Withering, who was concerned that his own botanical book not include any of Linnaeus's improper words and invented euphemisms such as "chives" and "pointals" for stamens and pistils, themselves only words of Latin and Greek origin meaning threads and columns.[23] Darwin on the

---

ed. Lady Smith, 2 vols. (London, 1832); and Smith's translations, *Reflections on the Study of Nature: Translated from the Latin of the Celebrated Linnaeus* (London, 1785), and *Sexes of Plants* (cit. n. 11).

[20] Botanical Society of Lichfield, *A System of Vegetables, According to Their Classes, Orders, Genera, Species, with Their Characters and Differences . . . translated from the 13th Edition of the* Systema vegetabilium *. . . and from the* Supplementum plantarum *of the Present Professor Linnaeus,* 2 vols. (Lichfield, 1783); and Botanical Society, *The Families of Plants, with Their Natural Characters . . . Translated from . . . the* Genera plantarum *of . . . Linnaeus* (Lichfield, 1787). On the Botanical Society see also *Letters of Erasmus Darwin,* pp. 109–111. Although there is little direct evidence beyond a few letters, it seems clear that Erasmus Darwin was the sole author of these catalogues. Yet Darwin always referred to the translations as the joint activity of the Botanical Society and the title pages of both books name only the society as author. Few contemporaries would have attributed them to Darwin alone.

[21] *Letters of Erasmus Darwin,* pp. 112–120.

[22] Stafleu, *Linnaeus and the Linnaeans* (cit. n. 19), pp. 199–240; and Patrick O'Brian, *Joseph Banks: A Life* (London: Collins Harvill, 1987). See also D. J. Mabberley, *Jupiter Botanicus: Robert Brown of the British Museum* (London: British Museum [Natural History]; Brunswick: J. Cramer, 1985).

[23] *Letters of Erasmus Darwin,* pp. 74–75; and William Withering, *A Botanical Arrangement of All the Vegetables Naturally Growing in Great Britain* (Birmingham, 1776). Withering's concern was reiterated by Curtis, *Linnaeus's System* (cit. n. 19), p. 2: "One chief aim in this translation, has been to convey to the English reader the Author's explanation of his system in terms the least exceptionable."

other hand, like James Lee, John Berkenhout, and even his brother Robert Waring Darwin, who had all published before him,[24] called them males and females, husbands and wives (see Figs. 1 and 2).

It is worth emphasizing here that it was Linnaeus who initiated this personification of the sexual relations of plants and that his more robust followers were merely accepting and extending the practice into English-language works. This use of personification allowed Linnaeus to write of plant sexuality as a "marriage" and the male and female organs as "husbands" and "wives"; he wrote of the petals (corolla) as the "marriage bed"; and he discussed the existence of monoecious and dioecious plants in terms of one or two different "houses." By coining the words *monoecious* and *dioecious* (derived from the Greek for one or two homes or houses), Linnaeus set up a system of metaphors through which plant sexuality could be made intelligible by being modeled on human society, in much the same way as La Fontaine's moral fables owed their dramatic force and piquancy to their location in the animal world rather than the human. Many translators saw the value of such metaphors. In *The Elements of Botany* (his translation of Linnaeus's *Philosophia botanica*), for example, Hugh Rose wrote: "The calyx then is the marriage bed, the corolla the curtains, the filaments the spermatic vessels, the antherae the testicles, the dust the male sperm, the stigma the extremity of the female organ, the style the vagina, the germen the ovary, the pericarpium the ovary impregnated, the seeds the ovula or eggs."[25]

However, Darwin ventured much further than Linnaeus in the bravura with which he maintained a policy of plain speaking in the translations. He believed that the English language had greater expressivity than Latin, and he consciously attempted to use English to display the inner meanings of Linnaeus's terms. In this he had, for a short while at least, the advice of the celebrated Samuel Johnson, also once resident in Lichfield.[26] Darwin spelled out his views in the preface to the Botanical Society's first translation: "The learned reader will perceive, that we have made a slight change in the construction of the sexual distinctions of the Classes on account of the greater delicacy of modern language; hence the words one male, and one female, are used in preference to one virility and one feminality."[27] In later years he referred to the Botanical Society's translations as having "rendered that translation of Linnaeus as expressive and as concise, perhaps more so, than the original."[28] In order to maintain such expression, Darwin went so far as to coin more than fifty new botanical words—for example, "stipule" for a lateral appendage often resembling a small leaf or scale[29]—and introduced a set of terms to describe the various physical juxtapositions of stamens,

[24] Lee, *Introduction to Botany*, pp. 10–11, 72–73; Berkenhout, *Clavis anglica* (both cit. n. 19); and Robert Waring Darwin, *Principia botanica: Or, a Concise and Easy Introduction to the Sexual Botany of Linnaeus* (Newark, 1787).

[25] Linnaeus, *Elements of Botany*, trans. Rose (cit. n. 19), p. 151.

[26] *Letters of Erasmus Darwin*, pp. 114, 172.

[27] Botanical Society of Lichfield, *System of Vegetables*, Vol. 1, p. v. The same sentiment is echoed in the society's *Families of Plants*, p. v: "The conciseness, the perspicuity, and the spirit of our author live, we hope, undiminished by the change of language." (Both cit. n. 20.)

[28] Darwin, *Loves of the Plants*, p. 130.

[29] Desmond King-Hele, "Erasmus Darwin, Man of Ideas and Inventor of Words," *Notes and Records of the Royal Society of London*, 1988, 42:149–180. Darwin's interest in language was further expressed in philosophical notes to the *Temple of Nature* (London, 1803), pp. 93–106, entitled "The Theory and Structure of Language."

CHARACTERS of CLASSES.

I. ONE MALE.
One hufband in marriage.
One ftamen in an hermaphrodite flower.
II. TWO MALES.
Two hufbands in the fame marriage.
Two ftamens in an hermaphrodite flower.
III. THREE MALES.
Three hufbands in the fame marriage.
Three ftamens in an hermaphrodite flower.
IV. FOUR MALES.
Four hufbands in the fame marriage.
Four ftamens in the fame flower with the fruit.
(if the two neareft ftamens are fhorter, it is referred to Clafs 14.)
V. FIVE MALES.
Five hufbands in the fame marriage,
Five ftamens in an hermaphrodite flower.
VI. SIX MALES.
Six hufbands in the fame marriage.
Six ftamens in an hermaphrodite flower.
(if the two oppofite ftamens are fhorter, it belongs to Clafs 15.)
VII. SEVEN MALES.
Seven hufbands in the fame marriage.
Seven ftamens in the fame flower with the piftil.
VIII. EIGHT MALES.
Eight hufbands in the fame marriage.
Eight ftamens in the fame flower with the piftil.
IX. NINE MALES.
Nine hufbands in the fame marriage.
Nine ftamens in an hermaphrodite flower.
X. TEN MALES.
Ten hufbands in the fame marriage.
Ten ftamens in an hermaphrodite flower.
XI. TWELVE MALES.
Twelve hufbands in the fame marriage.
Twelve ftamens to nineteen in an hermaphrodite flower.
XII. TWENTY MALES.
Generally twenty hufbands, often more.
Stamens inferted on the calyx (not on the receptacle) in an hermaphrodite flower.

XIII. MANY MALES.
Twenty males or more in the fame marriage.
Stamens inferted on the receptacle, from 20 to 1000 in the fame flower with the piftil.
XIV. TWO POWERS.
Four hufbands, two taller than the other two.
Four ftamens of which the two neareft are longer.
XV. FOUR POWERS.
Six hufbands, of which four are taller.
Six ftamens : of which four are longer, and the two oppofite ones fhorter.
XVI. ONE BROTHERHOOD.
Hufbands, like brothers, arife from one bafe.
Stamens are united by their filaments into one body.
XVII. TWO BROTHERHOODS.
Hufbands arife from two bafes, as if from two mothers.
Stamens are united by their filaments into two bodies.
XVIII. MANY BROTHERHOODS.
Hufbands arife from more than two mothers.
Stamens are united by their filaments into three or more bodies.
XIX. CONFEDERATE MALES.
Hufbands joined together at the top.
Stamens are connected by the anthers forming a cylinder (feldom by the filaments).
XX. FEMININE MALES.
Hufbands and wives growing together.
Stamens are inferted on the piftils, (not on the receptacle).
XXI. ONE HOUSE.
Hufbands live with their wives in the fame houfe, but have different beds.
Male flowers and female flowers are on the fame plant.
XXII. TWO HOUSES.
Hufband and wives have different houfes.
Male flowers and female flowers are on different plants.
XXIII. POLYGAMIES.
Hufbands live with wives and concubines.
Hermaphrodite flowers, the male ones, or female ones in the fame fpecies.
XXIV. CLANDESTINE MARRIAGES.
Nuptials are celebrated privately.
Flowers concealed within the fruit, or in fome irregular manner.

Figure 2. Darwin's translation of the chief characteristics of Linnaeus's twenty-four taxonomic classes, as given in The Families of Plants, *pages lxxviii–lxxix. Reproduced by permission of the Syndics of the Cambridge University Library.*

such as "confederate males," which are joined together at the base, or "brother-hoods," which mature in sets of three or five at different times in the life of the flower. Terms such as these may well have inspired further metaphors or trains of thought that eventually came together in his later botanical verses.

THE LOVES OF THE PLANTS

A close friend and confidante of Darwin's was the poetess Anna Seward, known to many as "The Swan of Lichfield" and at this time celebrated for her *Elegy on Captain Cook* (1780) and for a critical account of George Washington in her *Monody on Major André* (1781). Seward, who helped Darwin tend his botanic garden, encouraged him to turn his talent for light verse toward plants by presenting him with a short poem on the nymphs and gnomes in his Lichfield garden. Darwin, never averse to nymphs and goblins in his poetry as well as in his garden, thought that "the Linnean System is unexplored poetic ground, and an happy subject for the muse. It affords fine scope for poetic landscape; it suggests metamorphoses of the Ovidian kind, though reversed."[30]

Darwin wrote the bulk of *The Loves of the Plants* concurrently with the Botanical Society translations of Linnaeus. The poem, published in 1789, was begun

[30] Anna Seward, *Memoirs of the Life of Dr. Darwin, Chiefly during His Residence at Lichfield; with Anecdotes of His Friends and Criticisms on His Writings* (London, 1804), on pp. 125–131, quoting Darwin on pp. 130–131. Seward's verses, slightly altered by Darwin, were published as the "Exordium" to *Loves of the Plants.*

in 1779, and the two translations were issued in 1783 and 1787. There is a strong possibility that he intended *The Loves of the Plants* to be a reaffirmation of Linnaeus's insistence on plant sexuality in the face of increasingly numerous anti-Linnaean publications. Furthermore, it seems likely that the poem also represents a mild amendment of Linnaeus's ruling about the number of organs alone being the crucial factor. Darwin's personification of the stamens and pistils can in itself be seen as an attempt to introduce a real, physiological element into a highly abstract scheme; but he went further by also stressing the proportion, length, and arrangement of the organs within Linnaeus's numerical system. Darwin believed that the length of the male filaments or of the female style had a marked effect on the process of fertilization. Other botanists had demonstrated the way in which stamens bend over the stigma to pollinate it, some stamens even moving in turns, bending and retreating. Equally, the pistil in some plants bends to a set of stamens, and other pistils do not develop fully until the first has retreated.

Darwin emphasized these behavioral traits in his verses by accurately representing both the structure of each plant and its individual means of fertilization. He wrote of relative positions, of males and females bending to embrace each other, of sets of brothers, of knights and their squires, and so on. *Melissa,* the lemon balm, was defined just as much by its reproductive actions as by its structure:

> Two knights before thy fragrant alter bend,
> Adored Melissa! and two squires attend.[31]

The alpine flower *Draba* received much the same treatment from "four rival Lords" while "two menial youths attend," a comment on the differing maturation rates of the various stamens.[32] Later on, in his botanical book the *Phytologia,* published in 1800, Darwin set out this belief in the importance of the relative proportions and situations of the stamens, hoping to improve a little on Linnaeus's system while still expressing his sincere opinion that the numerical, sexual approach was unrivaled among taxonomies.[33] So the personification of stamens and pistils was perhaps Darwin's way of putting some organic functioning back into Linnaeus's artificial constructs, without conceding the game to French and British proponents of natural classification.

He had other aims as well, aims that were equally efficiently served by the sexual arrangements of flowers and the motif of human love and that were not so far removed from those attributed to Linnaeus. These aims, though apparently only nascent in Darwin's thoughts during the 1780s, soon emerged in his long poem *The Economy of Vegetation* and were thence elaborated in other books and writings, particularly the *Zoonomia* and *Phytologia.* Darwin wanted to demonstrate the fecundity of the natural world and to present his thesis that sexual

---

[31] Darwin, *Loves of the Plants,* canto 1, lines 59–60 (p. 6). All references are to the first (1789) edition, since Darwin added and altered later editions. Unfortunately, the printer made several mistakes in the line numbers (particularly in canto 1, lines 250–300). Line numbers are therefore followed by page references to the first edition.

[32] *Ibid.,* canto 1, lines 219–222 (p. 22).

[33] Erasmus Darwin, *Phytologia; or the Philosophy of Agriculture and Gardening: With the Theory of Draining Morasses, and with an Improved Construction of the Drill Plough* (London, 1800), pp. 564–578.

reproduction is the "chef d'oeuvre, the masterpiece of nature."[34] Nothing in nature could exist, he thought, without reproduction, and the purpose of existence was to reproduce. Central to this argument, as Roy Porter has recently emphasized, was Darwin's belief that sexual reproduction was the hidden force behind evolution and progress, since new organisms were introduced into the world through variations that arose in the offspring of sexual unions alone.[35]

Like others before him, Darwin was unsure exactly how variation came about or how much or what each parent might contribute to their progeny, but he seems to have held the view that both parents were involved and that the inherent irritability of living matter and the association of ideas led to adaptive responses in the embryo. However, it is difficult to know precisely what Darwin thought at the time he was writing *The Loves of the Plants*. In the first edition of his *Zoonomia* (1794–1796) he suggested that the male partner alone carried the formative influence, a view subsequently changed in favor of both partners in the *Phytologia* and the third edition (1801) of the *Zoonomia*.[36] In this later view, males and females provided different, complementary materials, a suggestion Darwin culled from Linnaeus's doctrine of plant reproduction, and individual variations were produced by a rearrangement of different quantities of the respective parental molecules or contributions.[37] Out of these individual differences there emerged a chain—or continuity—of forms, seen by Darwin as an evolutionary scale of nature progressing from the simplest to the most complex of living organisms.

He included plants in this evolutionary chain of being and applied his arguments about sexual reproduction to them with as much gusto as he did for the animal kingdom. Plants were given the attributes of sensation, movement, and a certain degree of mental activity, in order to provide a continuous scale between the lowest, simplest forms of living beings and the highest.[38] They possessed the same four classes of bodily actions itemized in the *Zoonomia*, that is, the properties of irritation, sensation, volition, and association, although to be sure Darwin

[34] Darwin, *Zoonomia* (cit. n. 5), Vol. I, p. 514; Darwin, *Phytologia*, p. 114; and Darwin, *Temple of Nature* (cit. n. 29), p. 36. The term *masterpiece* had more meanings than the obvious here; for cultured people in the eighteenth century the word was a euphemism for *vagina*, a hidden allusion that certainly reinforced Darwin's general meaning. See Peter Fryer, *Mrs. Grundy: Studies in English Prudery* (London: Dennis Dobson, 1963), p. 48.

[35] Darwin, *Phytologia* (cit. n. 33), p. 115: "But from the sexual, or amatorial generation of plants new varieties, or improvements, are frequently obtained"; see Roy Porter, "Erasmus Darwin: Doctor of Evolution?" in *History, Humanity and Evolution: Essays for John Greene*, ed. James R. Moore (Cambridge: Cambridge Univ. Press, 1989).

[36] King-Hele, *Doctor of Revolution* (cit. n. 2), p. 283; and Darwin, *Zoonomia* (cit. n. 5), Vol. I, pp. 478–533. See also *ibid.*, 3rd ed., 4 vols. (London, 1801), Vol. II, pp. 277–304. It was apparently Darwin's studies of the reproductive mechanisms of plants that led him to change his opinion: *ibid.*, Vol. II, p. 277.

[37] Darwin, *Phytologia*, pp. 91–131, esp. pp. 127–129, in which male organs are held to secrete fibrils or molecules with "formative" or "nutritive appetencies" and female organs secrete "formative" or "nutritive propensities." These mingle together, making an individual "resembling in some parts the form of the father, and in other parts the form of the mother, according to the quantity or activity of the fibrils or molecules at the time of their conjunction" (p. 130).

[38] Darwin, *Zoonomia* (cit. n. 5), Vol. I, pp. 101–107. Even in his earliest botanical writings, Darwin had ascribed such attributes to plants. In the Botanical Society of Lichfield *Families of Plants*, p. xix, he wrote, "For vegetables are, in truth, an inferior order of animals, connected to the lower tribes of insects, by many marine productions, whose faculties of motion and sensation are scarcely superior to those of the petals of many flowers, or to the leaves of the sensitive plant, the moving plant, and the Fly trap." See also Ritterbush, *Overtures to Biology;* and Delaporte, *Nature's Second Kingdom* (both cit. n. 7).

agreed that these were displayed to a lesser degree than in animals or humans.[39] Plants too indulged in the pursuit of pleasure and the avoidance of pain; they too sought gratification through sexual reproduction. Darwin's *Phytologia,* his subsequent paean to the vegetable kingdom, is crammed with examples of sensate plants, plants that move, plants that feel, plants that think or, at the very least, can tell the time of day. Sadly neglected by historians, this botanical text takes up all the themes we traditionally associate with the *Zoonomia* and the evolutionary poems, but applied by Darwin to plants in order to bring them fully into a comprehensive philosophy of nature.

The Loves of the Plants, then, may be seen as an early study in what was to be Darwin's lifelong commitment to the idea of transmutation. It was important for him to show plants as an integral part of animate nature, as organisms with the same attributes as animals in a degree appropriate to their place in the scale of organization, and important to show them as sexual beings able to contribute to the variability and progress of the natural world. His first public expression of these interests therefore took the form of identifying himself as a Linnaean who believed in the sexuality of plants. This was to be carried out by a sustained application of the simple metaphorical device of seeing plants as people.

THE IDEA OF A BOTANIC GARDEN

The Loves of the Plants has long been acknowledged as an extended didactic analogy between plants and humans; not wonderful poetry, by any means, but as Desmond King-Hele puts it, full of "glittering couplets" that led Wordsworth in his youth to write of the "dazzling manner of Darwin."[40] Samuel Taylor Coleridge, who famously condemned Darwin's extravagant diction ("I absolutely nauseate Darwin's poem" he wrote in 1796), dryly admitted that he had at least "accumulated and applied all the sonorous and handsome-looking words in our language."[41] Historians of science, accustomed to finding serious meaning only in Darwin's other poems and longer prose works, might justifiably ask how far the glitter and dazzle of *The Loves of the Plants* served merely to satisfy the usual requirements of story, meter, and rhyme. But a closer examination of the structure of Darwin's poem and the metaphorical framework shows that its author had several conscious aims that could best be expressed through this deliberately chosen vehicle.

The poem, which takes the form of a narrative delivered by a "Botanic Muse" who is described as having formerly guided Linnaeus, is loosely arranged to reflect the passing hours of a single day. After some prefatory advice from the author and others on what follows, the verses are divided into four cantos, each canto opened and closed by the narrator calling her nymphs back to her side, interspersed with dialogues between the poet and his bookseller about the metaphysics and characteristics of poetry. Beyond this, the verses have little narrative thread. Each plant, described as if it were a group of human beings according to the number of stamens and pistils it possesses, is presented in an anecdote designed to amuse the assembled nymphs as they dally in an Arcadian landscape (see Fig. 3).

[39] Darwin, *Zoonomia,* Vol. I, pp. 37–53.
[40] King-Hele, *Romantic Poets* (cit. n. 2), pp. 67–68.
[41] *Ibid.,* p. 136.

The setting is clearly a botanic garden in which exotic species intermingle with indigenous plants. For Darwin, as for other members of the intellectual leisured classes, reference to a botanic garden evoked a constellation of ideas and emotions that combined scientific purpose with recreational pleasure. Gardens glorified both the practical expertise of horticulturists and the serious activities of taxonomists and medical personnel. National pride was reflected in the breadth and variety of such collections, each plant representing geographical explorations in the past and the nation's political allegiances and commercial intentions. Gardens were also obvious repositories of "nature," a display of plants outside their usual geographic boundaries conjuring up notions of an untrammeled, fecund

Figure 3. *"Flora attired by the Elements," designed by Henry Fuseli and engraved by Anker Smith, 1791. The frontispiece to Erasmus Darwin,* The Economy of Vegetation, *Part I of* The Botanic Garden *(London, 1792). Reproduced courtesy of the Wellcome Institute Library.*

world—for some people almost literally a "garden of Eden," for others representative of organisms living in a state somehow beyond or outside the conventional limits and laws of nature.[42] Considerations like these were easily generalizable to the social and even the political world, should an author wish to do so. In Darwin's case it seems highly probable that he intended to make full use of this particular set of associated images. The motif of a botanic garden served to indicate that his verses dealt with plant species and their human analogues as if they were temporarily free of the usual constraints of the ordinary world.

But Darwin did not depend on these indefinable evocations alone. The structure of his poem was also closely tied to the idea of a botanic garden. As in a real botanic garden, the species were arranged or disposed according to their taxonomy or their useful attributes. Darwin deftly manipulated this metaphor to allow himself to group together species of plants that do not necessarily follow each other in strict botanical order, as in his canto 3 (on medicinal plants), and to give himself room to ignore other, less poetic, plants that would overload his delicate confection. He also capitalized on the chance to juxtapose extravagant imagery, appropriate to tropical exotics, with gentler, more pastoral allusions, providing the variety and ingenuity that his contemporaries would have expected and in which he came to excel. The poem's "garden" is full of profusion and confusion, all artfully ordered and cultivated by a knowing eye to give the impression of unadulterated nature, an impression central to eighteenth-century ideas about the picturesque and an integral element in the way in which Darwin and other gentlefolk thought about the natural world.[43] Darwin gave depth to the imagery here with the revelation that the garden loosely described in the poem was none other than his own in Lichfield, carefully laid out by himself, here translated from a form intended to delight the visual senses into the medium of poetry.

Thus the idea of a botanic garden in which to set the amours of flowers can be seen to be far more than a simple trope: it served as an organizing principle and as structure and metaphor. In addition, the botanic garden of the poem was a real garden in Lichfield, the poet's personal creation. The pictures painted by Darwin therefore possess meanings that went beyond the surface of the "gorgeous diction" that Coleridge so decried.

Darwin described only eighty-three species out of the many hundreds catalogued by Linnaeus. Each description included the numbers of stamens and pistils, in accordance with the Linnaean system, and ten or more lines of metaphorical, allusive poetry closely based on the appearance of the plant or its known attributes: for example, the grapevine is shown as a clinging, twining female; the poppy as a queen of sleep; the foxglove as a healing goddess bringing the drug digitalis; and so on. Lengthy footnotes, as in all Darwin's poems, explained these allusions. Other personifications took their cue from classical learning, though reversing the usual human-to-plant metamorphosis of classical myth. Linnaeus, like others before him, had laid great weight on the actual name of a plant or

---

[42] See esp. John Prest, *The Garden of Eden: The Botanic Garden and the Re-Creation of Paradise* (New Haven, Conn.: Yale Univ. Press, 1981, 1988).

[43] See McNeil, "Scientific Muse" (cit. n. 3), pp. 183–190; and Keith Thomas, *Man and the Natural World: Changing Attitudes in England, 1500–1800* (London: Allen Lane, 1983). Darwin's emphasis on an epistemology governed by the visual sense was partly based on his son's work. See Charles Robert Darwin, "On Ocular Spectra," in E. Darwin, *Zoonomia* (cit. n. 5), Vol. I, pp. 534–566; and E. Darwin, *Loves of the Plants*, pp. 128–129.

animal, stipulating that it should define the taxonomic relationships of the species and in a more traditional sense "encapsulate" the very essence of the species. He perceived the activity of naming as akin to religious baptism, almost as if the organism was not part of the Christian world until it possessed its own particular species name.[44] To a large extent, Linnaeus's nomenclature therefore reflected the ancient myths that had emerged around each species.[45] Erasmus Darwin, naturally enough, used the Linnaean names freely in his verses. More often than not, the classical allusions enshrined in Linnaeus's names were the motif on which Darwin's personifications were embroidered. These needed no explanation in the world of the classically educated eighteenth-century reader, and even women, Darwin's intended readers, who rarely had any formal training in ancient literature, would have been familiar with the gods and goddesses mentioned by Darwin.

Out of this rich mixture of allusions one tendency emerges clearly. Although Darwin was interested in describing accurately the reproductive structures and habits of plants, his poem focused largely on the sexual and social behavior of women. The characterizations of men and women were carefully matched to create an appropriate anecdote that would explain and define each chosen species; yet Darwin's efforts seem primarily directed toward creating a vivid picture of the *women* invoked in his verses: he gave the plant-women the central role in characterizing the behavior or story of each partnership, and the female personalities were allowed to carry the tone and impact of each stanza. The men—the stamens of Linnaeus's scheme—were not given the same attention or depth of characterization, even in some cases being sketched solely in terms of almost empty labels such as "swain" or "beau." In some sense this is a reversal of Linnaeus's system, in which the stamens—the males—defined the primary groups of plants (taxonomic classes) and could therefore be said to be more significant than the pistils, the females, which are merely secondary taxonomically (see Fig. 1). But Erasmus Darwin favored the idea of females taking a substantial part in reproduction, contributing actual molecules for the medulla of the offspring, not just a nutritive location for the growth of preformed seeds (although, as already mentioned, he did at a later stage question this interpretation in the first edition of the *Zoonomia*). In part, Darwin's literary sensibilities, in common with those of others of the same generation, whose taste was formed by the works of Fielding, Defoe, and Richardson, encouraged him to cast the poem essentially in terms of what women did and did not do. One hint given by Desmond King-Hele in his edition of Darwin's letters is also relevant here, that by 1778 Darwin was in love with Elizabeth Pole, the wife of another local resident.[46] *The Loves of the Plants*, begun in 1779 and composed intermittently

[44] Janet Browne, "Botany and Botanists," an essay review of E. L. Greene, *Landmarks of Botanical History*, ed. F. N. Egerton; and *Linnaeus: The Man and His Work*, ed. Tore Frängsmyr, *Hist. Sci.*, 1984, *22*:207–209.

[45] John L. Heller, "Classical Poetry in the *Systema naturae* of Linnaeus," *Transactions of the Proceedings of the American Philological Association*, 1971, *102*:183–216. Even in manuscript notes Linnaeus framed his identifications in terms of classical allusions: next to his written description of the species *Andromeda*, he drew a sketch of the girl Andromeda, chained to a rock with a dragon at her feet as in the Greek myth, juxtaposed with a hand-drawn picture of the plant itself. MS *Lachesis Lapponica*, fol. 87, Linnean Society, London.

[46] King-Hele, in *Letters of Erasmus Darwin*, pp. 76–78; and Henry Nidecker, "The Poetical Prelude of Erasmus Darwin's Second Marriage," in *Festschrift Gustav Binz . . . zum 70. Geburtstag am*

during the following decade, may at first have been intended as a kind of love song to Elizabeth Pole, hence Darwin's emphasis on women as the arbiters of masculine behavior. As luck would have it, Mrs. Pole was soon a widow and free to marry Darwin in 1781.

THE PERSONIFICATION OF PLANTS

Table 1 presents a synopsis of Darwin's poetic imagery relating to women and their sexual relations with men, ranged against the number of stamens and pistils as stipulated in Linnaeus's classification of plants. Putting it another way around, we can say that the table lists what might be called the "facts" of nature (the number of stamens and pistils) in conjunction with the social moral or metaphor that Darwin draws out of these "facts" when viewed in a human context. The characterizations in the table are necessarily brief but serve, it is hoped, to present an accurate version of each botanical image. Darwin's words have been used whenever possible and the key ideas checked against Darwin's own index, which was provided in a "Catalogue of the Poetic Exhibition" at the end of *The Loves of the Plants* for those who might have missed the point of his metaphors.

For clarity, the table has been divided into sections according to the relative number of stamens and pistils. In the first section there are fourteen anecdotes based on the sexual relationships of plants with one stamen and one pistil (one man and one woman). In this section the table follows Darwin and Linnaeus by including the Cryptogamia as sexually reproducing organisms that—as the name given to them by Linnaeus indicates—hide their activities from the eyes of naturalists. The Cryptogamia are marked by an asterisk in the table.

Otherwise, the number of plants with only one stamen and one pistil is very small. The vast majority possess five stamens and one pistil, although another biological quirk affects the figures slightly. Often the stamens are fused together in a tube, or the florets, male and female alike, are reduced in size and clustered together to make up a single flower head, as in the chrysanthemum, sunflower, or daisy—classed together by Linnaeus as the Polygamia. The table again follows Darwin's understanding of the scheme by including these in the section with five males and one female. When the numbers of males and females are both greater than one, they are given as Darwin presented them and not reduced to their lowest common denominator, since Darwin had different things to say about ratios of, for example, ten-to-ten from those he said of one-to-one.

Darwin himself took considerable artistic license and made use of only the more interesting or appropriate plants for his purposes. He followed Linnaeus's outline and gave at least one example of each of his classes and orders, though not necessarily in strict taxonomic series, as, for example, in the third canto, where he deals with medicinal and other useful plants together. In the table,

16 Januar 1935 von Freunden und Fachgenossen dargebracht (Basel: Benno Schwabe, 1935). The inference that *Loves of the Plants* was partly written with Elizabeth Pole in mind is wholly mine but is based on a poem of Darwin's addressed to her in 1775, in which Darwin, thinly disguised as a wood nymph from his botanic garden, begs that she should not proceed to lop any more trees in that garden. Certainly the garden metaphor played a significant role in their courtship, and *Loves of the Plants* was composed during the first years of their marriage. Together they raised a large second family, which cheerfully incorporated Darwin's two remaining sons from his first marriage and two natural daughters, Mary and Susan Parker, by another woman. See E. Posner, "Erasmus Darwin and the Sisters Parker," *History of Medicine*, 1975, 6(pt. 2):39–43.

Table 1. Images of women in *The Loves of the Plants*

No. of males (stamens)	No. of females (pistils)	Image
		I. One male and one female
1	1	A virtuous, timorous beauty (*Canna*, 1:39)
1	1	Disdained by husband, two beds divide (*Cupressus*, 1:73)
1	1	Betrayed by the appearance of progeny after clandestine relations (*Osmunda*, 1:93)*
1	1	Gentle, tender as a lamb (*Polypodium*, 1:247)*
1	1	Retiring, pursued by plighted swain (lichen, 1:293)*
1	1	Intrepid wife seeking her spouse (*Ulva*, 1:353)*
1	1	Hapless lover, killed by snow and cold (*Tremella*, 1:373)*
1	1	Sings of her secret loves (*Fucus*, 4:159)*
1	1	Awakened by enamored lover (*Muschus*, 4:259)*
1	1	Impatient for her lover (*Conferva*, 4:269)*
1	1	Chaste daughter who avows her love to husband (truffle, 4:297)*
1	1	Strikes a talisman that charms husband (*Caprificus*, 4:327)*
1	1	Blooming bride (*Byssus*, 4:357)*
1	1	Playful bride (*Conferva*, 4:363)*
		II. More than one male and one female
2	1	A pitying beauty who soothes in turns (*Collinsonia*, 1:51)
2	1	Tearful, calls her faithless lover (*Vallisneria*, 1:341)
2	1	Baleful queen-sorceress (*Circaea*, 3:6)
3	1	Has unjealous husbands (*Iris*, 1:71)
3	1	Two houses hold a fashionable pair (*Osyris*, 1:75)
3	1	Enthroned queen who grants gift of fame (*Papyrus*, 2:105)
4	1	Adored by 2 knights, attended by 2 squires (*Melissa*, 1:59)
4	1	Ambitious, soars and flies like an eagle (*Visca*, 1:225)
4	1	Revived from faint by attentive youths (*Dypsaca*, 1:307)
4	1	Blushing beauty, blending dye in cauldron (*Rubia*, 1:321)
4	1	Protected from the throng by her helpers (*Digitalis*, 2:419)
4	1	Flings poisoned darts and stings (*Urtica*, 3:191)
4	1	Modest virgin (*Trapa*, 4:169)
5	1	Laughing belle with a wanton air (*Meadia*, 1:61)
5	1	Cold and shy, an obdurate beauty (*Curcuma*, 1:65)
5	1	Reigns with charms despotic (*Chondrilla*, 1:97)
5	1	A plumed lady who leads a gaudy band (*Helianthus*, 1:191)
5	1	A fair lady with artless grace (*Lonicera*, 1:211)
5	1	A fair mechanic, lady balloonist (*Carlina*, 2:7)
5	1	Gentle timekeeper watching over the year (*Lapsana*, 2:163)
5	1	A bright lady with golden hair (*Calendula*, 2:164)
5	1	Priestess offering votaries to health (*Cinchona*, 2:343)
5	1	Frantic queen, avenges rejected love by killing infants (*Impatiens*, 3:131)

NOTE: The numbers in parentheses refer to the canto and the initial line number (see also n. 31). For clarity, the table has been divided in sections according to the relative number of stamens and pistils. The cryptogamia, in section I, are marked by an asterisk.

Table 1. (Continued)

No. of males (stamens)	No. of females (pistils)	Image
		II. More than one male and one female, *cont'd*
5	1	Her contagious breath brings death (*Lobelia*, 3:193)
5	1	Seductive harlot (*Vitis*, 3:287)
5	1	Gentle, grieving for dead baby (*Cyclamen*, 3:311)
5	1	Goddess with a train of cherubs (*Bellis*, 4:113)
6	1	Ensnares with harlot smiles and wily charms (*Gloriosa*, 1:119)
6	1	Folds her infant in her arms (*Tulipa*, 1:171)
6	1	A tall beauty who casts her shadow on distant lands (*Draba*, 1:219)
6	1	Playful beauty (*Galanthus*, 4:103)
8	1	Chaste, saintlike (*Tropaeolum*, 4:43)
10	1	Haughty maid wooed by brothers (*Genista*, 1:57)
10	1	Stalks with gloomy dignity (*Dictamnus*, 3:184)
10	1	A beauty guarded by fond brothers (*Cassia*, 3:343)
10	1	African beauty in transparent clothes (*Hedysarum*, 4:237)
20	1	Wild priestess/seer (*Laurocerasus*, 3:39)
20	1	Breathes her virgin vows (*Cerea*, 4:15)
100	1	Desdemona, won by sooty monster (*Plantago*, 1:77)
many	1	Gigantic nymph reigning over puny lovers (*Kleinhovia*, 1:157)
many	1	Queen of the coral groves (*Zostera*, 1:231)
many	1	Queen of the seraglio (*Mimosa*, 1:267)
many	1	Nymph encouraging factory operations (*Gossypia*, 2:85)
many	1	Fair (*Nymphaea*, 2:163)
many	1	Leads a sprightly troop (*Cistus*, 2:301)
many	1	Keeper of fragrant treasures (tea, 2:473)
many	1	Amazonian beauty (*Arum*, 4:187)
		III. One male and more than one female
1	2	Virgins smitten by beauty (*Callitriche*, 1:45)
		IV. Multiple males and females
2	2	Shepherdess sisters and wives (*Anthoxa*, 1:85)
3	2	Chaste sister-nymphs (*Avena*, 4:73)
4	2	Harlot-nymphs (*Cuscuta*, 3:259)
10	2	Burn with unallowed desires (*Dianthus*, 4:207)
12	2	Sister-nymphs (*Menispermum*, 2:227)
6	3	Blushing maids (*Colchica*, 1:181)
10	3	Harlot band (*Silene*, 1:131)
4	4	Sister-wives (*Ilex*, 1:143)
5	5	A queen with 4 sister-nymphs (*Drosera*, 1:199)
5	5	An inventor with 4 sister-nymphs (*Flax*, 2:67)
10	5	Wanton beauties in gay undress (*Lychnis*, 1:107)
many	many	Glittering throng of beaux and belles (*Anemone*, 1:263)
many	many	Gay sisters with seductive smiles (*Helleborus*, 2:199)
many	many	Sorceress, sofa'd on silk (*Papaver*, 2:265)
many	many	A hundred blushing virgins (*Adonis*, 4:387)

however, no useful purpose would be served by duplicating the miscellaneous order of Darwin's verses, and sections I to IV are consequently arranged solely by the numbers involved. Within each section the entries are tabulated in the order in which they appear in the poem, with the canto and initial line number given in parentheses.

Turning to the first section of the table, where the numbers of the sexes are equal, we see that Darwin depicted a wide range of possible situations encompassed by courtship and marriage. His opening scene concerning the canna lily is significant in that it shows the couple in an idealized, romantic light: the female is a "timorous beauty," fragile and tremulous, unaccustomed to the British climate, dreading the "rude blast of Autumn's icy morn"; the male is defensive and tender in his actions, clasping his bride in his arms. The reader is invited to see this as a love match, against which situations and behavior in the rest of the poem can be measured.

The following verses describe other forms of relationship, indicating that Darwin was well aware of the wide range of feelings that draw or hold people together. Of the married state itself, his images amply reflect what Lawrence Stone has called the companionate marriage, in which the relations between the sexes depended on a greater sense of equality and sharing than was common previously. Though Stone's taxonomy has been strongly criticized in recent years for its failure to cover fully the realities of marriage and family life in the seventeenth and eighteenth centuries, it perhaps remains a valid concept in discussing a possible *image* of marriage in Darwin's time.[47] These married women (or those who were otherwise possessed of only one partner) were described by Darwin in terms conventional to his time: they were "impatient" for their lovers, "playful," "chaste," "gentle," and "blooming"; they sought "talismans" to charm their husbands, or sang of their "secret love."

Extending the range of these conventional images, Darwin also mentioned in passing women with marital problems. One woman (*Ulva*) seeks her long-gone husband by sailing over the ocean, another is betrayed by a clandestine child (*Osmunda*). But among Darwin's characterizations of partnership some ideas that might have been expected on the strength of reading plays or novels of the time are missing: material benefits or possible financial incentives for marriage are never mentioned in the verses; divorce or separation hardly appears (although mutual dislike is represented by the plant *Cupressus*,[48] portrayed as a couple who share the same roof but occupy separate beds); adultery (apart from *Osmunda*) does not feature, either. Of course, it was hardly Darwin's intention to write of real life in the full sense. The point for historians here is rather that the presence or absence of certain features of eighteenth-century existence indicates just how completely Darwin was using the idealized pictures of his time in describing human relationships.

The next section of the table shows Darwin's descriptions of situations in which a single woman (pistil) coexists with more than one male (stamen). With small numbers of males, from say two to four, the female is shown by Darwin not as a wife this time but as a helpmate or associate, or as a figure not necessarily

[47] Lawrence Stone, *The Family, Sex, and Marriage in England, 1500– 1800* (London: Weidenfeld & Nicolson, 1977), pp. 325–404; and Linda A. Pollock, *Forgotten Children: Parent-Child Relations from 1500 to 1900* (Cambridge: Cambridge Univ. Press, 1983).

[48] Darwin, *Loves of the Plants*, canto 1, lines 73–74 (p. 8).

needed by the men at all, who may have other bonds such as those of scholarship
or brotherhood to support their personal life.

Toward the five-to-six mark, Darwin became more explicit about female sexuality and described the woman with this number of suitors as being seductive or
wanton in her charms. There is something of the sense of polite comedy or the
stage plot in this, for at a certain point in the story his readers would expect a
new kind of "character" to enter. In the poem, as in contemporary drama, the
scene was set for the entrance of a very different sort of woman. Like *Meadia,*
the American cowslip, she was a hoyden:

> Meadia's soft chains five suppliant beaux confess,
> And hand in hand the laughing Belle address;
> Alike to all, she bows with wanton air,
> Rolls her dark eye, and waves her golden hair.[49]

What counts here is not so much the predictable terms in which the "laughing
Belle" is described but the exact moment at which she appears. The species
Melissa and *Trapa,* in which the single pistil has four male associates, were not
characterized as wanton. *Vitis,* also with five males, and *Gloriosa,* with six, were
"seductive harlots." The transition from what might be called "acceptable" to
"promiscuous" behavior hence takes place at a ratio somewhere around five to
one, a point of view remarkable even in the eighteenth century for its perception
of female sexual activity as an essentially "natural" phenomenon.

If she is not described as a houri or a flirt, the woman with so many males is
shown as a person needing protection, with the males supplying the protection
rather than being the objects from which the lady needs to be saved. *Digitalis,*
the foxglove, invokes this kind of description: she has gifts of healing that are
preserved and treasured by her male companions, in order—in Darwin's lines at
least—to restore another, dropsical man to health:

> Divine Hygiea, from the bending sky
> Descending, listens to his piercing cry;
> Assumes bright Digitalis' dress and air,
> Her ruby cheek, white neck, and raven hair;
> Four youths protect her from the circling throng,
> And like the Nymph the Goddess steps along.—
> O'er him she waves her serpent-wreathed wand,
> Cheers with her voice, and raises with her hand,
> Warms with rekindling bloom his visage wan,
> And charms the shapeless monster into man.[50]

The female who is catalogued with eight or more males, however, leaves this
divalent imagery behind and takes on unambiguous metaphors of power and
command, being pictured as a saint, a reigning sovereign, a sorceress, a proto-
industrialist mixing vermillion dyestuffs, a priestess, and so on, through the Linnaean classes up to that of Icosandria, with twenty stamens (beyond which Linnaeus does not direct botanists to count), and on to Polyandria, where there are
from twenty to a hundred stamens in the same flower with the pistil. In this group

[49] *Ibid.,* canto 1, lines 61–64 (p. 6).
[50] *Ibid.,* canto 2, lines 419–428 (pp. 78–79).

there is a stern Amazonian beauty, the *Arum* or cuckoopint, who "trails her long lance, and nods her shadowy plumes," while

> Wolves, bears and pards forsake the affrighted groves,
> And grinning Satyrs tremble as she moves.[51]

And an inspired Pythian priestess, the "Lauro-cerasus" or cherry laurel:

> With maniac step the Pythian Laura moves;
> Full of the God her labouring bosom sighs,
> Foam on her lips, and fury in her eyes,
> Strong writhe her limbs, her wild dishevel'd hair
> Starts from her laurel-wreath, and swims in air.—
> While twenty Priests the gorgeous shrine surround
> Cincture'd with ephods, and with garlands crown'd,
> Contending hosts and trembling nations wait
> The firm immutable behests of Fate.[52]

Other females are seen as fairy sovereigns pledged to virginity, as leaders of a sprightly troop of choristers, and so forth, as indicated in the table.

Section III shows one male coexisting with two females, the only instance in botany of there being more pistils than stamens. Darwin's metaphor, which presents two women gently caring for one beautiful youth, is devoid of sexuality.[53] Powerful conventions govern the depiction of the women here, conventions running through the literature and drama of the period, in which women are seen as items of property, competed for but not duplicated. Rather than envisage the assemblage of one man and two women in a sexual context, Darwin chose to locate it in a neutral, possibly even familial relationship that secured the principals from any erotic connotation.

In the poem Darwin also worked through those parts of the Linnaean system where there are multiples of each sex. The fourth section of the table indicates that he was perhaps more interested in showing pastoral or mythological scenes than in characterizing individual men or women, but he still deployed images derived from the world of morals, as in his account of *Silene,* the catchfly, with three females and ten males in each flower, whose sticky nets for catching flies are likened to the deadly activities of three "dread sirens," skilled in destruction.[54] The poppy is seen as a sultry oriental queen surrounded by a helpless throng of enchanted young people, all made languorous and empty by continued opium eating.[55] Others are variously harlot-nymphs or gentle shepherdesses, blushing maids or sisters, with no apparent logic behind the imagery beyond the botanical properties of the plants themselves yet still presenting a fine succession of pictures of women in society.

Darwin's final scene endeavored to place all these varied pictures into a single frame. His choices of setting and the imagery used were evidently intended to provide the key to the way in which he hoped the poem would be considered and

[51] *Ibid.,* canto 4, lines 190, 205–206 (pp. 148–149).
[52] *Ibid.,* canto 3, lines 40–48 (pp. 92–93).
[53] *Ibid.,* canto 1, lines 45–50 (p. 4).
[54] *Ibid.,* canto 1, lines 131–142 (p. 14).
[55] *Ibid.,* canto 2, lines 265–290 (pp. 69–70).

remembered, perhaps even a veiled reference to the metaphysical position embodied within it. Darwin chose to describe the fertilization of plants belonging to
the Linnaean class Polyandria as if it were a Tahitian marriage ceremony, invoking the idea that human bonding was no more sacred than the purely physical
meetings of stamen and pistil. Based on a close reading of the scientific and
popular literature emanating from James Cook's and Joseph Banks's famous encounter with South Sea Island life, and steeped in an idealized view of Tahitians
as untarnished natural beings whose society functioned admirably according to
what Darwin perceived as 'natural,' non-Christian behavior,[56] Darwin's anecdote
served to remind readers that his poem was constructed around the idea that
human actions in the realm of love were, in reality, natural phenomena and not
owing to attributes bestowed by a creator. Darwin wrote of the Areoi people
engaging in one great nuptial ceremony:

> A hundred virgins join a hundred swains,
> And fond Adonis leads the sprightly trains;
> Pair after pair, along his sacred groves
> To Hymen's fane the bright procession moves;
>
> . . .
>
> As round his shrine the gaudy circles bow,
> And seal with muttering lips the faithless vow,
> Licentious Hymen joins their mingled hands,
> And loosely twines the meretricious bands.—
> Thus where pleased Venus, in the southern main,
> Sheds all her smiles on Otaheite's plain,
> Wide o'er the isle her silken net she draws,
> And the Loves laugh at all, but Nature's laws.[57]

This was the overall image intended to be left in the mind of the reader. Such
pronounced naturalism did not, however, lead Darwin to prescribe a sexual free-
for-all in England; complete subjugation to the animal (and plant) passions was
characteristic only of animals and plants, not humans. But he wished to set out
the materialist point that human love and feelings about sexual relations were
ultimately rooted in physiology not in Christianity. This idea was also to lie at the
heart of his deistic—possibly even atheistic—philosophies of nature and society
in *The Economy of Vegetation* and *The Temple of Nature*.

WOMEN IN ARCADY

Although Darwin hoped only to make Linnaean ideas about plant sexuality clear
and attractive to readers by modeling it on human society, he nevertheless provided a catalogue of his own social world which deserves wider historical attention. In his poem Darwin listed a procession of female images ranging from virtuous brides and tender mothers to attentive sisters, nymphs, and shepherdesses.
Laughing belles and wily charmers were followed by queens and amazons. Despite the robust sexuality and obvious insistence on 'natural' behavior, the overall impression is of an 'artificial' world far removed from real life. There are no

---

[56] Walter Veit, ed., *Captain James Cook: Image and Impact: South Sea Discoveries and the World
of Letters* (Melbourne: Hawthorne, 1972); and O'Brian, *Joseph Banks* (cit. n. 22); see also Harold B.
Carter, *Sir Joseph Banks 1743–1820* (London: British Museum [Natural History], 1988).
[57] Darwin, *Loves of the Plants*, canto 4, lines 287–390, 399–406 (pp. 164–165).

doubts or struggles with conscience in *The Loves of the Plants*. There are no sexual victims, no rape or violence of the kind found in Ovid or, for that matter, in some of Linnaeus's work.[58] There is little sexual jealousy, no murder, vice, abortion, prostitution, death or disease, no heartbreaks or abandoned lovers— except for the victims of the harlot band—and only one instance of a clandestine child. All is clean, healthy, and pastoral. Rather like the images of women on the Portland vase, itself a topic of much interest to Darwin, or in the frontispieces to each of the poems that together go under the title *The Botanic Garden* (Figs. 3 and 4), the world that Darwin was describing was the world imagined by classically educated gentlefolk of the late eighteenth century, in which Flora and Cupid gaily exchange the tools of their trade (see Fig. 4) and reality is temporarily forgotten in a rustic poetic paradise.[59]

Darwin took pains to explain some of his intentions in the prose interludes between cantos. There he put forward the theory that poetry consists of a series of pictures set in a landscape affording graceful and apposite imagery—a theory that he evidently followed closely in *The Loves of the Plants*. In the "proem" he explained:

> Whereas P. Ovidius Naso, a great Necromancer in the famous Court of Augustus Caesar, did by art poetic transmute Men, Women, and even Gods and Goddesses, into Trees and Flowers; I have undertaken by similar art to restore some of them to their original animality, after having remained prisoners so long in their respective vegetable mansions; and have here exhibited them before thee. Which thou may'st contemplate as diverse little pictures suspended over the chimney of a Lady's dressing-room, connected only by a slight festoon of ribbons.[60]

The poet, in Darwin's view, writes principally to the eye, in the sense that he or she creates pictures in the imagination.[61] Abstract thoughts and complex trains of reasoning that cannot be visualized are best expressed in prose writings; and Darwin followed his own recommendations by confining the philosophical comments and explanations of his botanical poetry to lengthy prose footnotes and interludes, and to his meticulously ordered scientific writings, the *Zoonomia* and *Phytologia*. Prose was the vehicle for what Darwin called the "strict analogies of philosophy" as opposed to the looser analogies with which he and other versifiers "dress out the imagery of poetry."[62] In setting *The Loves of the Plants* in a garden and personifying flower parts he deployed instantly recognizable and attractive metaphors, providing a mental landscape that stimulated readers to create their own personal pictures.

Darwin's description of the cantos as largely a display of poetic pictures makes

---

[58] See Karl Robert van Wikman, *Lachesis and Nemesis: Four Chapters on the Human Condition in the Writings of Carl Linnaeus* (Scripta Instituti: Donneriani Aboensis, 4) (Stockholm: Almquist & Wiksell, 1970); and Wolf Lepenies, "Linnaeus's *Nemesis divina* and the Concept of Divine Retaliation," *Isis*, 1982, 73:11–27; see also Sten Lindroth, "The Two Faces of Linnaeus," in *Linnaeus: The Man and His Work*, ed. Tore Frängsmyr (Berkeley/Los Angeles: Univ. California Press, 1983).

[59] Darwin discussed the Portland, or Barberini, Vase at length in the *Economy of Vegetation* (cit. n. 5), Additional Notes, pp. 53–59. In his opinion the figures represented scenes from the Eleusinian mysteries, consisting of an emblem of death in the first compartment and of immortal life in the second. The relief on the bottom of the vase he believed to be of a priestess, placed there as an emblem of secrecy or caution to the initiated. See also Irwin Primer, "Erasmus Darwin's *Temple of Nature*: Progress, Evolution, and the Eleusinian Mysteries," *J. Hist. Ideas*, 1964, 25:58–76.

[60] Darwin, *Loves of the Plants*, p. vi.

[61] See esp. McNeil, "Scientific Muse" (cit. n. 3).

[62] Darwin, *Loves of the Plants*, p. i.

him here the poetic equivalent of a genre painter or, more appropriately, a landscape gardener—like Humphry Repton, who created an air of natural harmony, balance, and beauty by encouraging the landscape to reveal its features from artfully selected viewpoints that "frame" the resulting "picture." The theory of the picturesque followed by Uvedale Price and modified somewhat by Repton was expressly intended to improve on natural scenery through a study of the best landscape pictures,[63] although Repton always maintained that nature and art should be recognized as different entities, following distinct sets of rules. Repton consciously used this distinction to create the pleasant tricks and confusions that emerge from a careful juxtaposition of real and cultivated nature.[64] His desire to hide the boundary of a park or lawn and the facility with which he maintained the illusion that grounds extended in every direction from a house show a commitment to the creative involvement of the imagination of the viewer that mirrors the philosophy of mind held by Erasmus Darwin: the two men provided the topography, the skill, and the imagery in order to arouse in the mind of the spectator a train of analogies that created satisfying pictures.[65] For them, as for others of the time, art was indeed artful; gardens and landscapes were graceful artifices that displayed nature at her best; cultivation did not signify the drudgery of the farmyard but rather the fostering of the gentle world of polite society. Darwin's poem, like Repton's sumptuously illustrated landscape designs, presented a series of views in which the subjects were carefully arranged to give the desired naturalistic and picturesque effect.[66]

The women that Darwin created were therefore entirely appropriate for the pastoral setting he envisaged. With one exception, there are no intellectual women in Darwin's verses, no educated poetesses like Anna Seward; no artists like Angelica Kaufmann (who is only mentioned in passing in one of the prose interludes);[67] no one like Maria Edgeworth, well known personally to Darwin as a girl;[68] no Mary Wollstonecraft or Madame de Staël. Even though there is some account of women with power or special knowledge, and of certain intrepid females such as the lady balloonist (*Carlina,* the thistle) and the nymph who turns the waterwheels for the cotton manufacturing industry on the river Derwent,[69]

---

[63] The key text here was William Gilpin, *Observations Relative to Picturesque Beauty* (London, 1786). See Ann Bermingham, *Landscape and Ideology: The English Rustic Tradition 1740–1860* (London: Thames & Hudson, 1987), for a full bibliography.

[64] Humphry Repton, *Observations on the Theory and Practice of Landscape Gardening* (London, 1805); and Repton, *Variety. A Collection of Essays* (London, 1788); see also *The Red Books of Humphry Repton: Facsimiles of the Red Books for Sherringham in Norfolk, Antony House in Cornwall, Attingham in Shropshire,* 4 vols. (London: Basilisk Press, 1976).

[65] There is a sizable literature on the interconvertibility of the pictorial, poetic, and landscape arts, and on the belief common during the 1780s and 1790s that this was an important philosophical movement. See esp. John Barrell, *The Idea of Landscape and the Sense of Place* (Cambridge: Cambridge Univ. Press, 1973); and John Dixon Hunt, *The Figure in the Landscape: Poetry, Painting, and Gardening during the Eighteenth Century* (Baltimore/London: Johns Hopkins Univ. Press, 1976).

[66] In *Loves of the Plants,* p. 40, Darwin wrote: "I am only a flower-painter, or occasionally attempt a landskip; and leave the human figure with the portraits of history to abler artists." Darwin's friend Anna Seward knew Humphry Repton personally; see Repton, *Variety* (cit. n. 64).

[67] Darwin, *Loves of the Plants,* pp. 45, 49. Three other women artists are mentioned: Mrs. Delaney, who prepared paper mosaic pictures of flowers according to Linnaeus's system (pp. 61–62); Mrs. North, the flower painter (p. 62); and Miss Emma Crewe, who drew the frontispiece (Fig. 4) and is praised by Darwin, pp. 70–71.

[68] *Letters of Erasmus Darwin,* pp. 338–339.

[69] Darwin, *Loves of the Plants,* canto 2, lines 7, 85 (pp. 52–55, 56–58).

there are no descriptions of intelligent, literary women of the kind prominent in his own life and in late eighteenth-century society as a whole.

The one apparent exception is the narrator herself, the goddess of botany, the didactic lecturer who speaks the whole poem. She is not only an expert botanist but also displays a deep and varied knowledge of contemporary science and the world about her. By choosing such a voice for his work, Darwin apparently demonstrated his genuine regard for educated women. But in fact this knowledgable goddess would not have been perceived in this way, for it was impossible for any reader of the time to have believed that the author was indeed a woman. Darwin may have been free to write a poem about sexual conduct, but his

Figure 4. *"Flora at play with Cupid,"* designed by Emma Crewe and engraved by S. Alken, 1791. The frontispiece to Erasmus Darwin, The Loves of the Plants, *Part II of* The Botanic Garden *(London, 1792). Reproduced courtesy of the Wellcome Institute Library.*

counterpart would not have dared to do the same. The disguise was clearly transparent: the botanic muse turns out to be a man.

The absence of educated women in the poem does not in itself betoken Darwin's dislike of them or any deeper views on sexual inequalities in nature. For one thing, his poetic Arcady had little room for intellectuals of either sex: if there was no Lady Hester Stanhope, there was no Dr. Johnson either; both would have had a hard time masquerading as shepherds. Darwin's intentions and the deliberate frame in which he cast his images provided only a limited, preconceived range of metaphor into which certain categories would not be allowed.[70] Furthermore, we know, for example, that Darwin argued for the better education of women, devised a progressive and liberal scheme for a girls' boarding school,[71] and endorsed an extraordinary Pygmalion plan carried out by his friend Thomas Day to educate a foundling girl to such a pitch that she would make a perfect Mrs. Day.[72] Yet (as this last project suggests) like most of the men of his time and social position who advocated a better education for women, Darwin saw it primarily in terms of the benefit to men. Education should produce "a good daughter, a good wife, and a good mother, that is, an amiable character in every department of life." Moreover, the female character "should possess the mild and retiring virtues rather than the bold and dazzling ones; great eminence in almost any thing is sometimes injurious to a young lady."[73] Entirely in accord with other male writers on women's education, Darwin wished to enlarge the world that women negotiated, yet the choices he wished women to make were still circumscribed and favored the maintenance of contemporary society and, in particular, the status quo of contemporary men. Similarly, the images in *The Loves of the Plants*, for all Darwin's progressive views, remained deeply polarized between the chaste, blushing virgin and the seductive predatory woman, the modest shepherdess and the powerful queen.

BOTANY FOR GENTLEMEN

In the end Darwin's personal attitude to women or their emancipation is less significant than the limited and entirely traditional nature of his images, which reflect more generally held views about women and the relations between the sexes. Given that Darwin was personifying a particular scientific classification scheme in order to make it attractive and easily memorable, it is only to be expected that he would choose metaphors instantly recognizable, familiar, and memorable in their own right. He presented pictures of women that were for many people reassuring stereotypes: the images that his contemporaries—both

---

[70] James Venable Logan, *The Poetry and Aesthetics of Erasmus Darwin* (Princeton Studies in English, 15) (Princeton, N.J.: Princeton Univ. Press, 1936), pp. 46–92; and Hassler, *The Comedian as the Letter D: Erasmus Darwin's Comic Materialism* (The Hague: Martinus Nijhoff, 1973).

[71] Erasmus Darwin, *A Plan for the Conduct of Female Education in Boarding Schools* (Derby/London, 1797). The school was run by Darwin's two natural daughters, Susan and Mary Parker; see *Letters of Erasmus Darwin*, pp. 270–271; and E. Posner, *Darwin and the Sisters Parker* (cit. n. 46).

[72] Seward, *Memoirs* (cit. n. 30), pp. 35–51. The girl, called Sabrina, did not rally to Day's Rousseauean ideals and was eventually placed in a boarding school in Sutton Coldfield, whence she married a friend of Day's. See also Keir, *Life and Writings of Thomas Day* (cit. n. 6), pp. 27–29.

[73] Darwin, *Female Education* (cit. n. 71), pp. 47, 10. The plan is explained as being designed to equip girls for life in polite society, especially if their male support should fail (pp. 52, 55).

male and female—were accustomed to finding in the romantic novels, pastoral poetry, and dramatic arts patronized by the landed gentry. It is in this sense that one might suggest that Darwin's scheme was basically patriarchal and that his botany was botany for gentlemen, rather than for ladies. Deliberately directed to "lady readers," *The Loves of the Plants* elaborated a series of views designed to reinforce women's roles as sexual partner, friend, wife, and mother, promoting the view that these stereotypes were in some sense "natural," built into the physiology or structure of women. Intentionally or not, the poem conveys a masculine view of what was considered appropriate feminine behavior.

To some extent it is therefore possible to locate Darwin's poem in the dark transformations in sexual feelings that Michel Foucault describes, from the "bright day" of seventeenth-century sexuality to the "monotonous nights" of the Victorian bourgeoisie.[74] For Foucault, it is the things left unsaid that point the way to a deeper understanding of the views expressed in a text, and such an approach is clearly helpful in assessing Darwin's position on sexual relations and women's role in society. *The Loves of the Plants* can be seen as avoiding those areas where contemporary fears might have jolted or outweighed the overall ideas being presented; as expelling unwanted forms of behavior; and as ignoring the physical and emotional results of sexual activity in the real world. Darwin's catalogue of the behavior of the plants can signify a form of sexual regulation among humans. Certainly it represents a particular point in the complicated process of "naturalizing" the way that society considered the body, particularly the female body, and of rethinking the relations between god and nature, a process that took place gradually over the early modern period.[75]

Darwin's contribution to this process was not, however, based on fear, as a reading of Foucault might lead some to suggest. It is true that new studies reveal how his mentor Linnaeus may have exorcised his fears about the body by putting sex at the heart of his classification system and thereby rendering it neutral, or at least turning it into a "scientific" and hence more manageable commodity.[76] But there was a world of difference between Linnaeus's and Darwin's personal life, the one a believer in divine retribution and a fierce, avenging, moralistic God, the other a liberal, freethinking deist with an obvious interest in the opposite sex. Rather than feeling anxious about sexual relations, Darwin undoubtedly relished them. Both his marriages were happy ones, by all accounts, and certainly fruitful: Darwin had three surviving (out of five) children by Mary Howard and seven by Elizabeth Pole. Nor did he, in the interval between marriages, feel any need to remain celibate. Living with Mrs. Parker, a widow of Lichfield, he fathered two natural daughters who continued to reside with him until fully grown. As an

[74] Michel Foucault, *The History of Sexuality*, trans. Robert Hurley, Vol. I: *An Introduction* (London: Allen Lane, 1979).

[75] See Ruth Bleier, *Science and Gender: A Critique of Biology and Its Theories on Women* (Oxford: Pergamon Press, 1984); Brian Easlea, *Science and Sexual Oppression: Patriarchy's Confrontation with Women and Nature* (London: Weidenfeld & Nicolson, 1981); Evelyn Fox Keller, *Reflections on Gender and Science* (New Haven, Conn.: Yale Univ. Press, 1985); Carol MacCormack and Marilyn Strathern, eds., *Nature, Culture and Gender* (Cambridge: Cambridge Univ. Press, 1980); and Ludmilla Jordanova, "Naturalizing the Family: Literature and the Bio-Medical Sciences in the Late Eighteenth Century," in *Languages of Nature* (cit. n. 3).

[76] Lepenies, "Linnaeus's *Nemesis divina*"; Lindroth, "Two Faces of Linnaeus" (both cit. n. 58); and Delaporte, *Nature's Second Kingdom* (cit. n. 7), pp. 139–140.

unidentified obituarist remarked in 1803, Darwin could never forsake the charms of Venus.[77] He fits more happily into the British tradition of "rational" thought, keen to disclose the basic "laws of nature," to show the identity between plants, animals, and humans; and to demonstrate that all living beings were governed by the same physiological processes and indeed, in Darwin's case, how they were all linked together by one unbroken evolutionary chain. The classification of women that emerges from his classification of plants is important precisely because Darwin took a range of female feelings and activities and deliberately lifted them out of the world of traditional Western morals in order to relocate them in nature, represented in his poetry by the non-Christian world of antiquity and the island of Tahiti and made explicit by his use of the imagery of a botanic garden. He made sexuality a normal feature of human life, love a "natural law."

By personifying plants, Darwin was therefore offering an interpretation of nature that operated on many levels. At its most obvious, *The Loves of the Plants* encouraged readers to think of plant species as sophisticated living organisms that enjoyed all the benefits of human existence, most notably sexuality. Even if for nothing else, *The Loves of the Plants* was significant in the history of botany for its emphatic restatement of Linnaeus's doctrine of the sexuality of plants and for bringing this concept to the forefront of natural science during the 1790s. Few readers—James Edward Smith, Joseph Banks, Robert Thornton, Samuel Taylor Coleridge, William Wordsworth, and Charles Darwin among them—could afterwards forget that garden flowers had a sex life. This personalized, sexualized picture remained vivid through the early years of the nineteenth century: so vivid that many of the efforts of women botanists such as Priscilla Wakefield were directed to rendering the subject in more neutral terms, suitable for the new wave of feminine enthusiasts emerging in the pre-Victorian period.[78] Successfully bowdlerized and sentimentalized, the image of plants as people lived on well past the turn of the century, particularly in literature directed toward women and children such as, for example, the well-known Flower Fairy Books, first published by Cicely Mary Barker in 1923.

Darwin also encouraged readers to see in his work a statement of the interconnectedness of the living world, a view first expressed in the translations of Linnaeus's *Species plantarum* and *Genera plantarum*: "For vegetables are, in truth, an inferior order of animals."[79] Plants were like animals because they possessed the same natural functions, different only in degree. *The Loves of the Plants* can therefore be seen as preliminary to, and closely intermeshed with, Darwin's later views on transformism and on the existence of an evolutionary chain of organisms stretching from molecules to man.

At another level entirely, Darwin's work took up views about human sexual and social behavior common to his personal intellectual circle and more generally to those of his class and wealth, and expressed them through the various images that the idea of personification generated. While it was not Darwin's intention to make great philosophical play with his metaphor, one consequence of this ex-

[77] King-Hele, *Erasmus Darwin* (cit. n. 2), p. 14.

[78] Ann B. Shteir, "Priscilla Wakefield's Natural History Books," in *From Linnaeus to Darwin: Commentaries on the History of Biology and Geology,* ed. Alwyne Wheeler and James H. Price (Papers from the Fifth Easter Meeting of the Society for the History of Natural History, 28–31 March 1983) (London: Society for the History of Natural History, 1985).

[79] See Botanical Society of Lichfield, *Families of Plants* (cit. n. 20), p. xix.

tended analogy was that as it became easier to think of plants as people so it became possible to think of human beings as plants. Like all metaphors in the history of science,[80] Darwin's idea of the personification of plants allowed the fruitful interplay of ideas between one realm (the human) and another (the botanical). We know we are not plants, but it is both amusing and informative to think about why we are not. Darwin invited his readers to consider whether humans were solely natural beings or whether there were also higher spiritual qualities inherent to mankind. Darwin's pictures revealed that he believed only in nature, and the poem's organizing structure of a botanic garden served to allude to the possibility of a world without the Christian church, a view made more explicit in *The Economy of Vegetation,* issued only two years after *The Loves of the Plants,* and in *The Temple of Nature.* Part of this manipulation and interplay of images was that women were plainly seen as "natural" beings, their function being primarily reproductive, their behavior seen through a wide range of stereotypes that themselves were presented as "natural" roles.

Linnaeus's classification scheme was thus being used to project an interconnected nexus of personal and communal views, commitments, and judgments, many of which were subsequently worked out by Darwin in his evolutionary verse and other writings, but which never came together again in quite the same evocative combination of philosophical and social values. Darwin turned the sexual system of Linnaeus to his own purposes and made it embody his metaphysical beliefs, his scientific commitments, his social world, and the intellectual preoccupations and assumptions of the wealthy, freethinking, professional class to which he belonged—and also those of his gender. So although Ann Shteir and David Allen are quite right to exhort us to think of the study of plants as a particularly feminine, female, occupation,[81] it would be a pity, in the continuing search for "Linnaeus's daughters," to overlook this other kind of botany, botany for gentlemen.

---

[80] See esp. Andrew E. Benjamin, Geoffrey N. Cantor, and John R. R. Christie, eds., *The Figural and the Literal: Problems of Language in the History of Science and Philosophy, 1630–1800* (Manchester: Manchester Univ. Press, 1987); Stanley Hyman, *The Tangled Bank* (New York: Atheneum, 1962); Jordanova, ed., *Languages of Nature* (cit. n. 3); Thomas, *Man and the Natural World* (cit. n. 43); and Robert M. Young, "Darwin's Metaphor: Nature's Place in Victorian Culture (Cambridge: Cambridge Univ. Press, 1985).

[81] Ann B. Shteir, "Linnaeus's Daughters: Women and British Botany" in *Women and the Structure of Society: Selected Research from the Fifth Berkshire Conference on the History of Women,* ed. Barbara J. Harris and JoAnn K. McNamara (Durham, N.C.: Duke Univ. Press, 1984), pp. 67–73; and David E. Allen, "The Women Members of the Botanical Society of London, 1836–1856," *British Journal for the History of Science,* 1980, *13*:240–254.

Goethe's Botany

Lessons of a Feminine Science

By Lisbet Koerner

I N 1831 EUROPE'S BEST-KNOWN MAN OF LETTERS, the eighty-two-year-old Johann Wolfgang von Goethe, coyly describing himself as a "middle-aged man of some reputation as a poet," began to rewrite his essay entitled "The Author Relates the History of His Botanical Studies." The central claim of this piece, first drafted in 1817, is astonishingly audacious. Goethe not only presents his life history as a recapitulation of contemporary developments in the discipline of botany. He also claims *personal* responsibility for this coincidence of one man's life with the history of the science. In his youth, he recalls, people interested themselves in tulips and other spring bulbs. "And if in addition to the usual fruit varieties, the apricots, peaches and grapes also turned out well, it was occasion enough for old and young to rejoice. No thought was given to exotic plants, still less to the idea of teaching natural history in the schools."[1] Goethe's own first interest in natural history was

This essay is dedicated to Moti Feingold and Juliet Fleming, with affection and gratitude.

[1] For the quotations see Johann Wolfgang von Goethe, "The Author Relates the History of His Botanical Studies" (hereafter cited as **Goethe, "History"**), in *Goethe's Botanical Writings*, trans. Bertha Mueller (1952; Woodbridge, Conn.: Ox Bow Press, 1989) (hereafter cited as **Goethe, Botanical Writings**), pp. 149–168, on pp. 149, 150; for the German see Goethe, "Geschichte meines botanischen Studiums" (hereafter cited as **Goethe, "Geschichte"**), in Goethe, *Werke*, 13 vols., Vol. 1: *Goethes naturwissenschaftliche Schriften*, ed. Rudolf Steiner (Weimar: Hermann Bohlau, 1890–1904) (hereafter cited as **Goethe, Naturwissenschaftliche Schriften**), pp. 63, 64. "Geschichte meines botanischen Studiums" was first published, in two parts, in 1817 in Goethe's pamphlet series *Zur Naturwissenschaft überhaupt, besonders zur Morphologie: Erfahrung, Betrachtung, Folgerung, durch Lebensereignisse verbunden* (Stuttgart/Tübingen: Cotta, 1817–1824) (hereafter cited as **Zur Morphologie**). It was then expanded and republished in an 1831 edition of *Versuch über die Metamorphose der Pflanzen*. I use Steiner's edition in this article. See also the shorter version of this personal history in Goethe, *Die Schriften zur Naturwissenschaft: Vollständige mit Erläuterungen verschene Ausgabe, heraugaben im Auftrag der deutschen Akademie der Naturforscher Leopoldina* (Weimar: Hermann Bohlaus, 1947–) (hereafter cited as **Goethe, Schriften zur Naturwissenschaft**, with series, volume, and year of publication added as necessary), I:9 (1954), pp. 15–19. For Goethe's claim of personal responsibility for the coincidence of his life with the history of botany see Goethe, "History," p. 151 (Goethe, "Geschichte," p. 66).

likewise a "mere uncertain, unsatisfied pondering."[2] Over time his botany matured, as did the discipline as a whole. But, Goethe argued in this autobiographical sketch, his development did not parallel other people's achievements. Rather, his *own* study of botany had transformed the discipline. Looking back from 1817, he identified as the key moment in this transformation his 1790 *Attempt to Explain the Metamorphosis of Plants*—"after almost thirty years finally accepted in scientific circles."[3]

At other times Goethe doubted his botanic theories. He eagerly searched for signs that the scientific community accepted them. For example, in the preface to an 1831 edition of *The Metamorphosis of Plants,* he listed all the scientists who had, however briefly, quoted him. And when Ludwig Reichenbach's 1828 introduction to Goethean botany was described in the May 1830 issue of *Bulletin des Sciences Naturelles* as "botanique pour les dames, les artistes et les amateurs des plantes, contenant une exposition du regne végétal dans ses métaphores et une instruction pour étudier la science et pour former des herbiers," Goethe was mortified, and he suspected that the metamorphosis of *metamorphosis* into *metaphor* was a "sarcastic allusion to the German manner of handling scientific subjects!"[4] If so, the French wit was misplaced. As Goethe had explained in his 1817 "History," "for more than half a century I have been known as a poet. . . . But the fact that I have busily and quietly

---

[2] Goethe, "History," p. 150 (Goethe, "Geschichte," p. 64). In his autobiography, *Dichtung und Wahrheit* (which he completed in 1831, but had worked on since 1809), Goethe describes his father's keen interest in gardening (see Goethe, *Schriften zur Naturwissenschaft,* II:9A [1977], p. 260) and how in 1765–1766 among his acquaintances "die Namen Haller, Linné, Buffon hörte ich mit großer Verehrung nennen" (*ibid.,* p. 262). *Dichtung und Wahrheit* also reports that Goethe attended medical lectures and socialized with medical students while in Straßburg in 1770 and 1771 (*ibid.,* p. 263). For the standard text of the autobiography see the "Hamburger" edition: Goethe, *Werke,* Vol. 9: *Aus meinem Leben: Dichtung und Wahrheit,* ed. Lieselotte Blumenthal and Erich Trunz (Autobiographische Schriften, 1) (Munich: Beck, 1974).

[3] Goethe, "History," p. 167 (Goethe, "Geschichte," p. 87). The *Metamorphosis of Plants* was first published as *J. W. von Goethe Herzoglich Sachsen-Weimarischen Geheimenraths Versuch die Metamorphose der Pflanzen zu erklären* (Gotha: Carl Wilhelm Ettinger, 1790). It was soon pirated, then legitimately reprinted as *Versuch über die Metamorphose der Pflanzen: Übersetzt von Friedrich Soret, nebst geschichtlichen Nachträgen* (Stuttgart, 1831). For its publication history see Goethe, *Schriften zur Naturwissenschaft,* II:9A, p. 534. For the standard German edition see *ibid.,* I:9, pp. 23–61. For a useful general discussion of Goethe's understanding of plant metamorphosis, particularly the role of the leaf, see *ibid.,* II:9A, pp. 602–605, editors' annotations.

Today Goethe's greatest achievements in science are considered his discovery of the intermaxillary bone in humans (J. W. von Goethe to J. G. Herder, 27 Mar. 1784, announcing the intermaxillary bone, *ibid.,* p. 602) and his coinage of the word *morphologie* (Goethe to Friedrich Schiller, 25 Sept. 1796, using the word for the first time, *ibid.,* II:9B [1986], pp. 419–420). Goethe's concept of metamorphosis was an all-encompassing principle of the natural world, as evidenced, e.g., by the 1796 manuscript *Die Metamorphose der Insekten, besonders der Schmetterlinge, wie auch ihre übrige Eigenschaften und Oekonomie betreffend;* on this see Goethe, *Schriften zur Naturwissenschaft,* I:10 (1964), pp. 176–193, and II:9B, pp. 447–456. For the use of the term *metamorphosis* in the 1790s see *ibid.,* II:9A, p. 539, editors' annotations.

[4] *Bulletin des Sciences Naturelles,* May 1830, 5:268 (see also Goethe, *Schriften zur Naturwissenschaft,* I:10, pp. 315–316); for the complaint see Goethe, "The Influence of My Publication," in Goethe, *Botanical Writings,* p. 210 (Goethe, from the preface to the 1831 edition of *Metamorphosis:* "Wirkung meiner Schrift: Die Metamorphose der Pflanzen und weiter Entfaltung der darin vorgetragenen Idee," in *Naturwissenschaftliche Schriften,* p. 214). See also the earlier version, the 1830 "Wirkung dieses Schrift und weitere Entfaltung der darin vorgetragenen Idee," in Goethe, *Schriften zur Naturwissenschaft,* I:10, pp. 297–318. On Ludwig Reichenbach (1793–1879), professor of natural history and medicine at the Dresden medical academy and director of the Dresden museums and botanical gardens, see Goethe, *Botanical Writings,* p. 207. Reichenbach's goal was a natural system of plant classification. The *Bulletin* referred to his *Botanik für Damen, enhaltend eine Darstellung des Pflanzenreichs in seiner Metamorphose* (Leipzig, 1828). "Der Verfasser widmet sein Werk Frauen, Künstlern und sinnigen Naturfreunden" and describes Goethe's as one of three schools of botany, along with those of Linnaeus and Antoine de Jussieu.

occupied myself with Nature . . . constantly and passionately pursuing seriously for-
mulated studies—this is not so generally known; still less has it been accorded any
attention." Indeed, "people were extremely astonished to find that a poet . . . could
turn for a moment from his path and in a cursory study achieve such an important
discovery [as the metamorphosis of plants]. It is to combat this mistaken idea that
the present essay has been written."[5]

If Goethe understood himself as a scientist, why did he react so anxiously to the
Bulletin's misprint—or joke? What did he mean by "the German manner of handling
scientific subjects"? Goethe helped formulate, if he also criticized, a particularly
German scientific methodology: *Naturphilosophie,* which drew on both a philosophical tradition, idealism, and an artistic movement, Romanticism.[6] In botany, the
Naturphilosophen were post-Linnaeans. They rejected the Enlightenment project of
mapping new species onto Linnaeus's self-consciously artificial taxonomy, which
groups plants according to the number, size, shape, and placement of stamens and
pistils. Instead, they hoped to identify a natural taxonomy, an aim they shared with
eighteenth-century French naturalists like Michel Adanson and Antoine-Laurent de
Jussieu.[7] But German Romantic botanists understood the road to a natural system
differently from their French counterparts. To Goethe and the *Naturphilosophen,*
plants' relations could be grasped through an understanding of the natural world that
was at once teleological and historicist. This understanding depended as much on a
scientist's personal sympathy with nature as on his systematic study of it.[8]

Importantly, this specifically German understanding of nature included "ladies,
artists, and botanic amateurs" in the scientific community. Within his methodology,
Goethe could truthfully assert both that he systematically studied nature and that the
secret of his scientific genius was that he was a dilettante. He understood his botany
as an alternate form of science that rejected the divisions between public and private,
amateur and professional, and contemplation and experience. Locating Goethe's botany between these various dyads, this essay discusses his scientific method and model
of cognition from the perspective of how he gendered his botany.

Scholars have suggested that Romantic botany, which validated amateur science
and was dialogic in its relation to the natural world, offered European gentlewomen
around 1800 a way to participate in science at a time when (it is concomitantly

---

[5] Goethe, "History," pp. 164–165 (Goethe, "Geschichte," p. 83). Goethe often felt misunderstood
in his science; see the undated fragment, probably from the late 1790s, in Goethe, *Schriften zur Naturwissenschaft,* II:9A, p. 254.

[6] Timothy Lenoir, "The Göttingen School and the Development of Transcendental Naturphilosophie
in the Romantic Era," *Studies in the History of Biology,* 1981, 5:111–205; and Lenoir, *The Strategy of
Life: Teleology and Mechanics in Nineteenth-Century German Biology* (Chicago: Univ. Chicago Press,
1989), p. 76 and *passim* on the related morphological tradition in German biology. See also Goethe to
Wilhelm von Humboldt, 22 Aug. 1806, for an ironizing critique of the *Naturphilosophen;* this is quoted
in Goethe, *Schriften zur Naturwissenschaft,* II:9B, p. 501. The critique ends with the statement "es ist
unangenehm, gerade diejenigen lassen zu müssen, die man so gern begleitete." These included Friedrich
Wilhelm Joseph von Schelling and his circle, Gustav Carus, and Lorenz Oken, as well as now-forgotten
professors at, e.g., the universities of Jena (Ludwig Reichenbach), Dresden (August Johann Georg Karl
Batsch), and Heidelberg (Franz Joseph Schelver).

[7] Linnaeus, too, shared this aim, but he understood the way to it to be through an artificial taxonomy,
allowing a universal nomenclature of botany and, hence, a scientific community with a comprehensive
language. Goethe arranged his own herbaria according to Jussieu's system. See his *Tagebuch,* 5 Feb.,
6 Feb., 9 Feb., and 9 Mar. 1800; quoted in Goethe, *Schriften zur Naturwissenschaft,* II:9B, pp. 167–
168.

[8] In this essay the term *Romantic* is used to characterize Goethe's critique of Enlightenment science
and not, obviously, his complex relation to the arts movement.

argued) they were marginalized from public roles.[9] One hypothesized cause of this process of female privatization is positivism. In turn, Goethe's botany could be hypothesized as an exception to the rule that sees modern science more as a male conspiracy than as a human achievement.[10]

Contemporary German women seem to have interested themselves more in Goethe than in academic post-Linnaeans. (So, however, did German men.) Possibly they did so because his botany rejected the boundaries between the life of the individual and that of society. Goethe's cult of individual genius promised, after all, to merge private thought and public significance. But even if it could be documented that Goethe's female audience did embrace his botany for this reason, that would not in itself be sufficient evidence that Goethe's botany was a feminist science—that it provided a sphere freeing women participants, and scientific discourse itself, from male domination. Moreover, whether or not contemporary women construed Goethe's botany as a romantic promise of personal liberation, it still failed to offer a viable alternative to nineteenth-century positivism.

Goethe celebrated a hybrid form of knowledge based on primal archetypes and on "lived experience"—what G. W. F. Hegel and Wilhelm Dilthey later termed *Erlebnis*. He believed that scientific knowledge cannot be transmitted solely through historical learning but must be lived through (*erlebt*) by each generation. The Goethean ontological certainty provided by experience and intuitive knowledge, with its concomitant experience of a liberation of self, understands itself as replacing master scripts of science with personal experience, verified through a (historically prefigured) notion of authenticity.

Within this framing assertion of intuitive certainty, Goethe postulated as a natural fact an "age-old association" between women and nature that, in the words of a latter-day adherent, "has persisted throughout culture, language, and history." Goethe's botany thus raised to a founding axiom what feminist historians have begun to read as a historical construct: the essentialist theory of gender complementarity that gathered force in the Romantic era.[11]

---

[9] For arguments on women's attraction to botany see, on the English context, Ann B. Shteir, "Botany in the Breakfast Room: Women and Early Nineteenth-Century British Plant Study," in *Uneasy Careers and Intimate Lives: Women in Science, 1789–1979*, ed. Pnina G. Abir-Am and Dorinda Outram (New Brunswick, N.J.: Rutgers Univ. Press, 1987), pp. 31–43; Sandra Harding, *The Science Question in Feminism* (Ithaca, N.Y.: Cornell Univ. Press, 1988); Londa Schiebinger, *The Mind Has No Sex? Women in the Origins of Modern Science* (Cambridge, Mass.: Harvard Univ. Press, 1989); and Nancy Tuana, ed., *Feminism and Science* (Bloomington: Indiana Univ. Press, 1989). For the argument that women's lives were privatized from the time of the late Enlightenment see Joan B. Landes, *Women and the Public Sphere in the Age of the French Revolution* (Ithaca, N.Y.: Cornell Univ. Press, 1988); and Schiebinger, *The Mind Has No Sex?* For an excellent critique of Landes, and of the argument itself, see Olwen Hufton's review in *American Historical Review*, 1991, 96:528–529.

[10] For one formulation see Harding, *Science Question in Feminism*, p. 125: "Why should we regard the emergence of modern science as a great advance for humanity when it was achieved only at the cost of a deterioration in social status for half of humanity?" A list of authorities and a quotation are invoked to establish the epistemological uncertainty of positivism: "Along with such mainstream thinkers as Nietzsche, Derrida, Foucault, Lacan, Rorty, Cavell, Feyerabend, Gadamer, Wittgenstein, and Unger, and such intellectual movements as semiotics, deconstruction, psychoanalysis, structuralism, archeology/genealogy, and nihilism, feminists 'share a profound skepticism regarding universal (or universalizing) claims about the existence, nature, and powers of reason, progress, science, language, and the "subject/self"'" (pp. 27–28).

[11] Carolyn Merchant, *The Death of Nature: Women, Ecology, and the Scientific Revolution*, 2nd ed. (New York: Harper & Row, 1990), p. xix; see also the bibliography section of Lorraine Anderson, ed., *Sisters of the Earth: Women's Prose and Poetry about Nature* (New York: Vintage, 1991), p. 386, which describes Susan Griffin's *Woman and Nature: The Roaring Inside Her* (New York: Harper &

But Goethe's linkage of intuitive botanic archetypes and "lived experience" into a liberation of self and the founding of a new science is not a feminist or gynocentric methodology. By locating botany specifically in a predefined private sphere, Goethe separated it from the public sphere that remains modern science's founding and necessary arena. Moreover, he particularly excluded women from sharing in his botany by casting them as audience, muse, and object to male scientific genius and sexual power. The essay thus argues that although Goethe's botanic practice could easily be emulated by polite women of his time, his botanic philosophy resisted extension into the public sphere of science. Finally, it claims that Goethe's botany's paratheory of the female—the interpenetration of his personal and botanical relations to women—shows that Goethe understood women as equipment to his thought rather than participants in his science.

I

Goethe began learning botany by helping a friend, an apothecary, grow a medical herb garden. He collected *naturalia* until his rooms resembled a curiosity cabinet, where minerals, plants, and fossils nestled beside stuffed animals, printed works of art, and human skeletons. He planted a garden with beehives and linden trees.[12] And he studied Linnaeus's three primers: *Fundamenta botanica* (1736), *Philosophia botanica* (1751), and *Termini botanici* (1762). "I want to confess that next to Shakespeare and Spinoza the greatest influence on me has been Linnaeus." He added: "and this precisely because he challenged me to oppose him."[13] For Goethe's botanic study culminated in the 1790 *Attempt to Explain the Metamorphosis of Plants,* a project very different from Linnaeus's, though its formal composition was modeled on *Fundamenta botanica*'s 365 aphorisms. In the winter of 1790 glass prisms captivated Goethe. Shifting his attention from botany to optics, he attempted to refute Newton's theory of the colored light spectrum, limiting his botanic research to more scattered readings, experiments, and observations, and to the study of the reviews of his *Metamorphosis*.[14]

In the 1817 "History of His Botanical Studies" Goethe described himself as a poet "whose time . . . appeared to be taken up by manifold interests and duties." As the word *appeared* indicates, he did in fact find "opportunity to devote a great part of

---

Row, 1978) as celebrating "woman-as-nature and nature-as-woman." On the essentialist theory of gender complementarity see Thomas Laqueur, *Making Sex: Body and Gender from the Greeks to Freud* (Cambridge, Mass.: Harvard Univ. Press, 1990); Schiebinger, *The Mind Has No Sex?* (cit. n. 9); and Catherine Gallagher, paper presented at the History of Science Society Annual Meeting, Seattle, 1990.

[12] Johann Wolfgang von Goethe, "Geschichte der botanischen Studien," in Goethe, *Schriften zur Naturwissenschaft*, I:10, p. 322; see also *ibid.*, II:9A, pp. 266–267. For garden annotations from Goethe's diary in the mid and late 1770s see *ibid.*, p. 268, e.g.: "Linden gepflanzt" (1 Nov. 1776) and "Mit den Bienen beschäftigt und sie zur Winterruh gebracht" (7 Nov. 1776).

[13] Goethe, "Geschichte," p. 68 (passage not included in Goethe, "History"). *Fundamenta botanica* (Amsterdam, 1736) was Linnaeus's first extended methodological treatise. *Philosophia botanica* (Amsterdam/Stockholm, 1751) became *the* methodological botany primer in the later eighteenth century, explaining both Linnaeus's sexual system of plant classification and his rules on how to generate plant names. *Termini botanici*, a 1762 Uppsala university dissertation, is a vocabulary, defining 673 botanic terms.

[14] See Goethe's two-part treatise: *Beyträge zur Optik* (Weimar: Verlag des Industrie-Comptoirs, 1791–1792) and *Zur Farbenlehre* (Tübingen: Cotta, 1812). On Goethe's continued attempts during the 1790s to verify experimentally the hypothesis of plant metamorphosis see Goethe, *Schriften zur Naturwissenschaft*, II:9B, p. 469.

life, with interest and passion, to nature studies." He also claimed, however, that "it was necessary that I pursue without interruption my other duties and recreations." For Goethe's awakening interest in botany coincided with his arrival in the Thuringian grand duchy of Saxe-Weimar-Eisenach in 1775. His friend Karl August, the hereditary prince of Weimar, made the twenty-seven-year-old author of the Romantic best-seller *Werther* privy councillor in 1776, manager of the ducal mines in 1784, and minister of state in 1815.[15] Goethe administered the castle gardens, the city theater, and the ducal forests and counted among his offices the ministries of finance, defense, and arts.

In 1786, leaving his administrative work behind, Goethe embarked on his first Italian journey. Even during those two years of travel, he protested that he found little time for botany. When crossing the Alps he lamented that the carriage rolled too quickly for botanical observations, "even though I carried my Linnaeus with me and had his terminology firmly stamped on my mind."[16] After he toured southern Italy, "regulated study [of botany] on my return to Rome was not to be thought of. Poetry, art, and antiquity all seemed to demand my entire attention, and seldom in my life have I experienced more arduous and laborious days."[17] Botany was relegated to a few moments. Goethe picked some plants on his daily walks. He raised a couple of date palms in the garden of Duchess Anna Amalia von Sachsen-Gotha of Weimar, and in the garden of the celebrated painter Angelica Kaufman he grew a pine tree, which she declared "a dear and most significant plant because it comes from a dear hand." One night, as if visited in his sleep by a botanic muse, he awoke to the sound of capsule pods he had collected exploding in a box. As for the botanical studies that he planned for his return to Weimar: "my busy life . . . interrupted and thwarted my good intentions."[18]

Like most women of his time who were interested in science, Goethe pursued science part time. He forewent what women were denied—university courses and full-time study. As he liked to explain, he was not interested in *Bibliothekarwissenschaft,* the science of librarians. He was suspicious of exact scientific methods, and turned to landscape drawing, the portraiture of nature. After studying mathematics briefly in 1787, he trained in the medium of watercolors (a common hobby of polite women). Through sketching and painting, he wished to "read directly off the pages of the book of nature." "How readable the book of nature has become to me I cannot tell you," he wrote to Charlotte von Stein. "My quiet joy is inexpressible."[19] Imi-

[15] Goethe, "History," pp. 149, 165, 160 (Goethe, "Geschichte," pp. 63, 83, 77). Goethe claims that the purpose of his botanical history is to show that he in fact found time to pursue his interest in nature studies. On Goethe's arrival in Saxe-Weimar-Eisenach see Friedrich Gundolf, *Goethe* (1916; Berlin: Georg Bondi, 1922), p. 261. *Die Leiden des jungen Werther* (Leipzig: Weygand) was first published anonymously in 1774; later editions gave the author's name. For the novel's publishing history see *Goethe: An Exhibition at the Houghton Library* (Cambridge, Mass.: Harvard College Library and Goethe Institute of Boston, 1982).

[16] Entry dated 8 Sept. 1786, "abends," Brenner, in Johann Wolfgang von Goethe, *Italian Journey,* trans. W. H. Auden and Elizabeth Mayer (San Francisco: North Point, 1982) (hereafter cited as **Goethe, Italian Journey**), p. 15; for the German see Goethe, *Italienische Reise,* 2 vols. (Taschenbuch, 175) (Frankfurt am Main: Insel, 1976) (hereafter cited as **Goethe, Italienische Reise**), p. 26.

[17] Goethe, "History," p. 162 (Goethe, "Geschichte," p. 80).

[18] Goethe, *Schriften zur Naturwissenschaft,* II:9A, p. 378, editors' annotations (date palms); Angelica Kaufman to Goethe, 23 July 1788, *ibid.,* p. 381; and Goethe, "History," pp. 163 (exploding pods), 164 ("busy life") (Goethe, "Geschichte," pp. 81, 82). Duchess Anna Amalia's villa was later bought by Ludwig I of Bavaria, who in 1829 reported to Goethe on his date palms.

[19] Goethe's remarks on sketching and painting are quoted in Hans Blumenberg, *Lesbarkeit der Welt*

tating Rousseau's botanic walks, he climbed Weimar's woody hills (later the site of KZ Buchenwald). He was unencumbered by the enlightened naturalist's parapher-nalia, bringing only Linnaeus's *Termini botanici* and *Fundamenta botanica,* supple-mented by a German-language commentary.[20] "All bound in a single slender volume, [these works] accompanied me into the highways and byways and today that same volume reminds me of the active, happy days when those precious pages opened up a new world to me."[21] Many years later, as an old man confined to his study, Goethe recollected the days when Duke Karl August flew Montgolfian balloons from the castle terraces. These newfangled aircraft spread themselves over the hills in the same manner, he continued, as just-introduced regional floras allowed plant studies, "previously limited to a narrow cloister garden, [to be] extended to the entire rich region."[22]

Goethe advertised his study of botany as an exercise in immediacy, a "free, joyous study of Nature" in "the freedom of out-of-doors, shedding new light for me on gardens and books." His botany was, however, a practice embedded in the social spaces of holiday outings, tourist trips, hunting parties, spa hotels, and public parks. On his first Italian journey Goethe collected "seeds and pressed leaves" on the beaches of Lido. He discussed plant taxonomies at the Weimar "skating grounds—at that time a gathering place of good society."[23] He examined trees with Karl August dur-ing the lulls of the hunt. And he mused on the Primal Plant while promenading, in the manner of Sicilian noblemen, at dusk in Palermo's public parks.

Goethe's 1785 spa holiday illustrates the circumstances of his botany. One summer morning he set out from Weimar toward Karlsbad (in the present-day Czech lands), his carriage drawn over the Erz Mountains by four horses, their coats matted with

---

(Frankfurt am Main: Suhrkamp, 1981), p. 216. See also Goethe to Charlotte von Stein, 15 June 1786, in Goethe, *Schriften zur Naturwissenschaft,* II:9A, p. 335.

[20] Goethe's *Botanical Writings* explains that the commentary was written by "Johann Gessner" and gives an English title. I have not been able to find an exact bibliographic reference to the work. In student dissertations of 1740 and 1741 Johannes Geßner (1709–1790), from 1738 a professor of math-ematics and natural history at Zurich, explains the system of *Fundamenta botanica.* These might be the texts Goethe refers to in his 1817 "History" as *Dissertationen zu Erklärung Linnéischer Elemente.* Geßner also wrote, among many other works, *Ersten Grundriß der Kräuterwissenschaft aus der characteris-tischen Pflanzentabellen des Dr. J. Geßner . . .* (Zurich: Salomon Schinz, 1775).

[21] Goethe, "History," p. 153 (Goethe, "Geschichte," p. 68). According to extant bills, Goethe had *Fundamenta botanica* and *Termini botanici* bound together in one volume in Jan. 1779, then in Mar. 1779 had bound (separately) Linnaeus's *Systema plantarum, Termini botanici* (he must have owned two copies), and another primer entitled *Nomenclator botanicus enumerans plantas omnes.* Goethe owned several explications of Linnaeus. E.g., at that time he also had bound Karl Friedrich Dietrich's *Das Pflanzenreich nach d. Linneschen System,* 2 vols. (Erfurt, 1770; Regensburg, 1775), and in June the same year an *Index Regni vegetabilis,* a 1770 work giving an overview of Linnaeus's sexual system. See Goethe, *Schriften zur Naturwissenschaft,* II:9A, pp. 270–271, for the bookbinder's bills.

[22] Goethe, "History," p. 153 (Goethe, "Geschichte," p. 69). For accounts of Karl August's Mont-golfian balloons see Goethe, "History," p. 153; and Goethe, *Schriften zur Naturwissenschaft,* I:10, p. 324.

[23] Goethe, "History," pp. 153, 160 (Goethe, "Geschichte," pp. 69, 77). I have modified the English translation, which uses "in the great out-of-doors" for "in der freien Welt." For "seeds and leaves" see entry dated 8 Oct. 1786, Venice, in Goethe, *Italian Journey,* p. 82 (Goethe, *Italienische Reise,* p. 119); on the skating grounds see Goethe, "History," p. 155 (Goethe, "Geschichte," p. 71). August Karl Batsch (1761–1802) was professor of natural history, philosophy, and medicine at Jena and author, among other works, of *Botanik für Frauenzimmer und Pflanzenliebhaber* (Weimar, 1795), which was translated into French, Danish, and Swedish. He had a son, August Wilhelm, also a natural historian, with whom Goethe discussed botany on the Weimar skating grounds. Both Batsches sought to establish a natural system of plants.

sweat. Trotting alongside the wheels, a peasant boy gathered flowers beside the road—
"whenever possible," as Goethe remembered in his 1817 "History,"

> handing them into my carriage on the spot and, in the manner of a herald, announc-
> ing the Linnaean designation, both genus and species, with happy conviction, if some-
> times with the wrong pronunciation. In this way I attained a new relationship to open,
> splendid Nature, while my eyes enjoyed her wonders, and at the same time the scien-
> tific designations of the individual plants reached my ears as though from a dis-
> tant study chamber.

In fact, Goethe informs us, the entire family of his boy servant, Friedrich Gottlieb
Dietrich, had taught itself Linnaean nomenclature in order to get these kinds of jobs.
Profiting from the late eighteenth-century botanic fashion, rural families who had
previously collected medical herbs for apothecaries now lived "chiefly from fur-
nishing teachers and students with so-called lessons, that is, bundles of plants blos-
soming each week in the vicinity." Goethe's companion belonged to the Dietrich
family from the village of Ziegenhain near Weimar, which "distinguished itself especially. The head of the family [*Stammvater*] had even attracted the attention of
Linnaeus and could display a holographic letter from this highly esteemed man—a
diploma, as it were, through which he justifiably felt elevated to the botanical nobility."[24]

Goethe thus was part of a burgeoning holiday industry for scientific tourists, who
purchased knowledge from local guides. Though these guides might, at best, be
authorized with a Linnaean "diploma" establishing them as "botanic nobles," they
remained their clients' social inferiors. Writing in 1817, Goethe acknowledged the
improved lot of the middle-aged man he called "young Dietrich": "to this day [Die-
trich] zealously and creditably heads the grand ducal gardens in Eisenach." He
and Karl August had helped to bring this about by paying for Dietrich's apprenticeship. Yet Goethe's 1817 memory of his 1785 holiday remained shaped by his
original experience of the boy as an object of desire, "a well-built youth of regular,
pleasant features," a "pretty country boy."[25] (Goethe's travel description survives
as the only memory: Dietrich's autobiography, known to have existed in manuscript, is lost.)[26]

Goethe at once lingered over, and sought to naturalize, the sexually charged reversal of roles in which a servant gave his master "lessons" by scripting Dietrich as
a being that at once *is* nature, in his beauty, youth, rustic origins, and autodidact
Latin, and helps his social superior *control* nature, by speaking its scientific nomenclature "as though from a distant study chamber." The boy remained outside,
excluded from Goethe's protected spaces: his carriage, his hotel, his study. Cast as
a living representative of nature, Dietrich negotiated the link between the actual out-

[24] Goethe, "History," pp. 154, 153, 154, 153–154 (Goethe, "Geschichte," pp. 70, 69). I changed
"Linné" to Linnaeus" in the English translation.

[25] Goethe, "History," pp. 155, 154 (Goethe, "Geschichte," pp. 71, 69 ["wohlgebauter Jüngling von
regelmäßig angenehmer Gesichtsbildung"], 70 ["schmucker Landknabe"]). On Goethe's and Duke Karl
August's contributions to the apprenticeship see Goethe, *Schriften zur Naturwissenschaft*, II:9A, p. 321,
editors' annotations.

[26] Goethe, *Schriften zur Naturwissenschaft*, II:9A, p. 321, editors' annotations. We now have only a
few extant notes from people who talked with Dietrich about Goethe (*ibid.*, pp. 320–323). Dietrich
authored a pamphlet on the plants he and Goethe examined during the Karlsbad holiday, but that, too,
is lost (see *ibid.*, p. 323, editors' annotations).

doors and the tourist's idealized wilderness. As master and servant crossed the Erz Mountains, "in hilly regions, always on foot, [Dietrich] zealously tracked down growing things."[27] The image turns servant boy into hunting dog.

Once the author was installed in a Karlsbad hotel, the boy

> was in the hills at sun-up, bringing abundant "lessons" to me before I had emptied my first mug at the spa. The hotel guests all participated. . . . They found their minds stimulated in the most charming way by the sight of a pretty country boy in a short little vest running about, exhibiting great bundles of plants and designating them by names of Greek, Latin, and barbaric origin. It was a phenomenon that excited much interest among the men—and apparently also among the women![28]

Goethe, then, practiced botany sociably. He exemplified how polite women could (and did) do botany. He also represented himself as choosing to be what women must remain: a dilettante. He enjoyed scientific texts for amateurs, which around 1800 often meant works addressed to women, such as Ludwig Reichenbach's 1828 *Botany for Ladies,* dedicated to "women, artists and thoughtful lovers of Nature," August Karl Batsch's 1795 *Botany for Women and Lovers of Plants,* and Jean-Jacques Rousseau's eight *Letters on Botany,* written between August 1771 and April 1773 to educate a four-year-old girl. Goethe first read Rousseau's "*ganz allerliebste*" letters in the summer of 1782. "They give me an occasion," he wrote to Karl August that June, "to recommend anew the beautiful realm of flowers to my beautiful women friends." He often returned to the letters, quoting with approval the author's self-characterization: "I am only a student in this field and am not thoroughly grounded in it. When I botanize, I am thinking more of diversion and pleasure than of instruction." Explaining his taste for Rousseau, Goethe remarked that "the dilettante likes to learn from the dilettante. This would of course be questionable . . . if experience did not show that dilettantes contribute a great deal to science."[29]

Goethe supported this claim not by examples but, paradoxically, by appeal to a scientific authority: Alexander von Humboldt, the founder of plant geography, who during his 1799–1804 voyage to South America had collected some sixty thousand plant specimens and authored thirty-four volumes of travel journals. "Poesie, too, can succeed in lifting the veil of Nature," he paraphrased a letter Humboldt had written him, adding: "And if *he* admits this, who would dare deny it?" Linnaeus, too, was mustered. In a 1798 letter Goethe noted that "Linnaeus was liberal enough

---

[27] Goethe, "History," p. 154 (Goethe, "Geschichte," p. 70).

[28] *Ibid.* The English version adds the exclamation mark. I have also changed the translation of "ein schmucker Landknabe im kurzen Westchen" from "a handsome jerkinclad country boy" to "a pretty country boy in a short little vest." Dietrich, for his part, may not have been as interested in the women as they were in him. See Goethe, *Schriften zur Naturwissenschaft,* II:9A, p. 323, where Dietrich remembers the names of the men present but groups the women as "einigen gebildeten Frauen."

[29] For references to Batsch and Reichenbach see Goethe, "History," p. 153 (Goethe, "Geschichte," p. 69); on Rousseau see Goethe to Karl August, 17 June 1782, in Goethe, *Schriften zur Naturwissenschaft,* II:9A, pp. 279–280; and Goethe, "History," p. 158 (Goethe, "Geschichte," pp. 73, 75). See Jean-Jacques Rousseau, *Essais élémentaires sur la botanique* (Paris, 1771), rpt. in *Collection complette des oeuvres de J. J. Rousseau,* 24 vols. (Deux-ponts: Chez Sanson, 1782). According to Goethe, *Schriften zur Naturwissenschaft,* II:9A, p. 280, editors' annotations, Rousseau's letters were published in a German translation as *Botanik für Frauenzimmer* (Frankfurt/Leipzig, 1781). Goethe's favorite edition of these letters would be the illustrated *Botanique de J. J. Rousseau, ornée de soixante-cinq planches, imprimées en couleurs d'après les peintures de P. J. Redouté* (Paris: Baudoin, 1821).

to name also the poet among those who can further science." His (unspecified) source was the last of *Fundamenta botanica's* sixteen categories of botanists, which includes poets, librarians, and theologians and is entitled "Anomali."[30]

As mentioned, Goethe mistrusted scientific tools. Yet he used them. For his mid-1790s experiments on plants he employed a camera obscura and plates of colored glass. In the 1780s he participated in dissections of human cadavers. Like his contemporaries, he greatly interested himself in monsters and curiosities. In June 1797 he wrote to Karl August to describe an oddly-colored ivory billiard ball, which a dog had swallowed "by chance." Visiting the Stuttgart natural history cabinet in September of that year, he viewed with fascination a fetus, "dry as leather," that a woman purportedly had carried for forty-six years. In November 1812 the carcass of a two-headed calf, delivered to Weimar's zoological cabinet, provided the occasion for a long letter to C. G. von Voigt.[31] Goethe also enthusiastically used microscopes to observe pollen and bacteria: in April 1785 the duchess Louise von Sachsen-Gotha of Weimar gave him a *"recht extra gutes"* one. That spring's microscope fashion was most intense: Johann Gottfried Herder jokingly warned his Weimar friends that if they were not careful they might themselves metamorphose into the small *"Infusions-Thierchen"* they so eagerly studied.[32]

II

Yet Goethe privileged contemplation over quantification. He felt that scientific instruments like microscopes were crutches—necessary, perhaps, in Germany, but not in Italy. Once he had crossed the Alps, he likened himself to a scientific Robinson Crusoe on a peninsular Arcadia, cast afloat without instruments and textbooks (excepting an old edition of Linnaeus's *Genera plantarum*). From his first Italian journey, he wrote: "Much that I could only conjecture and seek for with the microscope at home, I here see with the naked eye, with undoubtable certainty." The next letter repeated: "What I only assumed in the North I find obvious here."[33] Goethe's discovery of plant metamorphosis was made possible, he believed, by a particular natural landscape, the interpretation of which did not call for scientific instruments.

Goethe doubted the utility of science tools partly because of the problem of scale, a theme addressed in Jonathan Swift's *Gulliver's Travels* (1726). In the land of Brob-

[30] Goethe, "History," p. 185 (Goethe, "Geschichte," p. 141 [underlined in the original]); and Goethe to Neuenhahn, 14 Sept. 1798, in Goethe, *Schriften zur Naturwissenschaft,* II:9B, p. 151.

[31] Goethe's notebooks report on his use of the camera obscura and plates of colored glass; see Goethe, *Schriften zur Naturwissenschaft,* II:9A, pp. 225–227. On his other investigations see Karl Ludwig von Knebel to Henriette von Knebel, 3 Nov. 1781, *ibid.,* p. 277 (dissection of human cadavers); Goethe to Karl August, 12 June 1797, *ibid.,* 9B, p. 106 (billiard ball); Goethe's notes on the Stuttgart *Naturalienkabinett,* 5 Sept. 1797, *ibid.,* p. 116; and Goethe to C. G. von Voigt, [Nov 1812], *ibid.,* p. 355 (two-headed calf).

[32] *Ibid.,* 9A, p. 313 (on Goethe's microscope). Charlotte von Stein arranged for the purchase of the microscope. The annotations on the same page note that "Die Lust am Mikroskopieren ergriff die ganze Umgebung Goethes." In Apr. 1785 Goethe filled his letters with references to microscopes; see *ibid.,* pp. 316, 503, editors' annotations. For Herder's admonition see Herder to Knebel, [early May 1785], *ibid.,* p. 319: "Goethe guckt fleißig ins Glas und auf die Pflanzen; vielleicht tun Sie es auch. Macht aber nicht, lieben Leute, daß Euch die große Massenwelt fatal werde, weil die kleine Samen- und Baumwelt so niedlich ist, damit Ihr nicht gar Infusions-Thierchen werdet." Herder also participated in microscope studies; see Goethe to Charlotte von Stein, 10 Apr. 1786, *ibid.,* p. 332; and, on the same page, Goethe to Stein, 8 Apr. 1786, where he notes that "Ich lasse Infusionstierchen zeichnen."

[33] Goethe to Knebel, 18 Aug. and 3 Oct. 1787, *ibid.,* pp. 370, 373 (p. 374 records that he carried *Genera plantarum*).

dingnag, out-of-scale Gulliver cannot recognize his giant hosts. Instead he analyzes their agglomerations of features, abstracting them into mental notions of the human form. Overall comprehension eludes him even as he assembles an impressive number of local facts.[34] Though Goethe does not draw on Swift's tale, it nicely prefigures his thought. Gulliver's dilemma was that of Goethe, deluded by new technologies of vision. Rather than beholding nature enlarged by magnifying lenses, Goethe held, the scientist ought to read the world physiognomically and quasi-intersubjectively— straight from appearances, face-to-face and on a one-to-one scale.

For Goethe, we murder to dissect. Nature should neither be scale-distorted nor arbitrarily classified. The "born poet" and "true scientist," he argued, "always seeks to derive his terminology directly from the subjects themselves, each time anew."[35] Names cannot be invented willy-nilly; they must be patiently awaited as they emerge out of the experiential relation between man and object. Humboldt's "poesie" unveils nature when thought corresponds to world. Finding himself in an arbitrarily baptized world, Goethe desired a congruence between names and things, thought and reality. Martin Heidegger later took it upon himself to prove that this congruence was factual for German speakers, through German philology deployed upon German Romantic poetry. Goethe, less optimistic, at least could note on his first Italian journey that "thank God, Venice is no longer a mere word to me, an empty name, a state of mind which has so often alarmed me, who am the mortal enemy of mere words."[36]

When word and object coincide, language is immediate and untranslatable, like the primal onomatopoeia already ironized in Plato's *Cratylus*. Goethe's eight-line poem "The Worth of Words" lamented the loss of such linguistic immediacy. "Words are images to the soul—/Not an image! They are a shadow!"[37] However removed primal language was from ordinary speech, Goethe felt that it should be pursued. Hence his rejection of Linnaeus's artificial nomenclature. "Imagine that such a man [as Goethe] is now expected to commit to memory a ready-made terminology, a certain number of words, and bywords, with which to classify any given form. . . . A procedure of that sort always seemed to me to result in a kind of mosaic in which one completed block is placed next to another, creating finally the appearance of a picture from thousands of pieces; this was somewhat distasteful to me." Goethe felt betrayed. "I had devoted myself to him [Linnaeus] and to his theory with complete trust."[38]

Linnaeus expected artificially classified herbaria to enable a synthetic analysis of nature, which in turn would allow a natural plant taxonomy. Confessing that "I was by nature averse to classification and counting," Goethe instead sought the Primal

---

[34] Jonathan Swift, *Gulliver's Travels* (Harmondsworth, Middlesex: Penguin, 1985), pp. 130–131.

[35] Goethe, "History," pp. 159–160 (Goethe, "Geschichte," p. 76). The link between the poet and the scientist was the unifying factor of genius. That was also the manner in which some reviewers of the *Metamorphosis of Plants* understood the book; see review in the *Kaiserl. priv. Hamburger Neuen Zeitung*, 6 May 1791, quoted in Goethe, *Schriften zur Naturwissenschaft*, II:9A, p. 402: "mit Vergnügen sah er [the reviewer] eine neuen Beweis, daß ein Mann von Genie in jedem Fache groß sein kann."

[36] Entry dated 28 Sept. 1786, in Goethe, *Italian Journey*, p. 58 (Goethe, *Italienische Reise*, Vol. 1, p. 87).

[37] Johann Wolfgang von Goethe, "Wert des Wortes," in Goethe, *Gedichte*, selected with an introduction by Stefan Zweig (Universal-Bibliothek, 6782) (Stuttgart: Reclam, 1980) (hereafter cited as **Goethe, Gedichte**), p. 191: "Worte sind der Seele Bild—/Nicht ein Bild! Sie sind ein Schatten!"

[38] Goethe, "History," pp. 159–160, 159 (Goethe, "Geschichte," p. 76). In the first quotation I have altered the translation of "den Schein eines Bildes" from "a single picture" to "the appearance of a picture."

Plant. As he walked the public gardens of Palermo in April 1787, "where instead of being grown in pots or under glass as they are by us, plants are allowed to grow freely in the open fresh air and fulfill their natural destinies," he wondered: "Among this multitude might I not discover the Primal Plant? There certainly must be one. Otherwise, how could I recognize that this or that form *was* a plant if all were not built upon the same basic model?"[39]

Once the existence of the Primal Plant is realized, all other plant forms take on an immediate significance. Meanwhile, the fact that we can sort out our heterogeneous sense impressions proves the Primal Plant's existence. To grasp its form not only means to understand existing flora, but also to be able to construct an infinity of notional plants. As Goethe explained to Herder, with the discovery of the Primal Plant "it will be possible to go on for ever inventing plants and know that their existence is logical; that is to say, if they do not actually exist, they could, for they are not the shadowy phantoms of a vain imagination, but possess an inner necessity and truth."[40]

To Goethe the Primal Plant was thus the necessary and sufficient cause of all flora, in the sense that it functioned as a single explanatory principle. As he explained in Rome in June 1787 (under the heading "Some Questions about Nature which Intrigue and Perplex Me"), in Palermo "it came to me in a flash that in the organ of the plant which we are accustomed to call the *leaf* lies the true Proteus who can hide or reveal himself in all vegetal forms. From first to last, the plant is nothing but leaf, which is so inseparable from the future germ that one cannot think of one without the other." Once discovered, what he called the "key to all signs of nature" closes the gap between language and experience. Or more exactly, it does away with the need for a scientific language, in the sense of formally agreed-upon conventions. As Hans Blumenberg has reminded us, Goethe's Primal Plant is not an a priori Platonic image, "but emerges out of the fullness of our cognition."[41] It is not postulated or deduced. Therefore, it does not need to be translated into language. Rather, it is brought forth as the completion of the sympathetic observer's nontranscendental experience of nature.

Goethe hastily noted in an undated fragment, probably from the later 1790s, that this observer's task is to "contemplate nature's products in themselves without reference to their use or purpose without [their] relation to their initial creator only as a living whole [and] that precisely because it is alive [which] already [is] cause and effect in itself."[42] The Primal Plant is not wrestled forth by human reason alien to its object. It appears to meet the observer, once he realizes that in nature appearance and reality coincide and, therefore, ceases to mistrust personal experience.

Goethean botanic knowledge is thus acquired in a contemplative yet immediate cognition, and it takes the form of a conviction. Rather than pondering probabilities,

---

[39] Goethe, "History," p. 155; and entry dated 17 Apr. 1787, Palermo, in Goethe, *Italian Journey*, pp. 251–252 (Goethe, *Italienische Reise*, Vol. 1, p. 345).

[40] Goethe to Herder, 17 May 1787 (from Naples), in Goethe, *Italian Journey*, pp. 305–306 (Goethe, *Italienische Reise*, Vol. 1, p. 417).

[41] Entry dated 31 July 1787, Rome, *ibid.*, p. 363; Goethe, *Schriften zur Naturwissenschaft*, II:9B, p. 432; and Blumenberg, *Lesbarkeit der Welt* (cit. n. 19), p. 214. See also Gundolf, *Goethe* (cit. n. 15), pp. 267, 379.

[42] Goethe, *Schriften zur Naturwissenschaft*, II:9A, p. 254. The annotation to the text (p. 256) suggests that it might have been a sketch of a plan for publishing all of Goethe's science texts together, an idea he had in the spring of 1795.

the scientist becomes aware of his moral certainty. One March day in 1787, ambling along the seashore in Naples, Goethe remarks, "suddenly I had a flash of insight concerning my botanical ideas." Such moments, Goethe felt, came especially to empirics and amateurs. In 1815 he related how, recently, "an old court gardener in Dresden had discovered the metamorphosis of plants on his own, and he told [Goethe] with joy, how he noticed that also [Goethe] knew something about it." In the end, Goethe felt, all science is one truth. "*Ach Gott,*" he concluded his gardener's tale, "it is all so simple and always the same, it is truly easy to be our Lord God, once creation is in place, [since] it only entails one single thought."[43]

However convincingly Goethe proved to himself the theoretical necessity of the Primal Plant, empirical evidence eluded him. After that April morning's useless search along the Palermo garden's planted borders, he cried out: "Gone were my fine poetic resolutions—the garden of Alcinous had vanished and a garden of the natural world had appeared in its stead. Why are we moderns so distracted, why do we let ourselves be challenged by problems we can neither face nor solve!"[44] Torn between "poetic resolutions" and "the natural world," Goethe takes on the character of one of his fictional creatures, *Werther*'s flower-picking madman "in a shabby green coat," scrambling among river rocks on a cold and misty day in late November. "I have been looking for two days and can't find them. There are always flowers out here, yellow and blue and red. . . . But I can't find a single one of them."[45]

The madman seeks the Homeric garden Werther, languishing for Lotte, has given up. "Wretched man! And yet how I envy your melancholy, the confusion of your senses in which you are languishing. You start out hopefully to pick flowers for your queen—in winter. . . . And I—and I go out without hope, without purpose, and return home in the same purpose." The next day Werther learns that the madman's illicit passion for Lotte drove him insane. As he hears the story, he feels descending upon him the same madness, the cause of which is now revealed.[46] Werther's shock when he understands that the idiot is his alter ego is analogous to Goethe's shock when he realizes that "we moderns" can never grasp the primal order that inhabited classical antiquity's mythic parks, but not Palermo's 1787 *Weltgarten.*

Yet Goethe's disappointment fired rather than dampened his imagination. In a marvelous flight of fancy, the 1817 essay "Intuitive Judgment," he argued that sci-

---

[43] Goethe, *Italian Journey,* p. 212 (Goethe, *Italienische Reise,* Vol. 1, p. 288); and, for the story of the gardener, see Goethe, *Schriften zur Naturwissenschaft,* II:9B, p. 400 (in translating I changed the verb tense from present perfect to past perfect). This story is related by Sulpiz Boisserée in his diary for 2/3 Aug. 1815; its source is a discussion he had with Goethe. See also, however, Goethe's ironizing critique of the *Naturphilosophen*'s longing for all-encompassing unity in a letter to Wilhelm von Humboldt, 22 Aug. 1806, *ibid.,* p. 501: "Erfreulich ist es auf jenes wünschenswerte Zeil hingewiesen zu werden, daß aller Zwiespalt aufgehoben, das Getrennte nicht mehr als getrennt betrachtet, sondern alles aus Einem entsprungen und in Einem begriffen, gefaßt werden solle."

[44] Entry dated 17 Apr. 1787, Palermo, in Goethe, *Italian Journey,* pp. 251–252 (Goethe, *Italienische Reise,* Vol. 1, p. 345). King Alkinoos or Alcinous, mentioned in Homer, was the grandson of Poseidon. The *Odyssey* describes his palace gardens.

[45] Johann Wolfgang von Goethe, *The Sufferings of Young Werther,* trans. Harry Steinhauer (New York: Norton, 1970) (hereafter cited as **Goethe, Young Werther**), p. 69; for the German see Goethe, *Leiden des jungen Werther* (Taschenbuch, 25) (Frankfurt am Main: Insel, 1973) (hereafter cited as **Goethe, Jungen Werther**), pp. 118–119. The quotation is from Werther's diary entry for 30 Nov. 1772.

[46] Goethe, *Young Werther,* pp. 70, 71 (Goethe, *Jungen Werther,* pp. 120, 122); the quotation is from Werther's diary entry for 1 Dec. 1772.

ence ought to be modeled on §77 of the *Critique of Judgment* (1790), where Immanuel Kant writes: "In contrast to our own analytical intellect, we can conceive of an intuitive one which proceeds from the synthetically universal (the concept of the whole as such) and advances to the particulars."[47] Goethe admitted that Kant "seems to be referring here to godlike understanding." Yet he extended this notion to science, grandly choosing as his example his own botany:

> Since it is possible in the moral realm to ascend to a higher plane, drawing close to the Supreme Being through faith in God, virtue, and immortality, the same might well hold true in the intellectual realm. Through contemplation of ever-creative Nature we might make ourselves worthy of participating intellectually in her productions. I myself ceaselessly pressed forward to the archetype [of the Primal Plant], though at first unconsciously, from an inner urge.[48]

Faith, Goethe here argued, produces both the religious saint and the scientific genius.

Given this profoundly Romantic philosophy of science, Goethe was understandably "taken aback and somewhat irritated" by his famous 1794 discussion with Friedrich von Schiller. They met one August evening at the Jena Society for Scientific Studies:

> By chance we left the hall together, and began a conversation. . . . We had reached his house; the conversation lured me in. I gave a spirited explanation of my theory of the metamorphosis of plants with graphic pen sketches of a symbolic plant. He listened and looked with great interest, with unerring comprehension, but when I had ended, he shook his head, saying, "That is not an empiric experience, it is an idea." . . . The old antipathy was astir. Controlling myself, I replied, "How splendid that I have ideas without knowing it, and can see them before my very eyes."[49]

Schiller's reaction was particularly galling, since Goethe liked opposing his own "realism" to Schiller's "idealism." In the same spirit, Goethe later accused Hegel of contriving to "destroy the eternal reality of nature with a bad sophistical joke," by claiming that when a fruit appears its preceding flower is revealed as *"ein falsches Dasein der Pflanze."* "It is not possible to state something more monstrous," Goethe

---

[47] Johann Wolfgang von Goethe, "Intuitive Judgment," in Goethe, *Botanical Writings*, p. 232; for the German see Goethe, "Anschauende Urteilskraft," in Goethe, *Schriften zur Naturwissenschaft*, I:9, pp. 95–96. Goethe quotes from Kant's 1790 *Kritik der Urteilskraft*, a series of appendixes to the 1781 *Kritik der reinen Vernunft* and the 1788 *Kritik der praktischen Vernunft*: "Wir können uns einen Verstand denken, der, weil er nicht wie der unsrige diskursiv sondern intuitiv ist, vom synthetisch Allgemeinen der Anschauung eines Ganzen als eines solchen, zum Besondern geht, das ist, von dem Ganzen an den Teilen." "Anschauende Urteilskraft" was first published in Goethe's pamphlet series *Zur Morphologie*, Vol. I, No. 2 (1820). Note how in the generic title of his pamphlet series he already ties up his science with "life events" (*Lebensereignisse*). The translation of *anschauende* as "intuitive" does not quite capture the sense of how this "power" (*Kraft*) of judgment comes out of a willed inner gaze, rather than being merely spontaneous and unexpected. See Goethe, *Schriften zur Naturwissenschaft*, II:9A, p. 582, editors' annotations, for a brief analysis of this essay.

[48] Goethe, "Intuitive Judgment," p. 233 (Goethe, "Anschauende Urteilskraft," pp. 95, 95–96).

[49] Johann Wolfgang von Goethe, "On General Theory," in Goethe, *Botanical Writings*, p. 217; for the German see Goethe, "Glückliches Ereignis," in Goethe, *Schriften zur Naturwissenschaft*, I:9, pp. 81–82. This essay was first published in *Zur Morphologie*, Vol. 1, No. 1 (1817). For another account of the same meeting, also by Goethe, as told to an acquaintance in 1813 see Goethe, *Schriften zur Naturwissenschaft*, II:9A, pp. 430–431; this account differs only in details.

wrote a friend. "Comfort me, *mein Liebster,* if it is possible."[50] Yet, as Schiller's comment suggests, Goethe's botanic philosophy—his "passion"—was in a way solipsistic and antiempirical, even as it attempted to reflect "the eternal reality of nature." It was not solipsistic in George Berkeley's strict sense of the "I" believing that it and only it exists. But Goethe did doubt that his botanic theories were communicable; he explained to a visitor in 1813 that "it seems that most people simply don't have the organ to understand this [plant metamorphosis], therefore this theory is and will remain obvious, yet secret [*das offenbare Geheimnis*]."[51]

Stemming from an "inner urge," Goethe's botanic insights were (he believed) verifiable only through "passionate divination": an intuitive, sympathetic, intersubjective understanding modeled on sexual love. Goethe here at once received and transmitted the enduring German valorization of the private sphere gently satirized in Christian Morgenstern's 1910 poem "Science," where the comic hero Palmström famously encourages his companion, Kopf, after professors dismiss their homegrown science. "'Come comrade,' says Palmström, 'All that's finest stays—private!'"[52] Goethe, cultural icon ruling a neoclassical dukedom, and Palmström, tragicomic "little man" lost in an absurd modernity, are brothers at heart. They unite around a slogan importantly constitutive of German self-consciousness—and still powerful in the period just after World War II, when educated Germans wove into their conversations this magic apotropaic formula warding off public responsibility.

Palmström and Kopf reach their understanding of the natural world within the exclusive sphere of a lifelong male friendship. Goethe arrived at his method for sharing knowledge, "passionate divination," through a briefer love affair in the 1810s with Marianne von Willemer, an affair celebrated in their verse correspondence or "exchange of flowers and signs." (Goethe published their letters as his own work in the 1819 orientalist verse cycle *West-East Divan.*)[53] Passionate divination not only

---

[50] Goethe's account of his first meeting with Schiller appears in Goethe, *Schriften zur Naturwissenschaft,* II:9A, p. 430. See also *ibid.,* p. 506, editors' annotations; and, on Goethe's self-understanding as a realist in the 1780s, *ibid.,* 9B, p. 469. For the complaints see Goethe to Thomas Johann Seebeck, 28 Nov. 1812, *ibid.,* pp. 354–355. The passage from the preface of Hegel's *Phänomenologie des Geistes* that Goethe so disliked argues—I quote from Goethe's transcription—that "Die Knospe verschwindet in dem Hervorbrechen der Blüte, und man könnte sagen, daß jene von dieser widerlegt wird; eben so wird durch die Frucht die Blüte für ein falsches Dasein der Pflanze erklärt, und als ihre Wahrheit tritt jene an die Stelle von dieser."

[51] Goethe, "History," p. 162 (Goethe, "Geschichte," p. 80); and, for the "obvious, yet secret" theory of plant metamorphosis, Goethe, *Schriften zur Naturwissenschaft,* II:9B, p. 365 (which reproduces Johannes Daniel Falk's notes from a discussion with Goethe, 29 Mar. 1813). However well meaning, and however prepared to grant Goethe the status of universal genius, visitors often were confused by his conversations on natural history. See, e.g., *ibid.,* pp. 362–363, which reproduces Falk's notes from a talk with Goethe, 15 Mar. 1813.

[52] "'Komm,' spricht Palmström, 'Kamerad,/—alles Feinste bliebt—privat!'": Christian Morgenstern, *Galgenlieder, Palmström und andere Grotesken* (Munich: Piper, 1979), Vol. 1, pp. 140–141. For a fine explication of the concept of passionate divination see Geoffrey H. Hartman, *Saving the Text: Literature/Derrida/Philosophy* (Baltimore: Johns Hopkins Univ. Press, 1981), pp. 135–137.

[53] Hartman, *Saving the Text,* pp. 135–137. Maria Anna (Marianne) Katharina Therese Jung (1784–1860) was born to an unmarried Viennese actress. In 1800 she met a widowed Frankfurt banker and senator, Johann Jakob von Willemer, who educated his sixteen-year-old Eliza in modern languages and the polite arts and eventually married her. The verse cycle appeared as Johann Wolfgang von Goethe, *West-Östlicher Divan* (Stuttgart: Cotta, 1819). Goethe first titled these poems *Versammlung deutscher Gedichte mit stetem Bezug auf den "Divan" des persischen Sängers Mahomed Schemseddin Hafis,* since they were inspired by the 1812 German translation of Hafiz, a fourteenth-century Persian poet (Hafiz, *Der Diwan von Mohammed Schemsed-din Hafis: Aus dem Persischen zum erstenmal ganz übersetzt vor Joseph v. Hammer* [Stuttgart: Cotta, 1812]). In 1863 Herman Grimm revealed that Marianne von Willemer had proved that she had written parts of the *Divan.* Richard Friedenthal, *Goethe: Sein Leben und*

verified prior results; it also enabled the experience of nature that yielded the conclusions to be verified. Both in its most everyday forms and in its highest creation, nature was a sacred mystery. "When I kill a fly," a 1790 fragment in Goethe's hand declares, "I don't think and dare not think what organisation is destroyed. When I think of my death I dare not cannot think what organisation is destroyed." In Goethe's view, man relates to nature as to woman, approaching with the awe appropriate to that which always remains distinct yet inevitably draws him in. As the Chorus Mysticus sings in the famous closing lines of *Faust*, "The eternal-feminine/draw us on."[54]

Goethe grounded knowledge, whether literary or scientific, in what he called *Lebensereignisse* and what Hegel and Dilthey, following Goethe's lead, retermed *Erlebnis:* the privileged moment of experience. Unlike many later Germans, however, Goethe did not make a cult of experience for its own sake. Rather, he saw *Lebensereignisse* as the irreducible medium through which we grasp literature, the arts, and the sciences. That is why, as he set out toward the Alps on his first Italian journey, he looked forward to reading Tacitus in Rome, and why, having traveled through Sicily, he reread the *Odyssey*. "A word about Homer," he wrote Herder from Naples in May 1787: "The scales have fallen from my eyes. . . . Now that my mind is stored with images of all these . . . rocky cliffs and sandy beaches, wooded hills and gentle pastures, fertile fields, flower gardens, tended trees, festooned vines, mountains wreathed in clouds, eternally serene plains and the all-encircling sea with its ever-changing colours and moods, for the first time the *Odyssey* has become a living word to me."[55]

Even as he wrote these lines Goethe knew that, despite his vignette, the *Odyssey* could not be "a living word" to untraveled friends back in Germany. He believed that the mutual involvement of theory and experience profoundly complicates communication. We cannot reproduce experiences as we can experiments. Borrowing Walter Benjamin's terminology, we might liken Goethe's passionate divination to an original (as opposed to a mechanically reproduced) form of communication. Goethe at times even characterized his botanic ideas as "secret." When he wrote from Naples about his April 1787 "flash of insight," he added to that evening's letter: "Please tell Herder I am very near discovering the secret of the Primal Plant. I am only afraid that no one will recognize in it the rest of the plant world." The summer before he had written Charlotte von Stein: "If only I could share with someone my insight and my joy [in nature], but it is not possible. And [yet] it is no dream no fantasy."[56]

Goethe's *Naturphilosophie* was esoteric, like the body of knowledge that—along with botany, anatomy, geology, and optics—most interested him, alchemy. In Fras-

---

seine Zeit (Munich: Piper, 1963), p. 512, argues that without Grimm's testimony, "kein Forscher hätte es gewagt, eine ganze Reihe der schönsten Gedichte Goethes einer kleinen ehemaligen Tänzerin zuzusprechen." Gundolf, *Goethe* (cit. n. 15), p. 642, suggests that Willemer was more important as muse than as writer and goes on to analyze the poems as if they were written only by Goethe: "Wichtiger aber als ihre aktive Mitarbeit am westöstlichen Divan ist ihre passive: als 'Suleika' ist sie das letzte dichterische Frauenbild Goethes geworden."

[54] Goethe, *Schriften zur Naturwissenschaft*, II:9A, p. 140; and Johann Wolfgang von Goethe, *Faust: Gesamtausgabe* (Rotenburg an der Fulda: Insel, 1951), p. 483: "Das Ewig-Weibliche/Zieht uns hinan."

[55] Entry dated 17 May 1787, Naples, in Goethe, *Italian Journey*, p. 305 (Goethe, *Italienische Reise*, Vol. 1, pp. 416–417). Note that the former translates "lebendiges Wort" as "living truth." See also Goethe, *Schriften zur Naturwissenschaft*, I:9, unpaginated preface, editors' annotations.

[56] Entry dated 25 March 1787, Naples, in Goethe, *Italian Journey*, p. 212 (Goethe, *Italienische Reise*, pp. 288–289); and Goethe to Charlotte von Stein, 9 July 1786, in Goethe, *Schriften zur Naturwissenschaft*, II:9A, p. 336. See also Max Seiling, *Goethe als Esoteriker* (Melsbach: Silberschnur, 1988).

cati, in the autumn of 1787, he explained: "My theory [of the Primal Plant] is difficult to describe in any case, and, no matter how clearly and exactly it is written down, impossible to understand merely from reading." Eight months before, in January 1787, he had already suspected that the theory of plant metamorphosis he was working on might be incommunicable. "One may say what one likes in praise of the written or the spoken word, but there are very few occasions when it suffices. It certainly cannot communicate the unique character of any experience, not even in matters of the mind." Goethe's critique of the *Naturphilosoph* F. W. J. von Schelling, expressed in a 1798 letter to Schiller, was that instead of quietly pursuing his own philosophy, "the transcendental idealist . . . argues with other [scientific] beliefs, for one can't actually argue with [people who hold different] beliefs."[57]

Distrusting words, Goethe also doubted images. He suggested in the foreword to the 1810 *Theory of Colors* that scientific illustrations are "mere notions; they are symbolic resources, hieroglyphic modes of communication, which by degree assume the place of the phenomena and of Nature herself, and thus hinder rather than promote true knowledge." Twelve years later, he concluded that "a [scientific] idea cannot be demonstrated empirically, nor can it actually be proved." Even the titles Goethe and his botanic mentor Rousseau chose for their scientific writings suggest the quintessentially never-completed projects of Romanticism, the self-consciously partial vehicles of the inexpressible: letters, fragments, dialogues, prolegomena, plans. *Faust*'s Chorus Mysticus, concluding a play far too long to perform, again comes to mind: "The indescribable, /here it is done."[58]

In sum: Goethe lauded his amateur status. He distrusted scientific instruments and scientific nomenclature. His Primal Plant remained elusive. And he suspected that scientific information could be shared only in kindred souls' passionate divination. In effect, and in a complex way also by intent, Goethe *celebrated* his botanic methodology as incommunicable. In the autobiographical *Maxims and Reflections* Goethe wrote that "in New York there are ninety different Christian denominations, each of which professes in its own manner its belief in God without being disconcerted in any way by the others. In the sciences, yes, in every field of research, we must do the same."[59] Both as *activity* and as *philosophy,* Goethe's botany was unable to

---

[57] Entry dated 28 Sept. 1787, Frascati, in Goethe, *Italian Journey,* p. 387 (Goethe, *Italienische Reise,* Vol. 2, p. 524); entry dated 2 Jan. 1787, Rome, *ibid.,* p. 143 (Vol. 1, pp. 202–203); and Goethe to Schiller, 6 Jan. 1798, in Goethe, *Schriften zur Naturwissenschaft,* II:9B, p. 128. On Goethe's interest in alchemy see Ronald D. Gray, *Goethe the Alchemist: A Study of Alchemical Symbolism in Goethe's Literary and Scientific Works* (Cambridge: Cambridge Univ. Press, 1952).

[58] Johann Wolfgang von Goethe, *Theory of Colours,* trans. and annotated by Charles Lock Eastlake (1840; Cambridge, Mass.: MIT Press, 1970), p. xlix (for the German see Goethe, *Farbenlehre* [Tübingen: Wissenschaftliche Buchgemeinschaft, 1953], p. 173); Goethe, "Increasing Difficulty in Botanical Instruction," in Goethe, *Botanical Writings,* on p. 115 (Goethe, "Botanik," in Goethe, *Schriften zur Naturwissenschaft,* II:9, pp. 236–237); and Goethe, *Faust* (cit. n. 54), p. 483: "Das Unbeschreibliche,/ Hier ist es getan." "Botanik" was first published in *Zur Morphologie,* Vol. 1, No. 4 (1822). For examples of the self-consciously partial works of the period see Jean-Jacques Rousseau, *Fragmens pour un dictionnaire des termes d'usage en botanique* (published posthumously in 1782); Rousseau, *Lettres sur la botanique* (1771–1773); Goethe, *Versuch die Metamorphose der Pflanzen zu erklären* (1790); Goethe, *Beiträge zur Optik* (1791–1792); Goethe, *Versuch, die Elemente der Farbenlehre zu entdecken* (1793); and Goethe, *Aufsatze, Fragmente, Studien zur Naturwissenschaft in allgemeinen* (1817–1822).

[59] Max Hecker, ed., *Goethes Maximen und Reflexionen:* (Schriften der Goethe Gesellschaft) (Weimar: Hermann Bohlaus, 1907), no. 1181; also quoted in Susan Gustafson, "The Religious Significance of Goethe's 'Amerikabild,'" *Eighteenth-Century Studies,* 1990, 24:69–91, on p. 69. Goethe's celebration of his botanic methodology as incommunicable resembles Paul Feyerabend's idealization of multiple, incommensurable scientific communities.

bridge the gap between the public and the private by showing that such divisions were artificial. Balancing precariously between the solipsistic and the sociable, it instead at once reproduced and enclosed leisured women's everyday worlds.

For Goethe—man of letters, statesman—this hardly mattered. His ambitions for his botany were grand. But it only needed to serve what we now understand as his private desires and purposes. This was even more true for Rousseau, who from July 1767 to December 1769 signed himself Jean-Joseph Renou in an attempt to shed his literary fame and dedicate himself to botany. At its most basic, Goethe and Rousseau used botany as therapy. Rousseau preferred taxonomic drudgery; Goethe, botanic theory.[60] At its most elevated, Goethe made botany part of an individual's personal development, or *Bildung*. But as such it risked being displaced by literature and art. The quest was for self; the sciences, the arts, love, and even statesmanship were interchangeable means. Goethe remembered in Rome that his Weimar friends "mocked and wanted to restrain my passion for observing stones, plants and animals from certain definite points of view, and tried to make me give it up. Now my attention is fixed on the architect, the sculptor and the painter and in them, too, I shall learn to find myself." A 1786 letter to Herder makes the same point: "Now[,] my dear old friend[,] architecture and sculpture and painting is to me like [what] mineralogy botany and zoology [were]."[61]

Goethe, then, practiced botany *as* contemporary women did. His botanic philosophy also was *like* women's thought, in the sense that it celebrated as positive virtues what the West long has construed as archetypes of femininity, for example, intuition, spontaneity, holism, amateurism, a cult of poetics and interiority, and nature worship. But a feminine science—that is, one that meshes with the twentieth-century social construction of gender—does not equal a feminist science—that is, one committed to equal rights for men and women. Like most human endeavors, a feminist science must be a cooperative enterprise. Goethe's botanic project remained a private hobby. Moreover, as this essay now goes on to argue, his botany contains an account—or paratheory—of women's proper role within it, an account that renders Goethe's feminine science not only antifeminist, but also finally antifemale. For Goethe turned to botany not only as a solace for the specifically male self; he also used it to negotiate his personal relations to women.

III

In the summer of 1798, eight years after he had completed his scientific treatise *The Metamorphosis of Plants*, Goethe composed a ninety-seven-line poem with the same title and on the same topic. The verse "Metamorphosis" was first published by Schiller in autumn 1798, in the *Musen-Almanach* for 1799, and was reprinted in 1799, 1800,

---

[60] James F. Jones, Jr., rev. of Peggy Kamuf, *Signature Pieces: On the Institution of Authorship* (Ithaca, N.Y.: Cornell Univ. Press, 1988), in *Eighteenth-Cent. Stud.*, 1991, 24:372 (on Rousseau's pseudonym); and entry dated 17 Mar. 1787, Naples, in Goethe, *Italian Journey*, p. 202 (Goethe, *Italienische Reise*, Vol. 1, pp. 275–276) (on preferences within botany). It was clear to Goethe that philosophy was a healing agent superior to craft.

[61] Entry dated 2 Jan. 1787, Rome, in Goethe, *Italian Journey*, p. 143 (Goethe, *Italienische Reise*, Vol. 1, p. 203): "Ihr habt mir oft ausgespottet und zurückziehen wollen, wenn ich Steine, Kreuter und Tiere," etc. Note that *Italian Journey* translates "habt mir oft ausgespottet und zurückziehen wollen" as "You often used to make fun of." Goethe to Herder, 29 Dec. 1786, in Goethe, *Schriften zur Naturwissenschaft*, II:9A, p. 350.

1806, and 1817. Goethe wrote the piece when, as he put it, "an irresistible desire for country and garden life overtook people. Schiller bought a garden by Jena, and moved out there; [Christoph Martin] Wieland settled in Oßmannstedt," and Goethe himself bought a little manor house on the banks of the river Ilm.[62] Botanic art was the fashion, such as the drawings of "allegorical-symbolic-mystic plant metamorphoses" that the Romantic artist Philipp Otto Runge presented to Goethe in 1806.[63] So was botanic poetry that explained the latest scientific theories. Goethe's "Metamorphosis of Plants" was most immediately inspired by his January 1798 reading of a German translation of Erasmus Darwin's 1789 verse explication of Linnaeus's sexual system, *The Botanic Garden, Part II: Containing the Loves of the Plants.* Intended for "ladies and other unemploy'd scholars," it promised a racy read in its very first lines: "BOTANIC MUSE! who in this latter age/Led by your airy hand the Swedish Sage/Bade his keen eye your secret haunts explore."[64]

In a letter to Schiller of 26 January 1798, Goethe mercilessly mocked Darwin's heavy-footed poem.

> I only wish that I could show you this English work of fashion [*Modeschrift*] as it lies here in front of me, in a large quarto, bound in Morocco leather. It weighs exactly $5^1/_2$ pounds, as I made sure of myself yesterday. . . . It is sumptuously printed on shiny paper, with insanely allegorical copper etchings by [Johann Heinrich] Füßli . . . has introductions, lists of contents, footnotes, endnotes. . . . [But] among all these oddities the oddest, I think, is that in this botanic work you can find everything, except plants.

Goethe goes on, however, to outline in a more serious vein how *Loves of the Plants* is ordered: "Here then you have the scheme for a poem! This is how a didactic poem [*Lehrgedicht*] must look."[65]

[62] The passage about the passion for country life, referring to 1797, comes from Goethe's *Annalen;* it is reprinted in Goethe, *Schriften zur Naturwissenschaft,* II:9B, p. 127. This poem was first published in *Musen-Almanach für das Jahr 1799,* ed. Friedrich von Schiller (Tübingen: Cotta, [1798]). According to his diary, Goethe began writing the poem on 17 June 1798, at a time when he was also occupied in studying Schelling's *Weltseele.* For the standard version of "Metamorphosis of Plants" see Goethe, *Schriften zur Naturwissenschaft,* I:9, pp. 67–69; for the poem's printing history see *ibid.,* II:9B, pp. 476–480.

[63] The passage about Runge's drawings comes from Goethe's *Tischreden und Aphorismen,* 11 Dec. 1806; it is reprinted in Goethe, *Schriften zur Naturwissenschaft,* II:9B, p. 264. Goethe enthused about Runge's art as a starting point for the allegorization of the world of plants: "Die Nationen lassen sich auch mit Pflanzen, ihren Blüten und Früchten vergleichen. Die untern Stände sind die Kotyledonen und die daraus sich entwickelnden ersten Stengelblätter; die höhern Stände und die Kulturen derselben repräsentieren die fernern Blättern, Blüten, Früchte. Hier öffnete sich ein weites und artiges Feld für die Rungische allegorisch-symbolisch-mystische Pflanzenmetamorphose." The annotation on the same page in *Schriften zur Naturwissenschaft* remarks that when the artist gave Goethe these flower drawings, Johanna Schopenhauer commented: "der tiefe Sinn, der darin liegt, die hohe Poesie, das mystische Leben!"

[64] For an excellent treatment of Darwin's poem see Janet Browne, "Botany for Gentlemen: Erasmus Darwin and *The Loves of the Plants,*" *Isis,* 1989, *80:*593–621. The same topic has also been discussed by Londa Schiebinger, "The Private Life of Plants: Sexual Politics in Carl Linnaeus and Erasmus Darwin," in *Science and Sensibility: Gender and Scientific Enquiry, 1780–1945,* ed. Marina Benjamin (Oxford: Basil Blackwell, 1991), pp. 121–143. The description of the intended readership of the poem is quoted from a letter of Darwin's; see Browne, "Botany for Gentlemen," p. 595. For Goethe's diary annotations while writing "Metamorphosis of Plants" see Goethe, *Schriften zur Naturwissenschaft,* II:9B, pp. 130–131. According to his diary, Goethe read Darwin's work on 26 Jan. 1798. It is in a letter to Schiller written the same day that he mocks "diese englische Modeschrift" and notes that he "finde es wirklich unter meiner Erwartung." In a letter to Goethe dated 30 Jan. 1798 Schiller agrees that Darwin will not be a success in Germany.

[65] Goethe to Schiller, 26 Jan. 1798, in Goethe, *Schriften zur Naturwissenschaft,* II:9B, p. 130.

Inspired to write "Metamorphosis of Plants" as an improved *Loves of the Plants,* Goethe also intended it as part of a larger cycle of natural history poems, including verses on magnetism that apparently were never written and an undated fragment of hexameter verse entitled "Metamorphosis of Animals." "Metamorphosis of Plants" turned to older genres as well. Although Goethe abandoned his plan to pattern it strictly on Lucretius's *De natura rerum,* it belongs to the pastoral tradition both in its topic and in its meter, the mixed hexameter and pentameter characteristic of classical elegy.[66]

"Metamorphosis of Plants" had two audiences. Publicly, it was addressed to Goethe's

> women friends, who formerly had wanted to take me away from lonely mountains and from my study of lifeless stones, [and] again were far from satisfied with my abstract gardening [the theory of plant metamorphosis]. In their opinion, flowers and plants should have form, color and fragrance, whereas in my work they faded out into wraithlike figures. And so I attempted to entice the interests of these well-disposed ladies by an elegy.

In 1817, when Goethe explained the circumstances of the poem, he complacently reminisced that from these "*liebenswürdigen*" Weimar court women "I had much to suffer, for they wrote parodies of my metamorphoses, with droll and bantering allusions."[67] His poem thus served both as a verse popularization of a scientific tract and as a courtly vehicle for adult play, rephrasing gender stereotypes (e.g., man as abstract and woman as concrete) into controlling figures of an occasional poetry.

But "Metamorphosis of Plants" sought a second, private, audience as well. It was unofficially dedicated to a thirty-three-year-old former worker in Weimar's artificial flower factory, the woman who also inspired Goethe's 1788–1790 cycle of erotic poems, the *Roman Elegies.* Goethe met his lover and future wife, Christiane Vulpius, in July 1788, when she approached him as he was taking a walk and handed over a petition from her brother, an impoverished Romantic hack writer. During that first summer Goethe liked to think that his affair with Vulpius was secret. He even fancied, when he confessed to Charlotte von Stein, that only she knew. But for weeks the little town had busily gossiped about the *Herr Geheimrat*'s mistress, nestled in Goethe's summer residence, the "simple wooden cottage" in the English park of Goethe's own design. Unwilling or unable to handle a Weimar Faustina, Stein broke with Goethe. He cruelly suggested to her that her senses were overexcited by coffee drinking, and turned away from the cultured noblewoman in her forties to the semiliterate worker in her twenties.[68]

Goethe nicknamed Vulpius "the little Erotikon." His mother called her "the bed

---

[66] On the magnetism poem, mentioned in Goethe to Knebel, 16 July 1798, see *ibid.*, p. 477. On "Metamorphose der Tiere" see *ibid.*, pp. 481–487. Goethe printed one version of this poem in 1820 in the series *Zur Morphologie,* but it was probably first written in 1798 or 1799. In contrast to the "Metamorphosis of Plants," it does not explicate a previously written scientific prose text; instead, scientific findings are presented in verse from the start. On the initial plan to pattern his poem on Lucretius see Goethe, *Schriften zur Naturwissenschaft,* II:9B, p. 477, editors' annotations. Goethe owned a copy of *De rerum natura.*

[67] Goethe, "History," pp. 172, 174 (Goethe, "Geschichte," pp. 96, 99). The context makes it clear that Goethe here refers to a "Gesellschaft" specifically of women.

[68] For a facsimile edition of the *Roman Elegies* see Johann Wolfgang von Goethe, *Römische Elegien,* facsimile ed. and with commentary by Hans-Georg Dewitz (Frankfurt am Main: Insel, 1980). The quotation about the cottage comes from the *Encyclopaedia Britannica,* 1911 edition, s.v. "Weimar." On Goethe's first meeting with Vulpius see Friedenthal, *Goethe* (cit. n. 53), p. 279; on the break with Charlotte von Stein see *ibid.*, pp. 280–281. Vulpius (1765–1815) came from the *Kleinbürgertum,* but her family was downwardly mobile.

treasure." The person whom his 1797 will more prosaically called "my woman friend and housemate of many years" married Goethe on 19 October 1806.[69] Goethe was then fifty-six, the same age as Rousseau when, in 1768, he married his lover Thérèse Levasseur, a laundry maid. The two men married for the same reason: to reward their mistresses. Both were grateful for their lover's loyalty during hard times (Rousseau's political exile; Goethe's experience of the plunder of Weimar by Napoleon's troops in 1806. To commemorate the latter event, Goethe even predated the marriage rings five days, to 14 October, the day of the battle of Jena). In fairness to Goethe, however, the parallel between him and his botanic mentor ends here. Goethe and Vulpius reared their son at home, while Rousseau forced Levasseur to abandon their five children. "Maman" Warens, Rousseau's older mistress, compares badly to Goethe's Stein. And Goethe, though he may have engaged a prostitute in Rome, would hardly, like Rousseau in Venice, have purchased outright a twelve-year-old girl—to share, moreover, with a friend.

Goethe was ashamed of Vulpius, and never more than when Schiller first met her. Despite Schiller's respect for the great author, he could not disguise his surprise (the Romantic taste was for cultured, complex, and unconventional women from good families). Goethe himself thought Vulpius stupid. In August 1797 he warned her not to eat some zoological sea-mussel specimens dispatched from Hamburg: "Don't think that it's a snack." He could certainly be rude to her, as when she essayed that she feared his half-tame snake and he replied "Shut up, you!" and turned again to his visitor. During their first few years, however, Goethe also cherished Vulpius. His brief but affectionate letters from his travels arrived with packets of what he called "Jew tingle-tangle [*Judenkrämchen*]": lace, cloth, and small presents. By return post she sent him wine, cheese, and homegrown asparagus. He added kisses to "the little one," their first-born son August, and mourned with Vulpius when their five other children died at birth (from, it is thought, Rh-factor incompatibility).[70]

Goethe's letters to his "honest house-treasure" or "dear kitchen-treasure" were shot through with domestic admonitions. "Keep everything in good order." "Bring the house nicely in order." "Clean the hall for me really carefully." "I hope to find you well and the house in good order." There is something forlorn about the artless replies Vulpius sent to the man she in private called "treasure," "most beloved," and *Du* (the intimate form of address) and in public *Herr Geheimrat* and *Sie* (the formal form of address): "[I] only think about how I will bring everything in the household in order, to give you a little joy." "In the house and the garden you should find everything perfectly clean." "I love you and am hard-working and in all ways do my duties." Her rewards seem small, belated. The day after her marriage the forty-one-year-old *Geheimrätin* was introduced to Weimar society for the first time

---

[69] Friedenthal, *Goethe*, pp. 283, 282; and Johann Wolfgang von Goethe and Christiane Vulpius, *Wolfgang und Christiane: Goethes Ehe in den neunziger Jahren* (selected letters), ed. Siegfried Seidel (Weimar: Nationale Forschungs- und Gedenkstätten der klassischen deutschen Literatur in Weimar, 1989) (hereafter cited as **Goethe and Vulpius, *Wolfgang und Christiane***), p. 61. From Goethe's will of July 1797: "Ich setze nämlich den mit meiner Freundin und vieljährigen Hausgenossin, Christianen Vulpius, erzeugten Sohn August zu meinem Universalerben."

[70] Goethe to Vulpius, 15 Aug. 1797, in Goethe, *Schriften zur Naturwissenschaft*, II:9B, p. 111 (sea mussels); conversation among Falk, Goethe, and Vulpius in Goethe's garden, 30 June 1809, recorded by Falk, *ibid.*, p. 299 ("Schweig Du!"); and Goethe and Vulpius, *Wolfgang und Christiane, passim*. Gundolf, *Goethe* (cit. n. 15), p. 425, stresses how embarrassed Goethe was by Schiller's surprise on first meeting Vulpius. He casts their relationship as one of martyrdom for Goethe, saying in effect that the author of *Elective Affinities* should have abandoned her.

(as plain "Erotikon," she ate in the kitchen when there were guests). After hesitating for a few seconds, Johanna Henriette Schopenhauer graciously held out a cup of tea. Goethe, and perhaps Vulpius too, was deeply grateful. Nine years later, Vulpius's kidneys became acutely infected. Unnerved by her ceaseless screams, Goethe never entered the chamber where, for two days, she lay dying. As if in remembrance of what the *Hausfrau* had been, only one visitor came to say farewell—Karl August's mistress.[71]

The 1798 "Metamorphosis of Plants" aims to naturalize Goethe's and Vulpius's imperfectly socialized union, almost exactly ten years old when Goethe penned the poem. Styled at once as a "sacred riddle" and as an object lesson in Goethe's theory of the Primal Plant, the poem represents a movement from ignorance to knowledge. But there is no place in it for dialogue, for the inspired guesses and riddles that would characterize Goethe and Willemer's exchange of flowers and signs some fifteen years later. The poem is a determined monologue. It opens by narrating the woman reader's confusion on entering a garden and seeks to persuade her that "You are confused, beloved, by the mix of thousands/of flowers all over the garden./You hear many names, and always the one drives away/the next. Barbaric sounds all mingle in your ears./All forms are like each other, and none is like the/other."[72]

To the "beloved" names have become only sounds, and her sight is blurred. Her state of mind is presented as a simple fact, and it both provides a poetic starting point and sets a pedagogic task for the male writer-teacher. In fact, it is an impossibility: for the woman reader's state of mind to be as described, she would have to have given up voluntarily (as if such a volition was possible) what Maurice Merleau-Ponty has called the horizon of our thought, that is, the unconscious preordering of our sense impressions. She would have to have accepted the absurd premise that since she does not know science, a specific knowledge form, she literally cannot perceive the garden she has just entered.

It is an ironic expectation, given Goethe's philosophy of science—and also given Vulpius's personal taste. As noted, in the summer of 1788 she lived in the summer cottage. Later, when she had moved into the winter residence, the garden remained her chief joy. "I am busy now with the fruit," she wrote Goethe in October 1799. "When that's over it's about potatoes and apples." In 1794 she declared that "the garden gives me much joy, I can hardly pull myself away from it."[73] Vulpius was an interested amateur gardener, well versed in the everyday care of plants (Figure 1).

Yet, though at other times and for himself Goethe celebrated the privileged moment of the authentic experience, he did not allow Vulpius's garden *Erlebnis* to form

---

[71] Goethe to Vulpius, 27 Sept., 21 Aug., 9 Aug., 17 Aug. 1792, 10 July 1793, 2 Sept. 1795; Vulpius to Goethe, 17 or 18 June, 5 July 1793, 6 Oct. 1797: all in Goethe and Vulpius, *Wolfgang und Christiane*, pp. 10, 7, 6, 7, 24, 37, 20, 23, 86. For the private forms of address see *ibid., passim;* for the public forms see Friedenthal, *Goethe* (cit. n. 53), p. 282. On Vulpius's introduction to Weimar society and on her death see *ibid.*, pp. 283–284, 287.

[72] Goethe, "Metamorphosis of Plants," in Goethe, *Gedichte*, p. 129: "Dich verwirret, Geliebte, die tausendfältige Mischung/Dieses Blumengewühls über dem Garten umher;/Viele Namen hörest du an, und immer verdränget/Mit barbarischem Klang einer den andern im Ohr./Alle Gestalten sind ähnlich, und keiner gleichet der/andern." The translation is mine. Another translation, by Heinz Norden, is provided in Goethe, *Botanical Writings*, p. 172.

[73] Vulpius to Goethe, 2 Oct. 1799, 11 Apr. 1795, in Goethe and Vulpius, *Wolfgang und Christiane*, pp. 85, 33.

Figure 1. Woman with watering can in garden: believed to depict Christiane Vulpius in Goethe's city house garden. Pencil and touche drawing, 110 × 186 mm. Undated; executed after 1802. (From Johann Wilhelm von Goethe, Corpus der Goethezeichnungen, ed. by the Nationalen Forschungs- und Gedenkstätten der klassischen deutschen Literatur in Weimar [Leipzig: Seemann, 1958–1973], Vol. 4B [1968], drawing no. 72.)

the basis of a science. Her womanly experience is scrambled and discarded when, over the course of the poem, he overlays her garden with his own interpretative grid. Goethe probably determined to teach Vulpius his botany partly to prevent her teaching herself. As we noted, the German book market around 1800 offered many vernacular botany tracts addressed to unlearned women. But Goethe had a taste for carefully managed female ignorance. The author of the *Roman Elegies* even advocated a theory of asexual plant reproduction developed in Schelling's circle of *Naturphilosophen*, observing that

> for the instruction of young persons and ladies this new pollination theory will be extremely welcome and suitable. In the past the teacher of botany has been placed in a most embarrassing position, and when innocent young souls took textbook in hand to advance their studies in private, they were unable to conceal their outraged moral feelings. . . . Indeed, we recall having seen arabesques in which the sexual relations within a flower calyx were represented, in the manner of the ancients, in an extremely graphic way.[74]

Goethe cherished his oft-donned role as a ladies' teacher. During his 1785 Karlbad

[74] Johann Wolfgang von Goethe, "Pollination, Volatilization, and Exudation," in Goethe, *Botanical Writings*, p. 109; for the German see Goethe, "Verstäubung, Verdunstung, Vertropfung," in Goethe, *Schriften zur Naturwissenschaft*, I:9, pp. 214–215. The essay was first published in *Zur Morphologie*, Vol 1, No. 3 (1820). The theory of asexual plant propagation that Goethe took up was authored by Franz Joseph Schelver (1778–1832), professor of medicine in Jena, curator of Weimar's Grand-Ducal Botanical Garden, and author of *Kritik der Lehre von den Geschlechtern der Pflanze*, 3 vols. (Heidelberg, 1812, 1814, 1823). For Goethe's 1812 correspondence with Schelver on his work see Goethe, *Schriften zur Naturwissenschaft*, II:9B, pp. 346, 348. In this correspondence Goethe praises Schelver only cautiously.

visit he led afternoon walks where he "developed in thoughtful speeches the elements of his theory of [plant] metamorphosis to his beautiful female listeners." At the end of each day he distributed among his women admirers the raw materials of his scientific demonstrations, as elegantly arranged bouquets. In the spring of 1807 Goethe gave private lessons on botany and on Humboldt's South American voyage to Weimar "ladies."[75] A year later, in April 1808, he spontaneously lectured on a cancerous growth of a dwarf mountain pine to an assembly of women at Charlotte von Stein's. "The ladies dearly would have wanted to kiss his hands." Science provided Goethe with an elegant way to flirt. A few months afterward, taking the waters at Karlsbad spa in the summer of 1808, he taught botany in private sessions to the young Pauline Gotter (who later married Schelling). Unsurprisingly, the girl was ecstatic. It was, after all, agreed in her circle that Goethe's only equals were Homer and Shakespeare. "I could say, couldn't I, without bragging, that he especially cherished me . . . he often came early [in the morning] to give me botanic lessons, and a few times he took me, just the two of us, for long walks."[76]

It was thus as a ladies' man and science teacher that Goethe continued the "Metamorphosis of Plants," after having decreed the woman reader's need for instruction and established himself as the scientific authority. In some eighty lines he outlines his theory of plant metamorphosis, analogizing it to sexual love in excruciating detail and evasive terms. Then, toward the end of the poem, he again addresses the woman reader: "Turn now your eyes, beloved, to the many-colored profusion./It doesn't confuse you any more./Now every plant announces eternal laws to you./Every flower speaks louder and louder with you."[77]

In "Metamorphosis of Plants" Goethe's botany is at once a means of producing and a metaphor for describing a particular relation between the sexes. As the poem closes, it becomes clear that he aims to make the female reader simultaneously a passive recipient and an object of knowledge. For there he reveals that the woman reader's full comprehension of his verses must mean her acceptance of her inevitable seduction. "O think also then on how from the seed of our acquaintance/Little by little sweet habits grew in us./Inside ourselves friendship powerfully lodged itself/ And like Amor at last could show flowers and fruits."[78]

That Goethe made the science of botany part of his seducer's repertoire is also evidenced in some occasional verses from 1812, which he gave to Vulpius to memorialize their twenty-four-year-old relationship. In this brief poem, tellingly entitled "Found," Vulpius is not, as in "Metamorphosis of Plants," merely analogized to a

[75] Material on the Karlsbad afternoon walks is from a notation from 1895, based on Dietrich's account (now lost), and from Goethe's 1817 "History"; see Goethe, *Schriften zur Naturwissenschaft*, II:9A, p. 323. On the lessons that ended with the distribution of bouquets see *ibid.*, 9B, pp. 277–289. Goethe's diary records such lessons for 1 Apr., 15 Apr., 6 May, and 13 May 1785, usually with a terse "Um 10 Uhr die Damen" (15 Apr.; p. 279) or "Kam der Damenbesuch" (6 May; p. 180).

[76] Charlotte von Stein to F. von Stein, 22 Apr. 1808, in Goethe, *Schriften zur Naturwissenschaft*, II:9B, p. 288; and recorded conversation between Karoline Schelling and Pauline Gotter, *ibid.*, p. 289. See also Goethe's diary entries for 6 and 8 July 1808, *ibid.*, pp. 288–289.

[77] Goethe, "Metamorphosis of Plants," in Goethe, *Gedichte*, p. 131: "Wende nun, o Geliebte, den Blick zum bunten Gewimmel,/Das verwirrend nicht mehr sich vor dem Geiste bewegt./Jede Pflanze verkündet dir nun die ewigen Gesetze,/Jede Blume, sie spricht lauter und lauter mit dir." The translation is mine; cf. Norden's translation in Goethe, *Botanical Writings*, p. 173.

[78] Goethe, "Metamorphosis of Plants," in Goethe, *Gedichte*, p. 131: "Oh, gedenke denn auch, wie aus dem Keim der/Bekanntschaft/Nach und nach in unse holde Gewohnheit entsproß,/Freundschaft sich mit Macht in unserm Innern enthüllte,/Und wie Amor zuletzt Blüten und Früchte gezeugt." The translation is mine; cf. Norden's translation in Goethe, *Botanical Writings*, p. 173.

plant in her ability to "show flowers and fruits." She is *herself* metamorphosed into a plant. "Found" opens with Goethe walking in a wood aimlessly and alone: "I did not search for anything." Then he sees a small flower "in the shadow." This obvious simile alerts the reader to her humble origins. He drives home the point by confessing that his first impulse on finding this flower was to pluck it. The German "*es brechen*" translates literally as "to break it" (this sexual metaphor appears most famously in Goethe's popular poem "Heidenröslein"). Vulpius was not, in other words, a girl of good family, protected from sexual assault by the ongoing patriarchal supervision of older relatives. As his hand reaches down, the flower speaks: "Should I be broken, only to wither away?" Goethe then abandons his first strategy—to seduce and desert—as he recognizes the consequences for the "flower" (ostracization and poverty). Instead, he replants her in his garden house, where she "continues to bloom" but inhabits a "quiet place." "I dug out the little herb/With roots and all,/I carried her to the garden/To the pretty house//And replanted her/In this quiet place;/Now she is always silent/And continues to bloom."[79] Instead of seducing and deserting the "little herb," Goethe trades material protection, if not social recognition, for her ongoing sexual favors. Once this trade relation is set up, Vulpius reverts into the silence that, the poem suggests, ideally characterizes women and plants alike: "Now she is always silent." The poem closes there, because its narrativity depends on the flower momentarily entering the realm of history and speech. Its mere existence cannot form the basis of a narrative, since tales need a dialogic tension. Vulpius and the poem about her must fall silent simultaneously.

In this self-congratulatory poem about self-imposed limits of male power, Goethe dictates that Vulpius can express herself only in silence. He makes her what we most fear becoming: a "vegetable." The link between nature and women that American ecofeminists celebrate here recovers its original place within an attempted male appropriation of language. For when Goethe casts his female lover as an "always silent" plant, he involves himself in a peculiar perversion of the German eighteenth-century ideal of *Bildung,* the claim that regardless of social background, each member of society can achieve a public voice and a personal identity through a program of cultural self-improvement.[80] As we have seen, Goethe's botany was part of that quest for *Bildung*. In the botanic poems "Metamorphosis of Plants" (1798) and "Found" (1812) he reveals that he builds into the concept itself a notion that the male voice is dependent on women's silence.

Or, even, on women's death. "Found" models Vulpius onto Ottilie, the young and innocent heroine of Goethe's *Elective Affinities*. "You are in love with her, Goethe, I have already suspected it for a long time," Bettina von Arnim coquettishly accused him. Another acquaintance remembered a talk in 1821, when Goethe still fantasized about his Gothic Pygmalion. "The stars had risen; he talked of his relation

[79] Goethe and Vulpius, *Wolfgang und Christiane*, p. 4: "Ich ging im Walde/So für mich hin,/Und nichts zu suchen,/Das war mein Sinn.//In Schatten sah ich/Ein Blümchen stehn,/Wie Sterne leuchtend,/Wie Äuglein schön.//Ich wollt es brechen,/Da sagt' es fein:/"Soll ich zum Welken/Gebrochen sein?"//Ich grub's mit allen/Den Würzlein aus,/Zum Garten trug ich's/Am hübschen Haus.//Und pflanzt es wider/Am stillen Ort;/Nun zweigt es immer/Und blüht so fort." The translation is mine; double slashes indicate ends of stanzas.

[80] On the German notion of *Bildung* see, e.g., Henri Brunschwig, *Enlightenment and Romanticism in Eighteenth-Century Prussia* (1947), trans. Frank Jellinek (Chicago: Univ. Chicago Press, 1974).

to Ottilie, how he loved her, and how she made him miserable. He finally became almost mysteriously reverential as he talked on."[81]

Toward the end of the novel, Goethe allows his "dear Ottilie," a virgin who loves a married man she can never have, only one means of emotional expression: self-starvation. In the last chapter she lies dying—eyes sunken, lips drawn back, ribs and hips protruding. "The doctor had some bouillon brought up; Ottilie pushed it away in disgust, yes she almost had convulsions, when people brought the cup closer to her mouth." A maid confesses that Ottilie, speechless and self-immured in her rooms, "already for a long time had hardly eaten anything at all." As the novel's other characters gather around Ottilie's emaciated pubertal body, the farewells conventional to death around 1800 are only gestured by the mute and dying girl. Her death remains the only thing she herself can choose. At the precise and only moment she speaks—to make her beloved promise not to commit suicide and thus cheapen by imitation the single expression that Goethe allows as authentically her own—she dies.[82]

Perhaps instead of joining Goethe and his contemporaries in their cult of the dying Ottilie, we should consider the terse comment of a friend of the brothers Grimm, referred to in a letter from Wilhelm to Jakob in December 1809, when Goethe's newly published novel was the talk of Germany: "Steffens is of the opinion that the child dies like a dog." If we agree, Goethe's exasperated cry to an elderly male critic of *Elective Affinities* seems all the more chilling: "After all, I haven't written it for you, I have written it for young girls!"[83]

---

[81] Bettina von Arnim to Goethe, 9 Nov. 1809, in Goethe, *Wahlverwandtschaften* (Munich: Deutscher Taschenbuch, 1980), p. 288, annotations; and conversation between Goethe and Boisserée, 5 Feb. 1815, quoted from Boisserée's notes, *ibid.*, p. 263, annotations.

[82] Goethe, *Wahlverwandtschaften*, pp. 248–249, 249. Ottilie's last words are "Versprich mir zu leben!" Goethe's reference to "dear Ottilie" comes in a letter to Karl Friedrich von Reinhardt, 21 Feb. 1810, quoted *ibid.*, p. 262, annotations.

[83] Wilhelm Grimm to Jakob Grimm, 3 Dec. 1809, quoted in Goethe, *Wahlverwandtschaften*, p. 280, annotations. Wilhelm adds: "Und doch ist die Szene so rührend und die Nacht darauf recht mit der Stille dargestellt." Goethe's complaint to the elderly male critic is quoted *ibid.*, p. 262, annotations. The story comes from the diary of Varnhagen von Ense, 28 June 1843. He relates a story he heard at second hand, from the critic; von Ense was not a party to the conversation.

The American Career of Jane Marcet's *Conversations on Chemistry, 1806–1853*

By M. Susan Lindee

JANE HALDIMAND MARCET'S *Conversations on Chemistry* has traditionally claimed historical attention for its effect on the young bookbinder Michael Faraday, who was converted to a life of science while binding and reading
it. Marcet "inspired Faraday with a love of science and blazed for him that road
in chemical and physical experimentation which led to such marvelous results,"
in H. J. Mozans's romantic account. Or, as Eva Armstrong put it, Marcet led
Faraday to "dedicate himself to a science in which his name became immortal."[1]

In these accounts Marcet is important for her effect on one prominent male
scientist. But her influence was much wider: *Conversations on Chemistry* was
the most successful elementary chemistry text of the period in America. American publishers printed twenty-three editions of Marcet's text, and twelve editions
of an imitative text derived from it. Many young men and women had their first
serious exposure to chemistry through the lively discussions of Mrs. B., Emily,
and Caroline, the characters Marcet used to convey her ideas. The book was
widely used in the new women's seminaries after 1818. There is also evidence
that young men attending mechanics' institutes used Marcet's text, and medical
apprentices favored it in beginning their study of chemistry.[2]

The widespread use of Marcet's book in the early women's schools is of particular interest. Allusions to domestic applications of science and the spiritual insight it offered were commonly used to justify science instruction for women in
these new institutions. But did the texts and style of instruction bear out that
justification? If they did not, what might this suggest about the goals and intentions of those offering scientific training to young women?

I have compared Marcet's *Conversations* with other elementary chemistry
texts published in the United States between 1806 and 1853. My purpose is to
shed some light on the priorities of a poorly understood group: teachers and

Special thanks to Margaret Rossiter, Sally Gregory Kohlstedt, Jeffrey Sturchio, R. Lawrence
Moore, and L. Pearce Williams. This essay won the 1988 Schuman Prize.

[1] H. J. Mozans, [John Zahm], *Woman in Science* (New York: Appleton, 1913), pp. 372–373; and
Eva V. Armstrong, "Jane Marcet and Her *Conversations on Chemistry*," *Journal of Chemical Education*, 1938, *15*:53–57. L. Pearce Williams gives Marcet's work credit for leading Faraday to connect
his fascination with electrical phenomena with forces of "fundamental importance in the universe"—
chemical reactions; see Williams, *Michael Faraday: A Biography* (New York: Simon & Schuster,
1965), pp. 18–20.

[2] John K. Crellin, "Mrs. Marcet's *Conversations on Chemistry*," *J. Chem. Educ.*, 1979, *56*:459–
460.

Jane Haldimand Marcet (1769–1858). Courtesy of the Beckman Center for History of Chemistry.

administrators at intermediate or college-level women's schools in the first half of the nineteenth century. I show that while these educational reformers had numerous options, they favored a chemistry text that was theoretical and experimental: Marcet's *Conversations on Chemistry*. More "domestic" or practical chemistry textbooks, which were widely available, fared poorly, as did the less common textbooks emphasizing chemistry's spiritual lessons.

School administrators and instructors used domestic and religious justifications to increase the social acceptability of science education for women in the early nineteenth century. My work suggests, however, that the actual instruction at the women's schools promoted feminine interest in scientific theory at a level that exceeded that required for domestic efficiency or religious gratification.

MARCET AND HER SOCIAL CIRCLE

Jane Haldimand Marcet (1769–1858) was born in London of a prosperous Swiss family. When she was thirty years old, she married Alexander Marcet, a London physician and chemist. Her husband's social circle included J. J. Berzelius, Humphry Davy, the botanist Augustin de Candolle, the mathematician H. B. de Saussure, the writers Harriet Martineau and Maria Edgeworth, the political economist Thomas Malthus, the physicist and naturalist Auguste de la Rive, and the chemists Pierre Prevost and Marc Auguste Pictet.[3]

Such social connections gave Marcet access to new ideas and she translated this access into a long, productive writing career. After the success of *Conversations on Chemistry,* her first book, came *Conversations on Political Economy* (1817), *Conversations on Natural Philosophy* (1820), and *Conversations on Vegetable Physiology* (1829). She published anonymously until 1837, and for this and other reasons her works were often attributed either to other women writers, or (in America) to the male commentators whose names appeared on the title page. Marcet's use of "Mrs. B." as the instructor in these conversations led to speculation that the author was Margaret Bryan, a British popularizer of science already prominent when Marcet was a child. Marcet may, of course, have chosen "Mrs. B." as an allusion to Bryan. Marcet's *Conversations* was also attributed to other women writing about science, including Sarah Mary Fitton, who wrote *Conversations on Botany* in 1817.[4]

All of Marcet's later *Conversations* involved the characters—Mrs. B., Caroline, and Emily—introduced in *Conversations on Chemistry*. Caroline, an impetuous and skeptical student, was somewhat more interested in explosions than in fundamentals of science. Emily was serious and bright, and more likely to ask important questions. The two young women were thirteen to fifteen years old (at

---

[3] Alexander Marcet was a physician and, later, chemistry professor at Guy's Hospital, London. When his wife inherited a substantial fortune upon the death of her father in 1817, Marcet was able to give up medicine and devote himself to chemistry. He was the author of several scientific papers, and his work on the specific heats of gases was cited in other textbooks and in his wife's book. Marcet's social circle is briefly explored in Armstrong, "Jane Marcet" (cit. n. 1). See also Auguste de la Rive's obituary notice for Jane Marcet, "Madame Marcet," *Bibliothèque Revue Suisse et Etrangère,* 1859, N.S., *4*:445–468 (transcription and translation in the Edgar Fahs Smith Collection, University of Pennsylvania, Philadelphia). Alexander Marcet is mentioned in J. R. Partington, *A History of Chemistry* (London: Macmillan, 1972); see also John Read, *Humour and Humanism in Chemistry* (London: Bell, 1947), pp. 177–191, for some biographical details of Marcet's life.

[4] Bryan's works included *A Compendious System of Astronomy* (London, 1797), *A Comprehensive Astronomical and Geographical Class Book* (London, 1815), and *Lectures on Natural Philosophy* (London, 1806).

least, Emily's age was given in *Conversations on Natural Philosophy* as thirteen) and apparently not related. Caroline's father owned a lead mine in Yorkshire, and Emily's family background was not mentioned.[5]

They were young women of wealth, well educated and sensitive to social conventions. In her introduction to the chemistry text Marcet apologized for their intelligence: "It will no doubt be observed that in the course of these Conversations, remarks are often introduced, which appear much too acute for the young pupils, by whom they arc supposed to be made. Of this fault the author is fully aware." She explained that the unusual brightness of the pupils was necessary lest the work become "tedious."

In the opening conversation, Caroline claimed to be uninterested in the science of chemistry:

> *Caroline.* To confess the truth, Mrs. B., I am not disposed to form a very favourable idea of chemistry, nor do I expect to derive much entertainment from it. I prefer the sciences which exhibit nature on a grand scale, to those that are confined to the minutiae of petty details.
> *Mrs. B.* I rather imagine, my dear Caroline, that your want of taste for chemistry proceeds from the very limited idea you entertain of its object. . . . [Nature's laboratory] is the Universe, and there she is incessantly employed in chemical operations. You are surprised, Caroline; but I assure you that the most wonderful and the most interesting phenomena of nature are almost all of them produced by chemical powers.

When the conversation turned serious, Emily joined in, and the first lesson centered on "constituent" and "integrant" parts. The book then progressed in twenty-six conversations from simple to compound bodies, and from elements to living systems. Marcet included discussions of light and heat, electricity, oxygen and hydrogen, sulfur and carbon, metals, attraction, acidification, decomposition, and animal productions. A twenty-seventh conversation, on the steam engine, was added from 1830 on.

This range of topics indicates the parameters of early nineteenth-century chemistry. The field included—in some fashion—geology, mineralogy, electricity, fermentation, plant respiration, and animal growth. Chemists studied meteors, minerals, animal phosphorescence, medicinal cures, and soil samples. Marcet could, quite reasonably, turn her attention to "bones, teeth, horns, ligaments and cartilage" or to the "effects of Light and air on Vegetation."

But chemistry, however broadly defined, was not the only subject of the text. By the time Marcet wrote her chemistry book, she had already completed her *Conversations on Political Economy* (published later) and some of the themes from that volume made their way into her chemistry lessons. For example, she touched on problems of class. She had Mrs. B. proclaim that the "well-informed" were often too eager to adopt new technology, while the uninformed, "having no other test of the value of a novelty but time and experience" were sometimes able to "prevent the propagation of error." Mrs. B. also praised England's colliers, "digging out of the bowels of the earth one of the most valuable necessaries of life." She expressed disdain for scientific pretense, urging Caroline not to use the word *oxydate* rather than *rust,* "for you might be suspected of affectation."[6]

---

[5] Jane Marcet, *Conversations on Chemistry* (Hartford, 1839), p. 150 (Caroline's father). For information on the various editions, see Table 1.

[6] 1806 Philadelphia edition, p. 81 (on the well-informed); 1822 Hartford edition, p. 116, and 1839 Hartford edition, p. 125 (on colliers); and 1829 Hartford edition, p. 162 (on *oxydate* vs. *rust*).

Marcet also was aware of the sexual politics of her work and made frequent reference to her feminine readers and their presumed interest in science. In her preface she apologized for daring to publish a work on science, describing her apprehension that her work would be considered "unsuited to the ordinary pursuits of her sex." But the recent establishment of public institutions "open to both sexes, for the dissemination of philosophical knowledge," proved that "general opinion no longer excludes woman from an acquaintance with the elements of science."

She explained that her interest in chemistry was aroused by attendance at the public lectures of Humphry Davy, which she initially found confusing. When the basic concepts of the new chemistry were explained to her in "familiar conversations," Marcet said, she could enjoy Davy's lectures much more. "Hence it was natural to infer that familiar conversation was, in studies of this kind, a most useful auxiliary source of information, and more especially to the female sex, whose education is seldom calculated to prepare their minds for abstract ideas, or scientific language." Her book was written because "there are but few women who have access" to scientific friends, such as her own, willing to converse with them about theory. (John Crellin has suggested that Marcet's "scientific friend" was almost certainly her husband.[7])

At the same time, she did not promote an unseemly female participation in science. When Caroline mentioned pharmaceutical chemistry, Mrs. B. proclaimed that pharmaceutical work "belongs exclusively to professional men, and is therefore the last [branch of chemistry] that I should advise you to pursue."[8]

In its approach to chemistry, Marcet's book was theoretical rather than practical. She updated her treatment of important ideas in later editions, and at least on some topics her elementary text kept pace with scientific changes. She followed Antoine-Laurent Lavoisier's scheme of classification of the elements, as laid out in the 1796 English translation of his *Traité élémentaire de chimie*, considering light, electricity, and caloric "imponderable agents." She somewhat conservatively clung to the caloric theory, however, even after Davy had abandoned it. She used a Newtonian, corpuscular theory of matter, and she explained chemical reactions in terms of affinity, aggregation, gravitation, and repulsion.[9]

Marcet did not mention John Dalton's atomic theory until after 1819, and even then she expressed doubts about its validity. This in part reflected Davy's skepticism. But it was a skepticism widely shared; many other writers of chemistry texts, including Thomas Thomson, W. T. Brande and Andrew Ure, continued to question Dalton's theory (as an explanation for the fundamental nature of matter) until as late as 1841.[10]

---

[7] Crellin cites a December 1803 letter from the London physician John Yelloly to Alexander Marcet, in which Yelloly seems to imply that Alexander is responsible for the quality of Jane's (as yet unpublished) manuscript: Crellin, "Marcet's *Conversations on Chemistry*" (cit. n. 2).

[8] 1839 Hartford edition, p. 14.

[9] A.-L. Lavoisier, *Principles of Chemistry in a New Systematic Order*, trans. from the 2nd French ed. (1793) by Robert Kerr (Edinburgh, 1796). David Knight's comparison of Marcet's chemistry and that of Samuel Parkes, a contemporaneous writer who wrote for young men, emphasizes Marcet's theoretical assumptions; see Knight, "Accomplishment of Dogma" (cit. n. 4).

[10] Thomas Thomson, *A System of Chemistry of Inorganic Bodies* (Edinburgh, 1831), pp. 3–31; W. T. Brande, *A Manual of Chemistry* (London, 1841), pp. 234–238; and Andrew Ure, *A Dictionary of Chemistry* (London, 1831), pp. 443–445. I am indebted to J. R. Clarke for these useful comparisons, explored in his unpublished paper "Jane Marcet and Her *Conversations on Chemistry*" (J. R. Clarke, 15 Exeter Grove, Belmont, Victoria 3216, Australia).

Despite its elementary nature, Marcet's treatment of chemical theory compared favorably with that of the Scottish chemist Edward Turner in his much-admired college-level textbook *Elements of Chemistry*. Turner's book, first published in 1827, was about as widely used and imitated as Marcet's. He had at least three near-plagiarists in America: Lewis C. Beck, John Lee Comstock (one of Marcet's editors), and John Johnston all produced chemistry texts that depended heavily on Turner. They borrowed his organizational format, illustrations, charts, appendixes and, in many cases, his words. All acknowledged their debt to Turner, stating that their work was "on the basis of Turner's *Elements of Chemistry*" or that Turner was "used more freely than any other" author.[11] Both Turner's text and those drawn from it were popular in American men's colleges.

Marcet's handling of chemical theory was remarkably consistent with Turner's. In a point-by-point comparison of the two authors' treatment of heat as an "imponderable substance," the correlation of the subjects explored and scientists cited is very high. Both discussed William Herschel's studies of light and heat, John Leslie's work with the radiation of heat, Marc-Auguste Pictet's *Essai sur le feu*, Count Rumford on clothing and the conduction of heat, William Wells's theory of the formation of dew, the problem of cold as the absence of heat (rather than as a negative quality), the use of a pyrometer, and Pierre Prevost's studies of radiation.[12]

Marcet's text also kept pace with William Brande's *Manual of Chemistry*. Brande was professor of chemistry at the Royal Institution (appointed to replace Humphry Davy in 1812), and his text was intended as an advanced accompaniment to his three-month lecture course for men, which he taught with Faraday as his assistant. Brande's and Marcet's classification schemes were very similar, with Marcet's in some ways superior. She organized the elements on the basis of their presumed nature, and she attempted to construct a meaningful system that would help her students understand the processes of chemistry. Marcet's book would have covered Brande's course adequately, for the topics listed in his syllabus and those discussed by Marcet were largely the same.[13]

These comparisons suggest that Marcet's book was no collection of tips for homemakers and farmers, but an introduction to the most important chemical theories of her day. Its popularity in the new women's schools in America therefore raises questions about the goals and priorities of the educational reformers who taught there.

MARCET'S *CONVERSATIONS* IN AMERICA

Conversations on Chemistry was first published in London in 1806. The first American edition appeared later that same year. From 1806 to 1850, American publishers made twenty-three impressions of various editions of the work, at Hartford, Boston, Philadelphia, New Haven, and New York.[14] There were also

---

[11] See Lewis C. Beck, *A Manual of Chemistry* (Albany, N.Y.: Webster & Skinners, 1831); John Lee Comstock, *Elements of Chemistry* (New York: Robinson, Pratt, 1839); and John Johnston, *A Manual of Chemistry* (Philadelphia: Charles Desilver, 1861) (1st pub. 1848).

[12] Cf. Marcet, 1839 Hartford edition, conversations 2 and 3, with Edward Turner, *Elements of Chemistry* (1842), pp. 9–50.

[13] Clarke makes this comparison in "Marcet and Her *Conversations on Chemistry*" (cit. n. 10).

[14] The book was often printed more than once in a single year by competing publishers, e.g., by Increase Cooke of New Haven and James Humphreys of Philadelphia in 1809. Two or more runs of Marcet's book or the imitative Thomas P. Jones *New Conversations on Chemistry* were also produced by various American publishers in 1818, 1824, 1831, 1836, 1839, and 1844. Cornell's efficient

Table 1. Printings of American Editions of Marcet's *Conversations on Chemistry.*

1806 Jane Marcet. *Conversations on Chymistry*. Philadelphia: James Humphreys on Change Walk. "In which the elements of that science are familiarly explained and illustrated by experiments and plates."

1809 Jane Marcet. *Conversations on Chemistry*. New Haven: Increase Cooke & Co. "To which are added some late discoveries on the subject of the fixed alkalies, by H. Davy; A description and plate of the pneumatic cistern at Yale College—and a short account of artificial mineral waters in the United States."

Jane Marcet. *Conversations on Chemistry*. Philadelphia: James Humphreys. With "an appendix consisting of a description of the new hydro-pneumatic blow pipe . . . also of three disquisitions, one on dyeing, one on tanning and one on currying."

1813 Jane Marcet. *Conversations on Chemistry*. New Haven: Increase Cooke & Co.

1814 Jane Marcet. *Conversations on Chemistry*. New Haven: Sidney's Press.

1818 Jane Marcet. *Conversations on Chemistry*. Greenfield, Mass.: Denio & Phelps. "From the 4th and latest English edition, revised, corrected, and considerably enlarged. To which are added notes and observations: by an American gentleman" (Comstock).

Jane Marcet. *Conversations on Chemistry*. Philadelphia: M. Carey & Son. "Rev. and cor. by Thomas Cooper, M.D., from the 5th London, considerably enl."

1820 Jane Marcet. *Conversations on Chemistry*. Greenfield, Mass.: Denio & Phelps. Additions by "an American gentleman" (Comstock).

1822 Jane Marcet. *Conversations on Chemistry*. Hartford: O. D. Cooke & Co. "To which are now added explanations by J. L. Comstock."

1824 Jane Marcet. *Conversations on Chemistry*. Hartford: Cooke. Additions by Comstock.

Jane Marcet. *Conversations on Chemistry*. Philadelphia: Thomas DeSilver; Baltimore: J. E. Coale. 10th American ed. "Anonymous." Comments by William H. Keating.

1826 Jane Marcet. *Conversations on Chemistry*. Hartford: Cooke. "To which are now added, explanations of the text by J. L. Comstock, M.D., together with a new and extensive series of questions by Rev. J. L. Blake."

1828 Jane Marcet. *Conversations on Chemistry*. Hartford: Cooke. "To which are now added explanations of the text, directions for simplifying the apparatus, and a vocabulary of terms—together with a list of interesting experiments" by Comstock, with questions by Blake.

1829 Jane Marcet. *Conversations on Chemistry*. Hartford: Cooke. 11th American ed. from the 8th London ed. Additions by Comstock and Blake.

1830 Jane Marcet. *Conversations on Chemistry*. Hartford: Cooke. 12th American ed. Additions by Comstock and Blake.

1831 Thomas P. Jones. *New Conversations on Chemistry*. Philadelphia: John Grigg. "Adapted to the present state of that science; wherein its elements are clearly and familiarly explained. With one hundred and eighteen engravings . . . appropriate questions; a list of experiments, and a glossary. On the foundations of Mrs. Marcet's Conversations on Chemistry."

Jane Marcet. *Conversations on Chemistry*. Hartford: Cooke. 13th American ed. from the last London ed. Additions by Comstock and Blake.

1832 Thomas P. Jones, *New Conversations on Chemistry*. Philadelphia: Grigg.

NOTE: Numbers for the various editions are those given by the respective publishers.

Table 1—*continued*

1833 Jane Marcet. *Conversations on Chemistry*. Hartford: Cooke. 14th American ed. Additions by Comstock and Blake.

Thomas P. Jones. *New Conversations on Chemistry*. Philadelphia: Grigg.

1834 Thomas P. Jones. *New Conversations on Chemistry*. Philadelphia: Grigg.

1835 Jane Marcet. *Conversations on Chemistry*. Hartford: Beach & Beckwith. 15th American ed. Additions by Comstock and Blake.

1836 Jane Marcet. *Conversations on Chemistry*. Hartford: John Beach; New York: Collins, Keese & Co. 15th American ed. Additions by Comstock and Blake.

Thomas P. Jones. *New Conversations on Chemistry*. Philadelphia: Grigg.

1838 Thomas P. Jones. *New Conversations on Chemistry*. Philadelphia: Grigg.

1839 Thomas P. Jones. *New Conversations on Chemistry*. Philadelphia: Grigg.

Jane Marcet. *Conversations on Chemistry*. Hartford: Belknap & Hamersley. 15th American ed. Additions by Comstock and Blake.

1841 Jane Marcet. *Conversations on Chemistry*. Hartford: Belknap & Hamersley. 15th American ed. Additions by Comstock and Blake.

1842 Thomas P. Jones. *New Conversations on Chemistry*. Philadelphia: Grigg & Elliot.

1844 Jane Marcet. *Conversations on Chemistry*. Hartford: Belknap & Hamersley. 15th American ed.

1845 Thomas P. Jones. *New Conversations on Chemistry*. Philadelphia: Grigg & Elliot.

1846 Thomas P. Jones. *New Conversations on Chemistry*. Philadelphia: Grigg & Elliot.

1848 Thomas P. Jones. *New Conversations on Chemistry*. Philadelphia: Grigg & Elliot.

1850 Jane Marcet. *Conversations on Chemistry*. Hartford: Belknap & Hamersley. Additions by Comstock and Blake.

Thomas P. Jones. *New Conversations on Chemistry*. Philadelphia: Lippincott, Grambo.

---

twelve printings of a highly imitative American text, *New Conversations on Chemistry,* by Thomas P. Jones. Marcet's book was almost as popular in Britain, going through eighteen printings.[15] It was printed four times in Paris (perhaps more, since Marcet had at least one French plagiarist) and once in Geneva.[16] The book was a failure in Germany, where a single 1839 edition sold poorly.[17]

Olin Library staff assisted me in tracking down sixteen of these twenty-three American editions, held in various libraries in the United States, and four editions of the version by Thomas P. Jones. Thanks also to the libraries that made copies available: University of Pennsylvania, University of Michigan, Princeton University, New York State Library at Albany, and University of Minnesota.

[15] The book's popularity in Britain has been dismissed by some historians as a by-product of Humphry Davy's charisma; see Judit Brody, "The Pen Is Mightier Than the Test Tube," *New Scientist,* 14 Feb. 1985, p. 58. See also David Knight, "Accomplishment of Dogma: Chemistry in the Introductory Works of Jane Marcet and Samuel Parkes," *Ambix,* 1988, *33*:94–98, on p. 97.

[16] Jean Jacques has noted the anonymous publication in Paris in 1826 of *Entretiens sur la chimie après les méthodes of MM. Thenard et Davy,* virtually a direct translation of Marcet's text. Mrs. B. became Mme de Beaumont, Emily was transformed to Gustave, but Caroline remained Caroline. The same year A. Payen produced a version under the title *La chimie enseignée en vingt-six leçons).* Though he restyled portions of the text, he lifted the order of the conversations and many discussions directly from Marcet's text. Jean Jacques, "Une chimiste qui avait de la conversation: Jane Marcet (1796–1858)," *Nouveau Journal de Chimie,* 1986, *10*:209–211.

[17] A modern facsimile of the German edition was published in 1984, with an afterword by the historian of chemistry Otto Paul Krätz (*Unterhaltungen über die Chemie,* trans. F. F. Runge [Wein-

A contemporary commentator set American sales figures at 160,000 copies.[18]

Marcet did not intend her *Conversations* to be used as a textbook. In Britain, it was apparently used as she expected, as a guide to popular lectures on chemistry or natural philosophy. But in America it became the most successful elementary chemistry text of the first half of the nineteenth century. A succession of male editors reshaped it for classroom use through twenty-three pirated American editions over forty-seven years. Indeed, as noted earlier, the work was commonly attributed (in biographical dictionaries, catalogues and obituaries in the United States) to its male editors.[19] In the absence of international copyright law, Marcet received no income from these American editions, nor had she any control over the American commentaries and improvements.[20]

The American editors added study questions, dictionaries of terms, guides to the experiments, and critical commentaries. These amendments for the classroom were not a marketing strategy concocted by the book's American publishers, but the response of professional chemists and educators to the book's growing use as an introductory chemistry text. *Conversations on Chemistry* was widely adopted in the schools by 1818. It then attracted American editors, most of whom seemed to be disturbed by its popularity.

Marcet's American commentators included a minister and four professors of chemistry or chemical lecturers. They worried about questionable theories (Davy's) and dangerous experiments. They also attempted to promote American scientists—Robert Hare, whom Marcet neglected, and Benjamin Franklin, whom she misinterpreted.[21]

---

heim: Verlag Chemie, 1984]). Krätz concluded that the book failed in Germany because it discussed technologies, such as steam engines, unfamiliar to German readers. Karl Hufbauer has suggested that it may not have been successful because young German women had limited access to chemical education: see Hufbauer's review, *CHOC News*, Spring 1984, 2(1):7–8, on p. 8.

[18] It is difficult to determine exactly how many copies of Marcet's work sold in America. Sarah J. Hale provided the figure of 160,000, cited by several other historians, in Sarah Josepha Buell Hale, *Woman's Record; or, Sketches of Distinguished Women from the Creation to A.D. 1868, Arranged in Four Eras* (New York, 1874), p. 732. But Hale refers to 160 impressions of the book and assumes a print run of 1,000 copies per impression. I have managed to find records of only 32 impressions (counting Jones's version). Hale may have known of more, or perhaps print runs were much larger.

[19] Thomas P. Jones included Marcet's name in the frontispiece of all his editions, but the text was listed as his work in catalogues of texts used in chemical instruction in the women's academies. See Thomas Woody, *A History of Women's Education in the United States* (Lancaster, Pa.: Science Press, 1929), p. 553. Woody also lists John Lauris Blake, the Episcopalian minister who provided questions for numerous editions, as the author of a chemistry text that must have been Marcet's work: Blake's questions appeared in every American edition after 1828. (Woody again lists Blake as the author of a natural philosophy text that must have been Marcet's later *Conversations on Natural Philosophy*, to which he added similar questions.) Blake also appears as the author of *Conversations on Chemistry* in the *Dictionary of American Biography* (1936), Vol. XI, p. 343, and the *National Cyclopedia of American Biography* (1931), Vol. XXI, p. 172. This attribution of male authorship must have been the work of persons who had not read the book, since every edition carried Marcet's self-deprecating preface, which was clearly written by a woman, and was almost unchanged for forty-four years.

[20] Protection of copyright for American authors was established in 1790, but foreign authors were granted no such protection until 1891. The publication of Marcet's book in America was also influenced by the chaotic and competitive nature of early nineteenth-century book publishing. The "cutthroat" conditions of the era "forced many publishers to specialize in fields where competition was not so general and returns more stable," such as science: Henry Walcott Boynton, *Annals of American Bookselling, 1638 to 1850* (New York: John Wiley, 1932), p. 144. See also Warren S. Tryon and William Charvat, eds., *The Cost Books of Ticknor and Fields* (New York: Bibliographical Society of America, 1949)—a reprint of the publishing records of a major Boston publisher from 1832 to 1858.

[21] For a discussion of American envy of European chemistry see Robert V. Bruce, *The Launching of Modern American Science, 1846–1876* (New York: Knopf, 1987), pp. 14–28.

Marcet's most frequent editor was John Lee Comstock, who made his debut anonymously in the fourth edition of 1818 as "an American gentleman." His name first appeared in the 1822 Hartford edition. In 1826 O. D. Cooke produced an edition of *Conversations on Chemistry* with both Comstock's commentary and a series of numbered "study questions" provided by the Rev. John Lauris Blake. The combination of Comstock's criticisms and Blake's questions was the standard format for most American editions throughout the rest of the book's career.

Blake (1788–1857) was an Episcopalian minister in Boston. He had resigned his rectorship in 1822 to devote himself to "literary work," which included writing an introductory astronomy book and providing numbered study questions to both Marcet's *Conversations on Chemistry* and her later *Conversations on Natural Philosophy*. Blake must have been interested in the education of women: he started a girl's school at Concord, New Hampshire.[22] His questions in Marcet's chemistry book (1,456 of them in the 1836 Hartford edition) were printed at the bottom of each page and intended to aid in classroom instruction. On the title page Blake warned (in triple negative) that "no small portion of learners will pass over without study, all in which they are not to be questioned."

Blake's questions were not particularly thought provoking—they promoted rote learning—but they were apparently taken seriously by some students. In several copies of Marcet's text reviewed for this study, some long-ago student had dutifully penciled in the proper answers to these questions in the small space allotted on the page.

Comstock (1789–1858) was a self-educated surgeon who served in the Army in the War of 1812 and later settled in Hartford to write and edit textbooks on chemistry, natural history, botany, physiology, and mineralogy. Comstock's "original" work was apparently often borrowed from European authors. The *Dictionary of American Biography* credits him with the authorship of a *History of Gold and Silver,* a *History of the Greek Revolution* and a *Cabinet of Curiousities*.[23] He also wrote a highly derivative chemistry text: his *Elements of Chemistry* (1831) was a much-simplified and quite popular version of Turner's text of the same name. His 1822 *Grammar of Chemistry* was apparently also borrowed from another author. It was written "on the plan of David Blair," a pseudonym of R. Phillips, and "adapted to the use of schools and private students by familiar illustrations and easy experiments."[24] And Comstock's *Conversations on Natural Philosophy* was in fact Marcet's work, with his name on the title page as editor.

Two other American editors, the Philadelphia chemistry professors William H. Keating and Thomas Cooper (who produced one edition each), merely inserted a few mild footnotes clarifying Marcet's experiments or ideas.[25] But Comstock, her first and most persistent American editor, provided from 156 to 173 notes in

[22] *National Cyclopedia of American Biography* (1931), Vol. XXI, p. 172.

[23] On Comstock see *Dictionary of American Biography* (1872), p. 211; see also the extensive list of Comstock's publications in John F. Ohles, *Biographical Dictionary of American Educators* (Westport, Conn.: Greenwood, 1978), Vol. I, p. 295.

[24] John Lee Comstock, *A Grammar of Chemistry* (Hartford: S. G. Goodrich, 1822), title page.

[25] Keating was founder of the Franklin Institute and a chemistry professor at the University of Pennsylvania. Cooper was the son-in-law of Joseph Priestley, a professor of chemistry and mineralogy at the University of Pennsylvania, and, after 1821, a chaired professor of chemistry at South Carolina College.

his various editions, for a text averaging about 330 pages. In these notes he frequently disagreed with Marcet and sometimes implied that she was incompetent. When she explained the presence of so much "calcareous matter" as the "effect of a general combustion occasioned by some revolution of our globe," Comstock noted: "This idea is at random. We cannot account for the origin of carbonic acid in its native state, any better than we can for oxygen."[26] When Marcet suggested that it was highly unusual for three or more substances to combine without any of them being precipitated, Comstock noted that "such compounds are quite numerous." He characterized her explanation of volcanoes as "supposition piled on supposition."[27] When she attempted to explain the role of water in the life cycle of plants, he responded in a footnote: "The foregoing paragraph might mislead the student. Indeed, it seems to have been written without regard to proper authorities." When she suggested that "combustion is the result of intense chemical action," he responded: " 'Intense chemical action' neither explains the process, nor, indeed conveys to the mind any definite idea."[28] And when she said the concepts of negative and positive indicated "different quantities of the same kind of electricity," Comstock replied (with italics): "In this chapter, Mrs. B. has used these terms of the American philosopher [Franklin] improperly, for *plus* and *minus* were never meant to signify two sorts of electricity, but only its *presence* or *absence*."[29]

If Comstock disapproved of many of Marcet's proposals (both theoretical and experimental) why did he continue to edit the book vigorously for four decades? His introduction provides a partial explanation: Comstock was worried about the book's widespread use in the classroom. "Known and allowed facts are always of much higher consequence than theoretical opinions," he said in the "Advertisement of the American Editor" that introduced his editions. "A book designed for the instruction of youth, ought, if possible, to contain none but established principles."[30]

Keating and Cooper, while milder in their criticisms of Marcet, also expressed concern about the promotion of questionable theories to beginning students. Cooper edited the text "lest the young student should adopt as certainties many theoretical views which have hardly yet arrived at probability." He noted that Marcet had followed Davy where his contemporaries "have not yet dared to follow him." This adoption of Davy's ideas rendered the book "extremely interesting" but less than ideal for instruction in the fundamentals of chemistry.[31]

Marcet's editors also worried about her depiction of the use of hands-on laboratory experiment in the training of beginners. They found such a proposal extremely risky, and their concerns were not unwarranted. From Comstock's corrections of her experiments, it appears possible that Marcet did not actually perform all the experiments she described. Certainly her suggestion that elementary chemical instruction might include laboratory experiment was quite novel. Indeed, in 1822 her editor William Keating, of the University of Pennsylvania

[26] 1829 Hartford edition, p. 225.

[27] 1825 Hartford edition, pp. 13, 172.

[28] 1839 Hartford edition, pp. 281 (water), 234 (combustion).

[29] 1822 Hartford edition, p. 79.

[30] This introduction, essentially unchanged, is printed in every impression of Comstock's version, before the table of contents.

[31] Thomas Cooper, 1818 Philadelphia edition, preface; see also William H. Keating, 1824 Philadelphia edition.

and the Franklin Institute, was one of the first to apply this teaching method in an American college. It was not until after the Civil War that laboratory instruction for beginning students became the norm.[32]

While her other American editors merely inserted footnotes or study questions, Thomas P. Jones wrote a "new" text that followed Marcet's format precisely in terms of data presented, but eliminated the humor and personal commentary of the original. Jones, a professor of chemistry at Columbia College in Washington and a popular lecturer on chemistry and natural philosophy, was interested in filling the text with as many chemical facts as possible.[33] Publishing his first version of Marcet's book in 1831, he explained that while Marcet's text received "deservedly high praise" and had "contributed more than any other work to promote the study of chemistry," its original role as "companion for the parlour" had been superseded. The new role of textbook called for a different presentation. The digressions which gave the original work "variety and interest" in the "family circle" were now an impediment to the rapid assimilation of new facts, he said.[34] Jones's version, though lacking the entertainment value and "charm" which might be assumed to be one reason for Marcet's success, was relatively successful itself: it was reprinted twelve times, more frequently than most other chemistry texts of the era.

Marcet's American editors suggested that she went too far in her promotion of the latest chemical theories. Yet her discussions of theory may have been what academy-level instructors found so attractive. And the proposed experiments her editors found so risky may have made her work more valuable to instructors hoping to spark young women's interest in science.

SCIENCE IN THE WOMEN'S ACADEMIES

The antebellum women's academies have been a subject of increasing historical interest since 1979. Science instruction at these institutions was touched on in Thomas Woody's classic 1929 history of education for women. In 1979 Deborah Jean Warner examined more precisely the kinds of instructions and instructional materials that women's academies offered. Linda Kerber and Anne Firor Scott have explored their complex cultural role, suggesting that practice was not always in line with public rhetoric. Those promoting women's education for the sake of "republican motherhood" (the rearing of good male citizens who could defend the republic) may have had more radical intentions. And as Patricia Cline Cohen has shown, women's education in mathematics was predicated on the household applications of numerical reasoning (as in knitting or cooking), while actual instruction was much more advanced than these simple tasks required.[35]

---

[32] See Wyndham Miles, "William H. Keating and the Beginning of Chemical Laboratory Instruction in America," *Library Chronicle*, 1952/3, 29:1–34. See also the entry on Keating in *Dictionary of American Biography* (1872), p. 502.

[33] Wyndham D. Miles, "Public Lectures on Chemistry in the United States," *Ambix*, 1968, 15:129–153.

[34] Thomas P. Jones, *New Conversations on Chemistry* (Philadelphia: John Grigg, 1832), preface.

[35] Woody, *History of Women's Education* (cit. n. 19), Vol. I; Deborah Jean Warner, "Science Education for Women in Ante-Bellum America," *Isis*, 1979, 69:58–67; Linda K. Kerber, *Women of the Republic: Intellect and Ideology in Revolutionary America* (Chapel Hill: Univ. North Carolina Press, 1980); Anne Firor Scott, "The Ever-Widening Circle: The Diffusion of Feminist Values from the Troy Female Seminary, 1822–1872," *History of Education Quarterly*, Spring 1979, pp. 3–25; and Patricia Cline Cohen, *A Calculating People: The Spread of Numeracy in Early America* (Chicago: Univ. Chicago Press, 1982), pp. 134–149.

Certainly the historical picture of both the women's academies and the role of science therein is incomplete. Some sciences, including chemistry, were more widely taught in the women's academies than in boys' high schools of the early nineteenth century. And at least some women's schools, particularly Emma Willard's Troy Female Seminary, offered a greater range of sciences than contemporary men's colleges.[36]

Laboratories and observatories at the female colleges were not well funded, but they represented the single largest investment, excepting buildings, at many schools. A women's school in New York City, Abbott Collegiate Institute, claimed scientific apparatus "unsurpassed in character by that of any other Institution in our country." Astronomical equipment was particularly popular. Albany Female Academy and Packer Collegiate Institute each owned an orrery, a moving, mechanical representation of the solar system, made by a renowned Kentucky instrument maker.[37]

Such equipment, as Deborah Jean Warner has noted, proves nothing about the quality of science teaching. The paraphernalia was as important for promotional as for educational reasons. Yet she argues that other evidence suggests that the quality of the instruction in some sciences was relatively high. Some lecturers appearing at the women's schools were well known (Benjamin Silliman, Jr.; Elias Loomis), and some science teachers were extremely competent, among them Alonzo Gray, who taught at Brooklyn Female Academy, and Louis Agassiz, who with his wife Elizabeth ran a school for girls in Cambridge from 1855 to 1863.[38]

The availability of scientific apparatus and the high quality of some instructors suggest that science education at the women's academies was more than a public relations ploy. The selection of textbooks reinforces this conclusion. Those teaching chemistry to young women in this period had numerous options. Their choice of Marcet's text indicates their educational priorities. It suggests that their commitment to scientific instruction for women was not completely encompassed in their publicly stated goals. Textbooks conforming more properly to these stated goals were widely available before 1840. Most emphasized the practical applications of chemistry. But at least one important American chemistry text focused on the spiritual lessons it provided. This was the text of the American educator Almira Hart Lincoln Phelps, the sister of Emma Willard.[39]

Phelps should have had considerable insight into the instructional materials needed in the new women's schools. Yet her academy-level chemistry text, spe-

[36] See the discussion in Paul J. Fay, "The History of Chemistry Teaching in American High Schools," *J. Chem. Educ.*, 1931, 8:1533–1562, 1539–1540. For a valuable review of the state of chemical instruction in American colleges see Bruce V. Lewenstein, " 'To Improve Our Knowledge in Nature and Arts': A History of Chemical Education in the United States," *J. Chem. Educ.*, 1989, 66:37–44.

[37] Warner, "Science Education for Women" (cit. n. 32), pp. 59, 60 (quotation from Abbott Collegiate Institute, *Catalogue* [1854]).

[38] Warner, "Science Education for Women," p. 62.

[39] Almira Hart Lincoln Phelps, *Familiar Lectures on Chemistry* (New York: F. J. Huntington, 1838). Besides works cited in notes 11, 43, and 44, I have considered the following texts: William Henry, *An Epitome of Chemistry* (Boston, 1810); John White Webster, *Manual of Chemistry* (Boston, 1826); Edward Turner, *Elements of Chemistry* (New York, 1828); James Renwick, *First Principles of Chemistry* (New York, 1840); Benjamin Silliman, *First Principles of Chemistry* (Philadelphia/Boston, 1847); Edward Youmans, *A Class Book of Chemistry* (New York, 1851); and Youmans, *The Handbook of Household Science* (New York, 1853). For textbooks used in women's academies for other sciences see Woody, *History of Women's Education* (cit. n. 19), Vol. I, app.

cifically intended for the instruction of young women, was a failure: it was reviewed unenthusiastically and printed only twice, in 1838 and 1842.[40] Phelps's error may have been her assumption that chemical education was a form of religious instruction. While Marcet mentioned the relevance of chemical theory to religious faith only casually, Phelps's *Familiar Lectures on Chemistry* was metaphysical throughout. She said chemistry could provide lessons in humility— "Our own bodies are composed of a few elements of the same nature as those which form the very worm that crawls"—and in hubris: "There is a portion of ourselves which is beyond the scope of chemical science, which cannot be analyzed, because it is incapable of being separated into parts."[41]

While Phelps's primary interest was in the spiritual lessons of science, she also recognized that chemistry had a peculiarly practical aspect. She assigned her pupils to explain the "Chemical Principles involved in making bread" and informed them that chemistry had "an important relation to housekeeping . . . in the making of gravies, soups, jellies and preserves, bread, butter and cheese, in the washing of clothes, making soap, and the economy of heat in cooking, and in warming rooms."[42]

Other writers considered the utilitarian aspect of chemistry its chief value to potential students. The useful purposes these writers selected for discussion shifted with the intended audience. An author intending to address the problems of "household science" might discuss the relevance of chemical facts to the fermentation of bread, preservation of milk and butter, sources of impure air in the home, and properties of fuel used for artificial heating. Another, intending to reach workingmen, would focus on tanning leather, brewing wine, soil analysis, and medicine. John R. Coxe's translation of M. J. B. Orfila's *Practical Chemistry* (1818) contained little chemical theory, focusing instead on information useful to the pharmacist, farmer, or physician. Similarly, William Henry's *Elements of Experimental Chemistry* classified metals practically, rather than theoretically, and dealt solely with the relation of chemistry to the "practical arts." The American physician and Harvard chemistry professor John Gorham deemed even Henry's chemical text too experimental, and in his *Elements of Chemical Science* (the first original American chemistry textbook) simplified Henry's approach by eliminating virtually all laboratory work. John Lee Comstock's own text, *Elements of Chemistry*, first published in 1831, was an entirely descriptive and practical text that gave no attention to chemical theory. Even as late as 1867 J. Dorman Steele's popular *Fourteen Weeks in Chemistry* concerned only that "practical part of chemical knowledge" necessary in the "schoolroom, the kitchen, the farm and the shop." And a masculine version of Marcet's *Conversa-*

---

[40] Phelps's 1834 text for children, *Chemistry for Beginners,* was slightly more successful, and editions continued through the 1860s. Her most popular book was *Familiar Lectures on Botany,* which was reprinted dozens of times and had sold 230,000 copies by 1870. Phelps's biographer Emma Lydia Bolzau has attributed the failure of the chemistry texts to their derivative nature; see Bolzau, *Almira Hart Lincoln Phelps* (Lancaster, Pa.: Science Press, 1936), pp. 235–236.

[41] Phelps, *Chemistry for Beginners* (1867), p. 11. Marcet's most sustained discussion of chemistry and religious faith appeared in the closing paragraph of her book: "To God alone man owes the admirable faculties which enable him to improve and modify the productions of nature. . . . In contemplating the works of the creation, or studying the inventions of art, let us, therefore, never forget the Divine source from which they proceed; and thus every acquisition of knowledge will prove a lesson of piety and virtue." 1822 edition (e.g.), p. 327.

[42] Phelps, *Chemistry for Beginners* (1839), p. 5 (bread making); (1867), pp. 9–11 (housekeeping).

tions on Chemistry, the Rev. Jeremiah Joyce's *Dialogues in Chemistry,* featured conversations between a "Tutor" and two male pupils, Charles and James, on the relevance of chemistry to "agriculture, gardening and the arts of cooking and of making wine, beer and other fermented liquors."[43]

An introductory text that combined all these interests rather broadly was produced in 1822 by the New York educator Amos Eaton, a friend of both Almira Hart Lincoln Phelps and her sister Emma Willard. Eaton dedicated his *Chemical Instructor* to Willard because she was "the first in the interior of the Northern states to introduce experimental chemistry into [public] schools."[44] Eaton's text, written to replace Marcet's, which he disliked, was intended for the audience—academy chemistry instructors—that had already demonstrated its enthusiasm for her approach. Eaton interpreted that market as receptive to a practical treatment of the subject. He was unwilling to let a single chemical idea or principle pass without mentioning a practical application: he made special appeals to those engaged in the full-time management of a house. His intentions were egalitarian and democratic. He proposed simple, inexpensive experiments, recognizing that his readers might not have access to expensive chemical equipment or rare materials; part of his objection to Marcet was that she assumed her readers would have ample access to equipment and supplies.[45]

But Eaton's book was not widely used in the women's academies. Instead, many instructors of young women continued to introduce chemistry through Marcet's *Conversations on Chemistry,* a work that overlooked the domestic or practical applications Eaton and other American writers believed to be so important.[46]

CONCLUSION

By the 1820s "popular science" tailored to a female audience was a well-accepted social activity. From these public lectures and popular books women supposedly gained lessons in piety and useful household tips. School administrators at the women's academies transferred this reasoning to the formal educational setting. They offered their students those sciences promoted for women in popular lectures and books: natural philosophy, astronomy, chemistry, and botany.

But popular lectures and popular science books were casual entertainment,

[43] M. J. B. Orfila, *Practical Chemistry,* trans. from the French by John Coxe (Philadelphia: Thomas Dobson, 1818); William Henry, *The Elements of Experimental Chemistry* (Philadelphia: R. Desilver, 1822–1823); John Gorham, *The Elements of Chemical Science* (Boston: Cummings & Hilliard, 1819); Comstock, *Elements of Chemistry* (cit. n. 11); J. Dorman Steele, *Fourteen Weeks in Chemistry* (New York: Barnes, 1867); and Jeremiah Joyce, *Dialogues in Chemistry* (New York: James Eastburn, 1818). The quotation is from the third London edition, with "additional notes by an American professor of chemistry." In her third London edition Marcet stated that her format (a teacher and two students) was borrowed from a book entitled *Scientific Dialogues.* This was probably an earlier book by Joyce, who also wrote *Dialogues on the Microscope.* See Marcet's preface, 1809 (e.g.).

[44] Amos Eaton, *Chemical Instructor* (Albany: Websters & Skinners, 1822), dedication. The work went into four editions in Albany (1822, 1826, 1828, 1833). On the relationship between Eaton, Phelps, and Willard see Lois Barber Arnold, *Four Lives in Science: Women's Education in the Nineteenth Century* (New York: Schocken, 1984).

[45] Eaton, *Chemical Instructor* (1822), title page.

[46] As early as 1809 Marcet's New Haven publishers added a "description and plate of the pneumatic cistern of Yale College," a "short account of artificial mineral waters," and an appendix "consisting of treatises on dyeing, tanning and currying"; see 1809 New Haven edition.

essentially conservative, legitimated by the presumed domestic and religious applications of scientific knowledge. Education at the female academies entailed institutional approval of a sustained course of study of science, however elementary; there was often an implicit expectation that some students would pursue careers as teachers. While both activities were justified in similar ways, they reflected fundamentally different assumptions about female involvement in science. The conservative arguments that made sense of science education for women apparently had little impact on actual scientific instruction, which (at least in the case of chemistry) was often focused less on spiritual or domestic applications than on chemical theory and experiment.

Despite competition from dozens of other texts, Jane Marcet's *Conversations on Chemistry* dominated elementary chemical instruction in these academies. Administrators could have chosen texts that emphasized useful applications or spiritual lessons. They chose instead a presentation novel for both its attention to chemical theory and its advocacy of hands-on laboratory instruction for beginners. It was not simply a matter of teaching the principles of baking or soap making. Academy chemistry, at least in those schools that used Marcet's text, was serious chemistry for beginners: an up-to-date review of European chemical theory, illustrated by experiment, requiring an understanding of chemical terminology and facility in the manipulation of laboratory equipment and chemicals.[47]

The popularity of Marcet's book suggests that American educators wanted young women to understand the basics of theoretical and experimental science. Their reasons for this remain unclear. But certainly the instruction offered in the women's academies provided an important initial impetus for changes in the nature of women's participation in science. While the legacy of scientific training in the women's academies is difficult to measure, some women did become prominent scientists in the second half of the century. Wellesley College's first professor of physics, Sarah Frances Whiting, graduated from Ingham University for Women and taught at the Brooklyn Heights Seminary. The naturalist Lydia White Shattuck studied at Mount Holyoke Seminary. The botanist Graceanna Lewis attended the Kimberton Boarding School. The astronomer Maria Mitchell, her student and fellow-astronomer Mary Whitney, the chemist and educator Mary Lyon, the psychologist Christine Ladd-Franklin, and the chemist and home economist Ellen Swallow Richards were also products of this changing educational climate.[48]

The availability of serious scientific education in the new women's academies set the stage for increasing women's involvement in science. The access to introductory science instruction in a formal laboratory setting—rather than through a male family member, or a brother's tutor—legitimated feminine interest in scientific theory. And as the famous Faraday anecdote suggests, the young mind can sometimes reach grand conclusions from rather minor encounters.

---

[47] The Boston Girls' High School has been credited with being the first school to offer the teaching of chemistry with laboratory instruction, in 1865. By 1871 many high schools had chemistry laboratories. See Sidney Rosen, "The Rise of High School Chemistry in America (to 1920)," *J. Chem. Educ.*, 1956, *33*:627–633, on p. 628.

[48] Warner, "Science Education for Women" (cit. n. 35), pp. 65–66. See Margaret Rossiter, *Women Scientists in America: Struggles and Strategies to 1940* (Baltimore: Johns Hopkins Univ. Press, 1982), for a full discussion of this emergence of women scientists in the mid- and late-nineteenth century.

Parlors, Primers, and Public Schooling: Education for Science in Nineteenth-Century America

By Sally Gregory Kohlstedt

PARLORS, PRIMERS, AND PUBLIC SCHOOLING, if uncommon in our usual discussions in the history of science, are not trivial or irrelevant. Such topics are in fact fundamental to a more holistic understanding of individual and cultural incentives in the history of science in America, past and present. They help us understand a fundamental issue that underlies much of the recent research, teaching, and writing in the history of science in the United States: namely, the sources of the deep, if often ambiguous, public commitment that made science and technology a significant part of our national identity.

One trend in contemporary historical writing, influenced by movements from abroad and from other disciplines, is to tell more complex stories about the past and to use frameworks that involve multiple points of view. This trend can lead historians of science in new directions. When we study the history of science, we should inquire into the outlook on science taught to children. When we study the history of science, we should look to the initiatives of students and consider the objects of their curiosity and of their enthusiasms as well as the curriculum they were taught. When we study the history of science, we should know what texts, popular treatises, and scholarly monographs sold well, how they were written, and what they included. When we study the history of science, we should analyze the attitudes of those who sponsored and governed its activity. When we study the history of science we need, in short, to understand that the introduction, practice, sponsorship, and even critique of science are dynamic phenomena intimately connected to other dimensions of culture.

Looking at these aspects of American science may lead us to rethink the periodization customarily imposed on it, which now seems almost commonplace. Most of us have come to view the 1840s and 1850s as the formative years for modern American science. The title of Robert Bruce's recent Pulitzer Prize–winning volume, *The Launching of Modern American Science, 1846–1876*,

I want to express my gratitude to anonymous reviewers as well as colleagues from Syracuse University and Cornell University, particularly Janet Carlisle Bogdan and Margaret M. Rossiter, who sympathetically but critically read, listened to, and commented on an earlier version of this paper.

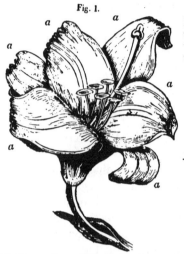

Fig. 1.

Parts of the flower.
Before you can learn the principles on which the classification of plants depends, it is necessary that you should become acquainted with the parts of a flower; for this purpose, you have here the representation of a Lily. (*See fig.* 1.) You know that at first this flower is folded up in a little green bud, and that by degrees, it expands and changes its colour; being in some kinds of lilies, white, in others, yel-low, orange, &c. , This is the picture of a white lily.

The part which you would call the blossom, is the *corolla;* this is composed of six pieces, each of which is a *petal*, as seen at *a.*

Fig. 2.

By examining the lily you will observe within the corolla six thread-like organs; these are called *Stamens.* Each stamen is composed of two parts, one long and slender, called the *Filament* (*Fig.* 2, *a*), the other part called the *An-ther* (*b*), is a kind of knob, like a little box, which, when the flower comes to maturity, opens, and throws out a colour-'ed dust, called the *Pollen.*

The central organ is called the *Pistil* (*c*); this consists of three parts, the top, which is called the *Stigma* (*d*), the slen-der filament which bears the stigma is called the *Style* (*e*), and the base is called the *Germ* (*f*).

In many flowers the corolla is surrounded by a kind of cup call-ed the *Calyx*, this is wanting in the Lily, but exists in the Pink.

The *Receptacle* (*g*), is the end of the stem, where all the other parts of the flower are inserted,

Figure 1. Almira Hart Lincoln Phelps designed her textbook, Familiar Lectures on Botany, *in order to teach young women and men the basics of plant taxonomy clearly and without embarrassment.*

makes the point directly. My own earliest research argued that the American Association for the Advancement of Science established an American scientific community between 1848 and 1860. Several scholars in the 1950s who pioneered the study of American science—such as A. Hunter DuPree, Nathan Reingold, and Edward Lurie—also concentrated much of their earliest research on the middle decades of the nineteenth century. A subsequent generation of dissertation writers followed their lead in searching out the nineteenth-century origins of scientific institutions. Consequently, we know a great deal about the National Academy of Sciences, state and federal surveys and agricultural bureaus, professional societies, and publication activities.[1]

It is time to reconsider the appropriateness of beginning in the 1840s while largely dismissing the previous decades and to reassess the years spent building the American republic, paying particular attention to scientific initiatives. The pioneering studies of I. Bernard Cohen, Raymond P. Stearns, and Brooke Hindle document the extent to which some extraordinary and other very earnest individuals pursued science in the late colonial period. Their accounts of trailblazing efforts also convey a certain sense of disappointment that there were so few outstanding figures like Benjamin Franklin and so few persisting institutions like the American Philosophical Society of Philadelphia. As John Greene has cogently put it, the Age of Jefferson was "not an age of brilliant scientific achievement judged by European standards, but an interesting and formative one, worthy of respect by all who value honest effort inspired by love of country and of science."[2] The curtain thus rises on American science with Enlightenment philosophes like Jefferson speaking enthusiastically but largely prophetically about universities and institutions to study an advanced science. In this standard drama of revolutionary-era science, the first scene closes with Jefferson's failure to persuade political colleagues to do much more than sponsor natural history as part of the Lewis and Clark Expedition.[3]

In the next scene, set in the early republic, characters act out often poignant personal histories in struggling institutions. The New York Lyceum of Natural History, founded in 1817, for example, saw its early ranks depleted by the career difficulties of Samuel Latham Mitchill and the eccentricities of Constantine Rafinesque. Later, members lost a new building at auction and failed to sustain regular publication of their *Annals* because they could not mobilize requisite

---

[1] Robert V. Bruce, *The Launching of Modern American Science, 1846–1876* (New York: Knopf, 1987); and Sally Gregory Kohlstedt, *The Formation of the American Scientific Community: The American Association for the Advancement of Science, 1848–1860* (Urbana: Univ. Illinois Press, 1976). Relevant literature is compiled by subject and with brief commentary in Marc Rothenberg, *The History of Science and Technology in the United States: A Critical and Selective Bibliography* (New York: Garland, 1982).

[2] John C. Greene, *American Science in the Age of Jefferson* (Ames: Iowa State Univ. Press, 1984), p. 419. Similarly, Brooke Hindle argued that "the most basic and permanent results of the cultural nationalism of the Revolution can be altogether missed if attention is limited to positive accomplishments. Strikingly, no scientific research of the time matched the importance of Franklin's colonial experiments in electricity"; see Hindle, *The Pursuit of Science in Revolutionary America, 1735–1789* (Chapel Hill: Univ. North Carolina Press, 1956), p. 383. See also I. Bernard Cohen, *Franklin and Newton: An Inquiry into Speculative Newtonian Experimental Science and Franklin's Work in Electricity as an Example Thereof* (Cambridge, Mass.: Harvard Univ. Press, 1966); and Raymond P. Stearns, *Science in the British Colonies of America* (Urbana: Univ. Illinois Press, 1970).

[3] Lillian B. Miller, ed., *The Selected Papers of Charles Willson Peale and His Family*, 5 vols. (New Haven, Conn.: Yale Univ. Press, 1983–1988) Vol. II, p. 390; and Daniel J. Boorstin, *The Lost World of Thomas Jefferson* (1948; Chicago: Univ. Chicago Press, 1985), pp. 217–218.

financial and intellectual resources.[4] Dispirited and frustrated by financial strain and small membership, some leading members explained their difficulties in terms of public apathy toward ideas in general and science in particular.

Nathan Reingold and others have already disabused us of some notions about American indifference by identifying the prevalence of science in visibly practical inquiries. Joseph Henry, a particularly apt and well-studied example, received an excellent education at the Albany Academy. In the 1820s Henry learned sufficient physics through courses on hydrography and surveying that he made major contributions to magnetic induction, was appointed to the Princeton faculty, and consulted with such notables as Samuel F. B. Morse; his contributions led to his prominence as secretary of the Smithsonian Institution in Washington, D.C.[5]

We are looking for national commitment to science in the wrong place, however, if we simply try to find outstanding scientists and piece together their experiences. We do know from the prosopographical studies of Clark Elliott, George Daniels, and Donald Beaver that there were a few individual "cultivators" and researchers in science and that the group was steadily growing between 1800 and 1850.[6] But quantitative measures tied largely to scientific publications indicate only the small numbers and intermittent successes of those men who constituted relatively exclusive learned societies. Their presence does not explain or describe more general cultural attitudes toward science, broadly defined. We need to dig deeper into the national record, into a range of private and public behaviors—independent study, conversations, group activities, popular publications, and educational institutions. The pluralism and often the transience of these phenomena defy any simple and orderly pattern, but their variety attests to pervasive public curiosity about scientific subjects and helps to explain the rapidity with which subsequent, permanent scientific institutions would be built in the last half of the nineteenth century.

In the decades after the Revolution, Americans sought to initiate a very self-conscious experiment in politics, industry, the arts, letters, and science. Charlotte Porter, for example, documents the efforts of Philadelphia naturalists who proposed a distinctively American system of zoological classification. In the founding charters of natural history societies and learned academies, rhetoric about patriotism and presumed republican values in science and technology was commonplace.[7] The theme was also underscored by the enthusiasm for science and technology in more general settings.

---

[4] Simon Baatz, *Knowledge, Culture, and Science in the Metropolis: The New York Academy of Sciences, 1817–1970* (Annals of the New York Academy of Sciences, 584) (New York, 1990).

[5] *The Papers of Joseph Henry,* ed. Nathan Reingold, with Arthur Molella and Marc Rothenberg, 4 vol. to date (Washington, D.C.: Smithsonian Institution Press, 1972–); see also Molella and Reingold, "Theorists and Ingenious Mechanics: Joseph Henry Defines Science," *Science Studies,* 1973, *3*:323–351; and Bruce Sinclair, *Philadelphia's Philosopher Mechanics: A History of the Franklin Institute, 1824–1865* (Baltimore/London: Johns Hopkins Univ. Press, 1974). On the indifference theme see Reingold, "American Indifference to Basic Research: A Reappraisal," in *Nineteenth-Century American Science,* ed. George H. Daniels (New York: Columbia Univ. Press, 1968), pp. 38–62.

[6] Clark A. Elliott, *Biographical Dictionary of American Science: The Seventeenth through the Nineteenth Centuries* (Westport, Conn.: Greenwood, 1979), esp. the detailed appendixes; George Daniels, *American Science in the Age of Jackson* (New York: Columbia Univ. Press, 1968); and Donald de B. Beaver, "The American Scientific Community, 1800–1860: A Statistical-Historical Study" (Ph.D. diss., Yale Univ., 1966).

[7] Charlotte M. Porter, *The Eagle's Nest: Natural History and American Ideas, 1812–1842* (Tusca-

It is therefore important to investigate places where science became embedded in personal inquiry and public discourse. Three such places seem particularly significant in the early nineteenth century. First, a basic familiarity with science was introduced *en famille,* practically and symbolically using the parlor where educated middle-class men and women entertained friends and neighbors. Second, supplementing and extending relatively private experience was the more public dissemination of ideas through primers, texts, magazines, newspapers, lectures, and museums. Third, ideas about science, presented in words and objects, were gradually incorporated into private academies, public schools, and colleges. Children were the explicit focus of much of this broadly educational activity, although the initiative came from adults who had relied on self-help measures for their own introduction to science.

Bernard Bailyn captured an important and inclusive perspective for historians in his now-classic formulation of education, which he defined not only as formal pedagogy but indeed as "the entire process by which a culture transmits itself across the generations." Moreover, ambitious educators quickly learn that while effective presentation is important, reception is essential. Ellen Lagemann's studies in education therefore take the point of view of the learner. Educational biography, she writes, records an interaction by which individual potential is "activated . . . and a change thereby produced in the self."[8] We must therefore study both intention and effect. While education is typically and significantly attached to childhood, nineteenth-century Americans already recognized it as a lifelong phenomenon.

PARLORS

Let me begin where the social and intellectual life of girls and boys began, namely, among family members in a home setting. By the late eighteenth century, as Lawrence Stone and other European historians argue, there had been some diminution of patriarchal authority. Companionate marriage had attained positive value. American historians concur: private correspondence and public speeches record more affectionate and complementary relationships between adult men and women. The Enlightenment model of parenthood also gave children a more important status, and child rearing gained new significance.[9] This interior-focused and emotionally connected family took shape at a time when the church and state seemed less attentive to the process of transmitting culture, and when community responsibility for children was weakening. What happened in and through the home therefore took on increased importance, as parents provided training that they hoped would enhance interpersonal ties even as it pro-

loosa: Univ. Alabama Press, 1986), esp. Ch. 4; and John F. Kasson, *Civilizing the Machine: Technology and Republican Values in America, 1776–1900* (New York: Penguin, 1977).

[8] Bernard Bailyn, *Education in the Forming of American Society: Needs and Opportunities for Study* (Chapel Hill: Univ. North Carolina Press, 1960), p. 14; and Ellen Condliffe Lagemann, *A Generation of Women: Education in the Life of Progressive Reformers* (Cambridge, Mass.: Harvard Univ. Press, 1979), pp. 4–5.

[9] Barbara Finkelstein, ed., *Regulated Children/Liberated Children: Education in Psychological Perspective* (New York: Psychohistory Press, 1979), pp. 1–9; Walter L. Arnstein, "Reflections on Histories of Childhood," in *Research about Nineteenth-Century Children and Books,* ed. Selma Richardson (Urbana-Champaign: Univ. Illinois Graduate School of Library Science, 1980), pp. 41–60; and Jay Fliegelman, *Prodigals and Pilgrims: The American Revolution against Patriarchal Authority, 1750–1800* (Cambridge: Cambridge Univ. Press, 1982), pp. 1–2.

vided skills and behavior appropriate for a productive economic and social life.[10] Leisure was linked to education and to family life by a society that emphasized cooperative use of time and resources. In this environment, interests and aspirations were nurtured; too often twentieth-century historians identify early family experience with neuroses rather than ambition.

Home life was facilitated by changing domestic architecture that included space designated for social interaction. Relatively few homes in the young republic had elaborate salons or drawing rooms. American historians have found little evidence of direct or self-conscious connection to the salon traditions in London, Paris, Vienna, or elsewhere, where conversations turned to science as well as literature and participants included men and women; still, there were similar outcomes. The parlor, a room increasingly common in middle-class homes, provided families with a location for more deliberate interaction, as well as an informal, domestic, and conversational setting for visitors. Although concepts of privacy were not yet well prescribed, architectural historians define the parlor as a semiprivate space, a meeting place that combined comfortable aspects of middle-class home life—soft chairs, subtle indoor lighting, floor-to-ceiling bookcases, and a fireplace for warmth—with social rituals of tea, coffee, and cake for welcoming visitors. The parlor was the site for conversations that were intermediate between ordinary routines of the family and the more formal requirements of church, courthouse, and other public forums.[11]

Historians of science should therefore not be surprised to find that the parlor was a site for scientific demonstrations, for cabinets of natural history specimens, and for discussions of the newest scientific books. Some of the activity was carefully planned and orchestrated. Charles Willson Peale used his Philadelphia living quarters to deliver lectures on natural history in order to promote his developing museum. Other times the routine was more intimate; Benjamin Silliman recalled that he and his siblings recited lessons on Sunday afternoons in their Connecticut parlor, which he described as "the best chamber in the house, which was also reserved for our guests."[12]

Ephemeral groups also met in homes for systematic discussions of science and technology. Unfortunately, we still know little about precisely *who* was involved in such activity, and we know even less about *how* the meetings functioned. In late eighteenth-century America it seems that wives, sons, daughters, and extended family members might well be part of such activity.[13] Sometimes men

---

[10] Daniel T. Rogers, "Socializing Middle-Class Children: Institutions, Fables, and Work Values in Nineteenth-Century America," in *Growing Up in America: Children in Historical Perspective,* ed. N. Ray Hiner and Joseph M. Hawes (Urbana: Univ. Illinois Press, 1985).

[11] Bonnie Anderson and Judith P. Zinsser, *A History of Their Own: Women in Europe* (New York: Harper & Row, 1987), Vol. II, pp. 103–166 (on the European evolution of this space and women's use of it see pp. 129–143); Clifford Edward Clark, Jr., *The American Family Home, 1800–1960* (Chapel Hill: Univ. North Carolina Press, 1986), pp. 40–42; and Katherine Grier, *Culture and Comfort: People, Parlors, and Upholstery, 1850–1930* (Amherst: Univ. Massachusetts Press for the Margaret Woodbury Strong Museum, 1988).

[12] Miller, ed., *Papers of Peale Family* (cit. n. 3), Vol. II, p. 259, n. 3; and Joy Day Bueland Richard Buel, Jr., *The Way of Duty: A Woman and Her Family in Revolutionary America* (New York: Norton, 1984), p. 196.

[13] Patricia Branca, *Silent Sisterhood: Middle-Class Women in the Victorian Home* (London: Croom Helm, 1975). On colonial Virginia see Rhys Isaac, *The Transformation of Virginia, 1740–1790* (Chapel Hill: Univ. North Carolina Press, 1981), p. 271. One group in Philadelphia met in each others' homes, sometimes in their shops, and also in a public house before they rented more perma-

alone met in the parlor or library—a model apparently followed for the early Wistar parties in Philadelphia and later by the Saturday Club of Boston—but other gatherings were more inclusive. It may well be that, following a pattern Steven Shapin uncovered in lives of late seventeenth-century members of the Royal Society, private practice of experiments and demonstrations at home preceded their more public presentations.[14] We still have much to learn about participation in these semiformal activities, which were undoubtedly important for men, but were for women often their essential access to scientific and other knowledge. Around these informal groups, such important resources as libraries, experimental apparatus, and cabinets of minerals, mounted insects, pressed plants, and anthropological materials were accumulated.[15] They were brought into systematic arrangement by mutual agreement and cooperation.

Science could be part of sociability. This is evident in the diary of Esther Edwards Burr, who had had access to the library of her father, Jonathan Edwards, while growing up. She could readily converse on various topics with friends and colleagues of her husband, Aaron Burr, president of Princeton. Amid themes of work, family, sisterhood, and religion, her letters reveal an interest in scientific thought. Trying to persuade her closest friend, Sarah Prince of Boston, to visit her in New Jersey, Esther Burr reported, "We have a very fine *Microscope* and *Telescope*. Indeed we have Two telescopes—and we make great discoveries tho' not yet any new ones that I know of."[16] Nor did that family ever make extraordinary scientific discoveries. Esther and Aaron Burr died relatively young, and their son ultimately turned to politics and acrimonious dueling; still, in her early years of family life Esther Edwards Burr found scientific information and apparatus readily accessible for cooperative amusement and learning.

Themes of domestic education are particularly evident when historians consider again those easily scanned first chapters in scientific biographies that recount years of upbringing by family and community. Too often early domestic experiences are passed over by biographers or subordinated to more readily documented and prestigious formal influences, which are identified as the basis for later significant achievement in science.

Nonetheless, the importance of family is demonstrated in "scientific genealogies" whose branches reach upward across generations and outward through kinship networks. Best recorded are immediate lineages between parent, especially father, and child: Jane Colden studied botany with her father; John Bartram taught his son William to be an observer and collector; Charles Willson Peale took pride in the collecting and preservation skills of his sons Titian and Linnaeus and his daughter Sophonisba; the émigré Patrick Kerr Rogers contributed four highly productive sons (James, Henry Darwin, Robert, and William) to

---

nent rooms; see Harry B. Weiss and Grace M. Ziegler, *Thomas Say: Early American Naturalist* (Springfield, Ill.: Charles Thomas, 1931).

[14] Steven Shapin, "The House of Experiment in Seventeenth-Century England," *Isis*, 1988, 79:373–404.

[15] Private libraries and collections served more than individual owners; see Joseph Ewan, "One Professor's Chief Joy: A Catalogue of Books Belonging to Benjamin Smith Barton," in *Science and Society in America: Essays in Honor of Whitfield J. Bell, Jr.*, ed. Randolph Shipley Klein (Philadelphia: American Philosophical Society, 1986), pp. 311–344.

[16] Carol F. Karlson and Laurie Crumpacker, eds., *The Journal of Esther Edwards Burr, 1754–1757* (New Haven, Conn.: Yale Univ. Press, 1984), p. 123.

Figure 2. *The Rollo series by the educator and writer Jacob Abbott featured ways in which family and friends could enlarge upon the presumed curiosity and energy of children, and several featured scientific activities. Many of the popular textbooks and self-help books for children featured illustrations depicting a tranquil domestic scene as the context for discussion of natural history and natural philosophy.*

the geological enterprise; William Mitchell trained his namesake son and his daughter Maria on the telescope.[17]

A mother's influence seems more subtle; certainly mothers' interests and aptitudes are less well remembered and recorded. Still, casual reference is not uncommon in scientific biographies. Charlotte Coues wrote on birds and flowers, reviewed visiting lyceum lecturers with stiletto precision in her local New Hampshire newspaper, and by example taught her son Elliott to do fieldwork in natural science. The mother of Josiah Willard Gibbs nourished similar interests in her children, although Josiah subsequently turned his efforts toward physical science. Siblings could be important, too, as in the case of Emma Willard, director of the Troy Female Academy, and her younger sister Almira Lincoln Phelps, whose textbooks in botany, chemistry, and geology sold in the tens of

[17] See H. W. Ricketts and Elizabeth Hall, eds., *The Botanic Manuscript of Jane Colden, 1724–1766* (Garden Club of Orange and Duchess Counties of New York, 1963); Helen Gere Cruickshank, ed., *John and William Bartram's America: Selections from the Writings of the Philadelphia Naturalists* (New York: Devin Adair, 1957); Miller, ed., *Papers of Peale Family* (cit. n. 3); Emma Rogers, ed., *Life and Letters of William Barton Rogers*, 2 vols. (Cambridge, 1896); and Sally Gregory Kohlstedt, "Maria Mitchell and the Advancement of Women in Science," in *Uneasy Careers and Intimate Lives: Women in Science, 1789–1979*, ed. Pnina Abir-Am and Dorinda Outram (New Brunswick, N.J.: Rutgers Univ. Press, 1987).

thousands. The Rogers brothers collaborated in various combinations on state geological surveys in Pennsylvania and Virginia. A grandmother taught Martha Maxwell to observe and preserve natural history.[18] These miscellaneous examples suggest the patterns of formative family guidance that recur in biographies, autobiographies, and diaries. The case could be overstated. Certainly there was later and independent discovery of science, and of course encouragement could fail in its effect; but a surprising number of biographies suggest, often in tangential ways, very early family and community influences that deserve further research.

Father-to-son occupational legacies were common in other work areas, but given the nature of science and the uncertainty of scientific employment, the connection here implies that family life involved parlor discussions, countryside excursions, books, and other shared activities. In one of the important recent studies to consider family life and science, Anne Shteir traces what she terms the "family nature of preprofessional botany." She documents the "centrality of women in botanical education" among prominent and not-so-well-known English naturalists of this period.[19] Scientific interests were not only initiated by but also pursued within families. It was particularly common for women trained in fine arts to illustrate and color the plates that accompanied the government reports and privately published texts credited to their husbands. According to Pnina Abir-Am and Dorinda Outram's *Uneasy Careers and Intimate Lives,* in such cases the efforts of wives and daughters provided a kind of "personalized patronage system."[20]

The domestic setting for scientific discussion reflects the permeability of private and public knowledge. In the early nineteenth century the kinds of insularity presumed in a later world of specialization and professional affiliations were hardly evident. In the era of the encyclopedists, natural philosophy and natural history were presumed to be and were made more or less accessible to men, to women, and at some level to children as well. President Jefferson deliberately arranged for his namesake and nephew, Thomas Jefferson Randolph, to live with Charles Willson Peale, where a stimulating home setting and a rich urban intellectual environment would complement the boy's studies with a Philadelphia schoolmaster.[21]

Peale was part of a modestly affluent community that had the resources to acquire collections and instruments, first in homes and increasingly in select learned societies. It is, however, too narrow an analysis that argues that "in 1800 science in Philadelphia was restricted to gentlemen-savants with wealth and lei-

---

[18] Paul Russell Cutright and Michael J. Brodhead, *Elliott Coues, Naturalist and Frontier Historian* (Urbana: Univ. Illinois Press, 1981), pp. 19–21; Lynde P. Wheeler, *Josiah Willard Gibbs: The History of a Great Mind* (New Haven, Conn.: Yale Univ. Press, 1962), p. 10; and Maxine Benson, *Martha Maxwell: Rocky Mountain Naturalist* (Lincoln: Univ. Nebraska Press, 1986). Maxwell subsequently expanded on this local knowledge during a year at Oberlin College in the "scientific course."

[19] Ann B. Shteir, "Botany in the Breakfast Room: Women and Early Nineteenth-Century British Plant Study," in *Uneasy Careers and Intimate Lives,* ed. Abir-Am and Outram (cit. n. 17), pp. 31–44, on p. 32; and Shteir, "Linnaeus's Daughters: Women and British Botany," in *Women and the Structure of Society: Selected Research from the Fifth Berkshire Conference on the History of Women,* ed. Barbara J. Harris and Jo Ann K. McNamara (Durham, N.C.: Duke Univ. Press, 1984).

[20] Sally Gregory Kohlstedt, "In from the Periphery: American Women in Science, 1830–1880," *Signs,* 1979, *4*:81–96; and Pnina Abir-Am and Dorinda Outram, "Introduction," in *Uneasy Careers and Intimate Lives* (cit. n. 17), p. 46.

[21] Miller, ed., *Papers of Peale Family* (cit. n. 3), Vol. II, P. 1121.

sure."[22] What is true is that the domestic and social circles of genteel society presumed that science was their domain. A young man traveling in Europe in the 1790s was urged to return after his studies to Philadelphia with a "mind well stored with useful and ornamental knowledge." To be an educated and refined gentleman, his aunt reminded him, was to have "regulated passion, improved intellect, cultivated taste, a love of Science—the fine arts and indeed all the embellishments of polished Society."[23]

Those not in "polished Society" and without a formal parlor, especially those antagonistic toward what they deemed aristocratic rank and privilege, sometimes viewed the knowledge confined to private homes and clubs with considerable skepticism. More than one public lecturer parodied learned treatises that recorded the "diameter of the antedeluvians" or discourse that explained "why a cedar tree does not bear apricots."[24] Irreverence toward the pretensions of the learned was expressed in 1819 in a satirical poem about his adopted city by the New York Knickerbocker, Fitz-Greene Halleck (brought to my attention by Simon Baatz). Halleck first satirized learned pretentiousness:

> I might say much about our lettered men,
> Those grave and reverent seigniors, who compose
> Our learned societies—but here my pen
> Stops short; for they themselves, the rumor goes,
> The exclusive privilege by patent claim
> Of trumpeting (as the phrase is) their own fame.
> And therefore I am silent.

Not silent for long, however. Halleck's next stanza turns to applaud public opportunities and relies on wit, not sarcasm, to make his point. He goes on:

> It remains
> To bless the hour the Corporation took it
> Into their heads to give the rich in brains
> The Worn-out mansion of the poor in pocket,
> Once the "old almshouse," now a school of wisdom,
> Sacred to S[cudder]'s shells and Dr. G[riscom].[25]

Halleck's commentary in these two stanzas distinguished between, on the one hand, the privileged few who pursued ornamental and abstract knowledge, and,

---

[22] Simon Baatz, "Philadelphia Patronage: The Institutional Structure in Natural History in the New Republic, 1800–1833," *Journal of the Early Republic*, 1988, 8:111–138, on p. 112.

[23] Philip J. Greven, *The Protestant Temperament: Patterns of Child-Rearing, Religious Experience, and the Self in Early America* (New York: Knopf, 1977), pp. 297–298.

[24] Quoted from Abraham Bishop, who was asked to present and then barred from delivering the Phi Beta Kappa address at Yale in 1789, and instead gave the talk in town; see Robert Walker, *Reform in America: The Continuing Frontier* (Lexington: Univ. Kentucky Press, 1985), p. 16. Other, later evidence is found in Taylor Stoehr, *Hawthorne's Mad Scientists: Pseudoscience and Social Science in Nineteenth-Century Life and Letters* (Hamden, Conn.: Archon, 1978); and Michele L. Aldrich, "New York Natural History Survey, 1836–1845" (Ph.D. diss., Univ. Texas, 1974).

[25] Quoted in Simon Baatz, *Knowledge, Culture, and Science* (cit. n. 4); see the poem "Fanny" in *The Poetical Writings of Fitz-Greene Halleck*, ed. James Grant Wilson (New York: Appleton, 1869), stanzas 67, 68, p. 120. Halleck was a friend of the apothecary Dr. William Langstaff and of James E. DeKay, according to Nelson Frederick Adkins, *Fitz-Greene Halleck: An Early Knickerbocker Wit and Poet* (New Haven, Conn.: Yale Univ. Press, 1930). Michele Aldrich ("New York Natural History Survey," cit. n. 24) cited a somewhat similar skeptical poem by Edgar Allan Poe, "Sonnet—To Science," in *The Complete Tales and Poems of Edgar Allan Poe* (New York: Modern Library, 1938).

on the other hand, those who, "rich in brains," could visit John Scudder's natural history museum or hear John Griscom's popular and practical lectures on chemistry. These public activities were, ironically, as Halleck suggests, available in the former almshouse in New York City's Battery Park. While the discussions and demonstrations about science commenced in private homes, they were scarcely confined there. Halleck was sarcastic about what he glimpsed in parlors and learned societies, but he was spirited in his enthusiasm for events in more public thoroughfares. Separatist and privileged study persisted, but alongside this, and in new and far-reaching ways, public presentations served as an important base for science in the new republic.

PRIMERS

This second and more public setting for informal scientific activities reveals a bewildering array that provided amusement, instruction, edification, and, for some, bemusement. Book hawkers, itinerant lecturers, museum entrepreneurs, and self-help organizations all vied for public attention. Science and technology were prominent subjects. The impact of the wars for American independence and the simultaneous advent of industrialization allowed technology, as John Kasson has argued, and I would add science, to appear as not merely the "agent of material progress and prosperity but the defender of liberty and the instrument of republican virtue."[26] A variety of printed materials, some pirated from abroad and others written here, could be used by those who wanted to review, debate, extend, and apply new knowledge. Literacy among the white population steadily increased, for boys more rapidly than for girls, in the first half of the nineteenth century. There was also more literature of every kind, with journals, newspapers, and books particularly important for science and technology.[27]

Middling-class families purchased general magazines that dealt with a wide range of topics, including natural history and natural philosophy. On winter evenings farmers pored over almanacs and agricultural journals that described geological formations and Copernican cosmologies even as they provided advice for planting and fertilizing; their children could read pages specially designed to help them identify local natural life or perform simple chemical experiments. A frontier settler was, at varying levels, inevitably some sort of naturalist. Diaries verify the extent to which men and women carefully observed and recorded weather, planting and harvesting dates, solar and lunar eclipses, comets, meteors, and unusually bright aurora borealis—and it is hardly surprising that these observers liked to hear from others who interpreted such phenomena.[28]

[26] Kasson, *Civilizing the Machine* (cit. n. 7), p. 8.

[27] Lee Soltow and Edward Stevens, *The Rise of Literacy and the Common School: A Sociological Analysis to 1870* (Chicago: Univ. Chicago Press, 1981). Literacy varied by region, race, gender, and class; it was undoubtedly highest in New England, where William Gilmore reports nearly universal literacy for men and, among the middling and upper classes, 80 percent for women; see Gilmore, "Elementary Literacy on the Eve of the Industrial Revolution: Trends in Rural New England, 1760–1830," *Proceedings of the American Antiquarian Society*, 1982, 92:114–126; and Gilmore, *Reading Becomes a Necessity of Life: Material and Cultural Life in Rural New England, 1780–1835* (Knoxville: Univ. Tennessee Press, 1989).

[28] Margaret Rossiter, *The Emergence of Agricultural Science: Justus Liebig and the Americans, 1840–1880* (New Haven, Conn.: Yale Univ. Press, 1975), Ch. 1. For examples from diaries see Rhys Isaac on Landon Carter's diary in *Transformation of Virginia* (cit. n. 13), p. 48; and Deborah Jean Warner, *Grace Anna Lewis: Scientist and Humanitarian* (Washington, D.C.: Smithsonian Institution Press, 1979), p. 13.

Less rural readers also subscribed to magazines that commonly included discussions of botany, mineralogy, and other sciences.[29] By the 1830s *Harper's Weekly* carried columns entitled "Scientific Intelligence," while the best-selling *Godey's Ladies Book* presumed that even women busy with homes and families had enough leisure to read at least short features on natural history, geography, and astronomy in addition to the usual fare of fiction, recipes, and household advice. Children's magazines such as *Youth's Companion* were similarly useful and didactic. Science seemed such a promising topic that two of the four children's magazines launched by Ticknor and Field's Old Corner Bookstore in the 1830s were *The Naturalist* and *Scientific Tracts*. Some magazines stressed the distinctively new "low science" orientation that Susan Sheets-Pyenson has identified with French and British periodicals. Here, as abroad, popular magazines elaborated empirical and experimental techniques for natural history observation and mechanical invention. The content was negotiated not only by sales but also by reader replies. The market was volatile, the failure rate of magazines high, but scientific publications held their own in this period of definition and competition. Into this marketplace Benjamin Silliman entered his specialized *American Journal of Science and Arts,* and he also recruited a diverse set of subscribers for its continuation.[30]

Newspaper circulation increased at twice the population rate, and the price of single issues dropped from six to one or two cents between the beginning and the middle of the century.[31] Donald Zochert argues in his study of newspapers of the period that Milwaukee papers relied heavily on articles borrowed from other newspapers. This usage constituted an informal news network and allowed the Wisconsin city's papers to offer a vigorous and sustained scientific coverage. Short news stories provided a vocabulary and general familiarity with scientific ideas that could be pursued in more detailed monographs.[32]

In recent years both the Library of Congress and the American Antiquarian Society have held symposia and published on the history of books. Reading itself, the participants tell us, was made easier thanks to commercially made eyeglasses and better home lighting. Even more direct technologies like typesetting machines, stereographed plates, lithography, and cheaper binding affected book publishing on every level from introductory primers to detailed monographs.[33]

[29] David Paul Nord, "A Republican Literature: A Study of Magazine Reading and Readers in Late Eighteenth-Century New York," *American Quarterly,* 1988, *40*:42–64. He discusses the list of subscribers and the content of *New-York Magazine,* in which science accounted for a small but constant percentage of the essays.

[30] W. S. Tryon, *Parnassas Corner: A Life of James T. Fields, Publisher to the Victorian* (Boston: Houghton & Mifflin, 1963), p. 63; Harriet R. Christy, "First Appearances: Literature in Nineteenth-Century Periodicals for Children," *Research about Nineteenth-Century Children and Books,* ed. Richardson (cit. n. 9), pp. 117–132; Susan Sheets-Pyenson, "Popular Science Periodicals in Paris and London: The Emergence of a Low Scientific Culture, 1820–1875," *Annals of Science,* 1985, *42*:549–572; Donald De B. Beaver, "Altruism, Patriotism, and Science: Scientific Journalism in the Early Republic," *American Studies,* 1971, *12*:10; and Simon Baatz, " 'Squinting at Silliman': Scientific Periodicals in the Early American Republic, 1810–1833" (unpublished manuscript).

[31] George R. Taylor, *The Transportation Revolution, 1815–1860* (New York: Rinehart, 1957).

[32] Donald Zochert, "Science and the Common Man in Ante-bellum America," in *Science in America since 1820,* ed. Nathan Reingold (New York: Science History Publications, 1976), pp. 7–32.

[33] For a general discussion of this interest see the special issue of *American Quarterly,* 1988, *40*(1), ed. Cathy N. Davidson, esp. Davidson's introduction, "Towards a History of Books and Publishers," pp. 7–17; see also Ronald Zboray, "Antebellum Reading and the Ironies of Technological Innovation," pp. 65–82. The American Antiquarian Society publishes *The Book: Newsletter of the Program in the History of the Book in American Culture* (1983–) and produced a bibliography: David D. Hall

Techniques for distribution became more diverse. Corner bookstores and itinerant book hawkers were supplemented by the mass advertising of growing firms who distributed by mail and other sales outlets.

Scientific publications changed dramatically. The firm Carey and Lea in Philadelphia pioneered in publishing scientific books using superb local illustrators and colorists. Initially such books were produced only for authors who could provide a list of subscribers. Some scientific authors, like John James Audubon, found it more profitable to publish abroad. A good many books were printed privately, like the one on local natural history produced by a collective of women in Wilmington, Delaware. Editions ran from one to five hundred copies.[34]

These small printed runs were quickly overshadowed by a burgeoning textbook market that included not only grammars, spellers, and copybooks but also books on special topics. As the major publishing firms came to rely on the steady income from texts, they experimented with new book genres and imposed themselves on the process of book writing in order to encourage, as in periodicals, the use of more popular and familiar language. By 1850 textbooks accounted for over one third of the book trade. The most popular, including Jane Marcet's chemistry text and Almira Hart Lincoln Phelp's botany text, sold in the hundreds of thousands. A few publishers like John Wiley and W. H. Appleton specialized in scientific texts.[35] Some of the books were produced by compilers, but many were written by men and women otherwise engaged in teaching. Faculty members like Benjamin Silliman and Denison Olmsted of Yale supplemented their incomes by writing texts variously directed at their students, at a more general audience, and even at small children. Books, like schools of this period, were not easily classified. One of the most often cited illustrations of the self-help nature of such books involves Michael Faraday, who, L. Pearce Williams reminds us, was an Englishman converted to scientific study while binding and reading Jane Marcet's popular *Conversations on Chemistry*. Indeed, as Susan Lindee has demonstrated, Marcet's popular text was initially intended as a parlor amusement, but American publishers reshaped it through numerous editions into a text widely used in schools.[36]

In fact, it may be that books for children particularly underscore the enthusiasm for science. Early children's books were rarely intended for amusement alone but rather pointed students back into observation of the world around them. John Newbery's famous children's books included *The Newtonian System*

and John Hench, eds., *Needs and Opportunities in the History of the Book: America, 1639–1876* (Worcester, Mass., 1987). See also [Fred Belliveau], *One Hundred and Fifty Years of Publishing, 1837–1987* (Boston: Little, Brown, 1987).

[34] William Charvat, "James T. Fields and the Beginning of Book Promotion, 1840–1855," *Huntington Library Quarterly*, 1944/5, 8:75–94.

[35] John Hammond Moore, *Wiley: One Hundred and Seventy-Five Years of Publishing* (New York: Wiley, 1982), p. 32. For some data on publication numbers see Samuel Austin Allibone, *Critical Dictionary of English Literature and British and American Authors* (1870), Vol. II, p. 1575. See also Charles A. Madison, *Book Publishing in America* (New York: McGraw-Hill, 1966), pp. 17, 31. Publication of technical monographs continued to be a problem, and in the last half of the nineteenth century the Smithsonian Institution became perhaps the major outlet for scientific publication—indeed, Wiley saw it as a competitor; see Moore, *Wiley*, p. 53.

[36] L. Pearce Williams, *Michael Faraday: A Biography* (New York: Basic Books, 1965), pp. 18–20; and M. Susan Lindee, "Gender Politics of a Textbook: The American Career of Jane Marcet's *Conversations on Chemistry, 1806–1853*," (unpublished paper; winner of the History of Science Society's Schumann Prize, 1988). For a citation of Faraday see, e.g., Charles Carpenter, *History of American Schoolbooks* (Philadelphia: Univ. Pennsylvania Press, 1963), p. 215.

of Philosophy, published in 1761 and purportedly written by Tom Telescope; James Secord has revealed in fascinating detail the book's evolution through numerous editions over the next seventy years toward an ever more adult and didactic text.[37] In more fanciful children's books like the *History of Little Goody Two-Shoes* (1765), Miss Margery rescues various animals and trains them to use their particular capacities to help her run her school. Animal fables persisted, but children's classics lost ground as the tendency to teach "very early science" and observation to children quickly crossed the Atlantic, pirated, as bibliophiles point out, in the absence of copyright laws and enforcement.[38] One of the earliest examples was Anna Letitia Barbauld's *Lessons for Children from Four to Five Years Old,* published shortly after the Revolution. Like many other American publications, it was derived in large measure from a similar English text but took its natural history examples from the American environment.

Initially such primers provided straightforward descriptions. As attitudes toward children changed, Maria Edgewood's injunction "not to lecture" but to "kindly instruct" led to a question-and-answer format or a conversational dialogue in order to engage young readers and their parents—but whatever their mode of presentation, authors increasingly linked general ideas about nature to the local, that is, American, environment of their readers.[39] The messages were inevitably multiple, invoking much more than specific knowledge of science. Values and responsibility were evident in the disquisitions on natural theology, orderly living, and neighborliness.

One of the most prolific authors of children's books, the educator-turned-author Jacob Abbott, produced a series of fourteen red-bound books about Rollo, perhaps the first truly American child in fiction. In a book entitled *Rollo's Museum,* the boy and his sister learned to observe, collect, and turn to experts for advice on the meaning of their local discoveries. Organizing a museum with neighborhood children was not easy, and the process taught Rollo valuable lessons about human nature as well as physical nature.[40] In many of the popular children's books, mothers were full of long explanations about the movements of the planets or the location of minerals. Extended families, too, offered a generous wellspring of facts and stimulating ideas, and maiden aunts, often teachers, were remarkable sources of information.[41]

Much of the public activity could be typed as adult education, and it provided

[37] James A. Secord, "Newton in the Nursery: Tom Telescope and the Philosophy of Tops and Balls," *History of Science,* 1985, *23*:127–151.

[38] For an oveview and bibliography see the unpaginated exhibition catalogue *Science in Nineteenth-Century Children's Books* (Harper Memorial Library, Univ. Chicago, 1966). See also Mary F. Thwaite, *From Primer to Pleasure in Reading: An Introduction to the History of Children's Books in England from the Introduction of Printing to 1914* (1963; Boston: Horn Book, 1972), pp. 181, 202; and Ruth Miller Elson, *Guardians of Tradition: American Schoolbooks of the Nineteenth Century* (Lincoln: Univ. Nebraska Press, 1964), pp. 14–40. Not everyone approved of this didacticism. Charles Lamb wrote to Coleridge, "Think what you would have been now, if you had been crammed with geography and natural history"; cited in Samuel F. Pickering, Jr., *John Locke and Children's Books in Eighteenth-Century England* (Knoxville: Univ. Tennessee Press, 1981), p. 60.

[39] Gail S. Murray, "Rational Thought and Republican Verities Children's Literature, 1789–1820," *J. Early Republic,* 1988, 8:159–177, on p. 162.

[40] Alice M. Jn, *From Rollo to Tom Sawyer and Other Papers* (Boston: Horn Book, 1948), pp. 21, 74. Abbott wrote over two hundred books for children and teachers.

[41] Anne Scott MacLeod, "Children's Literature and American Culture, 1820–1860," in *Society and Children's Literature,* ed. James H. Fraser (Boston: Godine, 1978), p. 25.

PART I.—DESCRIPTIVE.

FIG. 1.

HOLBROOK SCHOOL APPARATUS

COMMON SCHOOL SET.—PRICE, $20.

1. Orrery, ... $10.00
2. Tellurian, ... 6.00
3. Geometrical Solids, ... 1.25 Extra, $1.50
4. Terrestrial Globe, (5 inch.) 1.00
5. Numeral Frame, ...75 No. 2, .62½
6. Hemisphere Globe, ...75
7. Cube Root Block, (Extra,) . . . . . . .50 Double, .75
8. Text-book, ...37½ Cloth, .50
9. Magnet, ...25 .37 .50 & upwards.

MISCELLANEOUS ARTICLES.

10. Brass Mounted Orrery, ... $12.50 and $15.00
11. Brass Mounted Celestial Sphere, . . . 6.00
12. Lane's Mechanical Paradox, or Gyroscope, 2.50 3,50 5.00 & up'ds.
13. Pointing Rods, ...50 .75
14. Double Slates, No. 1, ...45 No. 2, .56
15. Holbrook's Noiseless Drawing Slates, No. 1, .20 No. 2, .25
16. Holbrook's Drawing Book, ...08
17. Holbrook's Nois'less High Sch'l Slate, single, .25
18.   "   "   "   " double, .56

Figure 3. Josiah Holbrook developed sets of instruments for demonstration and experimentation and made them available to teachers and parents through his School Apparatus Company.

opportunities to extend childhood experience or compensate for its gaps. Public lectures, for example, had predated the Revolution. The story of Franklin's conversion to electrical inquiry by a chemical lecturer is familiar. Itinerant speakers went from town to town, giving popular talks on physics, mechanics, chemistry, mineralogy, and other subjects that amused and challenged the imagination. By the 1810s major cities like New York had audiences sufficient to provide regular employment for the chemist John Griscom, in addition to the numerous itinerants.[42] Such speakers knew that it was essential to combine visual demonstrations with their lectures. They used apparatus imported or made by themselves. Audiences were diverse, determined by personal incentives as much as sponsors' intentions. Clerks, accountants, and small enterprising entrepreneurs heard lectures on mechanics, while the middling classes were more likely to attend dis-

---

[42] I. Bernard Cohen, *Benjamin Franklin's Experiments: A New Edition of Franklin's Experiments and Observations on Electricity* (Cambridge, Mass.: Harvard Univ. Press, 1941), pp. 49–54; and Edgar Fahs Smith, *John Griscom* (Philadelphia, 1925). On public demand for lectures see Ian Inkster, "Robert Goodacre's Astronomy Lectures (1823–1825) and the Structure of Scientific Culture in Philadelphia," *Ann. Sci.* 1978, *35*:353–363.

cussions on astronomy or mineralogy.[43] Some public lectures were actually short courses and had associated field trips. Information about such activities is scarce, preserved primarily in handbills and newspaper advertisements now in the archives of local history societies; these deserve further exploration. These ephemeral activities were of immense importance, however, in providing identity and coherence to scattered initiates and enthusiasts.

Many dissemination efforts were short-term, sporadic, and episodic. They were supplemented by a broad array of more systematic forms of activity in the literally hundreds of mechanics' institutes, young men's friendly societies, reading rooms, cabinets of natural science, lyceums, and public libraries. Such voluntary associations were sustained by members and philanthropists. In New York City, Frances Wright and Robert Owen called their workingman's association a "Hall of Science." A contemporary noted that besides regular presentations on the "more popular levels of science, [lecturers] generally embrace discussions, oral and written, on topics interesting for their practical bearing, but they judiciously avoid the profitless questions of theology, and party politics."[44] Science represented a noncontroversial, unifying force even as it provided factual knowledge, rational analysis, and practical applications.

Expository science, or popularized science, reached unprecedented and perhaps unsustainable levels in the first half of the nineteenth century. Audiences heard prominent lecturers speak in terms relevant to their situation, transposing themes of industry and technology with those of science without self-consciousness.[45] In the United States, at least, this ongoing public attention strengthened efforts to institutionalize science in that most pervasive antebellum social agency, the school. Social studies of science remind us that the status and authority of various sciences are intimately connected to the individuals and institutions that express and define them. It is therefore helpful to see schools as institutions that simultaneously established boundaries around knowledge and enabled individuals to push forward toward new personal knowledge.[46]

[43] Donald A. Scott, "The Popular Lecture and the Creation of a Public in Mid-Nineteenth-Century America," *Journal of American History*, 1980, 66:791–809; Frederick J. Antezak, *Thought and Character: The Rhetoric of Democratic Education* (Ames: Iowa State Univ. Press, 1985); and Elizabeth Barnaby Keeney, "The Botanizers: Amateur Scientists in Nineteenth-Century America" (Ph.D. diss., Univ. Wisconsin–Madison, 1985), p. 31. On England see Larry Stewart, "Public Lectures and Private Patronage in Newtonian England, *Isis*, 1986, 77:47–58.

[44] Quoted in Joseph J. McCadden, *Education in Pennsylvania, 1801–1835, and Its Debt to Robert Vaux* (New York: Arno, 1969), p. 74. Cf. the motto of *Youth's Companion*, "No sectarianism, no controversy." See also Hyman Kuritz, "The Popularization of Science in Nineteenth-Century America," *History of Education Quarterly*, 1981, 21:259–274. On systematic dissemination, besides literature cited above, standard accounts of various movements include Carl Bode's still-authoritative *The American Lyceum: Town Meeting of the Mind* (New York: Oxford Univ. Press, 1956). On Baltimore's early institutes and libraries see Philip Arthur Kalish, *The Enoch Pratt Free Library: A Social History* (Metuchen, N.J.: Scarecrow, 1969), pp. 29–46.

[45] For a discussion of how audiences reinterpreted public lecturers see Mary Kupiec Cayton, "The Making of an American Prophet: Emerson, His Audiences, and the Rise of the Culture Industry in Nineteenth-Century America," *American Historical Review*, 1987, 92:587–620. On popularization in other times and places see Terry Shinn and Richard Whitley, eds., *Expository Science: Forms and Functions of Popularization* (Dordrecht: D. Reidel, 1985).

[46] Anthony Giddens, *A Central Problem in Social Theory: Action, Structure, and Contradiction in Social Analysis* (London: Macmillan, 1979), p. 71; and Karen Knorr-Cetina and A. V. Cicourel, eds., *Advances in Social Theory and Method: Toward an Integration of Micro- and Macro-Sociologies* (Boston: Routledge & Kegan Paul, 1981).

EDUCATION

Schools of any kind were a small part of the terrain of childhood in revolutionary America, but by the 1840s a building boom dotted the countryside with one- and two-room frame, brick, stone, and even sod schoolhouses, consolidated enterprising villages around quasi-public academies, and situated literally hundreds of small colleges in towns and cities. Information on the curriculum of these schools is still limited, but there is considerable evidence to suggest that schooling directed at young people increasingly incorporated science. Political and legal action created state and city systems, but in the early years of educational reform they had less direct effect on curriculum than did the more idiosyncratic local concerns of teachers, parents, and administrators. The process of formalizing education created various types of schools, emphasized particular topics, and endorsed specific texts about science. The schools were a mechanism to both codify and transform the basic literacy and numeracy teaching that continued to be conducted in homes, churches, apprentice shops, asylums, and village meetinghouses.[47]

Pluralism among and within the diverse set of educational institutions blurred distinctions between public and private. This was particularly evident in the academies established in the late eighteenth and early nineteenth centuries. The academies operated both as preparatory schools for college and as capstones to the education of boys and some girls who might become surveyors, tradesmen, industrialists, teachers, and parents. Following a particularly American formulation, the academies were at once private, with self-perpetuating boards of trustees and tuition-paying students, and public, with income from lotteries and state and local subsidies. Such connections meant that the academies were, as Edward O'Neill terms it, "receptive vessels" for the values and aspirations of sponsoring communities.[48] The curriculum thus differed by location. In the seafacing merchant capitalist city of Salem the course of study was rich in geography (taught with a globe), bookkeeping, and navigation. In Albany, at the juncture of the Hudson River and the Erie Canal, students studied, along with standard subjects, geometry, surveying, geography, and hydrography.[49]

The curricular innovation extended to coeducational and women's academies as well. As Linda Kerber and Margaret Rossiter have argued, the status of girls as future mothers of citizens of the republic gave a patriotic rationale for educators to provide girls with basic skills and to teach them scientific subjects as well. Patricia Cohen's study of numeracy in the period finds no hint of math anxiety; both boys and girls studied arithmetic. Deborah Warner notes that certain sciences were actually more available to girls than to boys, and Elizabeth Barnaby Keeney suggests that academy botany preceded college botany by a decade or more.[50] Such freedom and opportunity impressed (or dismayed) foreign visitors.

---

[47] On elementary schools see Barbara Finkelstein and Kathy Vandell, "The Schooling of American Childhood: The Emergence of Learning Communities, 1820–1920," in *A Century of Childhood, 1820–1920,* ed. Mary Lynn Stevens Heininger *et al.* (Rochester, N.Y.: Margaret Woodbury Strong Museum, 1984), pp. 65–95; and David Tyack, Thomas James, and Aaron Benavot, *Law and the Shaping of Public Education, 1785–1954* (Madison: Univ. Wisconsin Press, 1987).

[48] Edward Herrin O'Neill, "Private Schools and Public Vision: A History of Academies in Upstate New York, 1800–1860" (Ph.D. diss., Syracuse Univ., 1984).

[49] Bernard Farber, *Guardians of Virtue: Salem Families in 1800* (New York: Basic Books, 1972), pp. 168–172.

[50] Linda K. Kerber, *Women of the Republic: Intellect and Ideology in Revolutionary America*

The Englishwoman Harriet Martineau, while visiting in Cincinnati, reported, "I attended the annual public exhibition of [a girls'] school and perceived with some surprise that the higher branches of science were among the studies of the pretty creatures assembled there."[51]

If academies were in the vanguard of the movement for education in science, there were parallel efforts based in both elementary schools and colleges. Multiple and overlapping initiatives came from teachers, students, and patrons.

Teachers and directors of private schools necessarily created their programs with attention to the aspirations of parents who wanted stimulating and relevant course work. Mary Peabody recalled her lessons in astronomy, chemistry, and natural philosophy, which, she later remembered, "my mother took great pleasure in teaching and for which, for want of text-books, she wrote voluminous works in the form of dialogues, which her pupils read aloud, and which she illustrated to the best of her ability with very scanty materials." This education was supplemented with visits to a neighbor, Dr. Nathaniel Bowditch, who invited Mary and her sister to look through his telescope and taught them more about astronomy and geography. Once she became a teacher herself, Mary Peabody discovered that students were easily lured into her own favorite study, natural history. Being an innovative educator like her sister Elizabeth Peabody, she took students into the field to collect specimens and to Boston to visit the small museum of the Boston Society of Natural History when it was still housed in a lawyer's office. Her justification for such scientific study emphasized self-improvement and natural theology, while the texts she used and her systematic instruction attended quite directly to the subjects at hand.[52]

Self-consciously innovative teachers of the period were influenced by such educational philosophers as Johann Heinrich Pestalozzi and Fredrich Froebel. Their theories emphasized stages of childhood development as these related to readiness for instruction. Their pedagogy involved object study, arguing that all the senses (hearing, touching, and smelling as well as seeing) contributed to learning. The wealthy Scottish immigrant, geographer, and social reformer William Maclure visited Pestalozzi's school and reported to fellow Americans on the use of objects for drawing and model building and the enthusiasm of the children for collecting and field trips. Maclure implemented these ideas by hiring Madame Fretegot to come to the United States in the 1820s and organize a school at the utopian community of New Harmony. His "boat load of naturalists" from Philadelphia ensured a generous supply of specimens for study and for research in this private experiment.[53]

In publicly sponsored schools the issue of curriculum was often negotiated. Even if teachers could prepare special subjects, the local community might resist

---

(New York: Norton, 1966); Margaret W. Rossiter, *Women Scientists in America: Struggles and Strategies to 1940* (Baltimore: Johns Hopkins Univ. Press, 1982); Patricia Cline Cohen, *A Calculating People: The Spread of Numeracy in Early America* (Chicago: Univ. Chicago Press, 1982); Deborah Jean Warner, "Science Education for Women in Antebellum America," *Isis,* 1978, *69*:58–67; and Elizabeth Barnaby Keeney, "The Botanizers" (cit. n. 43), p. 51.

[51] Frances Trollope, *Domestic Manners of the Americans* (1839; London: Century Publishing, 1984), p. 68.

[52] Mary Peabody Mann, "Reminiscences of School Life and Teaching," *American Journal of Education,* 1882, *32*:744.

[53] Recent historians of education have paid less attention to educational philosophy and theory than to political and cultural questions, but see Thomas A. Barlow, *Pestalozzi and American Education* (Boulder, Colo., Este Es Press, 1977).

innovation. In the handbook for teachers, Jacob Abbott reminded them that they were responsible to constituents. His example was an eager young woman who wanted to introduce the study of botany into her classroom. The parents and school board objected. They wanted the children to confine their attention exclusively to the elementary branches of education: "We want them to read well, to write well, and to calculate well, and not to waste their time in studying about pistils and stamens and nonsense."[54] Abbott, sympathetic to a curriculum with science, suggested that the teacher might try to persuade parents and school committee members—but warned that the final authority resided in their hands. Abbott's illustration is instructive about process. Curricula were negotiated and, increasingly, science was part of the bargain struck, although not necessarily the science subjects proposed by teachers.

Another evidence of the educators' interest in science was the growing business created by the use of visual illustrations and models. The public school reformer Henry Barnard was probably preaching to those already converted by popular lecture series when he advocated apparatus for common schools. Private school advertisements had already emphasized books and equipment along with teachers' qualifications in the late eighteenth century. Illustrations (usually pictures or diagrams) were said to ensure precision and permanence of knowledge, as were models of globes, geometrical solids, and anatomical figures.[55]

As in book publication and distribution, an early import business was rapidly transferred to domestic instrument makers. The boom in demand for precision instruments and scientific apparatus in Georgian England by a public that used them for amusement, for instruction, as things of beauty, or as status symbols moved across the Atlantic.[56] Deborah Warner indicates that the American entrepreneurs initially "identified themselves as philosophical, enjoying the status inherent in this term," but they soon realized that education represented a "vastly greater market than did research." Josiah Holbrook, an avid student of Silliman who is best remembered as the founder of the American lyceum movement, organized a School Apparatus Company. He advertised simple and inexpensive boxes of specimens in natural history, and his set of familiar mechanical devices illustrated the lever, pulley, inclined plane, screw, axle, and wedge. Chemistry equipment included various glass holders as well as a pneumatic cistern and a compound blowpipe. Just how the school supplies were utilized remains somewhat a mystery, but a record of tax dollars spent indicates that there is much more to be discovered about the school curriculum, especially in regard to science and technology.[57]

Advocates for school science represented various positions. The educational reform leaders in Massachusetts, New York, and Philadelphia all included essays about the value of science in their publications. The workingmen's parties of the

[54] Jacob Abbott, *The Teacher; or, Moral Influences Employed in the Instruction and Government of the Young; Intended Chiefly to Assist Young Teachers in Organizing and Conducting Their School* (Boston: William Peirce, 1836), p. 237. Abbott's use of a woman here is unusual, for most references to teachers are masculine in this handbook.

[55] Bode, *American Lyceum* (cit. n. 44), pp. 186–187; and Perry Miller, *The Life of the Mind in America from the Revolution to the Civil War* (New York: Harcourt, Brace & World, 1965), p. 272.

[56] Roy Porter *et al.*, *Science and Profit in Eighteenth-Century London* (London: Whipple Museum of the History of Science, 1985), p. 3, offers a detailed discussion of mathematical, optical, and philosophical instrument makers and their wares.

[57] Deborah Jean Warner, "Commodities for the Classroom: Apparatus for Science and Education in Antebellum America," *Ann. Sci.*, 1988, *45*:387–397, on p. 390.

1830s, undoubtedly inspired by ideas of Robert Dale Owen, lobbied for a more thorough, practical, and scientific education in common, publicly funded schools with a standardized curriculum.[58] The results of considerable activity and agitation were impressive in northern states. New York, a pioneer, reported in 1829 that it had fifty academies and nine thousand school districts. The majority at that point were one-room schools serviced by a minister, a vacationing college student, or the increasingly available farm daughter. Children studied by day, while adults met at night in local school buildings for meetings of agricultural societies, mechanics institutes, and itinerant lecturers. The written word supplemented and to some extent subordinated the spoken sociability of community education.[59]

Noah Webster's assertion that education has an "inseparable connection with morals and a consequential influence upon the peace and happiness of society"[60] underscores the reality that the chief business of schools established during the burgeoning education movement was citizenship, broadly defined. Advocacy for more science teaching was often couched in terms of moral sensibility and community responsibility. Religion monitored but did not oppose science. There was still a "place of grace" in antebellum evangelical Protestantism where knowledge and faith were complementary.

By mid century, advocates for more science in the schools shared the ideas of the editor of *Popular Science Monthly,* Edward Youmans. Science was an essential mental discipline as well as a part of the culture demanded by modern life. The push for modernity impelled colleges toward more science in the curriculum and more facilities for such study. The pattern was not predictable: apparatus and specimens sometimes preceded the institution of regular courses, while at other times popular courses led college administrators to acquire such resources. Here, too, lectures and specimens were often accessible to local community members.[61] Education was never confined to the schools, and the work carried out in the schools involved a rich variety of activities and participants that have yet to be examined in detail.

REFLECTIONS AND CONCLUSION

Looking back through the lens of history at scientific activity in the young republic is like aligning and realigning a kaleidoscope with its glittering bits of evidence that change with each refocus of the lens. Irregular fragments emerge and recede, creating ever-changing patterns that highlight marginal and working classes or more educated groups, women or men, lecturers or their audiences, northerners or southerners. Still, the historical alignments prove to be not simply random. Particular components of science tended to be advocated by an upper class, while others, ignored or disdained by them, were apparently attractive to

[58] Stephen Stimpson's *Working Man's Manual,* cited in McCadden, *Education in Pennsylvania* (cit. n. 44), pp. 95–96.

[59] Lawrence A. Cremin, *American Education: The National Experience, 1783–1876* (New York: Harper & Row, 1980).

[60] Noah Webster, *On the Education of Youth in America,* rpt. in *Essays on Education in the Early Republic,* ed. Frederick Rudolph (Cambridge, Mass.: Belknap Press of Harvard Univ. Press, 1965), pp. 43–67, on p. 43.

[61] Sally Gregory Kohlstedt, "Curiosities and Cabinets: Natural History Museums and Education on the Antebellum Campus," *Isis,* 1988, 79:405–426.

working families. Still different combinations of scientific information were pursued by middling groups who were in the process of defining a new, unprecedented professional identity in scientific occupations and used local academics as a mechanism to transfer practical scientific knowledge to their children.

Gender has also been a significant variable. Women's enthusiasm for science, their publication of texts, and their involvement in education as teachers and as learners are important because men and women became more self-conscious about the appropriateness of women studying science. The texts by women were sometimes published under pseudonyms or included disclaimers about the limits of the author's expertise. While Marcet, for example, did not cater to women by using domestic examples, she did feel impelled to explain her reasons for writing the text; Phelps, by contrast, drew on such female experience.[62] There was not equality of opportunity or condition, and we need to remember that it was possible but not probable that women could pursue scientific interests. Essentially no black people could, a fact that has had consequences for the twentieth century and for science. The pursuit and support of particular sciences were, these variables seem to suggest, intimately related to status considerations, technological implications, health and other public concerns, and geographic location. The stage lights illuminating the drama of American science need to focus beyond the exceptional events that have heretofore monopolized historians' attention. At no point in the history of science is that more important than in these formative years of national history.

By the 1840s scientific activity gained visibility in more formal settings, and its advocates presented new and largely unprecedented claims for "pure science." Science taught in conjunction with moral values, or even technical applications, seemed inappropriate, even embarrassing. General, diffuse, and sometimes demanding enthusiasm for science became channeled by those in a position to constrain, advance, and direct it. Private letters and printed pamphlets document the ways in which a generation with employment in science was able to coordinate a like-minded community. The goal was very explicitly to raise up a new generation in the established group's image and to disseminate ideas about science on terms it defined. The institutionalization of science was taking its now-familiar shape: scientific activities by state and federal governments were being gradually transformed into permanent agencies, colleges presumed that science courses should be in their catalogues, and specialists created their own sections of the AAAS, produced journals, and moved toward separate societies.

By the time this authority was established, there was also a legacy of commentary and criticism by those who would debate and contest the boundaries and definitions given to science. Conflict between "closet" or "armchair" botanists and field observers was often sharp, for example, and ran parallel to the increasing distinction that separated self-styled amateurs from employed or institutionally connected naturalists. Debates about method and expertise were possible because science and technology had become familiar topics susceptible to critique, and public meetings like those of the AAAS were occasionally jarred by challenges from members reluctant to rely on an exclusive and inner-circle style of national science. Outside scientific groups, too, thoughtful observers raised

---

[62] Lindee, "Gender Politics of a Textbook" (cit. n. 36), p. 28.

questions about, for example, applications of science, such as those used to buttress racial distinctions.[63]

In the process of professionalization, many of the early popular efforts were eliminated or were subordinated—sometimes only after serious controversy, as in the case of the Dudley Observatory—to the particular goals of the scientists.[64] The process narrowed participation in formal and public institutions to include some men but very rarely the women who had successfully popularized science. Nonetheless, some traditional activity by men and women persisted, usually defined as amateurism or as staff support and little acknowledged by contemporaries or by historians until very recently.[65]

My purpose has been to present the obverse of the coin of research science by considering family life, popularization, and education as part of the process by which scientific activity is initiated and scrutinized.[66] The National Science Foundation is currently emphasizing, under new funding programs, support for more science-related teaching in the formative years of elementary, middle, and secondary schooling. Professional educators have argued that these are the years when most children establish lifelong interests and attain fundamental skills. Reports like *A Nation at Risk* decry the country's slippage in science.[67] Historical research can evaluate such assessments and also document both the private and the institutional processes that have shaped study and research in science.

There is also something to be learned from a comparison of the formative years of scientific development in this country with those of other European, colonial, and even non-Western countries. The multiple and diverse initiatives taken during the first seventy years of the republic established public anticipation for greater scientific activity. Education—whether acquired in informal parlor discussions, through primers and popular lectures, or under the guidance of teachers—shaped and was in turn shaped by a cultural outlook in which the study of science gained a fundamental place. If the teaching and learning lacked uniformity and consistency, their very diversity expressed an orientation toward science that included but was not limited to the search for new knowledge. The process incorporated personal, interactive, and eventually intrusive techniques. All together, they created an ambiguous, inconsistent, but nonetheless pervasive commitment to science in American culture that has persisted for two centuries.

[63] Leonard R. Stanton, *The Leopard's Spots: Scientific Attitudes toward Race in America, 1855–1859* (Chicago: Univ. Chicago Press, 1960).

[64] Mary Ann James, *Elites Conflict: The Antebellum Clash over the Dudley Observatory* (New Brunswick, N.J.: Rutgers Univ. Press, 1987).

[65] Margaret W. Rossiter, "Women and the History of Scientific Communication," *Journal of Library History*, 1986, *21*:39–59.

[66] The need for more discussion of such topics is indicated by Sara Delamont in "Three Blind Spots? A Commentary on the Sociology of Science by a Puzzled Outsider," *Social Studies of Science*, 1987, *17*:163–170.

[67] National Commission on Excellence in Education, *A Nation at Risk: The Imperative for Educational Reform* (Washington, D.C., 1983).

Science Education for Women in Antebellum America

By Deborah Jean Warner

WOMEN TOO SHARED in the popular enthusiasm for science which emerged in America in the second third of the nineteenth century. Schools for women placed a new emphasis on natural history and natural philosophy. Books about science directed specifically at women proliferated, as did scientific articles in the general women's magazines, and public scientific lectures attracted large numbers of women. The message presented through these various media was a cultural one: the efforts of scientists should be supported, their achievements appreciated. The audience was encouraged to become "cultivators" of science, not necessarily "practitioners."[1] Once interested, however, some women—indeed a substantially larger number than is generally recognized—went on to pursue science on their own. By 1860 the foundations were securely planted for women's involvement in America's scientific enterprise.

Public education for women, practically nonexistent in the colonial and early republican periods, began to flourish in the 1820s, and within a few decades dozens of academies, seminaries, and colleges were established. Most of those admitting women were for women only, but some were coeducational; most were in the Northeast, but a good number were in the South and West.[2]

Many educators sought to establish curricula for women similar to that available to men. Evidence for this is easily multiplied, but for now let suffice two typical declarations of purpose. The Elmira Female College was founded "with the design of affording a superior education to young ladies, with all the advantages furnished by the best [male] Colleges in the country." Packer Collegiate Institute in Brooklyn aimed to furnish "all the advantages for thorough and complete education that are enjoyed by the other sex in our best appointed colleges."[3] To be sure, Emma Willard's

---

[1] Nathan Reingold, "Definitions and Speculations: The Professionalization of Science in America in the Nineteenth Century," in Alexandra Oleson and Sanborn Brown, eds., *The Pursuit of Knowledge in the Early American Republic* (Baltimore: Johns Hopkins University Press, 1976), pp. 33–69.

[2] The most comprehensive account is still Thomas Woody, *A History of Women's Education in the United States,* 2 vols. (New York: The Science Press, 1929). Catalogues and announcements for more than four dozen antebellum women's schools can be found at the Library of Congress and in the Sophia Smith Collection at the Smith College Library.

[3] Elmira Female College, *Catalogue* (1862–1863), p. 13. Packer Collegiate Institute, *Circular and Catalogue* (1861), p. 5. Recent work in women's studies tends to stress the differences between the experiences of men and women, often to the exclusion of the similarities; see, e.g., Keith Melder, "Masks of Oppression: The Female Seminary Movement in the United States," *New York History,* 1974, 55:261–279. The propaganda of the Ivy League sister schools has largely obscured earlier educational advantages, as can be seen in James Monroe Taylor, *Before Vassar Opened* (Boston: Houghton Mifflin, 1914), and all standard histories of American education which even mention schooling for women. Articulate feminist reformers who were sufficiently well educated to appreciate the opportunities, yet who

Troy Female Seminary (1821) offered studies "as different from those appropriate to the other sex, as the female character and duties are from the male," and Catherine Beecher took a similar tack at her Hartford Female Seminary (1823). But even these professedly feminine schools offered an academic rather than an ornamental education.[4]

In the 1830s, as Guralnick has shown, schools for men began to establish science as a prominent component of the required studies and to acquire expensive and often sophisticated scientific apparatus.[5] A similar trend occurred in schools for women. The Albany Female Academy promoted the natural sciences as "essential to a proper education." Even Bascomb Female Seminary at Granada, Mississippi, which advertised itself as a finishing school, offered semi-weekly "Chemical and Philosophical Lectures, accompanied with experiments." Among the annual awards at the Baltimore Female College, the highest was for Greek and Latin, the second highest for "eminence in Mathematics and Natural Sciences."[6]

The educational advantages enjoyed by American women were duly noted by foreign visitors. Fredrika Bremer of Sweden found American schools for women generally much superior to European ones. She was particularly impressed by the opportunities for advancing "as far as young men" in the sciences. As a result, women were distinguishing themselves in "mathematics, algebra, the physical sciences . . . and many other hitherto interdicted branches of learning."[7] American commentators often made similar remarks, and visiting committees charged with examining students at the end of each school year repeatedly emphasized how proficient girls were in mathematics and science.[8]

Since none of the women's schools before Vassar (1865) enjoyed an endowment at all commensurate with those of the leading men's schools, their expenditures for science could scarcely have been as extravagant as the schools considered by Guralnick. Within the limits of their resources, however, they did remarkably well. Scientific paraphernalia often represented the largest single investment aside from the actual buildings and furnishings and so were treated as a prime attraction. Rutgers Female Institute, in New York City, insisted that in chemistry and natural philosophy few colleges could "boast of greater facilities for instruction." Abbot Collegiate Institute, also in New York, obtained a large and expensive collection of chemical

---

denied them, presented a highly distorted picture: e.g., Elizabeth Cady Stanton, *Eighty Years and More, Reminiscences 1815–1897* (New York: European Publishing Co., 1898; reprint, New York: Schocken, 1971). Girls from better-off families, "polished at boarding schools in the social amenities and the esthetic arts of fancy embroidery and painting on velvet," described by Merle Curti, *The Growth of American Thought* (New York: Harper, 1951), p. 384, represented a minority of the women educated in the antebellum period.

[4] Emma Willard, *An Address to the Public, Particularly to the Legislature of New York, Proposing a Plan for Improving Female Education* (Middlebury, Vt., 1819). Joan Burstyn, "Catherine Beecher and the Education of American Women," *New England Quarterly*, 1974, 47:386–403.

[5] Stanley Guralnick, *Science and the Ante-Bellum American College* (Memoirs of the American Philosophical Society, 109) (Philadelphia: American Philosophical Society, 1975).

[6] Albany Female Academy, *Catalogue* (1861), p. 5. Bascomb Female Seminary, *Catalogue* (1852), p. 10. Baltimore Female College, *Annual Catalogue* (1859), p. 13. Unlike Baltimore, most colleges offered women science in lieu of classical languages. Woody, *A History of Women's Education*, Vol. I, p. 418, recognized this, but as a point of criticism, for knowledge of Greek and Latin was his standard for academic excellence.

[7] Fredrika Bremer, *America of the Fifties*, ed. Adolph Benson (New York: The American-Scandinavian Foundation, 1924), pp. 72–73, 285.

[8] E.g., "Ingham University," in *Scientific American*, 1864, *11*:54. Troy Female Seminary, *Catalogue* (1851–1852), p. 25.

and philosophical apparatus, "unsurpassed in character by that of any other Institution in our country." The obvious exaggeration of such claims should not obscure their intent. Abbot aimed explicitly to attract the brightest students, and to "elevate the standard of education for young Ladies."[9]

Of all the sciences, astronomy was the most extravagantly outfitted. As early as 1828 the Young Ladies' High School at Boston owned a tellurian, a cometarium, and an orrery which "for the number and correctness of the motions represented," the school claimed, was "probably one of the best instruments of the kind in the country." The Seminary for Female Teachers at Ipswich, Massachusetts, opened with one building, books worth $25, and "a pair of very valuable globes." Around mid-century the Albany Female Academy owned a large vertical orrery, and Packer Collegiate Institute had a very large horizontal one made by the renowned Thomas Barlow of Kentucky.[10] Elsewhere students learned the geography of heaven and earth by means of maps, charts, and magic lantern slides.

For symbolic impact nothing outmatched the astronomical observatory. The most famous observatory at a women's school, Vassar's $15,000 installation, was antedated by at least six others, and by as long as twenty years. In 1846 the Sharon Boarding School near Darby, Pennsylvania, spent $4,000 on equipment which included an equatorial refractor of 6¼ inches aperture made by Merz u. Sohn of Munich, a meridian circle, and a sidereal clock.[11] Through the beneficence of the local academy of sciences, the Elmira Female College had access to an observatory housing a transit instrument and a $2,000 8½-inch-aperture equatorial refractor made by Henry Fitz. Both Mount Holyoke Seminary and Packer Collegiate Institute had 6-inch Fitz telescopes. Observatories were built at the Albany Female Academy and the Ohio Female College, and the prospectus of the Chesapeake Female College at Hampton, Virginia, called for an observatory "with a very superior refracting telescope."[12] The telescopes, many of research quality, were used mainly for viewing celestial details and the transit instruments for determining times and locations, which was also the practice at most men's schools.

Students learned natural philosophy—such subjects as mechanics, pneumatics and hydrostatics, electricity and magnetism, and optics—by means of lectures with demonstrations. According to their catalogues, each women's school owned some physics apparatus, sometimes simply a standard set offered by dealers such as N. B. Chamberlain of Boston or Benjamin Pike of New York. But some schools had much more: the Buffalo Female Academy had over $2,000 worth of philosophical apparatus; Abbot Collegiate Institute purchased "several thousand dollars' worth of new and beautiful apparatus, American and European, chiefly made to order." The

[9]E. Porter Beldon, *New York, Past, Present and Future* (New York, 1849), p. 110. Abbot Collegiate Institute, [Catalogue] (1854), p. 13.

[10][Boston] Young Ladies' High School, *Catalogue* (1829), p. 19. *Catalogue of the Officers and Members of the Seminary for Female Teachers at Ipswich, Mass. for the Year Ending April 1839.* Albany Female Academy, *Catalogue* (1851 [?]), pp. 5–6. Packer Collegiate Institute, *Circular and Catalogue* (1861), p. 11.

[11]Elias Loomis, *Recent Progress of Astronomy* (New York: Morningside Heights, 1856), pp. 156–158. Eleanor Wolf Thompson, *Education for Ladies, 1830–1860* (New York: King's Crown Press, 1947), pp. 64–65.

[12]Elmira Female College, *Catalogue* (1858–1859), p. 29, and (1860–1861), p. 26. Sarah Stow, *History of Mount Holyoke Seminary* (Mount Holyoke, 1887), p. 193; and Henry Fitz, Notebook (Nov. 15, 1853), in National Museum of History and Technology. Packer Collegiate Institute, *Circular and Catalogue* (1861), p. 11. Albany Female Academy, *Catalogue* (1851 [?]), pp. 5–6. Ohio Female College, *Plan, Reasons and Encouragement for its Permanent Endowment* (1851). Chesapeake Female College, *Circular* (1856), p. 7.

Columbia Female Institute in Tennessee boasted "a beautiful working model in brass" of a steam engine, suggesting that technology too was an appropriate subject for feminine interest.[13]

Chemistry was taught in laboratories as well as in lecture halls. The Holston Conference Female College at Ashville, North Carolina, had apparatus for two hundred different experiments. The Elmira Female College laboratory was "appropriately arranged and furnished," and students were required "to take part themselves in the experiments." Buffalo Female Academy equipped its laboratory "with every appliance for a thorough and systematic course of instruction in Chemistry and Natural Philosophy," a course lasting two years and comprising about a hundred lectures.[14]

Collections for the study of natural history were especially commonplace. The Brooklyn Female Academy advertised a "Cabinet of Minerals and Shells," the State Normal School at Bridgewater, Massachusetts, a "cabinet of minerals and geological specimens," the Tuskegee Female College a "Cabinet of Minerals, Fossils and Curiosities." The Pittsburgh Female College stocked its cabinet with specimens from the Smithsonian. The Columbia Female Institute had a botanic garden and a fine cabinet of minerals largely collected by local women; Mount Holyoke Seminary had a garden and herbarium. Several schools bought compound microscopes, costing up to $400 each, for close scrutiny of specimens. With a solar microscope, like that at the Young Ladies' High School at Boston, details of flora and fauna could be projected on the wall for all to see.[15]

Educated Victorian women, reputedly too modest even to mention body parts, had relatively easy access to information about human biology. Indeed, courses in this subject were stressed in schools for women to a much greater degree than in schools for men. Nor was this just a development of the second half of the century, when in ever increasing numbers women attended the female medical colleges in Philadelphia, New York, and Boston and the male medical colleges which occasionally admitted them. As early as the 1830s Dr. Clemence Lozier had initiated formal courses in physiology, anatomy, and hygiene at her private school in New York City. At Mount Holyoke Seminary Mary Lyon made a special point of teaching students about their bodies; a manikin illustrated Edward Hitchcock's lectures inaugurating a course there in human anatomy and physiology in 1844. A "life size French manikin and a large number of exceedingly fine anatomical plates" enhanced the lessons at the Sharon Boarding School. Students at the Esther Institute in Columbus, Ohio, enjoyed perhaps the best array of teaching aids of all: "an entire set of 'Cutter Anatomical Plates', a skeleton, and another full size natural anatomical preparation so dissected as to show the vital organs, principal arteries, and veins, nerves, muscles,

[13]Buffalo Female Academy, *Circular and Catalogue* (1859–1860), p. 9. Abbot Collegiate Institute, [Catalogue] (1854), p. 13. Thomas C. Johnson, *Scientific Interests in the Old South* (New York: Appleton-Century, 1936), p. 122.

[14]Holston Conference Female College, *Catalogue and Circular* (1856), p. 13. Elmira Female College, *Catalogue* (1860–1861), p. 25. Buffalo Female Academy, *Circular and Catalogue* (1859–1860), p. 9.

[15]Brooklyn Female Academy, *Circular* (1846), pp. 10–11. [Bridgewater] State Normal School, *Catalogue and Circular* (1861–1862), p. 6. R. E. Ellison, *History of Huntington College* (University, Ala.: University of Alabama Press, 1954), p. 15. Pittsburgh Female College, *Catalogue* (1865–1866), p. 28. Johnson, *Scientific Interests in the Old South*, pp. 121–122. Arthur C. Cole, *A Hundred Years of Mount Holyoke College* (New Haven: Yale University Press, 1940), p. 62. [Boston] Young Ladies' High School, *Catalogue* (1829), p. 19.

etc., in their true positions, together with separate natural preparations of the heart, brain, lungs, liver, eye, ear, etc."[16]

In itself scientific paraphernalia proves nothing conclusive about the level and quality of science teaching—especially since such equipment had an intrinsic advertising value distinct from any actual use to which it was put. More directly indicative of what science a student might have learned are the people who taught it. Prominent scientific "practitioners," still few in number, seldom served on the regular faculty of the early women's schools. Nevertheless, the science courses for the most part were in competent hands. Alonzo Gray, professor of natural sciences at the Brooklyn Female Academy, is a good example. A graduate of Amherst College and Andover Theological Seminary, Gray had had experience teaching science to the men of Phillips Academy and Marietta College, and he had written popular texts on chemistry, geology, natural philosophy, and agriculture.[17]

Special lectures frequently enriched the routine classwork. The Albany Female Academy made certain that students took advantage of the scientific resources in the state capital such as the Dudley Observatory and the state geological collections; these attracted "numbers of literary and scientific gentlemen" from whom the Academy, in turn, obtained the "most important services." The Washington Female Institute promised students "the advantages of attending the lectures of the SMITHSONIAN INSTITUTE." In Indiana, students of St. Mary's Academy heard science lectures presented by Notre Dame University professors. Benjamin Silliman, Jr., gave a series of chemistry lectures at the Spingler Institute in New York in 1849, and Elias Loomis lectured there the following year. Ebenezer Snell of Amherst advised Mount Holyoke Seminary on its acquisition of philosophical apparatus, supervised installation, and then gave a series of lectures accompanied by experiments. Even students at the Troy Female Seminary, where Mrs. Willard stressed "studies appropriate to the female character," attended science lectures presented by Amos Eaton and his colleagues, at the seminary, at the Troy Lyceum, and at the Rensselaer School.[18]

Besides the many scientists who lectured to women occasionally, a few devoted considerable time and energy to women's education. From a practical point of view many men recognized the economic advantages of teaching women, both for their individual pockets and for the scientific establishment at large. As citizens, albeit disenfranchised ones, educated women might support professional and amateur scientific enterprises. Louis Agassiz and his wife Elizabeth ran a school for girls in Cambridge from 1855 to 1863, the professor himself taking charge of the instruction in physical geography, natural history, and botany, "giving a lecture daily, Saturdays excepted, on one or other of these subjects, illustrated by specimens, models, maps and drawings." Income from this school was a welcome contribution to the Aggasiz's household budget. Years later, alumnae donated $4,050 to Agassiz's Museum of Comparative Zoology in an expression of "affection for their old teacher, as well as their interest in his work." Oliver Payson Hubbard, professor of chemistry, mineralogy, and geology at Dartmouth, and his wife Faith, a daughter of Benjamin Silliman,

[16]"Mrs. Clemence S. Lozier, M.D.," in James Parton, ed., *Eminent Women of the Age* (Hartford, 1869), pp. 517–522. Cole, *A Hundred Years of Mount Holyoke,* pp. 61–62. Thompson, *Education for Ladies,* p. 65. Esther Institute, *Catalogue* (1855), p. 13.
[17]"Alonzo Gray," in Appleton's *Cyclopaedia of American Biography,* Vol. II, p. 728.
[18]Albany Female Academy, *Catalogue* (1861). *Boyd's Washington and Georgetown Directory* (1860). St. Mary's Academy, *Catalogue* (1863), p. 3. Abbot Collegiate Institute, [Catalogue] (1854), p. 13. Cole, *A Hundred Years of Mount Holyoke,* pp. 61–64. Ethel McAllister, *Amos Eaton, Scientist and Educator* (Philadelphia: University of Pennsylvania Press, 1941), pp. 484–485.

Sr., ran a women's school at Hanover from 1852 to 1865 and another at New Haven from 1865 to 1873.[19]

For many people, however, simple economic self-interest does not explain their commitment to women's education. Elizabeth Agassiz was later to become a prime mover behind the Harvard Annex, the forerunner of Radcliffe. F. A. P. Barnard aimed to promote the "higher education of women, equal in all respects to those provided for men"; to this end Barnard College, coordinate with Columbia, was established. John Locke—physician, scientist, and inventor—established the Cincinnati Female Academy in 1822 and directed its operations for thirteen years. The customarily critical Mrs. Trollope remarked on Locke's "liberal and enlarged opinions on the subject of female education," noting especially that he taught his students "the higher branches of science." Amos Eaton initially tried to establish Rensselaer on a coeducational basis. After failing at that he set up what amounted to a female branch and kept it going through the 1830s. Courses for ladies, "similar to the courses proposed for gentlemen," were managed by Laura Johnson, Eaton's sister-in-law and protégée. In the nine-week courses in chemistry and natural philosophy designed for teachers, all the students, male and female together, attended daily lectures and then, for the experiments, went off to separate but equal laboratories. Among the students in this program was Mary Lyon, founder of Mount Holyoke Seminary.[20]

The early private coeducational and women's schools were by no means the sole avenue of scientific education open to women in antebellum America; in fact, the emphasis on science in the schools is perhaps less significant in and of itself than as a reflection of prevailing educational ideals and values. Among the popular lectures sponsored by lyceums, mechanics' institutes, and similar groups, those on scientific topics attracted especially large turnouts, female as well as male. As early as 1803 a large class of ladies in New York City enjoyed George Chilton's lectures on natural philosophy. In 1817 Amos Eaton's seven-week course in botany at Northampton, Massachusetts, attracted some forty women and twenty-five men, and his evening lectures on chemistry, mineralogy, and geology drew fifty-five women. When the elder Silliman took to the road in the 1830s he found it remarkable how well his lectures on geology and chemistry were attended and appreciated by women. In the South women attended chemistry lectures at the University of Alabama and special courses in botany at Charleston. Female doctors, stressing common-sense prevention rather than heroic cures, gave health and hygiene instruction through the auspices of the so-called physiological societies and also to their private patients.[21]

In the latter regard, the prevalence and importance of private instruction in the sciences must be stressed, even in this period when schools were multiplying so

---

[19] Elizabeth Cary Agassiz, ed., *Louis Agassiz, His Life and Correspondence* (Boston, 1893), pp. 526–531. "Oliver Payson Hubbard," in *National Cyclopaedia of American Biography,* Vol. IX, p. 557.

[20] William Fletcher Russell, *The Rise of a University* (New York: Columbia University Press, 1937). Frances Trollope, *Domestic Manners of the Americans,* ed. Donald Smalley (1832, New York: Knopf, 1949), p. 82. McAllister, *Amos Eaton,* pp. 484–490.

[21] James G. Kelso, "The Lyceum and the Mechanic Institutes, Pre-Civil War Ventures in Adult Education" (Ph.D. dissertation, Harvard University, 1953), pp. 173–174, stresses the role of the lyceum in opening up educational opportunities for women. Marvin Fisher, *Workshops in the Wilderness* (New York: Oxford University Press, 1967), p. 100, states that at least two thirds of the audience at the Lowell lyceum lectures were mill girls. John C. Greene, "Science and the Public in the Age of Jefferson," *Isis,* 1958, *49*:13–25, points out the large attendance of women at popular scientific lectures of a slightly earlier period. Obituary of George Chilton in *American Journal of Science,* 1836–1837, *31*:421–424. McAllister, *Amos Eaton,* p. 181. George P. Fisher, *Life of Benjamin Silliman* (New York, 1866), Vol. I, pp. 355, 359, 360, 362–363, 373. Johnson, *Scientific Interests in the Old South,* p. 123. Rev. H. B. Elliot, "Woman as Physician," in Parton, *Eminent Women of the Age,* pp. 513–550.

rapidly. Theodore Strong, professor of mathematics at Rutgers, tutored his five daughters in a wide variety of subjects, not only history and languages, but also astronomy, mathematics, and geology. One daughter recalled his holding up the distinguished British science writer Mary Somerville "as an example of what woman may achieve in the domain of science." Eliza Youmans, sister and collaborator of the scientific publicizer Edward Livingston Youmans, studied chemistry in New York with Thomas Antisell, a graduate of the Dublin School of Medicine and student of the most celebrated chemists of Paris and Berlin. Benjamin Peirce of Harvard taught geometry to the young women of Cambridge.[22]

Yet another avenue to science education was the printed word, employed either as an adjunct to formal schooling or autodidactically. While American consumption of advanced scientific treatises remained small during the antebellum period, the market blossomed for introductory books aimed at adult amateurs and young students. Some of these books, though indistinguishable in content from those written for male or mixed audiences, addressed women expressly. *Conversations on Chemistry,* set as a dialogue between "Mrs. B." and "Caroline," was offered "particularly to the female sex" by its author, Jane Marcet; the first American edition appeared in 1806, and by mid-century 160,000 copies had been sold. Mrs. Marcet's *Conversations on Natural Philosophy* went through some three dozen American printings before the Civil War, and her *Conversations on Vegetable Physiology* did equally well.

In the wake of Mrs. Marcet's success, American women began turning out books about science. Botany was the most popular subject, attracting at least ten female writers before the Civil War, from Jane Welsh (*Botanical Catechism,* 1819) to Mrs. C. M. Badger (*Wild Flowers, Drawn and Colored from Nature,* 1859). Almira H. L. Phelps was the most successful author: her *Familiar Lectures on Botany,* first published in 1829, went through at least twenty-three printings by 1860; by 1872, when the ninth edition appeared, sales had topped 275,000. In other areas of science there were, for instance, Mrs. Redfield's *Zoological Science* (1858) and accompanying chart of the animal kingdom, Mrs. Phelps' popular *Chemistry for Beginners* (1834) and her *Natural Philosophy* (1838), Lydia Folger Fowler's *Familiar Lessons on Physiology* (1847), and Hannah Bouvier Peterson's *Familiar Astronomy* (1857), highly commended by George Airy and John Herschel.

Men too wrote science books for women. These were especially common in astronomy. Elijah Burritt's *Astronomia . . . Designed for the Amusement and Instruction of Young Ladies and Gentlemen* (1821) was followed by Montgomery Robert Bartlett's *Young Ladies' Astronomy* (1825) and Denison Olmsted's *Letters on Astronomy Addressed to a Lady* (1840). Samuel McCulloch dedicated his *Map of the Visible Heavens* (1840) "to those Young Ladies, who at various times, have composed Classes in Astronomy, under the care of their friend and instructor, the Author."

Popular periodicals, such as the *Guardian,* "devoted to the cause of female education on Christian principles," and the *Ladies Repository,* frequently included articles about science in their regular fare. *Godey's Lady's Book,* ostensibly dedicated to feminine fashion and fiction, ceaselessly championed advanced education for women; for many years it ran a column on "Chemistry for the Young," describing

---

[22]"Theodore Strong," in National Academy of Sciences, *Biographical Memoirs,* Vol. II, p. 10. John Fiske, *Edward Livingston Youmans* (New York, 1894), pp. 54 and 61. "Recent Movements in Women's Education," *Harper's New Monthly Magazine,* Dec. 1880, 103.

experiments which could be performed at home. *Scientific American* repeatedly and pointedly encouraged its female readers to study science and involve themselves in its pursuit.

In this supportive context, women did participate in science. Unlike the unique and curious Colonial botanist Jane Colden, the astronomer Maria Mitchell became a popular heroine in the middle years of the nineteenth century, symbolizing the ambitions and achievements of a growing number of American women. The Van Duzee sisters, who took "much interest in mathematics and astronomy," obtained a very large Fitz telescope in 1861.[23] Mary Van Duzee studied at the Buffalo Female Academy, taught at the Wheaton Female Seminary, and held membership in the American Association for the Advancement of Science. Eunice Foote presented a paper to the AAAS in 1856 and again in 1857—"Circumstances Affecting the Heat of the Sun's Rays" and "On a New Source of Electrical Excitation"—both of which were published in the *American Journal of Science*.[24] Another woman who presented a paper to the AAAS was Margaretta Hare Morris, who had been reporting her studies of insect pests to the Philadelphia Academy of Natural Sciences since the 1840s and was elected to membership in that organization in 1859.[25] In the field of botany, Mary Chase sent a collection of three hundred New York flowers, thoroughly described and classified, to the London Crystal Palace and received "gratifying testimonials" for her work;[26] and Rachel Littler Bodley, a graduate of the Cincinnati Wesleyan Female College, published a catalogue of the plants in a local herbarium. Representative of the women who earned recognition as scientific illustrators were Lucy Say, first female member of the Philadelphia Academy of Natural Sciences,[27] and Maria Martin, who supplied the floral paintings on which many Audubon birds sat.[28]

Many women educated in the antebellum period became successful practitioners in the 1860s and 1870s. Sarah Frances Whiting, the first professor of physics at Wellesley College, had graduated from Ingham University for Women and then taught at the Brooklyn Heights Seminary. The naturalist Lydia White Shattuck studied at Mount Holyoke Seminary. Graceanna Lewis, highly acclaimed in the 1870s for her botanical and zoological studies, had attended the Kimberton Boarding School; the teacher who most influenced her, Abigail Kimber, was a friend of William Darlington and had contributed to his *Flora Cestrica*. Rebecca Pennell Dean, graduate of the Lexington Normal School in Massachusetts, taught most of the science courses at Antioch College during its early years. In 1857 she was joined by Lucretia Crocker, a graduate of the State Normal School at West Newton, who served as Professor of Mathematics and Astronomy. After the Civil War, Crocker helped organize the Women's Educational Association, headed the science depart-

[23]"The New Comet," *Sci. Amer.*, 1862, 7:122.

[24]Eunice Foote, "Circumstances Affecting the Heat of the Sun's Rays," *American Journal of Science*, 1856, *22*:382–383, and *Philosophical Magazine*, 1857, *13*:167–172; "On a New Source of Electrical Excitation," *Am. J. Sci.*, 1857, *24*:386–387, and American Association for the Advancement of Science, *Proceedings*, 1857, 123–126. See also, "Scientific Ladies," *Sci. Amer.*, 1856–1857, *12*:5.

[25]For a bibliography of her work see Max Meisel, *A Bibliography of American Natural History: The Pioneer Century, 1769–1865*, 3 vols. (Brooklyn: The Premier Publishing Co., 1924–1929).

[26]Phoebe A. Hanaford. *Daughters of America* (Augusta, Maine, 1882), pp. 241–242, 263–264.

[27]*Who Was Who in America*, Historical Volume 1607–1896.

[28]For Jane Colden, Maria Mitchell, Rachel Littler Bodley, and Maria Martin see *Notable American Women* (Cambridge, Mass.: Belknap Press, 1971).

ment of the Society to Encourage Studies at Home, worked closely with the Teachers'
School of Science at the Massachusetts Institute of Technology, and, following her
election to the School Committee in 1873, "virtually revolutionized science teaching
in the Boston schools."[29]

In the mid-nineteenth century, when science was still somewhat of a cottage
industry, women often served as secretaries and research assistants for their husbands. My examples are selected from among the early members of the National
Academy. Louis Aggasiz's American wife, Elizabeth Cabot Cary, was adept at
imparting scientific knowledge "accurately and with . . . animation and authority."
On her own she wrote *Actaea, a First Lesson in Natural History*; with her husband, *A
Journey in Brazil*; and with her stepson, Alexander, *Seaside Studies in Natural
History*, an introductory guide to marine zoology. Elizabeth Agassiz was hardly the
sole refutation to "the myth of the stereotypic Victorian lady of manners in a male-
dominated society."[30] Alexander Bache's wife Nancy actively participated in the work
at his magnetic observatory and is said to have ultimately "given up her life to her
husband, and was part of all his labors."[31] Jane Gray was Asa Gray's "constant
companion, and established that familiarity with his work and his associates that
made her a constant help and delight."[32] An accomplished artist, Orra White Hitchcock prepared many of the illustrations for her husband Edward's reports.[33] John
Torrey's wife Eliza was known to have a "strong and scholarly mind."[34] William
Sterling Sullivant's second wife was "a lady of rare accomplishments, and, not the
least, a zealous and acute bryologist, her husband's efficient associate in all his
scientific work until her death. . . ." Her botanical work is commemorated in
Hypnum Sullivantiae, an Ohio moss.[35]

In assessing the role of women in science, the contributions of anonymous amateurs and of philanthropists must also be recognized. In 1822 Eaton estimated that
more than half the botanists in New England and New York were women. Speaking
before the Milwaukee Lyceum in 1840, Increase Lapham suggested that a knowledge
of plants and shells "seems now to be almost indispensable to an *accomplished* lady."
Banded together in the Caroline Herschel Association, women in Georgia raised over
$500 and purchased a Dollond telescope in 1842. A group of women led by Elizabeth
Peabody bought Maria Mitchell a splendid Alvan Clark telescope.[36] Clara Jessup
Moore, a donor of the Jessup scholarship fund at the Philadelphia Academy of
Natural Sciences, gave additional money in 1893 to support "young women who may
desire to devote the whole of their time and energies to the study of the natural
sciences." Although a writer by profession, Mrs. Moore recalled with special pleasure

[29] See *ibid.* for Sarah Frances Whiting, Lydia White Shattuck, Rebecca Pennell Dean, and Lucretia
Crocker; also "Graceanna Lewis" in *National Cyclopaedia of American Biography*, Vol. IX, pp. 447–448,
and John W. Harshberger, *The Botanists of Philadelphia* (Philadelphia, 1899).

[30] Edward Lurie, *Louis Agassiz: A Life of Science* (Chicago: University of Chicago Press, 1960), pp.
108–109. "Elizabeth Agassiz" in *Notable American Women*.

[31] Merle Odgers, *Alexander Dallas Bache* (Philadelphia: University of Pennsylvania Press, 1947), 203.
Joseph Henry, "Eulogy on Professor Alexander Dallas Bache," *Smithsonian Institution, Annual Report*
(Washington, D.C., 1870), p. 95.

[32] A. Hunter Dupree, *Asa Gray* (New York: Athenaeum, 1968), pp. 177–184.

[33] William J. Youmans, *Pioneers of Science in America* (New York, 1896), p. 299.

[34] Dupree, *Asa Gray*, 38, 175.

[35] Asa Gray, "William Sterling Sullivant," National Academy of Sciences, *Biographical Memoirs*, 1877,
Vol. I, p. 281.

[36] Amos Eaton, *Manual of Botany* (3rd ed., Albany, 1822), preface, quoted in McAllister, *Amos Eaton,*
p. 229. Donald Zochert, "Science and the Common Man in Ante-Bellum America," *Isis*, 1974, 65:465.
Johnson, *Scientific Interests in the Old South*, p. 123.

having attended Silliman's lectures during her student days at a girls' boarding school in New Haven.[37]

In all such cases what is important is not some qualitative assessment of scientific results, but rather what such activities suggest about general ideals and values. For women themselves, as for men, science was thought to confer a multitude of benefits. It developed mental acuity. Its heroes—truth-seekers defying irrational authority—inspired good republican citizenship. Study of science easily merged into an appreciation of nature and became an avenue to natural theology. For Maria Mitchell it brought "calm to the troubled spirit, and a hope to the desponding."[38] Esther Lewis justified sending her daughters to a course of scientific lectures inasmuch as it would enlarge their ideas "of the Great Creator of all these wonderful works."[39]

As the nineteenth century progressed, women assumed ever greater responsibility for the maintenance of cultural and moral standards, both within the home, as men moved ever more into the world of industry, and outside, as church members and as teachers in the rapidly expanding schools systems. Thus the personal benefits of science were not confined to individual women but passed through them to their family and community. In short, in the antebellum years, before it had become a recondite professional specialty, science played an important and wide-ranging role in American culture. As members of that culture women were encouraged to learn about science and to involve themselves in its pursuit.

[37]"Clara Jessup Moore" in *Notable American Women*; Moore family papers at the Philadelphia Academy of Natural Sciences.
[38]Mitchell quoted in Eve Merriam, ed., *Growing Up Female in America* (Garden City, N.Y.: Doubleday, 1971), p. 91.
[39]Quoted in manuscripts in a private family collection, Media, Penn.

Science, Women, and the Russian Intelligentsia

The Generation of the 1860s

By Ann Hibner Koblitz

T HE 1860S AND 1870S could be called the golden age of Russian science. In addition, the period witnessed a rise in women's consciousness and the beginnings of the Russian revolutionary movement. Largely as a result of this combination of circumstances, these years marked a time of substantial gains in scientific opportunities for Russian women. Encouraged by the social philosophy of the nihilists and by the new women's consciousness, young women of the gentry and intelligentsia sought education and careers in the natural sciences and medicine. Russian women were among the first women in the world to receive doctorates in mathematics, physiology, zoology, chemistry, and other scientific fields.

INTRODUCTION AND BACKGROUND

The death of the conservative Tsar Nicholas I in 1855 and the succession of Alexander II (who was at first thought to be more liberal than his father) were viewed by many members of the intelligentsia and nobility as events that heralded a new age.[1] The educated elite hoped that discussions of serf emancipation and educational reform would be accompanied by an improvement in the position of women, a democratization of the tsarist government, and a general liberalization of Russian society.

The mood was optimistic, especially among younger members of the intelligentsia and nobility. They called themselves "realists" or "new people" or "children of the sixties" (the 1860s having started for them with Nicholas's death in 1855).[2] After the well-known writer Ivan Turgenev called the hero of his 1862

The author gratefully acknowledges the support of the International Research and Exchanges Board (IREX), the Institute for Advanced Study in Princeton, the National Endowment for the Humanities, and the National Science Foundation.

[1] For background on Nicolaevan Russia see Nicholas Riasanovsky, *Nicholas I and Official Nationality in Russia, 1825–1855* (Berkeley: Univ. California Press, 1959).

[2] Much has been written about the ideas of the 1860s. In novel form, the best sources are Nikolai Chernyshevskii's *What Is To Be Done?* (serialized in 1862) and Ivan Turgenev's *Fathers and Children* (1862). The memoir literature on the period is extensive. Of particular interest are the following: N. V. Shelgunov, L. P. Shelgunova, and M. L. Mikhailov, *Vospominaniia,* 2 vols. (Moscow: Izdatel'stvo Khudozhestvennoi Literatury, 1967); A. Ia. Panaeva, *Vospominaniia* (Moscow: Izdatel'stvo Khudozhestvennoi Literatury, 1956); L. F. Panteleev, *Vospominaniia* (Moscow: Izdatel'stvo Khudozhestvennoi Literatury, 1958); and A. M. Skabichevskii, *Sochineniia,* 2 vols. (St. Petersburg: Er-

novel *Fathers and Children* a "nihilist," a certain segment of the young people adopted that term as well.

Turgenev seems to have intended to insult the younger generation by calling them nihilists. He wanted to depict them as mindless iconoclasts—young rebels who attacked all the institutions of tsarist society merely for the pleasure of shocking their elders. But some of the young people, far from being offended by Turgenev's charges, exuberantly accepted them. They *were* nihilists, they declared. Nothing in the society of their parents was worth saving; Russia had to be rebuilt completely.[3]

It was an exaggeration on the part of the nihilists to maintain that they denied the value of everything in tsarist Russia. And certainly Turgenev's fear that they respected nothing was ill founded. In fact, the children of the sixties had great faith in the natural sciences and the power of education, strongly believed in the equality of women, admired the traditional peasant commune and certain other rural institutions, and desired to be of use in some capacity to the masses of ordinary people in Russia.[4]

The nihilists confidently looked forward to the social revolution they considered inevitable and felt that the best way to help it along was through intensive study of the natural sciences.[5] There were several reasons why the young people of the 1860s found the sciences so attractive. It is almost a truism that in Europe the nineteenth century was an age of faith in science; this faith had filtered into Russia in spite of the attempts of Nicholas I and his reactionary advisers to keep it out.

To the nihilists, science appeared to be the most effective means of helping the mass of people to a better life. Science pushed back the barriers of religion and superstition and "proved" through the theory of evolution that (peaceful) social revolutions were the way of nature. Science was perceived as virtually synonymous with truth, progress, and radicalism; thus the pursuit of a scientific career

---

likh, 1903). E. N. Vodovozova, *Na zare zhizni,* 2 vols. (Moscow: Izdatel'stvo Khudozhestvennoi Literatury, 1964), contains delightful accounts of herself as a young girl exposed to the circles of the 1860s. The secondary literature is too extensive to enumerate, but Franco Venturi, *Roots of Revolution,* trans. Francis Haskell (1952, in Italian; New York: Grosset & Dunlap, 1966); and T. A. Bogdanovich, *Liubov' liudei shestidesiatykh godov* (Leningrad: Academia, 1929), are helpful to start on, as is the classic history of the intelligentsia in this period: V. R. Leikina-Svirskaia, *Intelligentsia v Rossii vo vtoroi polovine XIX veka* (Moscow: Mysl', 1971). The introduction to Alexander Vucinich, *Science in Russian Culture, 1861–1917* (Stanford, Calif.: Stanford Univ. Press, 1970) is especially relevant because of his emphasis on a scientific world view and faith in the natural sciences as integral parts of the intellectual ferment of the 1860s.

[3] See, e.g., D. I. Pisarev's analysis of *Fathers and Children,* "Bazarov," in Pisarev, *Sochineniia* (Moscow: Izdatel'stvo Khudozhestvennoi Literatury, 1955), Vol. II, pp. 7–50. The people most eager to call themselves nihilists were the scientists and science students. Possibly this was because Bazarov, the hero of *Fathers and Children,* was himself a medical student and a keen admirer of the natural sciences. Members of the movement who were less interested in the sciences were more likely to shy away from what they apparently saw as the connotation of brashness and heartlessness in the word.

[4] For a summary of the beliefs of the period and how they fed into the philosophies of later revolutionary movements see Isaiah Berlin's introduction to Venturi, *Roots of Revolution* (cit. n. 2), pp. vii–xxx.

[5] Vucinich, *Science in Russian Culture* (cit. n. 2), p. 14; and Peter Kropotkin, *Memoirs of a Revolutionist* (1900; New York: Grove, 1968), p. 115. Prince Kropotkin, besides being a noted anarchist, was also a naturalist of some reputation. His study of Siberian fauna, combined with his political beliefs, eventually led him to propose the theory of "mutual aid" as an alternative to the struggle for existence and natural selection: see Daniel P. Todes, "Darwin's Malthusian Metaphor and Russian Evolutionary Thought, 1859–1917," *Isis,* 1987, *78*:537–551, esp. pp. 546–548.

was viewed as in no way a hindrance to social activism. In fact, it was seen as a positive boost to progressive forces, an active blow against backwardness.[6]

NIHILISM, THE TSARIST GOVERNMENT, AND SCIENTIFIC EDUCATION

From the beginning of the movement the tsarist government was uneasily aware of the nihilist equation of science and progressivism. Nevertheless, Russian officials felt that they had no choice but to encourage the reform of scientific education and the training of scientific specialists. There was an urgent need for technical personnel of every conceivable type, and the military and medical establishments insisted on the necessity of increased attention to the sciences. The tsarist government decided on a crash program to train scientists and sent numerous young men abroad in the late 1850s to study the sciences in German and Swiss universities.[7]

This was an exciting time for Russian science. The nihilists' enthusiasm for and faith in science, coupled with the government's decision to devote material resources to scientific development, produced rapid results: the 1860s and 1870s were a "golden age" of Russian science.

The scientific specialist or historian of science is well acquainted with the names of Ivan Sechenov and Kliment Timiriazev in physiology; Dmitri Mendeleev, Aleksander Butlerov, and Vladimir Markovnikov in chemistry; Ilya Mechnikov and Aleksander Kovalevskii in embryology; Ivan Pavlov in neurophysiology; and Vladimir Kovalevskii in paleontology. What is perhaps less well known is that to varying degrees all of these scientists were involved in the ferment of the 1860s, and most considered themselves nihilists.

Ivan Sechenov, for example, was fully conscious of the social and political implications of his famous treatise *Reflexes of the Brain*. He knew that his research on automatic and semiautomatic responses in the brains of animals gave scientific support to materialist attacks on religion; and he was proud that his *Reflexes* had been confiscated by the censors and was often cited as a "nihilist tract."[8] In like manner, the Kovalevskii brothers were especially enthusiastic about their embryological and paleontological research, because they saw it as a defense of Darwinian materialist views of nature against vitalist and anthropocentric biological systems. Other Russian specialists felt the same about their own studies.[9]

This connection of nihilism and the natural sciences had far-reaching conse-

---

[6] Interesting in this connection are the works of D. I. Pisarev (1840–1868), particularly "Bazarov" (cit. n. 3) and "Progress v mire zhivotnikh i rastenii" (1864), in Pisarev, *Polnoe sobranie sochineniia*, Vol. III (St. Petersburg: Pavlenkov, 1904), pp. 309–496.

[7] Vucinich, *Science in Russian Culture* (cit. n. 2), p. xi and *passim*.

[8] I. M. Sechenov, "Refleksy golovnogo mozga," *Meditsinskii vestnik*, 1863, No. 47–48, pp. 461–484, 493–512. The tsarist censors prohibited publication of this work in *Sovremennik*, one of the most popular and progressive of the so-called thick journals (prestigious periodicals often the size of small books). For Sechenov's reaction to this censorship see Sechenov, *Avtobiograficheskie zapiski* (Moscow: Izdatel'stvo Akademii Meditsinskikh Nauk, 1952), pp. 186–187; and M. G. Yaroshevsky, *Ivan Sechenov*, trans. into English by Michael Burov (Moscow: Mir, 1986), esp. pp. 86–127.

[9] See, e.g., *Pis'ma A. O. Kovalevskogo k I. I. Mechnikovu, 1866–1900* (Moscow: Akademiia Nauk, 1955); "V. O. i A. O. Kovalevskie" [letters], *Nauchnoe nasledstvo* (Moscow: Akademiia Nauk, 1948), Vol. I, pp. 185–423; *Bor'ba za nauku v tsarskoi rossii* [correspondence of and reminiscences about the Kovalevskiis, Mechnikov, Sechenov, and others] (Moscow: Gosudarstvennoe Sotsial'no-Ekonomicheskoe Izdatel'stvo, 1931); and K. A. Timiriazev, *Nauka i demokratiia: Sbornik statei 1904–1919 gg.* (1920; Moscow: Izdatel'stvo Sotsial'no-Ekonomicheskoe Literatury, 1963).

quences. Many young scientists were at the forefront of movements for popular education and social reform. These children of the sixties were not full-time revolutionary activists, as some of their students would later become. But by their teachings and example, they set the stage for subsequent revolutionary movements. Throughout their lives many of them continued to show interest in radical politics, and they sometimes gave practical assistance to members of the revolutionary underground.[10] Moreover, the large numbers of progressives in scientific professions in Russia in the 1860s and 1870s had another effect: they provided a hospitable climate in their disciplines for the women of the sixties, many of whom felt encouraged to take up technical studies.

WOMEN UNIVERSITY AUDITORS IN THE EARLY 1860s

In the early 1860s women formed a significant part of an amorphous group of unofficial auditors who wandered in and out of the lecture halls in St. Petersburg's university and specialized technical academies. The professors and students welcomed them, and initially the authorities tolerated them, since there had been a certain liberalization in the university structure after the Crimean War and the emancipation of the serfs in 1861.[11] During this time the St. Petersburg Medical-Surgical Academy began to admit women as medical students on a semiofficial basis. They could not receive degrees, but they could informally take exams and work in the laboratories of the institution, and they were eagerly encouraged by the progressive professors on the faculty.

Among the women enrollees during this period two of the older and better-educated were Nadezhda Suslova and Maria Obrucheva (later Bokova, then Bokova-Sechenova). These women were members of the first Land and Freedom organization, which was one of the early groups dedicated to the spread of radical propaganda. They were in close contact with all aspects of the progressive

---

[10] See, e.g., I. S. Dzhabadari, "Protsess 50'ti," *Byloe*, 1907, No. 8/20, pp. 1–26, on p. 17; Ann Hibner Koblitz, *A Convergence of Lives: Sofia Kovalevskaia: Scientist, Writer, Revolutionary* (Boston: Birkhäuser, 1983), pp. 164–166; and German Smirnov, *Mendeleev* (Moscow: Molodaia Gvardiia, 1974), pp. 232–235.

[11] Panteleev, *Vospominaniia* (cit. n. 2), p. 215; and S. Ashevskii, "Russkoe studenchestvo v epokhi 60-kh godov," *Sovremennyi Mir*, 1907, No. 7–8, pp. 19–36. See also Daniel R. Brower, *Training the Nihilists: Education and Radicalism in Tsarist Russia* (Ithaca, N.Y.: Cornell Univ. Press, 1975). For information on the status of women in Russia see Dorothy Atkinson, "Society and the Sexes in the Russian Past," in *Women in Russia*, ed. Atkinson, Alexander Dallin, and Gail Warshofsky Lapidus (Stanford, Calif.: Stanford Univ. Press, 1977), pp. 3–38; and E. O. Likhacheva, *Materialy dlia istorii zhenskogo obrazovaniia v Rossii (1086–1901)*, 6 vols. (St. Petersburg: Stasiulevich, 1893–1901). For information on specific aspects of women's education and political activities see Vladimir Stasov, *Nadezhda Vasil'evna Stasova: Vospominaniia i ocherki* (St. Petersburg: Merkushev, 1899); Vera Broido, *Apostles into Terrorists: Women in the Revolutionary Movement in the Russia of Alexander II* (New York: Viking, 1977); Richard Stites, "Women and the Russian Intelligentsia: Three Perspectives," in *Women in Russia*, ed. Atkinson, Dallin, and Lapidus; pp. 39–62; Stites, *The Women's Liberation Movement in Russia: Feminism, Nihilism, and Bolshevism, 1860–1930* (Princeton, N.J.: Princeton Univ. Press, 1978); Ekaterina Zhukovskaia, *Zapiski* (Leningrad: Izdatel'stvo Pisatelei, 1930); and Barbara Alpern Engel, *Mothers and Daughters: Women of the Intelligentsia in Nineteenth Century Russia* (Cambridge: Cambridge Univ. Press, 1983). For information on the education of girls in mid-century Russia see Likhacheva, *Materialy*; Vodovozova, *Na zare zhizni* (cit. n. 2), Vol. I; V. Ia. Stoiunin, "Obrazovanie russkoi zhenshchiny (Po povodu dvadtsatipiatiletiia russkikh zhenskikh gimnazii," *Istoricheskii Vestnik*, 1883, 12:125–153; E. S. Nekrasova, "Zhenskie vrachebnye kursy v Peterburge: Iz vospominanii i perepiski pervykh studentok," *Vestnik Evropy*, 1882, 6:807–845. For complaints about the poor quality of education in the girls' schools at mid century see E. Ts—skaia [Vodovozova], "Chto meshaet zhenshchine byt' samostoiatel'noi? (Po povodu romana Chernyshevskogo 'Chto delat'?')," *Biblioteka dlia Chteniia*, 1863, No. 9, pp. 1–19.

and revolutionary movements of their day and were under police surveillance most of their lives.

Suslova and Bokova-Sechenova became the heroines of their generation. Young girls whispered about Suslova in their institutes and resolved to emulate her. Bokova-Sechenova was immortalized in Nikolai Chernyshevskii's *What Is To Be Done?* (1862) and thus became the model for numerous women of the sixties.[12] The two women's seriousness and devotion to the "new ideas" inspired others, so that the years 1860 through 1863 saw a steady increase in the number of women attending lectures unofficially or enrolled semiofficially as auditors.

The women students were optimistic that their capabilities and commitment to their studies would convince the tsarist government to admit them on an equal basis with men. They hoped that the new university statute, which was being worked out at the time, would codify their entry into institutions of higher education throughout the Russian empire. One proposed draft of the statute, approved by all universities except Moscow and Dorpat, did suggest that women be admitted with the same rights as men.[13] If this version of the university statute had been adopted, Russia would have been the first country in all of Europe to grant women the same degrees and status as men. Unfortunately, the fearful tsarist government intervened.

THE TSARIST GOVERNMENT REACTION

Alexander II issued his proclamation emancipating the serfs in March 1861. The publication of the document was followed almost immediately by a wave of student demonstrations protesting its inadequacies and injustices. The protests continued throughout the year and broadened to include issues of university life. The students demanded the firing of incompetent or unfair professors, permission to hold meetings (which were forbidden), and the right to join together in reading and lodging cooperatives (also banned).[14] Although Alexander II had emancipated the serfs and tolerated somewhat more discussion and dissent than his father had, this was going too far. The student uprisings were put down with force, and many participants were arrested or exiled to Siberia.

For women of the intelligentsia the most unfortunate result of the student unrest was that the universities closed to them. In purely numerical terms there were not many women involved in the protests. There were far fewer women auditors than male, and most of the women feared the suspension of their university privileges should they participate in demonstrations. But a few women had high visibility in the student movement and caught the attention of the tsarist police.[15]

---

[12] On Suslova as a role model see *Odna iz mnogikh: Zapiski nigilistki* (St. Petersburg: Mettsig, 1881), p. 29. For details of Bokova-Sechenova's life and influence see the preface and Section 3 of Bogdanovich, *Liubov' liudei* (cit. n. 2); and S. Shtraikh, "Geroinia romana 'Chto delat'?' v ee pis'makh," in *Zven'ia* (Moscow: Academia, 1934), Vol. III–IV, pp. 588–616. There has been some debate about whether Bokova-Sechenova was the only model for Chernyshevskii's Vera Pavlovna, but recent accounts argue that she was certainly the main model; see Yaroshevsky, *Ivan Sechenov* (cit. n. 8), pp. 57–59; and Ann Hibner Koblitz, "The Heroine of *What Is To Be Done?* in Fiction and Fact: Vera Pavlovna and Maria Bokova-Sechenova," forthcoming.

[13] A. N. Derevitskii, *Zhenskoe obrazovanie v Rossii i zagranitsei* (Odessa: Isakovich i Beilenson, 1902), p. 3.

[14] Ashevskii, "Russkoe studentchestvo" (cit. n. 11), No. 7–8, pp. 19–36; No. 9, pp. 48–85; No. 10, pp. 48–74.

[15] Nadezhda Suslova, e.g., was implicated because her organization Land and Freedom was vo-

The tsarist government made what officials saw as an obvious connection between women's desire for university education and a revolutionary world view.
The proposal to allow women to matriculate officially was abandoned, and
women were not mentioned at all in the revised university statute of 1863. Moreover, institutions of higher education were ordered to close their doors to all but
regularly enrolled students, in the hope that "subversive" elements could be
more easily controlled.

By 1863 these measures had had the effect of excluding women from most
university-level study in Russia. The Medical-Surgical Academy of St. Petersburg held out for a while longer, accepting women until 1864. But in May of that
year the War Ministry, under whose auspices the Academy functioned, issued a
directive banning women from the institution. The only exception to the order
was Varvara Kashevarova-Rudneva. She wanted to study medicine expressly to
help Muslim women of the Bashkir nationality who, because of their faith, could
not visit male physicians. The Orenburg provincial government considered a female doctor so much of a necessity that it made a special petition to the tsar, and
Kashevarova-Rudneva was permitted to continue her studies in spite of the ban
on women.[16]

In an attempt to mitigate the effects of this ban, a series of women's pedagogical courses was started in St. Petersburg. Initiated by the biologist E. K. Brandt
(Medical-Surgical Academy) and the physicists F. F. Petrushevskii (Artillery
Academy) and N. N. Tyrtov (Naval Academy), the lectures concentrated on
technical subjects. But the courses soon raised a furor because of rumors that
"immoral" and "materialist" propaganda was being given out in the guise of
anatomy and physiology instruction. In 1865, fearing government shutdown of
the whole program, the founders canceled most of their scientific lectures.[17]

RUSSIAN WOMEN GO ABROAD

The most determined and committed (and, often, the most financially secure) of
the women decided to try their luck abroad and traveled to Western Europe. The
majority of them went to the university and technical school in Zurich, because
the Swiss universities would admit foreign women without entrance examinations or gymnasium certificates. (A Russian, Maria Kniazhnina, determined this
in 1864.[18]) This was an important point, because most of the Russian women had
had little or no formal schooling, and virtually none of them possessed the precious gymnasium *attestat*. Moreover, tuition and living expenses were reputed to
be relatively low in Zurich, and there was already a large colony of Russian male

---

cally opposed to the terms of the emancipation proclamation. Several other women either spoke up at
demonstrations or were easily identifiable among the protesters: Bokova-Sechenova's brother V. A.
Obruchev (who later became an eminent ethnographer-geographer and a member of the Academy of
Sciences) was a leader of the uprisings, as was the brother of the later revolutionary Maria Mi-
khaelis-Bogdanova. See Panteleev, *Vospominaniia* (cit. n. 2), pp. 215, 317n.

[16] J. M. Meijer, *Knowledge and Revolution: The Russian Colony in Zurich (1870–1873)* (Assen,
Netherlands: Van Gorcum, 1955), p. 23; and Jeanette E. Tuve, *The First Russian Women Physicians*
(Newtonville, Mass.: Oriental Research Partners, 1984), pp. 48–50. Ironically, when Kashevarova-
Rudneva finished her studies, she was not permitted to work in Bashkiria because the only hospitals
there were military ones, and the military would not license her.

[17] Likhacheva, *Materialy* (cit. n. 11), pp. 176–185; and L. A. Vorontsova, *Sofia Kovalevskaia*
(Moscow: Molodaia Gvardiia, 1957), pp. 73–74.

[18] Derevitskii, *Zhenskoe obrazovanie* (cit. n. 13), p. 7; and Meijer, *Knowledge and Revolution* (cit.
n. 16), p. 25.

students and political exiles to make the women feel at home. In the years from 1865 to 1873 Zurich would see first a trickle, and then a torrent, of Russian women in pursuit of degrees in the natural sciences and medicine.

To a large extent women's higher education in continental Europe was pioneered by this first generation of Russian women. They were the first students in Zurich, Heidelberg, Leipzig, and elsewhere. Theirs were the first doctorates in medicine, chemistry, mathematics, and biology.[19] Ironically, the women of the sixties were often as surprised at their pioneer status as most people are today when they learn of the fact. Many of the nihilist women were full of idealism about the West and about the level of democracy and equality they assumed Western Europe to have achieved. They felt inferior, coming as they did from "backward" Russia, and they confidently expected to be joining the ranks of numerous European women already engaged in serious study.[20]

To the Russian women's astonishment, they discovered that their ideas and attitudes, their eagerness for education, and their determination to succeed in spite of obstacles in some respects put them in the forefront of the European women's movement. Especially in the early years of the Russian student colony in Zurich, they encountered few women of other nations. Even Swiss women tended not to take advantage of their country's opportunities for advanced education, perhaps in part because the Swiss, unlike foreigners, were required to have a gymnasium diploma.

The cumulative statistics for Zurich University show that 203 women were enrolled as auditors or students between winter 1864–1865 and summer 1872. There were 23 English, 10 Swiss, 10 Germans, 6 Austrians, 6 Americans—and 148 Russians. Later, when women were admitted to other universities in Switzerland, notably Bern and Geneva, the percentages would be much the same. Moreover, almost all of the women studied the natural sciences and medicine. Of the sixty-seven degrees given to women, mostly Russians, in Geneva from 1876 to 1883, for example, there were thirty-five in the natural sciences, thirty-one in medicine, and only one in letters. The proportions of degrees awarded in other Swiss universities were similar.[21]

Given this preponderance of Russian women in the female student communities of continental Europe (few Russian women studied in England), it is not surprising that they tended to band together. Although some of the women were supported by wealthy parents or husbands, many found themselves in straitened circumstances. They economized by forming eating, lodging, and studying cooperatives. They shared textbooks and laboratory animals and in extreme cases took turns with winter coats and boots.[22]

---

[19] Here "doctorate" is meant in the modern sense—that is, a degree granted for some original piece of scientific research, after at least a three-year period of study.

[20] The idealism about America was even greater than that about Western Europe. See, e.g., remarks in P. N. Arian, ed., *Pervyi zhenskii kalendar' na 1899 god*, pp. 86–88; and S. Panteleeva, "Iz Peterburga v Tsiurikh," in *Pervyi zhenskii kalendar' na 1912*, pp. 20–31 (second pagination), on p. 25. On *Pervyi zhenskii kalendar'* see n. 37.

[21] For the enrollment figures and for degrees granted in various fields see V. Bёmert, *Universitetskoe obrazovanie zhenshchiny* (St. Petersburg: Merkulev, 1873), pp. 8–9; and Marie Goegg, "Switzerland," in *The Woman Question in Europe*, ed. Thomas Stanton (New York: Putnam, 1884), pp. 388–389.

[22] For an excellent treatment of the Zurich colony in all its aspects see Meijer, *Knowledge and Revolution* (cit. n. 16). Additional information about the lives of the women there can be found in E.

Though the Russian students generally met with politeness, if not enthusiasm, from their professors, their habits and attitudes led to problems in staid Swiss society. Most of the nonacademic Swiss disapproved of the Russian students, especially the women. From the start most Swiss citizens could not understand the women's desire for education, particularly in such supposedly unwomanly areas as the sciences and medicine. The Swiss disliked the politics of the majority of the Russian students as well, and they were scandalized by the comradely, unceremonious relations of the Russian women with their male counterparts. For the middle-class citizens of Zurich, the nihilist women's short hair, simple dress, unstudied mannerisms, and liking for cigarettes were shocking. Add to that the Russians' tendency to talk in each others' rooms until all hours, walk around the town in mixed-sex groups, and treat each other with informality, and it is clear that many proper, conventional Swiss would have viewed them as little better than prostitutes. The women found it difficult to obtain satisfactory rooms and were often discriminated against and ridiculed in the shops and markets.[23]

Another annoyance was the attitude of the women's non-Russian fellow students. The mathematician Elizaveta Litvinova recalled that she had to be careful to avoid looking any of her classmates in the eye or even glancing in their direction for too long, lest she give them an excuse to accost her. Since Litvinova was studying mathematics at the Polytechnic Institute rather than at the university, with most of the other women, her situation was particularly difficult: she was often the only woman in a class of 150 men. But although the other women had more female companionship, they still seem to have experienced problems similar to those of Litvinova. At best, they could expect the male students to be "correct, but somewhat stupidly puzzled" by the women's desire to learn, as the nihilist physiologist Kliment Timiriazev disparagingly described Heidelberg sentiments toward the mathematician Sofia Kovalevskaia. Frequently, the women were insulted and harassed.[24]

FIRST WOMEN DOCTORATES IN THE SCIENCES AND MEDICINE

The first women to obtain their degrees were those who had been the best prepared before they left Russia—that is, those who had audited classes at the Medical-Surgical Academy and university in St. Petersburg in the early 1860s and those whose families had been both financially able and willing to give their daughters advanced tutoring. Among this group were the two heroines of the period, Nadezhda Suslova and Maria Bokova-Sechenova.

The first woman in nineteenth-century Europe to obtain a doctorate in medicine fully equivalent to men's degrees was Nadezhda Suslova (1843–1918).[25]

El' [Elizaveta Litvinova], "Iz vremen moego studenchestva," *Zhenskoe Delo,* 1899, No. 4, pp. 34–63; Panteleeva, "Iz Peterburga v Tsiurikh" (cit. n. 20); and Vera Figner, *Studencheskie gody, 1872–1876* (Moscow: Golos Truda, 1924).

[23] Meijer, *Knowledge and Revolution* (cit. n. 16), pp. 56–61.

[24] El', "Iz vremen moego studenchestva" (cit. n. 22), p. 40; Timiriazev, *Nauka i demokratiia* (cit. n. 9), p. 24; Meijer, *Knowledge and Revolution* (cit. n. 16), pp. 56–57; and Panteleeva, "Iz Peterburga v Tsiurikh" (cit. n. 20), pp. 22–23.

[25] At the time, Suslova's degree was hailed as the first doctorate of medicine granted to a woman. But "firsts" in the history of women in science are often problematic, as historians of the subject are beginning to recognize. In this case, Suslova's claim to being first would probably depend on one's definition of "doctorate of medicine."

Suslova was one of the first of the nihilist women to go abroad; she left Russia almost as soon as the Medical-Surgical Academy closed its doors to women in 1864. Suslova enrolled at Zurich University as an auditor in 1865 and became the first official female student there in 1867 (she was given credit for the courses she had taken in St. Petersburg). On 14 December 1867 she received her medical degree, producing a dissertation on reflexes in the hearts and lymph glands of frogs.[26] She obtained permission to practice in Russia on her return to St. Petersburg in 1868, after Ivan Sechenov and other nihilist scientists vociferously urged the unprecedented licensing of a woman. In succeeding years Suslova practiced medicine, wrote articles on physiology, gynecology, and hygiene, and published several short stories as well.[27] She was one of the first to note that eye inflammation and blindness in infants could be caused by exposure to gonorrhea during passage through the birth canal.[28]

Maria Bokova-Sechenova (1839–1929) soon followed Suslova abroad and was equally successful. She received her doctoral degree in medicine from Zurich in 1871 and was licensed as a doctor in St. Petersburg in December of that year. But Bokova-Sechenova was interested in research on eye diseases and so did not immediately practice medicine. She returned to Western Europe to specialize in opthalmology in Vienna and subsequently worked as a researcher and oculist in laboratories of the Russian Academies of Sciences and Medicine. Bokova-Sechenova's work concerned color blindness and color vision and the possibility of treating eye conditions with various colored lenses. After inheriting a small estate from her grandmother, she became interested in agronomy, and in the late 1880s she was quite proud of the wheat yields her experimental farm produced.[29]

Suslova and Bokova-Sechenova were the first of many Russian women who obtained their doctorates in medicine at foreign universities in the 1860s, 1870s, and 1880s. During those years Russia boasted more licensed women physicians than any other country in Europe. Like their male colleagues, many of the women carried out research as well as practiced medicine.[30] A sizable number of

[26] Suslova's dissertation, *Beiträge zur Physiologie der Lymphe* (Zurich, 1868), was acclaimed by her professors as a brilliant achievement; see Tuve, *First Russian Women Physicians* (cit. n. 16), p. 21. Ivan Sechenov, who was her unofficial adviser and worked closely with her while she did her research, called Suslova's dissertation "magnificent": Ivan Sechenov to Maria Bokova-Sechenova, [late October 1867], Archives of the Academy of Sciences of the USSR, Moscow (hereafter **Arch. Akad. Nauk**), fond 605, opis' 3, no. 25, pp. 20–21.

[27] See, e.g., N. Suslova-Erisman, "O vospitanii detei v pervye gody zhizni," *Arkhiv Sudebnoi Meditsiny i Obshchestvennoi Gigieny,* December 1870, 6(4):21–31. Among Suslova's stories are "Sashka" and "Fantazerka," both of which appeared in *Sovremennik*, in 1862 (*90*:489–520) and 1864 (*103*:169–219), respectively. Many of the women scientists and physicians wrote stories (often didactic or semiautobiographical) and essays for the popular press; see, e.g., the list of Litvinova's publications in Koblitz, "Elizaveta Fedorovna Litvinova," in *Women of Mathematics: A Biobibliographic Sourcebook,* ed. Louise Grinstein and Paul Campbell (New York: Greenwood, 1987), pp. 129–134, on pp. 133–134. In part their literary endeavors were inspired by the nihilist desire to instruct and popularize. More important in many cases, however, was the need to supplement the slim income of a laboratory assistant or teacher or country doctor.

[28] "Nadezhda Prokof'evna Suslova," in *Pervyi zhenskii kalendar' na 1901,* ed. Arian, pp. 377–383; and Tuve, *First Russian Women Physicians* (cit. n. 16), pp. 12–33. Suslova's first scientific article appeared in *Meditsinskii Vestnik* in 1862, when she was nineteen.

[29] Sechenov, *Avtobiograficheskie zapiski* (cit. n. 9), pp. 197–229; and letters from Sechenov to Bokova-Sechova, 1862–1863 and 1890, in Arch. Akad. Nauk, fond 605, opis' 3, nos. 24 and 30.

[30] Meijer, *Knowledge and Revolution* (cit. n. 16), p. 155; Tuve, *First Russian Women Physicians* (cit. n. 16). The training for European medical degrees was much more rigorous than that for degrees

Russian physicians of both sexes took the time to organize and report their observations on public health issues and the course of disease in the Russian countryside. This unusual combination of general medical practice with zealous research and the compilation of public health statistics has come to be seen as a trademark of the Russian medical profession in the second half of the nineteenth century.[31]

Russian women were the first in other fields besides medicine. Inspired by the ideas of the 1860s, Sofia Kovalevskaia (1850–1891) studied in Heidelberg and Berlin and obtained her doctorate in mathematics *summa cum laude* from Göttingen University in 1874.[32] She was the first woman in modern times to receive her doctorate in mathematics; the second and third women to be awarded the Ph.D. in this field were also Russian.[33]

Iulia Lermontova (1846–1919), Kovalevskaia's best friend, had also been affected by the ideas of the 1860s. After an unsuccessful attempt to obtain permission to study at a Moscow institute of agriculture and forestry, Lermontova joined Kovalevskaia in Heidelberg and later Berlin. She studied physical and organic chemistry and in 1874 received her degree *magna cum laude* from Göttingen, with a dissertation on the analysis of methyl compounds. Lermontova was the first woman in the world to obtain a doctorate in chemistry. Her advances in techniques for the separation of platinum alloys (a problem suggested to her by Mendeleev) and her work on oil distillation procedures have gained her a secure place in the history of chemistry in Russia.[34]

These first women scholars set the tone for their generation, and they were followed by many others. In 1873, at the height of the student colony in Zurich,

---

from most American medical schools at the time. The normal instruction period was four to five years, the natural sciences were studied intensively, and the candidate was required to present a dissertation based on original research.

[31] Nancy Mandelker Frieden, *Russian Physicians in an Era of Reform and Revolution, 1856–1905* (Princeton, N.J.: Princeton Univ. Press, 1981), especially Pts. 1 and 2. Frieden does an impressive job of chronicling the professionalization of Russian physicians, but the book contains little on changing medical practice and virtually nothing on women physicians.

[32] For details of Kovalevskaia's life see Koblitz, *Convergence of Lives* (cit. n. 10); for detailed exposition of her mathematical works see Roger Cooke, *The Mathematics of Sonya Kovalevskaya* (New York: Springer, 1984). Kovalevskaia and her adviser, Karl Weierstrass, decided that she should present three works to Göttingen in support of her application for the doctorate, on the grounds that, as a woman, she would need particularly good credentials. Her three papers were "Zur Theorie der partiellen Differentialgleichungen," *Crelle's Journal*, 1875, *80*:1–32; "Über die Reduction einer bestimmten Klasse von Abel'scher Integrale 3-en Ranges auf elliptische Integrale," *Acta Mathematica*, 1884, *6*:393–414; and "Zusätze und Bemerkungen zu Laplace's Untersuchung über die Gestalt des Saturnringes," *Astronomische Nachrichten*, 1885, *111*:37–48. Kovalevskaia's Prix Bordin paper (on the "Kovalevskaia top") was published as "Sur le problème de la rotation d'un corps solide autour d'un point fixe," *Acta Math.*, 1888–1889, *12*:177–232.

[33] Here, as elsewhere, I am ignoring the claims of several eighteenth-century Italian women (e.g., Laura Bassi, Maria Gaetana Agnesi) with regard to these "firsts" in women's education. Their achievements have not as yet been well enough documented (studies of them are needed), and it can be argued that they did not receive doctorates in the modern sense of the word.

[34] Iu. V. Lermontova, "Vospominaniia o Sofe Kovalevskoi," in S. V. Kovalevskaia, *Vospominaniia i pis'ma* (Moscow: Akademiia Nauk, 1951), pp. 378–379; for further details on Lermontova's life see Iu. S. Musabekov, *Iulia Vsevolodovna Lermontova 1846–1919* (Moscow: Nauka, 1967). See also Julie von Lermontoff, *Zur Kenntniss der Methylenverbindungen* (Göttingen: E. A. Huth, 1874). Lermontova was also the first (at least in Russia—"firsts" in the history of applied chemistry can be as problematic as those in the history of women in science) to study the alkylation of olefins by halogen derivatives and the first to obtain 1,3-dibromobutane and dimethylacetylene and to demonstrate the structure of 4,4-diaminohydrazobenzene.

there were over a hundred Russian women studying at the university and poly-technical institute. The women comprised more than 40 percent of the Russian degree aspirants in Zurich and approximately 85 percent of all women at the university. Their proportions were analogous in institutions of higher education elsewhere on the Continent.[35]

It would be difficult to overemphasize the extraordinary nature of the achievement of this first group of Russian women in science. They came from one of the most reactionary countries in Europe and approached the European educational establishment with determination. Sofia Kovalevskaia single-handedly persuaded Heidelberg University to break its prohibition against women. Thanks to her forcefulness, university officials there admitted not only her, but Iulia Lermontova and the future revolutionary Natalia Armfeldt as well. Nadezhda Suslova accomplished a similar mission in Zurich, and Vera Goncharova was among the first group of four women who were permitted to study at the Faculty of Medicine in Paris.[36]

The Russian women studied at the foremost centers in Europe, under some of the most famous scientists of their day. Kovalevskaia and Lermontova attended the lectures and assisted in the laboratories of Gustav Kirchhoff, Wilhelm Bunsen, Hermann Helmholtz, Leo Konigsberger, and Paul and Emil Du Bois-Reymond. Kovalevskaia wrote her dissertation under the world-famous mathematical analyst Karl Weierstrass and, as a graduate student, worked with the eminent mathematicians Leopold Kronecker and Hermann Schwarz (Elizaveta Litvinova's adviser). Other Russian women studied with equally prominent specialists.

SCIENCE OR REVOLUTION?

Despite the Russian women's triumphs at Western European universities, their subsequent paths in their chosen disciplines were by no means easy. Of the hundreds of women who went abroad to study during the early 1870s, only a small percentage actually completed their programs. And of the women who did graduate, not all continued in their specialties.

There were many reasons for this attrition, both before and after the granting of the degree. Naturally, the women experienced the stresses encountered by a woman student of the sciences in any age: pressure from family, the internalized conviction that her career choice is essentially "unwomanly," lack of appropriate role models, the subtle and overt prejudice of professors and male students, and so forth. But there were also certain factors peculiar to the situation of the Russian women of the 1860s.

The women of this group had decided to pursue higher education, diplomas, and careers in the sciences for several reasons. It seems, however, that love of science for its own sake was not the most important of them. As the radical feminist P. N. Arian pointed out, "One had to have immense bravery and energy to go against society, family, and friends and set out alone for far-off places in

---

[35] Meijer, *Knowledge and Revolution* (cit. n. 16), p. 1; Bĕmert, *Universitetskoe obrazovanie zhenshchiny* (cit. n. 21), pp. 8–9; and Goegg, "Switzerland" (cit. n. 21), pp. 387–389.

[36] Caroline Schultze, *La femme-médecin au XIX siècle* (Paris: Ollier-Henry, 1888), pp. 12–13. I am indebted to Joy Harvey for bringing Schultze's doctoral dissertation to my attention.

search of science."[37] The women who did so were motivated by their social philosophy as well as their interest in the sciences per se. They wanted to prove that women were capable of sustained intellectual effort and success in the so-called male professions. More importantly, they wanted to educate themselves so that they could be of use to the Russian masses—the recently freed serfs who had never seen a doctor, knew nothing of modern soil research and cross-breeding experiments, had not the slightest idea of proper nutrition and sanitation. And, along with the concrete use their professional skills would bring, many wanted to take revolutionary propaganda to the peasantry.

This sociopolitical motivation for their choice of the sciences and medicine was noted again and again by the women of the sixties. Sofia Kovalevskaia had at first intended to become a doctor for political exiles in Siberia and had planned to study her beloved mathematics only in her spare time. The revolutionary Vera Figner wrote that she and her comrades had learned medicine and the sciences in Zurich "in order to have in our hands *weapons for social activism*." Although she later became committed to her medical career, in the beginning Nadezhda Suslova wanted an official degree just to prove that a woman could stay the course. Her real goal had been to bring revolutionary propaganda to her Muslim women patients on the Siberian steppe.[38] It is clear that without the "new ideas" of the 1860s, the women of the intelligentsia and nobility would never have flocked to Zurich and elsewhere in such numbers. But the radical ideas also contributed to the high attrition before and after graduation.

As time went on, an increasing number of students began to question whether it was right to concentrate on scientific education. Perhaps it would be better to help the revolution along in a more active manner, some women suggested.[39] The 1870–1871 debates in the "Heidelberg women's commune," as Sofia Kovalev-skaia and Iulia Lermontova called their apartment, illustrate the dilemma of the women abroad. Lermontova and Kovalevskaia lived with the aspiring writer Anna Korvin-Krukovskaia (Kovalevskaia's older sister), the law student Anna Evreinova, and the mathematics student Natalia Armfeldt in Heidelberg, where all except Korvin-Krukovskaia attended the university. All were staunch nihilists; all looked upon their presence in Heidelberg as a political as well as intellectual act of rebellion.

Yet Korvin-Krukovskaia and Armfeldt were not content with what they increasingly saw as a passive role in radical politics. They argued with the others, saying that the time to help the people was now; education could be obtained later, they claimed. Kovalevskaia, Lermontova, and Evreinova, on the other

[37] P. N. Arian, "Russkaia studentka zagranitsa," in *Pervyi zhenskii kalendar' na 1912*, p. 57. *Pervyi zhenskii kalendar'* (The First Woman's Calendar) was an annual periodical published from 1898 to 1916 in St. Petersburg by P. N. Arian. It is a mine of information on Russian women in the late nineteenth century and contains numerous data on women in scientific and technical fields. (Arian's papers, which are in the Tsentral'nyi Gosudarstvennyi Arkhiv Literatury i Iskusstva in Moscow, fond 1018, indicate that she was instrumental in organizing pharmacological, agricultural, and engineering courses for women.)

[38] Figner, *Studencheskie gody* (cit. n. 22), p. 72 (emphasis in original); and Sechenov to Bokova-Sechenova, 27 Sept. and 29 Oct. 1867, Arch. Akad. Nauk, fond 605, opis' 3, no. 25, pp. 6, 23. According to Sechenov, Suslova tried unsuccessfully to persuade him that his and Bokova-Seche-nova's interests should be more centered on the spread of revolutionary propaganda.

[39] These debates are described in detail in Meijer, *Knowledge and Revolution* (cit. n. 16), pp. 120ff.; and Engel, *Mothers and Daughters* (cit. n. 11), pp. 141–147. For background on the theoretical

hand, felt that it would be absurd to abandon their studies. They would do more for the cause of women and the Russian masses by following through with their career plans and becoming fully qualified specialists, they maintained.

The Heidelberg commune split apart on this question. Korvin-Krukovskaia abandoned her literary aspirations to participate in the Paris Commune of 1871 and the First International Working Man's Association. (Kovalevskaia, Bokova-Sechenova, and several other Russian women students also took part in the Commune, though in less important capacities than Korvin-Krukovskaia.) Armfeldt gave up her mathematical studies, returned to Russia to join an underground revolutionary organization, and eventually died in detention in Siberia. Kovalevskaia, Lermontova, and Evreinova continued their education, and each became the first woman to earn a doctorate in her respective field.[40]

The same tensions and debates later surfaced in the Zurich women's colony, with similar results. Some of the women felt that it was self-indulgent to continue their education. They argued that Russia needed full-time revolutionaries more than it needed agronomists and doctors, however committed these specialists might be to radical ideas. Several women decided to return to Russia to engage in political agitation, abandoning their scientific studies to do so. Most of the Zurich students elected to remain and finish their work, however, although many felt guilty about their desire to continue their studies.[41]

Ironically, the tsarist government did its share to swell the number of women students in the revolutionary underground when it attempted to destroy the Zurich women's colony. Government officials had been watching events abroad uneasily. Through the information of spies, they knew that the students mingled freely with such radical émigrés as the anarchist Mikhail Bakunin and the populist Peter Lavrov, and it was clear that many of the young women were influenced by these political thinkers. Alarmed by the outright revolutionism espoused by a significant fraction of the students, the tsarist officials decided to call a halt to what they saw as a threat.

In June 1873 the Russian government issued a proclamation recalling all women students from Zurich. Any who remained after 1 January 1874 would not be hired in any capacity by the government, could never take any licensing or qualifying exams, and would be barred in the future from all Russian institutions of higher education. In addition, the government insulted the women in the crudest of ways by insinuating that so many were medical students in order to be able to perform abortions on one another.[42]

The women were shocked and offended by both the content and the tone of the proclamation, but some of them decided that they had no choice but to leave Zurich. The poorer students could not afford to endanger their chances of ob-

---

origins of the debates (the Mikhail Bakunin versus Peter Lavrov controversy) see Philip Pomper, *Peter Lavrov and the Russian Revolutionary Movement* (Chicago: Univ. Chicago Press, 1972).

[40] On participation in the Commune see notes from Bokova-Sechenova to Sechenov, Jan. 1871, Arch. Akad. Nauk, fond 605, opis' 2, no. 26, pp. 1–2; for more information on the Heidelberg group see Koblitz, *Convergence of Lives* (cit. n. 10), Chs. 4, 5, pp. 81–111.

[41] Engel, *Mothers and Daughters* (cit. n. 11), pp. 140–144.

[42] Likhacheva, *Materialy* (cit. n. 11), pp. 559–561; Meijer, *Knowledge and Revolution* (cit. n. 16), pp. 1, 142; and Panteleeva, "Iz Peterburga v Tsiurikh" (cit. n. 20), pp. 29–30. Panteleeva notes ironically that most of the Russian women medical students could not possibly have performed abortions because they had not yet had their gynecology course.

taining employment in Russia. The government was their main hope for a job as doctor, scientist, or teacher, so for the most part they felt forced to accede to the conditions of the proclamation. Moreover, some women were naive enough to believe that they would soon be admitted to licensing exams and Russian universities if they obeyed the government order. A few of these—the youngest and most innocent of the Zurich women—abandoned Western Europe and returned hopefully to Russia.[43]

Most women studying in Zurich attempted to find ways of circumventing the effects of the proclamation. Some of the more advanced students tried to compress their remaining studies into the seven months before the government's deadline. At least one woman, Anna Kleiman, finished her work in time and received her medical degree but was so exhausted and weakened by the pace she had set herself that she became seriously ill (her later fate is uncertain). Her dissertation, on the causes of mortality in children, was still being cited by Swiss doctors years later.[44]

Other women looked with contempt upon the proclamation and did not believe that the government would enforce it. Serafima Panteleeva (born 1846), for example, felt that the order could only play upon the fears of the immature and the inexperienced. For someone like her, who had learned progressive political ideas at her father's knee and shared Siberian exile with her husband, the proclamation was something to be flouted. Her adviser, a Professor Hermann, suggested that she ignore the order and continue with her physiological work. Panteleeva and her friend Aleksandra Ivanova complied with this advice, and Panteleeva eventually went on to work in the physiology laboratory of I. R. Tarkhanov in St. Petersburg.[45]

Elizaveta Litvinova (1845–1919) was also urged by her adviser, the mathematical analyst Hermann Schwarz, not to succumb to the intimidation of the tsarist government. She completed her master's dissertation at Zurich in 1876 and received her doctorate in mathematics from Bern University in 1878. Litvinova had some trouble when she returned to Russia, however. In Zurich she had earned a certificate of competence that licensed her to teach in the upper grades of boys' gymnasia. Generally, the Russian Ministry of Education permitted men with such a certificate to take the Russian licensing exam and teach in gymnasia at the senior level. By contrast, Litvinova was forbidden to take the examination for over ten years, and during the whole of her life she was deprived of the pension rights and vacation privileges enjoyed by her male counterparts. She was an excellent mathematics teacher, though, and by her example attracted many women to the subject, including the famous Bolshevik Nadezhda Krupskaia.[46]

---

[43] Panteleeva, "Iz Peterburga v Tsiurikh," pp. 30–31; and Meijer, *Knowledge and Revolution*, pp. 208–212.

[44] E. Litvinova, "Malen'kaia," in *Pervyi zhenskii kalendar' na 1912*, pp. 112–116. Litvinova claims that Kleiman died that summer, but reference is made to her practicing several years later in Engel, *Mothers and Daughters* (cit. n. 11), pp. 193–194.

[45] Panteleeva, "Iz Peterburga v Tsiurikh" (cit. n. 20), pp. 29–31. In later years Panteleeva became a scientific translator and popular science writer.

[46] For a biographical sketch of Litvinova and a list of her publications see Koblitz, "Elizaveta Fedorovna Litvinova" (cit. n. 27), pp. 129–134. Litvinova's numerous articles on mathematical pedagogy and her biographies of scientists were very popular in late Imperial Russia. For her dissertation see E. F. Litvinova, *Lösung einer Abbildungsaufgabe* (St. Petersburg: Buchdruckerei der Kaiserlichen Akademie der Wissenschaften, 1879). This was Litvinova's only research paper in mathematics.

Most women decided to obey the letter of the proclamation by leaving Zurich, while attempting to continue their studies elsewhere. Some, including the later famous revolutionaries Sofia Bardina and the Subbotina sisters, Maria and Evgenia, enrolled at the Sorbonne. Others, among whom were the radicals Vera Figner and Olga Liubatovich, tried to settle down at the small young university in Bern. Still others traveled to Geneva, where there was a large Russian émigré colony and a university that opened its doors to women in 1876. One woman journeyed to Pisa, where she obtained her medical degree in the late 1870s, became a surgeon, and participated in Italian radical movements. Two others studied in Boston and Philadelphia.[47]

Life at other universities was even more difficult than it had been in Zurich. The situation in Paris was especially bad. The tuition and laboratory fees were higher than they had been in Switzerland, so the women were hard-pressed to find the money to eat. Moreover, the tsarist government decided that several of the refugees from Zurich were so dangerous that it harassed them into leaving the Sorbonne as well. Discouraged, impoverished, and afraid that their government would hound them out of whatever foreign university they chose, some of the women gave up, returned to Russia, and entered the revolutionary underground.[48]

SOME SUCCESSFUL WOMEN SCIENTISTS

In spite of the normal attrition, hardship, and the barriers placed in the way of the women of the sixties by the tsarist government, a significant number of the women of the student colonies abroad received their degrees. Complete statistics are not available, but it would seem that in the period 1865–1890, some 500 to 700 women went abroad for study. If that estimate is correct, then about 20 to 30 percent of the women who studied in Switzerland and elsewhere eventually obtained some kind of advanced scientific or medical degree.[49] Considering the large and varied problems the women of the sixties faced, even the lower estimate is impressive.

Of the women who finished their degrees, not all continued in their specialties. Some had families to contend with, and some wanted to bear children while they were still reasonably young. Some women became discouraged because their diplomas did not immediately lead to offers of employment commensurate with their qualifications. There were formidable obstacles for women wishing to become university teachers, laboratory experimenters, and doctors in Russia. The tsarist government persisted in regarding the women as dangerous and would not hire them for most positions in the university, civil service, or hospital systems. Moreover, the government reneged on its implied promise of access to licensing

---

[47] Dora d'Istria, "Italy," in *Woman Question in Europe,* ed. Stanton (cit. n. 21), p. 327; Tuve, *First Russian Women Physicians* (cit. n. 16), pp. 34–44; and S. L. Chudnovskii, "Iz dal'nykh let," *Byloe,* 1907, No. 10/22, pp. 218–240, on pp. 226–227.

[48] Engel, *Mothers and Daughters* (cit. n. 11), pp. 144–145. For descriptions of the Russian women's lives in other university towns see Figner, *Studencheskie gody* (cit. n. 22); I. S. Dzhabadari, "Protsess 50'ti," *Byloe,* 1907, No. 9/21, pp. 169–192; and Koblitz, *Convergence of Lives* (cit. n. 10), pp. 81–103.

[49] Meijer, *Knowledge and Revolution* (cit. n. 16), p. 155, writes that slightly less than 25 percent of the Zurich University women finished their studies. For those who enrolled in the German and other Swiss universities the percentage appears to have been higher.

exams for women. As a result, jobs in the small private sector were closed to them as well.

In spite of these hindrances, however, a number of women of the sixties distinguished themselves in their chosen fields. The majority were doctors and so-called learned midwives. Nadezhda Suslova and others had successful practices in St. Petersburg and elsewhere, many women worked for the public health system in the countryside, and several women physicians won the admiration of their male colleagues for their excellent work in field hospitals during the Russo-Turkish War.[50]

Other women chose to work in biological or chemical laboratories or to concentrate on the research side of medicine.[51] Among these were Maria Bokova-Sechenova, Serafima Panteleeva, and Iulia Lermontova. Bokova-Sechenova made advances in the treatment of eye diseases, and Panteleeva was interested in skin ailments. Lermontova contributed substantially to the chemistry of petroleum distillation and wrote papers with Markovnikov and other eminent chemists of the day before returning to her first loves—agronomy and food production—and devoting herself to the study of cheese making.[52]

Elizaveta Litvinova, although barred by the tsarist government from university-level teaching, became an influential pedagogue and proponent of advanced methods of mathematical instruction. Moreover, in 1887 she became the first woman in Russia to obtain the right to teach in the higher grades of a boys' gymnasium.[53]

Zoology could also boast a successful woman of the sixties—Sofia Pereiaslavtseva (1849–1903), a distant relative of Sofia Kovalevskaia. Pereiaslavtseva graduated from Zurich University with a doctorate in zoology and for ten years was the director of the Sevastopol Biological Station. She wrote a series of works on Black Sea fauna, one of which won the Kessler Prize of the Russian Society of Natural Scientists, and she directed the work of at least two women students of her own.[54]

By far the most important contributions to science of the women of this generation were made by the mathematician Sofia Kovalevskaia. She was the first woman outside of eighteenth-century Italy to be awarded a chair at a research university, the first woman to be on the editorial board of a major scientific journal (*Acta Mathematica*), the first woman to be elected a corresponding

[50] Bëmert, *Universitetskoe obrazovanie zhenshchiny* (cit. n. 21), p. 20; and Tuve, *First Russian Women Physicians* (cit. n. 16), pp. 65–67.

[51] Interestingly, there does not appear to have been an appreciable difference in the research topics chosen by female and male scientists and physicians. I address this question further in "Gender and Science in the Russian Context," now in preparation for "Women in Russia," a conference to be held in Akron and Kent, Ohio, Aug. 1988.

[52] Bogdanovich, *Liubov' liudei* (cit. n. 2), p. 427; Musabekov, *Iulia Vsevolodovna Lermontova* (cit. n. 34), p. 48; and Arch. Akad. Nauk, fond 603, opis' 3, nos. 1, pp. 1–17, and 12, pp. 1–4. Lermontova even attempted to set up a women's chemistry laboratory but was discouraged by A. M. Butlerov's coworker Mikhail Dmitrievich L'vov; see L'vov to Lermontova, 6 Feb. 1881, Arch. Akad. Nauk, fond 603, opis' 3, no. 12, pp. 1–2.

[53] I. G. Zenkevich, "Elizaveta Fedorovna (Ivashkina) Litvinova," in Zenkevich, *Sud'ba talanta (Ocherki o zhenshchinakh-matematikakh)* (Briansk: Pedagogicheskoe obshchestvo RFSFR, 1968), pp. 33–37, on p. 36; and Koblitz, "Elizaveta Fedorovna Litvinova" (cit. n. 27).

[54] M. Kozhevnikova, "Sofia Mikhailovna Pereiaslavtseva," in *Pervyi zhenskii kalendar' na 1905*, pp. 383–390. Pereiaslavtseva's best-known works include *Protozoa Chernogo Moria, Zapiski Novo-rossiskogo Obshchestva Estestvennoispytatelei*, 1886, Vol. X; and *Etudes sur le devéloppement des Amphipodes* (Moscow: Société des Naturalistes de Moscou, 1888).

Sofya Vasilyevna Kovalevskaia (1850–1891)

member of the Russian Imperial Academy of Sciences. In some sense, Kovalevskaia was the first woman in modern times who could be considered a *professional* mathematician, accepted as such by male mathematicians all over the
world.

Kovalevskaia and her friends frequently acknowledged their debt to the sociopolitical creed of the radical Russian youth of the 1860s. They knew that without
the supportive, unprejudiced, free atmosphere that attended nihilist and scientific
circles when they were young women they probably would never have broken
away from the traditionalism of patriarchal Russian society.

THE END OF THE MOVEMENT

As was mentioned above, the tsarist government saw the connection between
progressive ideas and women's striving toward the sciences as clearly as did
Kovalevskaia, Suslova, Bokova-Sechenova, and their contemporaries. But Russian officials were faced with a dilemma. On the one hand, Russia needed medical and scientific personnel, and thus many administrators strongly urged the
training of women. On the other hand, many of those women who had studied
the sciences at home or abroad acquired police records in connection with their
political activities, and some were at the very heart of the revolutionary struggle.
Of the eight women on the eighteen-member executive committee of the militant
populist organization People's Will, for example, two had studied in Zurich, and
at least two more had had medical training in Russia.[55]

The autocracy took steps to curtail or at least control women's access to careers in the sciences and medicine. Besides issuing the Zurich proclamation, the
government seems to have ordered its embassies to harass some women students

---

[55] Engel, *Mothers and Daughters* (cit. n. 11), pp. 173–179. The executive committee of People's
Will was responsible for assassinating Tsar Alexander II in 1881.

in other European cities. Concurrently, however, it allowed the formation of so-called higher women's courses and courses for learned midwives in St. Petersburg, Moscow, and other large cities of the empire.[56]

The Russian women's movement considered the establishment of these courses a great triumph. The courses for learned midwives that opened in St. Petersburg in 1872 were the first medical courses exclusively for women anywhere in Europe. The instruction there soon reached a level equal to that at the Russian men's medical courses, although the women graduates were not allowed to call themselves "doctors" (as opposed to "learned midwives") until 1880.[57]

But it can be argued that these courses represented a step down for women in science in Russia. While the medical courses were on a reasonably high level, other higher women's courses were not of the caliber of the courses that the Russian women had taken in Western Europe. There was a problem of status as well. Learned midwives, after all, were not doctors, and a degree from Zurich was regarded as immeasurably superior to the certificate of competence one could earn from the Russian higher women's courses.

Moreover, even at the women's courses, while women could be classroom supervisors or laboratory assistants, they could not become professors. The women who had received their degrees abroad were fully qualified to teach in the courses, and several were more than willing to do so. The Ministry of Education, however, refused to accept their applications to take the necessary qualifying exam. For example, Kovalevskaia, whose application had the full approval of the mathematics faculty and the rector of Moscow University, was told by the ministry that both she and her daughter "would have a chance to grow old before a woman would be allowed to teach in a Russian institution of higher education."[58] So the first wave of Russian women scientists was not in a position to provide satisfactory role models for the next generation of women.

The government imposed strict controls on the higher women's courses, deliberately preventing them from achieving the status of the male universities. In the 1870s and 1880s, although the quality of women's higher education in Russia was not noticeably inferior to that offered for men (many of the same science professors taught in both institutions and commented favorably upon the women's level),[59] women graduates were not given the rank and privileges accorded the men. And every petition of a woman to enter the male universities met with the same response: study at the higher women's courses.

In addition, the government had so taken fright at the numbers of women scientific and medical students who involved themselves in progressive activities

[56] On harassment see *ibid.*, p. 145. For the history of the various higher women's courses see Likhacheva, *Materialy* (cit. n. 11); S. N. Valk, ed., *Sankt Peterburgskie vysshie zhenskie (Bestuzhevskie) kursy (1878–1918)* (Leningrad: Leningradskii Gosudarstvennyi Universitet, 1973); Stasov, *Nadezhda Vasil'evna Stasova* (cit. n. 11); A. N. Sheremetevskaia, "Stranitsa iz istorii vysshego zhenskogo obrazovaniia," *Istoricheskii Vestnik*, 1896, *65*:171–184; and Nekrasova, "Zhenskie vrachebnye kursy" (cit. n. 11).

[57] Tuve, *First Russian Women Physicians* (cit. n. 16), pp. 43–44; Barbara Alpern Engel, "Women Medical Students in Russia, 1872–1882: Reformers or Rebels?" *Journal of Social History*, 1979, *12*:394–414; and Christine Johanson, "Autocratic Politics, Public Opinion, and Women's Medical Education During the Reign of Alexander II, 1855–1881," *Slavic Review*, 1979, *38*:426–443.

[58] S. V. Kovalevskaia to A. O. Kovalevskii, [October 1880], in Kovalevskaia, *Vospominaniia i pis'ma* (cit. n. 34), p. 254.

[59] See, e.g., Sechenov, *Avtobiograficheskie zapiski* (cit. n. 8), pp. 240–241; and Stasov, *Nadezhda Vasil'evna Stasova* (cit. n. 11), pp. 338–339.

that it closed the courses at the least sign of student unrest, even if the women were not directly involved.[60] The medical courses stopped taking new students in 1881, after the assassination of Alexander II by People's Will, and they closed completely in 1887. Only in 1895 was a new women's medical institute approved.[61]

Similarly, in 1881 the government closed the St. Petersburg higher women's courses. When the courses reopened in 1889, it was discovered that decisive steps had been taken to prevent women from entering the sciences. The physical-mathematical and natural sciences faculties had been greatly reduced, their laboratories had been closed, and most natural science courses had been canceled.[62] Study languages or art, the government in effect was saying—that ought to be harmless enough.[63] Only after many years were the natural science and physical-mathematical faculties at the higher women's courses permitted to grow naturally again.

And so ended the first episode in the history of women in science in Russia. The women had flourished during the golden age of Russian science, which not coincidentally was the golden age of Russian nihilism as well. But they were largely prevented from educating a new generation of women, and circumstances led some of them into the revolutionary underground. Not until well into the twentieth century would there again be a number of women of scientific attainments comparable to those of Kovalevskaia, Pereiaslavtseva, Lermontova, Litvinova, Suslova, Bokova-Sechenova, and other women of the first generation of Russian women scientists.

[60] The tsarist government also periodically closed male institutions perceived to be sources of antigovernment activity. The St. Petersburg Medical-Surgical Academy and the technical institutes were especially prone to closures; see Leikina-Svirskaia, *Intelligentsia v Rossii* (cit. n. 2), esp. pp. 107–146.

[61] Johanson, "Autocratic Politics" (cit. n. 57), pp. 442–443.

[62] Valk, ed., *Sankt Peterburgskie vysshie zhenskie (Bestuzhevskie) kursy* (cit. n. 56), pp. 12–13, 21, 118–120. See also Sheremetevskaia, "Stranitsa iz istorii" (cit. n. 56), pp. 176–177, on the popularity of physics and mathematics among the early women students.

[63] This government policy could very well have been partly responsible for the flourishing of radical art movements (which included a large number of women) in Russia in the 1890s and 1900s. I am indebted to Linda Nochlin for suggesting this connection to me.

Women Astronomers in Britain, 1780–1930

By Peggy Aldrich Kidwell

"With a true eye and a faithful hand"

Between 1780 and 1930, that is from the time of Caroline Herschel to that of Cecilia Payne, women gained a foothold in British astronomy. Astronomical societies admitted them, universities let them attend relevant courses, and a few observatories hired them as computers. A brief survey of the careers of these women reveals patterns that persisted from the nineteenth into the twentieth century. At the same time, it shows that they had fewer opportunities, especially for independent astronomical research and teaching, than women elsewhere. These patterns merit further study in a more complete collective biography.

The careers of the Hanoverian-born singer Caroline Herschel and the Scottish-born author, wife, and mother Mary Somerville offered two early models for British women drawn to astronomy. Caroline Herschel (1750–1848) was the exemplary helpmate. She joined her brother William in England in 1772, and initially shared in his career as a musician. When William Herschel began to manufacture telescopes and observe the stars, she assisted. After William Herschel discovered Uranus in 1781, the British government awarded him a pension that allowed him to devote all his time to astronomy. Caroline Herschel also turned to astronomy, not only polishing lenses and recording observations, but discovering eight comets on her own. In 1787, she received her own pension of £50 annually. When William Herschel married in 1788, she moved to separate lodgings but continued to work in his observatory. After his death in 1822, she returned to Hanover, spending part of her long retirement preparing a catalogue of nebulae for the use of William Herschel's son John. In 1835, Caroline Herschel was elected an honorary fellow of the Royal Astronomical Society, one of the first two women so honored.[1]

The second woman elected to the RAS in 1835, Mary Somerville (1780–1872), entered astronomy by another path often used by later women. She wrote books for a popular audience, synthesizing contemporary discoveries in the physical sciences. Although Somerville had no formal training in mathematics and astronomy, she prepared an English "rendition," with extensive commentary, of

This research was carried out under a Smithsonian Institution fellowship. I thank Deborah Warner for her advice and direction and David DeVorkin, Karl Hufbauer, John Lankford, Margaret Rossiter, Marc Rothenberg, Elizabeth Patterson and two anonymous reviewers for their comments.

[1] See M. C. Herschel, *Memoir and Correspondence of Caroline Herschel* (London: John Murray, 1876); M. A. Hoskin, "Caroline Lucretia Herschel," *Dictionary of Scientific Biography*, Vol. V (New York: Scribners, 1972), pp. 322–323; and M. B. Oglivie, "Caroline Herschel's Contributions to Astronomy," *Annals of Science*, 1975, *32*:149–16l.

the first four books of Pierre-Simon Laplace's *Mécanique céleste*. The book, published in 1831 under the title *The Mechanism of the Heavens,* was widely acclaimed and served for a time as a textbook at Cambridge University. Three years later Somerville wrote *On the Connexion of the Physical Sciences.* Here she discussed new-found interconnections between gravitation, heat, light, electricity, and magnetism. When writing these and later books, Somerville consulted a distinguished group of contemporary natural philosophers. Her achievements and political connections brought rewards; in 1835 she was awarded an annual government pension of £200 that was raised to £300 two years later.[2]

Herschel and Somerville were accepted members of the British scientific community. Women also had more modest successes. In the first two decades of the nineteenth century, Margaret Bryan, a teacher and proprietor of a school for girls, wrote elementary texts on astronomy, geography and natural philosophy. During the 1830s and 1840s, Janet Taylor published lunar tables, ran a Nautical Academy, and sold an assortment of mathematical and navigational instruments at a business inherited from her husband George.[3] Finally, an anonymous "lady" devised cards showing the constellations, with stars represented by small holes. The size of the hole varied with stellar magnitude, so that when a card was held in front of a light, the dots of light varied in brightness like the stars in the sky. These cards, sold under the name Urania's Mirror, were issued in at least four editions.[4]

During the second half of the nineteenth century, changes both in the place of women in British society and in the astronomical community affected women's participation in astronomy. As the number of girls' schools increased, many young women came to receive a more systematic education than that provided by relatives, independent reading, scientific acquaintances, or glances through devices like Urania's Mirror. The growing number of schools prompted a demand for well-trained teachers that was met, in part, by colleges for women such as Queen's College (1848) and Bedford College (1849). A few ambitious women soon established houses of residence for women who wished to fulfill the requirements for Oxford and Cambridge degrees. By 1881, each of these universities had two women's halls or colleges.[5]

[2] "Rendition" is Somerville's term for her version of Laplace's work. On her life, see Martha Somerville, ed., *Personal Recollections from the Early Life and Old Age of Mary Somerville* (London: John Murray, 1874); Elizabeth C. Patterson, "Mary Somerville," *British Journal for the History of Science,* 1969, 4:311–339; and Patterson, *Mary Somerville, 1780–1872* (Oxford: for Somerville College, 1979); and Patterson, *Mary Somerville and the Cultivation of Science, 1815–1830* (The Hague: Martinus Nijhoff, 1983).

[3] For information about Janet Taylor, see E. G. R. Taylor, *Mathematical Practitioners of Hanoverian England* (Cambridge/London: Cambridge Univ. Press, for the Institute of Navigation, 1966). On Margaret Bryan, see the *Dictionary of National Biography* (*DNB*), Vol III, p. 154.

[4] Examples of Urania's Mirror are in the Rare Book Room, Smith College, Northampton, Mass., and at the National Museum of American History, Smithsonian Institution, Washington, D.C. Mention of the anonymous lady and contemporary comments on the cards are found in Jehosaphat Aspin, *A Familiar Treatise on Astronomy* (1825; 4th ed., London: M. A. Leigh, 1834), written "expressly to accompany Urania's Mirror."

[5] Women who completed degree requirements were awarded certificates; they did not receive degrees until 1920 at Oxford and 1948 at Cambridge. On women's education in Britain, see H. C. Barnard, *A Short History of English Education from 1760 to 1944* (London: Univ. London Press, 1947), pp. 182–195; Lee Holcombe, *Victorian Ladies at Work: Middle-class Women in England and Wales, 1850–1914* (Hamden, Conn.: Archon Books, 1973), pp. 21–66; and E. W. Jenkins, *From Armstrong to Nuffield: Studies in Twentieth-Century Science Education in England and Wales* (London: John Murray, 1979), esp. pp. 170–214. On specific schools, see Barbara Stephen, *Girton College, 1869–1932* (Cambridge: Cambridge Univ. Press, 1933); Ann Phillips, ed., *A Newnham Anthology* (Cambridge: Cambridge Univ. Press, 1979); and Vera Brittain, *The Women at Oxford: A Fragment of History* (New York: Macmillan, 1960).

The opening of women's colleges did not immediately make it possible for women to become professional astronomers. According to well-established tradition, rising British astronomers studied mathematics and mathematical physics at Cambridge. They then worked as assistants at major observatories or as directors of colonial observing stations until more suitable positions became available. Graduate degrees did not become common until the late 1920s. For example, in 1903 H. H. Turner described the newly appointed Superintendant of Indian Observatories as having "a sound knowledge of mathematics and some physics, his chief work having been on the flight of the boomerang." The appointee was to learn astronomy through reading, visits to observatories, and work on the spot.[6]

This system of recruitment affected women in three important ways. First, they had to persuade mathematics and physics professors to let them attend their courses. Some professors rejected women students. The eminent physicist James Clerk Maxwell firmly warned one assistant in 1877 that "it is a rule of the Cavendish Laboratory that the students thereof be of the male sex." Maxwell did relent to the extent of allowing lectures for women while he was on vacation in Scotland. By the turn of the century, the Cavendish had dropped such restrictions, and only a handful of Cambridge professors refused to have any women students whatsoever. However, the practice of having women sit in a separate row in classes continued at the Cavendish into the 1920s.[7]

Second, the British practice of teaching astronomy as part of mathematics and physics restricted employment opportunities for women interested in the subject. When the new women's colleges hired women to assist students with their courses, they hired mathematicians and physicists, not astronomers. This situation is in marked contrast to that in the United States, where Vassar, Smith, Mt. Holyoke and other schools developed departments of astronomy that not only trained women students, but employed women to teach and to run their observatories.[8]

Third, the British emphasis on acquiring professional credentials through paid work in observation rather than study put women at a further disadvantage. Women who wanted to work at telescopes outside university campuses had few choices. Some had access to instruments owned or used by their relatives, some had their own telescopes, and at least one woman, Agnes Clerke, visited a government observatory as the guest of the astronomer in charge. Elizabeth Iris Pogson Kent did work at the Madras Observatory from 1873 to 1896, first as assistant astronomer and then as meteorological observer. She began working for her father, Norman Robert Pogson (1829–1891), who was director of the

[6] H. H. Turner to G. E. Hale, 28 Jan. 1903, G. E. Hale Papers, Mt. Wilson and Palomar Observatories Library, Pasadena, California (quoted by permission of the Carnegie Institution of Washington) (henceforth Hale Papers).

[7] See J. C. Maxwell to Sedley Taylor, 31 May 1877, Add. MS. 6259, no. 88, Sedley Taylor Collection, University Library, Cambridge, U.K. (quoted by permission of the Cambridge University Library); I thank Paul Theerman for this reference. On later practices at the Cavendish, see J. G. Crowther, *The Cavendish Laboratory, 1874–1974* (London: Macmillan, 1974), p. 74; and C. Payne-Gaposchkin, *Cecilia Payne-Gaposchkin: Her Autobiography and Other Recollections,* ed. K. Harumdanis (Cambridge: Cambridge Univ. Press, 1984). In 1896, Isabel Maddison reported that the only mathematician at Cambridge who closed his lectures to women was Dr. Glaisher, the only physicist G. G. Stokes; see Maddison, *Handbook of British Continental and Canadian Universities, with Special Mention of Courses Open to Woman* (New York: Macmillan, 1896), p. 79.

[8] On women astronomers at American colleges, see Deborah J. Warner, "Women Astronomers," *Natural History,* May 1979, *88*(5):14–26; Pamela Mack, "Women in Astronomy in the United States, 1875–1920" (B.A. diss., Harvard Univ., 1977); and Margaret Rossiter, *Woman Scientists in America: Struggles and Strategies to 1940* (Baltimore/London: Johns Hopkins Univ. Press, 1982), esp. pp. 12–28, 160–217.

observatory. Significantly, there is no evidence that she was paid.[9] Such observing did not provide the accreditation required for better jobs. Again, the situation for women was somewhat different on the Continent and in the United States, where graduate studies played an increasing role in the training of astronomers.

Moreover, positions in computing, which allowed several women in the United States to enter astronomy, if at a low level, were rarely available to women in Britain. Most of these jobs went to young men just out of school. As Herbert Hall Turner, Savilian Professor of Astronomy at Oxford, explained: "to spend a whole life in such computing has manifest disadvantages, both for the individual and for the institution which must pay him an increasing salary with little corresponding return. On the whole the method of using the few years between 14 and 17 or 18 in the life of a boy who has no immediate prospects seems the least wasteful."[10] Despite such opinions, a few nineteenth-century observatories did hire women to carry out routine measurements and calculations. Even in the eighteenth century, one Mary Edwards worked for the Nautical Almanac Office. Observatory reports and other accounts reveal that a Miss Hardy worked at the Cambridge University Observatory from 1876 to 1879; she was succeeded by a Miss Walker, who stayed until 1903. Alice Everett and Annie Scott Dill Russell (later A. S. D. Maunder; 1868–1947) worked at Greenwich for a few years in the 1890s. In 1892 they observed the opposition of Mars with the 12.8-inch and 10-inch telescopes there, but this was an unusual exception to their routine work as computers. Finally, Edith Bellamy had begun doing computations for the Oxford University Observatory by 1899. Of these five women, only Everett and Maunder were university educated.[11]

Because of these restrictions, in the late nineteenth century the major focus of activity for women interested in astronomy was the scientific society. After 1838, women interested in astronomy could attend meetings of Section A (Mathematics and Physics) of the British Association for the Advancement of Science. Meetings of the BAAS were at most an opportunity to hear astronomy discussed; four papers by women were read to section A in the nineteenth century, but all were on physics.[12] Women's participation in Britain's Royal Astronomical Society was also limited. This organization, formed in 1820, admitted women as honorary fellows from 1835, but in the nineteenth century accorded

[9] On Elizabeth I. P. Kent, see *Monthly Notices of the Royal Astronomical Society,* 1920, *80*:336. For other women who in fact made observations at British observatories, see on Alice Everett and A. S. D. Maunder below. I do not know what formal regulations governed the employment of women at British observatories.

[10] H. H. Turner, *Annual Report of the Savilian Professor of Astronomy . . . for 1909 to 1910,* p. 6.

[11] Pamela Mack found 35 women who worked as computers at American observatories before 1900, 8 of whom had college degrees; see Mack, "Women in Astronomy," pp. 134–148. On Edwards see Dorothea Klumpke, "La femme dans l'astronomie," *Bulletin de la Société Astronomique de France,* 1899, *13*:162–215, p. 167. Hardy and Walker are listed in F. J. M. Stratton, "The History of the Cambridge Observatories," *Annals of the Solar Physics Observatory, Cambridge,* 1949, *1*:1–26, p. 25. On Everett and Maunder see A. J. Meadows, *Greenwich Observatory,* Vol. II: *Recent History, 1836–1975* (London: Taylor & Francis, 1975), pp. 9–15; and F. J. Sellers and P. Doig, eds., "The History of the British Astronomical Association," *Memoirs of the British Astronomical Association,* 1948, *36*(2):88. On Edith Bellamy's employment, see Turner, *Annual Report of the Savilian Professor,* from 1900 onward.

[12] On the role of women at early meetings of the BAAS, see Jack Morrell and Arnold Thackray, *Gentlemen of Science: Early Years of the British Association for the Advancement of Science* (Oxford: Clarendon Press, 1981), pp. 148–157. In 1858 W. S. Ayrton read a paper written by Rosina Zornlin to Section A; in 1895 Hertha Ayrton and Dorothy Marshall presented papers; Hertha Ayrton read another paper in 1897.

this honor only to Caroline Herschel, Mary Somerville, and Anne Sheepshanks (1789–1876). Sheepshanks was not a practicing astronomer. She was elected an honorary fellow in recognition of her donation to the RAS in 1857 of several instruments of her late brother, the astronomer and mathematician Richard Sheepshanks.[13]

By the end of the century, some members believed that women should join the RAS as regular members. In 1886 Elizabeth Isis Pogson Kent of the Madras Observatory was nominated as an ordinary fellow. When the society's legal advisor saw no bar to her election, the governing council sought another opinion. The second attorney concluded that those who had written the RAS charter had not meant to allow women as ordinary fellows, and Kent's nomination was withdrawn. Six years later, Annie Scott Dill Russell and Alice Everett of Greenwich were proposed for fellowship, as was the distinguished amateur Elizabeth Brown. This time the council decided that the women should be allowed to stand for election. The candidates did not receive the required three quarters of the votes cast, however, and hence did not become fellows.[14]

Women found a warmer welcome in newer astronomical organizations. The research of the Herschels, Lord Rosse, William and Margaret Murray Huggins (see below), and others laid the foundation for a new "physical astronomy," concerned not with mathematical calculations of the positions of the heavenly bodies but with the nature of these bodies as made manifest by their surface features, changes in brightness, distribution in space, and spectra. Data collected by amateurs was of great use to the new astronomy; Agnes Clerke commented in 1887 that "there is no one 'with a true eye and a faithful hand' but can do good work in watching the heavens."[15] Both regional astronomical organizations and a national society created specifically for amateurs carried out this new type of research.

From 1882 the Liverpool Astronomical Society sought to guide those with true eyes and faithful hands who wished to watch the heavens. Dues were low and women were welcome. To foster useful observations, the society was divided into sections devoted to the study of the sun, the moon, meteors and comets, star colors, double stars, variable stars, and planets.[16] Each section was headed by an experienced observer who could advise beginners and collect and collate results. Elizabeth Brown (died 1899), the head of the solar section, made and collected sunspot observations. She thought that drawing sunspots was a particularly appropriate activity for women, not only because ladies often had ample time and were skillful in the use of the pencil, but because no exposure to the night air was required. (Brown did not explain precisely what the adverse effects of night air were.) During her years with the LAS, Brown also participated in solar eclipse expeditions to Russia and to the West Indies; she published two short books describing her travels.[17]

[13] On Anne Sheepshanks, see A. M. Clerke, "Richard Sheepshanks," *DNB*, Vol. XVIII, p. 10.

[14] See J. L. E. Dreyer et al., *History of the Royal Astronomical Society, 1820–1920* (London: Wheldon & Wesley, 1923), pp. 233–234.

[15] Agnes M. Clerke, *A Popular History of Astronomy During the Nineteenth Century* (Edinburgh: A. & C. Black, 1887), pp. 6–7.

[16] Recent discussions of British amateur astronomy at the turn of the century have focused on studies of the planets; see Norris S. Hetherington, "Amateur vs. Professional: The British Astronomical Association and the Controversy over Canals on Mars," *Journal of the British Astronomical Association*, 1976, 86:303–308; John Burnett, "British Studies of Mars: 1877–1914," *ibid.*, 1979, 89:136–143; and John Lankford, "Amateurs versus Professionals: The Controversy over Telescope Size in Late Victorian Science," *Isis*, 1981, 72:11–28.

[17] On sunspot drawings, see Elizabeth Brown, "Solar Section," *J. Brit. Astron. Ass.*, 1891, 1:172. Brown's books, published anonymously, were *In Pursuit of a Shadow* (London: Turbner, 1887) and *Caught in the Tropics* (London: Griffin, Farran, Okeden, & Welsh, 1890).

In 1890 financial and administrative problems at the LAS led to the resignation of Brown and other members and prompted demands for a new, national association of amateur astronomers. That fall, the British Astronomical Association began meeting in London. Like the LAS, it provided amateurs with an opportunity to make observations under the direction of skilled section leaders. Women were welcomed as among those "practically excluded from becoming Fellows" of the RAS, and they took an active role as members of the council of the BAA, directors of sections, librarians, and, in a few cases, officers.[18]

Women also joined enthusiastically in eclipse expeditions organized by the BAA at the turn of the century. For example, in 1898 the BAA sent two parties to India to observe the solar eclipse. One group, ten men and five women headed by the amateur John M. Bacon, set up camp in imperial style at Buxar. Bacon's daughter Gertrude took four photographs during the eclipse. Two other women made observations with a crude slitless spectroscope; two more joined a group that sketched the solar corona. A smaller, more professional party, headed by Edward W. Maunder of Greenwich, went to Talni. The former Annie Scott Dill Russell, now Maunder's wife, had received a research grant from Girton College, Cambridge, for a photographic survey of the Milky Way, and she promptly set about this work. On the day of the eclipse, she made photographs of the solar corona that showed the largest coronal extensions yet observed. Neither the observations of these six women nor any results of the expedition were major contributions to astronomical knowledge. The episode does illustrate, however, that women had been incorporated into one part of the British astronomical community, even receiving small grants in aid of their own specific research projects.[19]

By the late nineteenth century, then, British women were attending classes at major universities, doing some routine work at observatories, and making observations and going on eclipse expeditions with the blessing of the BAA. They also continued on older paths, helping male relatives make observations and writing expository books and articles on astronomical subjects. To see how women made use of these opportunities at the turn of the century, I have examined the astronomical careers of 258 women active in British astronomy between 1901 and 1930. Of course, people active in this period often had begun to work in astronomy some years earlier. The women included were members of the BAA or other amateur organizations for five years or more, fellows of the Royal Astronomical Society, authors of books and articles on astronomical topics, computers at observatories, or assistants acknowledged in observatory reports, particularly reports published in the *Monthly Notices of the Royal Astronomical Society*.[20]

[18] Quoting *J. Brit. Astron. Ass.*, 1890, *1*:19. Sellers and Doig, eds., "The History of the BAA" (cit. n. 11), lists officers, section directors, and some councillors; on the difficulties of the LAS and the founding of the BAA, see *ibid.*, pp. 7–9, and *The English Mechanic*, July 1890–Dec. 1891, *51–52*.

[19] E. W. Maunder, *The Indian Eclipse, 1898* (London: Hazell, Watson, & Viney, 1899). On A. S. D. Maunder's grant, see Stephen, *Girton College* (cit. n. 5), p. 89.

[20] My sources include the annual *List of Members of the British Astronomical Association;* the notices of election of fellows in the *Mon. Not. Roy. Astron. Soc.;* lists of members in "annual" reports of the LAS for 1904, 1924, 1925, and 1928; and the list of members published annually in the *Journal of the Astronomical Society for Wales and Monmouthshire* (after 1908 *The Cambrian Journal*). To find women's publications, I consulted the *International Catalogue of Scientific Literature* for 1901–1914 and the *Astronomische Jahresbericht* for 1915–1930. In addition to observatory reports in the *Mon. Not. Roy. Astron. Soc.*, I looked at annual reports from the Greenwich Observatory, the Edinburgh Royal Observatory, the Cambridge Observatory, and the Oxford University Observatory. While further search would undoubtedly turn up other women, the sample used suggests both the opportunities and the difficulties for women in British astronomy at the turn of the century.

Table 1. Women active in British astronomy 1891–1930, by decade.

Decade	Number
1891–1900	(107)
1901–1910	150
1911–1920	131
1921–1930	131

Table 1 gives the number of women active in British astronomy during each decade between 1890 and 1930; data for the 1890s are included for comparative purposes. Bibliographical sources are most complete for the years 1901–1910, so that the figure for that period may be too large.[21] Nonetheless, it seems safe to say that women's participation in British astronomy did not increase greatly over the first thirty years of the twentieth century.

As in the nineteenth century, in the twentieth century British women participated in astronomy most frequently as members of astronomical societies. Over three quarters of those in the sample were members of amateur societies, primarily the BAA.[22] The matter of admitting women as fellows of the RAS arose again in December, 1914, when two well-established amateur astronomers, Mary Blagg and Fiammetta Wilson, were proposed as fellows. When the nominations came before the council, it proposed to petition the king for a supplemental charter that would allow for the admission of women fellows. The proposal was approved at the February 1915 meeting of the RAS. By the following January the legal steps necessary for the election of women had been taken. Blagg, Wilson, Ella K. Church, A. Grace Cook, and Margaret T. Meyer were elected fellows. By 1930 forty-two British women had been elected to the RAS.[23]

Women fellows of the RAS were not necessarily actively interested in astronomy. Only half were members of the BAA. This might suggest that once women could join the older society, they felt less need for the newer one. However, some women fellows simply had little interest in observational astronomy. Of the twenty-one fellows of the RAS not in the BAA, only Edith Bellamy published extensively in astronomy; at least seven other women had a greater interest in physics or mathematics.

While one should not underestimate the importance of astronomical societies to women, one also should not overestimate their role in these overwhelmingly male organizations. The RAS published annual reports of the directors of public and private observatories in the British empire. From 1915 to 1930, no woman filed such a report. Moreover, women were not chosen as officers of the society or as members of the council, and they were not invited to the Royal Astronomical Society Club, an elite group of fellows that entertained important visiting astronomers. Even at the BAA, women remained a small minority. Between 1901 and 1930, they made up at most 8.5% of the membership, and were rarely officers. During World War I Fiametta Wilson and A. Grace Cook were acting directors of the meteor section, and from 1921 to 1923 Cook directed the section alone. In 1930 Mary A. Evershed began her tenure as director of the historical section. Otherwise, the directors of sections were men. Probably the

---

[21] Since the literature search did not cover the 19th century, the figures for 1891–1900 are probably too low. The figures for 1915–1930 will also be somewhat low, since the *Astronomische Jahresbericht* did not list popular works, included in the *International Catalogue of Scientific Literature*.

[22] Of the 258 women, 185 were in the BAA; of the 73 not in the BAA, 10 were members of the Astronomical Society for Wales and 1 was in the LAS.

[23] In addition, Agnes Clerke and Margaret Huggins were elected honorary fellows in 1903; see Dreyer et al., *History of the RAS*, pp. 233–234.

Table 2. Highest degree earned by women active in British astronomy, 1901–1930.

Degree	Number
Ph.D. or D.Sci.	6
M.A. or M.Sci.	9[1]
B.A. or B.Sci.	11
A.R.C. Sci.	1
None (no university training)	13
Unknown	218

[1] One degree is honorary.

woman who did most for the BAA in these years was A. S. D. Maunder, who had first edited the *Journal of the British Astronomical Association* for a few years in the 1890s and took up this duty again from 1917 to 1930. Once again, institutional changes of the nineteenth century did not prompt an outburst of astronomical activity by twentieth-century women.

Information about the education of women in the sample is limited, with available data summarized in Table 2. Women who received certificates from Oxford or Cambridge are counted as having degrees (see note 5 above). In general, the membership lists of astronomical societies give the degrees attained by members and publications list the credentials of authors. Hence I suspect that most of the women for whom I have no information did not attend any university.

Most women active in the British astronomical community at the turn of the century were unmarried. Of those in the sample, 166 (64%) had not married, while 92 (36%) were married or widowed for at least part of the time they were active in astronomy. Among those paid for astronomical work, marriage was much less common. To be sure, 9 of 27 authors, translators and lecturers were married. However, only 2 of 26 computers and 2 of 14 teachers were or had been married. Further research is needed to determine whether formal regulations as well as social pressures produced these patterns.

Information is available about the employment of 85 (33%) of the 258 women in the sample; these data are summarized in Table 3. Of the fifteen women who had no paid employment, eleven worked as unpaid assistants, often in their husband's observatories. Foremost among the helpmates was Margaret Lindsay Huggins (1849–1915). Margaret Lindsay Murray had made drawings of the constellations and of sunspots as a child, and she went on to build a small spectroscope on a plan described in the magazine *Good Words*. In 1875 she married the former businessman and devoted astronomical spectroscopist William Huggins. They worked together for over thirty years, preparing an atlas of stellar spectra and making special studies of binary stars, nebulae, and Wolf-Rayet stars. Margaret Huggins took and developed photographic plates, classified spectra, and wrote and edited articles. When William Huggins was knighted in 1897, she was cited as his "gifted wife." After Sir William died in 1910, she accepted a government pension of £100 per year in honor of her service to science in collaboration with her husband.[24]

---

[24] On Margaret Lindsay Huggins, see the obituaries by Sarah F. Whiting in *Science*, 1915, *41*:853–855; and H. F. Newall in *Mon. Not. Roy. Astron. Soc.*, 1915, 76:278–282; see also R. S. Richardson, *The Star Lovers* (New York: Macmillan, 1967), pp. 157–162. Margaret Huggins wrote George Ellery Hale that she was pleased that the pension was granted for her service to astronomy and not (as was usual for widows) because she was in straitened circumstances; see Huggins to Hale, 11 Apr. 1910, Hale Papers (cit. n. 6).

Table 3. Employment of women in British astronomy, 1901–1930

Occupation	Number
No paid employment	15
Author, translator, or lecturer	27
Computer	26
Teacher or professor	14
Librarian or administrator	3

Other wives worked on a more modest level. In 1909 William Doberck, retired director of the Hong Kong Observatory, reported to the Royal Astronomical Society on recent research at his private observatory in Surrey. He noted that "with the cooperation of Mrs. Doberck, 939 measures of double stars had been secured by the end of the year." The wealthy amateur Isaac Roberts even persuaded Dorothea Klumpke (1861–1942) to leave her post at the Paris Observatory and marry him, in order to further Robert's research at his own observatory. Klumpke, an American citizen, had obtained her D.Sc. from the Sorbonne and worked for several years in Paris before she married in 1901. Isaac Roberts died three years afterward, leaving Dorothea Klumpke Roberts to complete the publication of his astronomical papers.[25]

Women also assisted in public and university observatories on a voluntary basis. A. S. D. Maunder helped her husband to reduce Greenwich sunspot observations while she was raising their two children. She is most unusual in that she then worked at Greenwich in a paying position from 1917 to 1920. Mary Acworth Evershed (1867–1949) published observations of solar prominences that she made with her husband John Evershed, director of the Kodaikanal Observatory in India.[26] Some professional astronomers even found unpaid help outside the family circle. For over twenty years, Mary Blagg (1858–1944) aided H. H. Turner of Oxford with his reduction of accumulated observations of variable stars. Another women of independent means, Elizabeth Williamson, worked for at least three years as an honorary assistant at the University College Observatory of the University of London.[27]

Of course, women of sufficient wealth and leisure also observed on their own, in the tradition of Elizabeth Brown. The most enthusiastic of these observers in the early twentieth century was Fiammetta Wilson (1864–1920), a trained musician who happened to attend a series of lectures given by the astrophysicist Alfred Fowler in 1910. Greatly impressed, Mrs. Wilson joined the BAA and began observing planets, comets, the zodiacal light, and, most especially, meteors. Between 1910 and 1920, she made over 10,000 observations of meteors. She joined A. Grace Cook as acting head of the meteor section during World War I, becoming "the brightest ornament and the most exhilarating presence in

---

[25] Quoting William Doberck, "Dr. Doberck's Observatory," *Mon. Not. Roy. Astron. Soc.,* 1909, *69*:281–282. On Klumpke see A. J. Meadow's biography of J. Norman Lockeyer, *Science and Controversy* (London: Macmillan, 1972), p. 281; Robert Aitken's obituary in the *Publications of the Astronomical Society of the Pacific,* 1942, *54*:217–222; *National Cyclopaedia of American Biography* (New York: James T. White, 1944), pp. 405–406; and Katherine Bracher, "Dorothea Klumpke Roberts: A Forgotten Astronomer," *Mercury,* 1981, *10*:139–140.

[26] On A. S. D. Maunder, see *J. Brit. Astron. Ass.,* 1947, *57*:238; as well as her published papers. On Mary Acworth Evershed, see obituaries by P. J. Melotte in *ibid.,* 1950, *60*:86–87; and A. D. Thackeray in *Mon. Not. Roy. Astron. Soc.,* 1950, *110*:128–129.

[27] Mary Blagg also took an active interest in lunar nomenclature. For an account of her life, see *Mon. Not. Roy. Astron. Soc.,* 1945, *105*:65–66. Elizabeth Williamson's work at the University of London observatory is mentioned in *ibid.,* 1926, *86*:203; 1927, *87*:278; and 1928, *88*:277.

the little community of meteor observers." In July 1920 she was appointed to the E. C. Pickering Fellowship, a one-year research position for women established at the Harvard College Observatory in 1916. Unfortunately, she died that very month and never learned of this honor.[28]

Some sixty-five women made money from activities connected with astronomy. Of these, twenty-seven sought to popularize astronomy as authors, translators, and lecturers. By far the most distinguished of these cultivators of astronomy was Agnes Mary Clerke (1842–1907). Educated at home and through travel abroad, Clerke chronicled the development of astronomy past and present in a steady stream of articles for periodicals like the *Edinburgh Review, Observatory,* and *Knowledge and Science News.* She also wrote entries on astronomers for the *Dictionary of National Biography* and the *Encyclopaedia Britannica.* Moreover, she synthesized her individual studies in books such as *A Popular History of Astronomy in the Nineteenth Century* (1887), *The System of the Stars* (1890), *Problems in Astrophysics* (1903), and *Modern Cosmogonies* (1906). Agnes Clerke was highly respected by contemporary astronomers. Her correspondents included William and Margaret Huggins; Norman Lockyer; David S. Gill, whom she visited at the Cape of Good Hope; E. S. Holden of the Lick Observatory; and E. C. Pickering of Harvard. Indeed, at the time of the opening of the Mt. Wilson Observatory in California, Turner persuaded her to outline an observing program for the new institution. She was elected an honorary fellow of the RAS, and a lunar crater is named in her honor.[29]

Other cultivators of astronomy sought a broader audience. Some, like Agnes Giberne and Geraldine Mitton, wrote for children. Another, Mary Proctor (1862–1944), turned out a stream of popular books for adults. Finally, there were women like Gertrude Bacon, mentioned earlier as a member of the BAA's expedition to India for the eclipse of 1898. Bacon and her father, the Rev. John M. Bacon, gave joint lectures describing their travels and balloon ascents. When the elder Bacon died in 1904, his daughter filled his speaking engagements. Apparently she was a success, for she continued to write and lecture on astronomy and aeronautics into the 1920s.[30]

The number of women working as paid computers in the British empire increased considerably during the first three decades of the twentieth century. British colonial observatories in Australia and South Africa hired at least fifty-six women for this work, and twenty-six others found positions in Britain itself.[31] About two thirds of the British posts were filled under policies established by Frank W. Dyson, first when he was Astronomer Royal for Scotland and then after he became the Astronomer Royal at Greenwich in 1910.[32] The computers

[28] Quoting from W. F. Denning's obituary of Fiammetta Wilson in *ibid.,* 1921, *81:*266–269; see also the obituary of Wilson by A. Grace Cook in *J. Brit. Astron. Ass.,* 1920, *30:*330–331.

[29] On cultivators of science, see Nathan Reingold, "Definitions and Speculations: The Professionalization of Science in America in the Nineteenth Century," in Alexandra Oleson and Sanborn C. Brown, *The Pursuit of Knowledge in the Early American Republic* (Baltimore/London: Johns Hopkins Univ. Press, 1976), pp. 33–69. On Clerke's life, see Margaret Huggins, *Agnes Mary Clerke and Ellen Mary Clerke: An Appreciation* (n. p.: printed for private circulation, 1907); there are also numerous contemporary book reviews and obituaries. Clerke mentions in the prefaces those she consulted in preparing her books. For her suggestions for Hale, see A. M. Clerke to H. H. Turner [March, 1903] filed under "Turner," Hale Papers.

[30] Gertrude Bacon, *Memories of Land and Sky* (London: Methuen & Co., 1928).

[31] For the names of the colonial women, see the sections of the International Photographic Chart of the Heavens (Carte du Ciel) for Melbourne, Perth, and the Cape of Good Hope, as well as the annual reports of directors of observatories in these cities.

[32] Dyson may even have consciously set out to hire women, for his daughter said that he favored women's suffrage; see Margaret Wilson, *Ninth Astronomer Royal: The Life of Frank Watson Dyson* (Cambridge: Heffer & Sons, 1951), p. 163.

were hired for temporary positions, often funded by government grants. About one fifth of them had attended college; two had advanced degrees. Both Dyson and other directors of observatories continued to hire young men just out of school to fill similar posts. Indeed, the men may have gained more from their employment, for they were often trained to make observations as well as to measure photographic plates and perform calculations.

Computers rarely made a career of their work; at least half of the women kept their jobs less than five years. However, a few people made computing a lifetime occupation. I have mentioned Edith Bellamy (died 1960), who did calculations for her uncle F. A. Bellamy, an assistant at the Oxford University Observatory, from the late nineteenth century. From the 1890s Bellamy worked at home, paid by government grants to reduce data for the astrographic catalogue and to calculate coordinates of the stars in the path of the asteroid Eros. By 1909 she was doing clerical work at the Oxford Observatory and, in the words of her employer, H. H. Turner, had "become a valuable member of the establishment." On his recommendation, she received a permanent appointment in 1912 and set about supervising computations for the Vatican Observatory's portion of the astrographic catalogue. In 1920 the Pope awarded her a silver medal in recognition of this work. During the next few years, Turner became most interested in seismology, and Bellamy took up the work of collecting and collating earthquake records from hundreds of stations around the world. These appeared quarterly in the *International Seismological Summary*, a publication Bellamy edited until shortly before her retirement in 1947. In addition to the Vatican medal, Bellamy's faithful service brought her election to the RAS in 1920 and an honorary M.A. from Oxford in 1935.[33]

The length of Bellamy's stay at Oxford, the honors she received, and the fact that some of her results were published under her own name are all most unusual. S. A. Falconer's brief career is more representative. Falconer worked at the Edinburgh Observatory from 1909 until 1916, measuring plates taken at Perth for the International Photographic Chart of the Heavens. At the outbreak of the war, she left to do volunteer work in France on a temporary basis; there is no record of her subsequent career.[34]

Computers received low pay to do tedious calculations, worked on problems selected by those who hired them, and were rarely acknowledged in publications. It is not surprising that several well-educated women interested in astronomy chose to teach school instead. At least 14 of the 258 women in the sample were teachers. They worked in secondary schools, in women's colleges, and as instructors at coeducational universities. Six published articles on astronomical topics, and a seventh wrote her dissertation on celestial mechanics. However, none of them published more than two articles before 1930, and, as might be expected under the British system, they taught mathematics or physics, not astronomy.

The careers of three of the six women in the sample who had doctorates exemplify this pattern. Maud O. Saltmarsh, a Cambridge graduate who received her Ph.D. from London University in 1924, was cited in the astronomical literature for her investigations of the spectra of excited states of phosphorous. After she completed her degree, she remained at London University, teaching physics at Bedford College. Like Saltmarsh, Dorothy Wrinch (1894–1976) at-

[33] On Edith Bellamy, see Turner, *Annual Report of the Savilian Professor* (cit. n. 10), from 1899/1900 onward, quoting here 1908/1909, p. 8; see also the obituary by H. H. Plaskett in the *Quarterly Journal of the Royal Astronomical Society*, 1960, 2:121–123.

[34] On S. A. Falconer's work, see the *Annual Report of the Astronomer Royal for Scotland*, from 1909/1910, No. 20, onward.

tended Cambridge and carried out advanced work at London University. Wrinch was elected a Fellow of the Royal Astronomical Society in 1920 and published a paper on the theory of relativity two years later. However, she was primarily interested in mathematics and philosophy. During the 1920s and 1930s Wrinch had a temporary teaching appointment at Oxford, and she used a series of grants to pursue research at both Oxford and Cambridge. She became an X-ray crystallographer and eventually emigrated to the United States, where she taught biochemistry at Smith College in Massachusetts. A third Cambridge graduate, Bertha Swirles (later Jeffreys) completed a Cambridge Ph.D. in 1929 and went on to publish on mathematical physics and theoretical astrophysics. Swirles held a variety of lectureships in mathematics, eventually becoming the Director of Studies in Mathematics at Girton College, Cambridge. Her career was largely overshadowed by that of her husband, the astronomer and geophysicist Harold Jeffreys.[35]

British women also pursued astronomy overseas. Some, like those who stayed in Britain, were helpmates and computers. In the nineteenth century, Marie Rümker worked with her husband K. L. C. Rümker at the Hamburg Observatory, while Annie Jacoby assisted her husband Harold Jacoby at Columbia University. Preeminent among the computers was the Scotswoman Williamina Fleming (1859–1911) who emigrated to Massachusetts, became a maid in the home of Edward C. Pickering, and eventually rose to be a mainstay of the corps of computers at the Harvard College Observatory. Rümker, Jacoby, and Fleming apparently had no thoughts of astronomy when they emigrated; others were more professionally minded. Alice Everett went to Vassar College in 1898 to assist in astronomy and later worked for a time in the Potsdam Observatory. Early in this century, Frances Lowater (1863–1957) and Gertrude Longbottom (1876–1935) did graduate work at Bryn Mawr College in Philadelphia; Lowater obtained her doctorate. Both women eventually returned to England and worked as assistants.[36]

The most distinguished of these emigrants, and one whose career exemplifies the difficulties women astronomers still faced in Britain in the 1920s, was Cecilia Helena Payne (later Gaposchkin). Payne (1900–1979) took full advantage of the opportunities secured for women in the nineteenth century. She studied physics and astronomy at Cambridge, joined the BAA, and was elected to the RAS. In 1923 Payne left England to do postgraduate research at the Harvard College Observatory. During her second year at Harvard, she completed her doctoral dissertation, a monograph on the temperature and composition of stellar atmospheres. The book was widely praised in both England and the United States, and Payne hoped that her work might earn her a research position at home. She gave a paper at the September 1925 meeting of the British Association "from motives of pure advertisement," and she was warmly received by such well-established figures as Frank Dyson, Horace Lamb, H. H. Turner and Oliver Lodge.[37]

[35] On Maud O. Saltmarsh, see the *Calendar of the University of London,* 1922/1923–1939/1940; see also the *List of Members of the Institute of Physics.* On Dorothy Wrinch, see Marjorie Senechal, "A Prophet without Honor: Dorothy Wrinch, Scientist, 1894–1976," *The Smith Alumni Quarterly,* April 1977, pp. 18–23. On Bertha Swirles, see "Bertha Jeffreys," *Who's Who in British Science* (London: Leonard Hill, 1953), p. 148; I thank Karl Hufbauer for this reference.

[36] On Williamina Fleming, see obituaries by Annie Jump Cannon in *Astrophysical Journal,* 1911, *34*:314–317; and in *Science,* 1911, *33*:987–988. On Gertrude Longbottom, see the obituary in *Mon. Not. Roy. Astron. Soc.,* 1936, *96*:295.

[37] On Payne's career, see Payne-Gaposchkin, *Cecilia Payne-Gaposchkin* (cit. n. 7); on her 1925 trip to England, see C. H. Payne to Margaret Harwood, 1 Sept. 1925, Margaret Harwood Papers, Schlesinger Library, Radcliffe College, Cambridge, Mass. (quoted by permission of the Schlesinger Library).

British scientists, whatever their good will, found no place for Payne in their midst. She won a National Research Council Fellowship, returned to America, and began a long and distinguished career at Harvard. By 1930, women could and did participate in most British astronomical institutions. Nonetheless, for a woman passionately devoted to her own research, the path to eminence led out of England.

"A Lab of One's Own"

The Balfour Biological Laboratory for Women at Cambridge University, 1884–1914

By Marsha L. Richmond

ABSTRACT

The Balfour Biological Laboratory for Women was established at Cambridge University in 1884 to prepare the students of Newnham and Girton Colleges to sit the Natural Sciences Tripos, first opened to women in 1881. For thirty years, until its closure in 1914, the Balfour Laboratory served as the central conduit for biological instruction for the women of Cambridge, introducing them to the new program of experimental biology developed by the physiologist Michael Foster and the embryologist Francis Maitland Balfour. Directed by distinguished women graduates, the Balfour Laboratory became recognized as the leading center for women's biological instruction in Britain. Its significance, however, extends beyond its nominal status as a teaching laboratory. It provided university positions for able scientists who otherwise would not have been placed, offered advanced students the opportunity to engage in independent research, and, most important, formed the locus for the scientific subculture created by women at Cambridge to compensate for their exclusion from the social community of science. Drawing upon college and university archival records, this essay offers institutional, social, and biographical research that broadens our understanding of the experience of the first generation of women to pursue a higher education in the life sciences at one of the world's premier universities.

Support for this project came in part from a summer grant from Wayne State University. I thank the Department of History and the Women's Studies Program at Oklahoma State University, along with Elizabeth Williams and Robert Mayer, for giving me the opportunity to present an earlier version of this essay. I am indebted for research assistance to Elisabeth van Houts, former archivist, and Carola Hicks, archivist, of Newnham College; Joan Bullock-Anderson, deputizing archivist, and Kate Perry, archivist, of Girton College; Elizabeth Leedham-Green, archivist of Cambridge University; and Perry O'Donovan, Darwin Correspondence Project. Anna McClean Bidder kindly granted me an interview concerning her career in zoology and that of her mother, Marion Greenwood Bidder. Discussions with Kostas Gavroglu, Anne Secord, Soraya de Chadarevian, Joan Mason, and Joe Lunn greatly assisted my understanding of the scientific subculture of Cambridge women. I also thank Margaret Rossiter, Joy Harvey, and several anonymous referees for their helpful suggestions about how to revise "an impossibly long" article.

T HE OPENING OF HIGHER EDUCATION IN THE SCIENCES to women after 1870
marks a major watershed in the history of women in science.[1] Only when women were
given access to a university education in the sciences—to the laboratory as well as to the
lecture room—could they become an increasingly integral part of the scientific community.

In Great Britain, the movement for the higher education of women achieved its first
success in 1869, when the Syndics of Cambridge University (one of the governing committees appointed by the university senate) granted girls the right to sit the Higher Local
Examinations, which served as university entrance examinations and were previously open
only to boys. This "educational experiment" led to the establishment of the first British
women's colleges: Girton, which opened in Hitchin in 1869 and moved to Cambridge in
1871, and Newnham, which opened in 1871 and removed to its present site in 1875. At
Oxford, the first colleges for women, Somerville and Lady Margaret Hall, were founded
in 1879.[2]

The opening of the women's colleges, although a major victory, was but the first step
in a long process of gaining for women the same educational advantages open to university
men. The higher education of women reached another milestone in 1880, when Cambridge
granted women the right to sit the university honors examinations, the tripos, generally
taken by students at the end of three years of study. But it would be 1948 (1920 at Oxford)
before Cambridge admitted women to university membership and conferred on them degrees. Several earlier proposals to grant women degrees (1887, 1897, and 1921) ended in
failure and polarized the university with respect to the "woman question." It is against this
backdrop of almost eighty years of struggle for full equality within Cambridge University
that the entry of women into academic science must be set.

Although there exists a burgeoning biographical literature on early university-educated
women in science, much less is known about the social milieu in which they worked,
particularly the subculture they fashioned to support their work in science. In this regard,
the archives of the women's colleges at Cambridge are treasure-houses, offering, in the
records pertaining to the Balfour Biological Laboratory for Women established by Newnham College in 1884, a glimpse of the social nexus created by early Cambridge women
in science.[3] For thirty years, from 1884 until its closure in 1914, the Balfour Laboratory

---

[1] See Margaret W. Rossiter, *Women Scientists in America: Struggles and Strategies to 1940* (Baltimore: Johns
Hopkins Univ. Press, 1982). Although no comparable study exists for Great Britain, see Carol Dyhouse, *No
Distinction of Sex? Women in British Universities, 1870–1939* (London: UCL Press, 1995); and Elizabeth Sey-
mour Eschbach, *The Higher Education of Women in England and America, 1865–1920* (New York: Garland,
1993).

[2] See Barbara Stephen, *Emily Davies and Girton College* (London: Constable, 1927); M. C. Bradbrook, *"That
Infidel Place": A Short History of Girton College, 1869–1969*, rev. ed. (Cambridge: Girton College, 1984);
Blanche Athena Clough, *Memoir of Anne Jemima Clough* (New York/London: Arnold, 1897); Alice Gardner,
A Short History of Newnham College (Cambridge: Newnham College, 1921); Mary Agnes Hamilton, *Newnham:
An Informal Biography* (London: Faber & Faber, 1936); Annie M. A. H. Rogers, *Degrees by Degrees: The Story
of the Admission of Oxford Women Students to Membership of the University* (Oxford: Oxford Univ. Press,
1938); Vera Brittain, *The Women of Oxford: A Fragment of History* (London: Harrap, 1960); and Pauline Adams,
Somerville for Women: An Oxford College, 1879–1993 (Oxford: Oxford Univ. Press, 1996).

[3] Primary material on the Balfour Laboratory is in the Newnham College Archive (hereafter **NCA**) in the
following dossiers: Newnham College Report (**NCR**); Balfour Laboratory: Papers and Correspondence (**BL:
P&C**); Balfour Laboratory Reports, 1884–1919 (**BLR**); Account Ledger Sheet; and Bathurst Fund, 1879–1919.
The Girton College Archive (**GCA**) holds relevant material in the Minute Books, Executive Committee, Vols.
6 (1879–1881) and 7 (1881–1883), and in the Gamble Prize Records. Published sources on Cambridge women
in science include Roy MacLeod and Russell Moseley, "Fathers and Daughters: Reflections on Women, Science,
and Victorian Cambridge," *History of Education*, 1978, 8:321–333; Rita McWilliams-Tullberg, *Women at Cambridge: A Men's University—Though of a Mixed Type* (London: Gollancz, 1975); McWilliams-Tullberg, "Women
and Degrees at Cambridge University, 1862–1897," in *A Widening Sphere: Changing Roles of Victorian Women,*

served as the central conduit for biological instruction and the locus of the scientific subculture for women at Cambridge.

Cambridge, as Mary Creese has noted, "occupied a special position in the education of the first generation of British women scientists, producing many of the most distinguished." This was no accident. Although by 1895 all universities in Britain, with the notable exception of Cambridge and Oxford, awarded degrees to women, "in the minds of most English people . . . a university education meant matriculation through Oxford or Cambridge, not a new university, or a provincial university, or an examining body such as the University of London." In the life sciences Cambridge was unexcelled, attracting the best and the brightest of those seeking to read the natural sciences. The men and women who came up to Cambridge in the 1870s were attracted by the new program in biological instruction that emerged, as Gerald Geison has shown, under the able leadership of Michael Foster, recently appointed prelector in physiology at Trinity College, Cambridge, and his protégé, Francis Maitland Balfour, a recent fellow of Trinity. The two established the first university biological laboratories in England, in which students were given firsthand experience with investigating the form and function of living organisms.[4]

The Balfour Laboratory for Women was one of the first such university teaching laboratories. As a designated women's laboratory in a mixed-sex university, it provides an interesting contrast to the situation in the United States, where women often received education in the life sciences in single-sex colleges, as Toby Appel has recently described for physiology.[5] At Cambridge women benefited from the teaching of some of Britain's most talented biologists. Moreover, physiology there was always understood as a biomedical science, not in terms of personal hygiene.

As Carol Dyhouse has recently noted, "there has been little research into the development of British universities as social institutions, let alone into patterns of gender differentiation within their walls" or the "gender-based subcultures" developed by university women. This is particularly true in science. Cambridge women had to create their own scientific subculture—lectures, laboratories, coaching, scientific clubs, research fellowships, and lectureships—as a prerequisite to engaging in scientific study and research. The barriers women faced in science are well known. Since women were not formally students at Cambridge until 1948, they were unable to participate fully in all aspects of the scientific community. Thus, reaping the benefits of pedagogical innovations and dynamic research programs was not always straightforward for women. The story of the creation and eventual discontinuation of the Balfour Laboratory exposes many of the barriers that nineteenth-

---

ed. Martha Vicinus (Bloomington: Indiana Univ. Press, 1977), pp. 117–145; and Christopher N. L. Brooke, *A History of the University of Cambridge*, Vol. 4: *1870–1990* (Cambridge: Cambridge Univ. Press, 1991), Ch. 6. However, none of these mentions the Balfour Biological Laboratory for Women. But see Paula Gould, "Women and the Culture of University Physics in Late Nineteenth-Century Cambridge," *British Journal for the History of Science*, 1997, *30:*127–149, which discusses women within the laboratory culture of Cambridge.

[4] Mary R. S. Creese, "British Women of the Nineteenth and Early Twentieth Centuries Who Contributed to Research in the Chemical Sciences," *Brit. J. Hist. Sci.,* 1991, *24:*275–305, on p. 299; and Eschbach, *Higher Education of Women* (cit. n. 1), p. 124. Thomas Henry Huxley and William Thiselton-Dyer spoke of Cambridge's superiority in the natural sciences in 1880. See Joseph Lester, *E. Ray Lankester and the Making of Modern British Biology* (Oxford: British Society for the History of Science, 1995), pp. 55–56. As the zoologist Anna McClean Bidder explained concerning her decision in 1922 to go up to Cambridge from University College London, for those reading biology Cambridge was still *the* university (interview with Marsha L. Richmond, July 1993). On the establishment of the first university biological laboratories in Britain see Gerald Geison, *Michael Foster and the Cambridge School of Physiology* (Princeton, N.J.: Princeton Univ. Press, 1978).

[5] Toby A. Appel, "Physiology in American Women's Colleges: The Rise and Decline of a Female Subculture," *Isis,* 1994, *85:*26–56.

century women had to surmount in order to pursue a higher education in the life sciences.[6] When we examine this laboratory's background—its creation, organization, direction, teaching, and research activities—a rich and complex picture begins to emerge of the institutional, social, and intellectual life of the first generations of women to pursue university-level studies in biology at one of the world's leading scientific establishments.

This story also carries a historiographic message. If it is accepted that for many years women were excluded from full participation in professional science—from gaining degrees, full-time paid positions, and membership in scientific societies—perhaps then historians need to revise the means by which the scientific accomplishments of women have traditionally been measured. By ignoring the full range of experiences of women in science, we risk marginalizing them even further. Exploring "female subcultures" in science—investigating the social systems women created apart from as well as in connection with the male-dominated scientific community—sheds light on women whose scientific work was serious but comparatively modest. It also broadens our understanding of how women in the late nineteenth and early twentieth centuries accommodated themselves to the mainstream culture of science in order to pursue biological study and research.

BIOLOGICAL INSTRUCTION AT CAMBRIDGE CIRCA 1880

After their founding in the early 1870s, Girton and Newnham colleges assisted the few women who opted to read the natural sciences by obtaining permission for them to attend the lectures of those university instructors who were sympathetic to the higher education of women. Among these were Michael Foster, lecturer in physiology, and Frank Balfour, lecturer in embryology and animal morphology. Foster introduced the first university course in elementary biology to Cambridge in 1873. It was modeled on the course that his mentor, Thomas Henry Huxley, had offered the two previous summers at South Kensington to school science teachers, for which Foster served as demonstrator. As Gerald Geison has noted, "perhaps no one, not even Huxley himself, did so much to make elementary biology a systematic and integral part of his institution's biological curriculum."[7]

This one-term course in elementary biology was intended to serve as the gateway to the study of anatomy and physiology at Cambridge. As advertised in the *Cambridge University Reporter,* the course entailed a short half-hour lecture "followed by practical work for about 1 1/2 or 2 hrs." that included "instruction in the use of the microscope and the art of dissection." In 1875, Foster's lectures were complemented by those of Balfour, a recent graduate who had taken a first class in the 1874 Natural Sciences Tripos. Balfour (see Figure 1) followed Foster's lead in 1875 by offering the first university course in animal morphology, in which he approached questions relating to animal form from an evolutionary perspective that was in stark contrast to the traditional systematic lectures given by the professor of zoology and comparative anatomy, Alfred Newton. The new program in experimental biology introduced by Foster and Balfour thus revolutionized the teaching

---

[6] Dyhouse, *No Distinction of Sex?* (cit. n. 1), pp. 5–7. The "women's subculture" that I discuss is therefore necessarily different from that described by Appel in "Physiology in American Women's Colleges," which was based on a disciplinary and instructional organization that was absent in Cambridge. Nonetheless, there are similarities in the formation at Cambridge of a "female network" among the students of biology and in the mentoring of students by staff members.

[7] Geison, *Foster and the Cambridge School of Physiology* (cit. n. 4), pp. 117–118. See also Graeme Gooday, " 'Nature' in the Laboratory: Domestication and Discipline with the Microscope in Victorian Life Science," *Brit. J. Hist. Sci.,* 1991, 24:307–341.

Figure 1. *Portrait of Francis Maitland Balfour, executed on stone from a photograph after his death by E. Wilson of the Cambridge Scientific Instrument Company. (Reproduced from the frontispiece to Volume 1 of* The Works of Francis Maitland Balfour, *edited by Michael Foster and Adam Sedgwick [London: Macmillan, 1885].)*

of biology at Cambridge, marking "the beginning of a new epoch in the teaching of biology in the English universities."[8]

This much has long been known. What has not hitherto been noted, however, is that men were not the only beneficiaries of this revolution in biological instruction. So too were the growing numbers of natural science students of Girton and Newnham colleges. Both Foster and Balfour were supporters of the higher education of women. Each served as an advisor to the women's colleges, Foster primarily to Girton and Balfour to Newnham, to which he was connected by both familial and collegial ties. His sister Eleanor Mildred Sidgwick, later principal of Newnham College, was married to Henry Sidgwick, a founder

---

[8] *Cambridge University Reporter,* 1873, p. 19, cited in Geison, *Foster and the Cambridge School of Physiology,* p. 117—where the remark about the "new epoch" also appears. On Balfour's Cambridge career see Michael Foster, "Introduction," in *The Works of Francis Maitland Balfour,* 4 vols. (London: Macmillan, 1885), Vol. 1, pp. 1–24; Mark Ridley, "Embryology and Classical Zoology in Great Britain," in *A History of Embryology,* ed. T. J. Horder, J. A. Witkowski, and C. C. Wylie (Cambridge: Cambridge Univ. Press, 1985), pp. 35–67; Frederick B. Churchill, "Francis Maitland Balfour," in *Dictionary of Scientific Biography,* ed. Charles C. Gillispie, 16 vols. (New York: Scribner's 1970–1980), Vol. 1, pp. 420–422; Brian K. Hall, "Francis Maitland Balfour," in *New Dictionary of National Biography* (Oxford: Oxford Univ. Press, in preparation); and Geison, *Foster and the Cambridge School of Physiology,* pp. 124–130. Alfred Newton's lectures are in the Newton Collection, Manuscript Room, Cambridge University Library (hereafter **MSS Room, CUL**).

of Newnham and a former fellow of Trinity College. By admitting women students to their courses and the practical classes that accompanied them, Foster and Balfour provided them the same opportunity as men to benefit from the first laboratory-based university biology classes offered in Britain.[9]

As advertised in the 1882 *Girton Review,* "the student will find that advantages in scientific work are enjoyed in Cambridge which could hardly be obtained elsewhere. It is probably true that nowhere in England can Physiology be studied so profitably as in Cambridge; certainly no course of lectures is more justly celebrated than Dr. Michael Foster's elementary Biology and Physiology." One of the women who attended lectures of Foster, Balfour, and the professor of anatomy George Murray Humphry between October 1875 and December 1878 later recalled the experience in a letter published in the *British Medical Journal:*

> Michael Foster allowed us women to sit up in a gallery overlooking his big lecture room, full of men, among whom were not a few greyheads [i.e., older academics]. He did indeed draw on the blackboard in many coloured chalks—then leant his back against it and transferred the colours to his lecture gown! May I add a word about Mr. Frank Balfour? I attended his lectures on embryology one May term, in a tiny room, where men and women were squeezed together . . . and the demonstrator, Mr. Adam Sedgwick, sat on a table dangling his legs, and the lecturer towered above us, in height, as in everything else. I remember specially his long, delicate hands and beautiful manipulation of sections, but also his marvellous and stimulating teaching, not hindered by hesitating speech, and his power of making his pupil feel—oh, *so* small! To this very day the teaching of all three has been an inspiration and a guide.[10]

Owing both to the growth of the Cambridge medical school and to the great demand for teachers in zoology in the secondary schools, by the early 1880s Foster's and Balfour's classes were overflowing with students. This created a general crisis within the university, prompting a debate over how to provide lecture and laboratory facilities for the rapidly expanding experimental sciences in an ancient university with a decentralized collegial system.[11] Under the circumstances, it was urgent for the women's colleges to provide their

---

[9] "During the 1880s the numbers of Cambridge women reading for the NST grew steadily, reaching an average of 14 each year in the 1890s": MacLeod and Moseley, "Fathers and Daughters" (cit. n. 3), p. 326. At Oxford, the fledgling School of Animal Morphology, under the direction since 1881 of Henry Nottidge Mosely, was closed to women. In 1888, when Mosely fell ill and was temporarily replaced by Halchett Jackson, two women were permitted to attend his lectures and laboratory demonstrations. See Adams, *Somerville for Women* (cit. n. 2), p. 41; and Lester, *Lankester and the Making of Modern British Biology* (cit. n. 4), pp. 115–116.

[10] *Girton Review,* Dec. 1882, p. 6; and "Science Teaching at Cambridge," *British Medical Journal,* 9 Oct. 1920, p. 572 (I thank Mark Weatherall for bringing this letter to my attention). The date range is suggested by Arthur E. Shipley, *"J.": A Memoir of John Willis Clark* (London: Smith, Elder, 1913), p. 277, who stated that Balfour's 1878 course was attended by "two or three ladies," and by the fact that Adam Sedgwick first served as Balfour's demonstrator in 1878. The author may have been Margaret Emily Pope (d. 1924), who took a second class in the 1878 Natural Sciences Tripos and taught at a number of girls' schools before going out to India as a missionary in 1898. See *Girton College Register, 1868–1946* (Cambridge: Girton College, 1948), p. 7.

[11] "In 1870 Cambridge had a small Medical School; by 1883–4 the intake of 90 was second only to St. Bartholomew's among the medical schools in Britain; from then it remained among the largest in the land": Brooke, *History of the University of Cambridge,* Vol. 4 (cit. n. 3), p. 166. See also Humphry Davy Rolleston, *The Cambridge Medical School: A Biographical History* (Cambridge: Cambridge Univ. Press, 1932), pp. 25–30. In the Michaelmas term of 1881, seventy students were attending lectures in elementary physiology and fifteen in advanced physiology; fifty-five were attending lectures in elementary morphology and twenty in advanced morphology. These students were fitted into rooms designed to hold half that number. Foster and Balfour thus petitioned the university "to take steps to increase our accommodation": Cambridge University Register (hereafter **CUR**) 39.36: Professor of Animal Morphology, Registry, Cambridge University Archives, MSS Room, CUL. See Shipley, *"J."* (cit. n. 10), pp. 281, 307; and Brooke, *History of the University of Cambridge,* Vol. 4, Ch. 6.

own facilities for students requiring practical instruction in biology in preparation for the Natural Sciences Tripos.

THE CREATION OF THE BALFOUR BIOLOGICAL LABORATORY FOR WOMEN

On 24 February 1881, the women's colleges at Cambridge achieved a major victory when the university senate passed three graces (amendments to the university charter) giving women the right to sit for the university honors degree examinations, the tripos, including the Natural Sciences Tripos (NST).[12] Eleven students of Girton and Newnham colleges had taken the NST since 1874, but on each occasion a personal application had to be submitted to the council of the senate. Now that women had secured the right to sit the honors examinations, it was especially incumbent on the women's colleges to meet the needs of students reading for the NST.

A central requirement of the reformed NST was the completion of a two-day examination testing a candidate's proficiency in practical laboratory techniques. This provision posed a dilemma for the women's colleges. Space limitations increasingly forced many instructors, even those sympathetic to the "cause," to reconsider allowing women, who were not officially members of the university, into classrooms that were already overflowing. Although Girton and Newnham both had college laboratories, these facilities were small and mainly outfitted for work in chemistry.[13] New laboratory facilities for biology were urgently required. Indeed, the very day after the university's decision was announced, both women's colleges began making plans to establish a biological laboratory for the women students at Cambridge.

From the beginning, the positions of the two colleges regarding the relationship of the proposed laboratory to the university were at variance, reflecting their philosophical differences over women's higher education. Emily Davies of Girton College, who believed that women should have the same rights and responsibilities as Cambridge undergraduates, wanted the university to establish and manage a separate laboratory for women, with the stipulation that the women's colleges "would provide an adequate contribution to the cost." The Girton College Executive Committee approached Michael Foster for his opinion concerning this proposal. Foster demurred from speaking for the university but expressed his personal view that there should be but one laboratory for all the women, that it should be located as near as possible to the university science buildings, and that it should be managed by the lecturers and demonstrators.[14] In May Girton formed a committee—comprising Foster, John Willis Clark (superintendent of the Museum of Zoology and Comparative

[12] John Roach, *Public Examinations in England, 1850–1900* (Cambridge: Cambridge Univ. Press, 1971), pp. 124–127.

[13] On the reformed NST see Roy MacLeod and Russell Moseley, "The 'Naturals' and Victorian Cambridge: Reflections on the Anatomy of an Elite, 1851–1914," *Oxford Review of Education,* 1980, 6:177–195. Ida Freund said of the first Girton laboratory that "no one could tell whether it was the post-office box, a safe, or a draught-cupboard": *Girton Review,* May 1909, p. 9. The Newnham laboratory was likewise inadequate, particularly for microscopical study: *Newnham College Report,* Dec. 1884, p. 15, NCA. "I still quiver with cold," one early student recalled, "as I remember those raw days in the laboratory barely tempered by a little grate fire in one corner": Mary Ann Willcox, "The Sidgwicks in Residence," in *A Newnham Anthology,* ed. Ann Phillips (Cambridge: Newnham College, 1979), pp. 13–16, on p. 14.

[14] Minute Books 3.7, Executive Committee, Vol. 6 (1879–1881), 25 Feb. 1881, GCA (Davies); and A. F. Bernard to Henry Sidgwick, 18 Mar. [1881], BL: P&C, NCA (Foster's view). The different approaches to women's education taken by Emily Davies, founder of Girton College, and Henry Sidgwick, prime founder of Newnham, were well known to contemporaries and have been fully described in the secondary literature (see note 2, above).

Anatomy), and Sydney Howard Vines (reader in botany)—charged with collecting funds for such a laboratory. However, the action of Newnham College soon made the task of this committee redundant.

True to the college's founding principles, the leaders of Newnham took a more pragmatic approach, deciding to move forward in providing the required women's laboratory rather than waiting on yet another protracted university debate over the question, especially since a legal challenge was pending over the passage of the recent graces. The students concurred. In March 1881, a group of twenty-two natural sciences students of Newnham presented a memorial to the governing council urging the college to take immediate action. Citing the hindrance they experienced owing to the cramped conditions in the current college laboratory, the want of light for undertaking microscopical work, and the time they lost in "walking from lecture to laboratory, and again back to lecture," they presented the college with over £200 they had raised.[15] Acceding to the students' concerns, in April the Newnham College council appointed a subcommittee consisting of Vice-Principal Eleanor Sidgwick as head and Principal Anne Jemima Clough, along with Frank Balfour and Coutts Trotter of Trinity College, another longtime supporter of women's education, "to carry on preliminary enquiries as regards the proposed laboratory."[16]

Eleanor Sidgwick's role in spearheading the Newnham drive for a women's laboratory was not simply a matter of nepotism. Prior to her marriage to Henry Sidgwick in 1876, she had herself been preparing to sit the Higher Local Examination in mathematics, and her tutor was of the opinion that "she would have been a high Wrangler, if she had read for the Tripos." Moreover, she was currently engaged in scientific research, having assisted the experimental work of her brother-in-law, John William Strutt, Lord Rayleigh, Cavendish Professor of Experimental Physics at Cambridge, since 1879. Their work on the determination of standards for electrical units resulted in several joint publications that were highly regarded. Rayleigh's successor, J. J. Thomson, thought highly of Sidgwick's research work.[17]

On 13 April 1881 Eleanor Sidgwick sent a letter to Emily Davies informing Girton of developments concerning the new laboratory. A "very favourable opportunity" had arisen when the subcommittee located an abandoned Congregational chapel in Downing Place, near the New Museums site, that appeared suitable for their purposes. "We should like therefore to know," she wrote, "whether in the event of the university refusing the application that you propose, Girton would still be willing to co-operate with us—say on some terms as these:—that the laboratory should be conveyed to trustees and managed by a

---

[15] Student Petition to the Council of Newnham College, n.d. [Mar. 1881], NCA. The loss of time mentioned by the students was not an inconsiderable issue. Even though Newnham was far closer to the university science buildings (the New Museums site) than Girton, located three miles from the university, it was still a long way from the university lecture rooms to the college laboratory, especially before the introduction of the bicycle in the 1890s.

[16] Eleanor Sidgwick to Caroline Croom Robertson (secretary of Girton College), 12 Apr. 1881, BL: P&C, NCA. A member of the Association for Promoting the Higher Education of Women in Cambridge, the Trinity physicist Coutts Trotter served as vice-president of the Council of Newnham College and was particularly interested in promoting laboratory-based science instruction. See Michael Foster, John Willis Clark, and Sedley Taylor, *Coutts Trotter: In Memorium* (Cambridge: Macmillan & Bowes, 1888).

[17] Ethel Sidgwick, *Mrs. Henry Sidgwick: A Memoir by Her Niece* (London: Sidgwick & Jackson, 1938), p. 66. Eleanor Sidgwick once told a friend that "mathematics especially appealed to her in early youth because she thought a future life would be much more worth living if it included intellectual pursuits" (p. 66 n 1). Lord Rayleigh and Eleanor Sidgwick's studies were published in the *Philosophical Transactions of the Royal Society of London*, 1882, *174:*173–185, 1884, *175:*411–460. See Ethel Sidgwick, *Mrs. Henry Sidgwick*, pp. 71–73; and Helen Fowler, "Eleanor Mildred Sidgwick, 1845–1936," in *Cambridge Women: Twelve Portraits,* ed. Edward Shils and Carmen Blacker (Cambridge: Cambridge Univ. Press, 1996), pp. 7–28. Thomson's opinion was reported by his student, Ernest Rutherford. See Margaret Rossiter, *Women Scientists in America: Struggles and Strategies to 1940* (cit. n. 1), p. 324 n 34.

Committee consisting e.g. of some or all of the teachers of the respective branches of science and the principals of the two Colleges." Should Girton be disinclined to collaborate at present, she noted, Newnham planned to "proceed independently in the negociation for the site." In May, Sidgwick initiated legal proceedings for the purchase of the building.[18]

After a meeting early in May between the committees of Girton and Newnham, the different positions of the two colleges became more entrenched. The fears of the Newnham College committee about pressing the university for further concessions to women's education in the wake of the recent victory were well expressed in a 11 May letter from committee member Coutts Trotter to Emily Davies. Reiterating the major points of their conversation of the previous day, Trotter stated that "we probably attached much less intrinsic importance to the question of formal University management than the Girton executive" and further admitted "that some of us might, as members of the University, feel some doubt as to whether it would be wise for the University to accept the trust at present." Indeed, given its current outlay for laboratories, it was highly unlikely that the university could accept final responsibility for any shortfalls the proposed laboratory might incur.[19]

Trotter's concern about financial reservations the university might have in acceding to the women's colleges' request for a university-managed women's laboratory was not a specious argument, given Foster's and Balfour's recent requests for increased university funding for biological facilities. With the science instructors petitioning for a substantial augmentation of the university funds allocated for their subjects, it was indeed highly unlikely that monies would be forthcoming to provide facilities for those who were not officially students of the university.[20]

Despite such reservations, the Girton College Executive Committee continued to maintain its original position. Disheartened by this intransigence, Newnham moved forward on its own, informing Girton, however, that the new laboratory would be made "as fully available for the use of Girton Students as if a more complete combination had been effected" and that Newnham would keep open the possibility of transferring control to the university should the opportunity arise. Over the course of the next three years, Newnham College raised more than £2,000 for the new laboratory, with considerable sums coming from Eleanor Sidgwick herself and her sister, Alice Blanche Balfour. Girton College also contributed toward equipping the laboratory.[21]

---

[18] Minute Books 3.7, Executive Committee, Vol. 7 (1881–1883), 13 May 1881, GCA; and correspondence between Eleanor Sidgwick and Mr. Bond, 1881, BL: P&C, NCA. The final sale of the building, however, went through only in 1883.

[19] Coutts Trotter to Emily Davies, 11 May 1881, Minute Books, 13 May 1881, GCA.

[20] The "crisis" in accommodating the growing numbers of students in the natural sciences is a frequent topic in the *Cambridge University Reporter* in the 1880s. See also the numerous petitions from the life sciences instructors in CUR 39.36 and CUR New Museums, Vol. 2, MSS Room, CUL.

[21] Girton College Executive Committee meetings, 13 May 1881, 2 June 1881; and letter from M. G. Kennedy (secretary of Newnham College), 16 May 1881, Minute Books, 27 May 1881, GCA. The building itself cost £1,400: Account Ledger Sheet, Balfour Laboratory, NCA. Subscriptions for the new laboratory came from a number of sources, including £50 from the journalist and amateur physiologist George Henry Lewes, husband of Marianne Cross (George Eliot), who had already donated £200 for books for the Newnham library; see Ethel Sidgwick, *Mrs. Henry Sidgwick* (cit. n. 17), pp. 88–89. Alice Balfour, who had assisted Frank Balfour's scientific work by making drawings of his specimens, contributed £710, and Miss S. Bathurst £200. Although no exact figures are given in the Balfour Laboratory ledger as to how much Eleanor Sidgwick contributed, it was reported in the *Newnham College Club: Cambridge Letter* for 1884 that "Newnham College was enabled by a generous donation from Mrs. Sidgwick to purchase [the old chapel] towards the end of last year (1883)" (p. 9). Edith Rebecca Saunders, last director of the Balfour Laboratory, reported in a brief retrospective that "the Building and site were presented to the College by members of the Balfour family (?Mrs Sidgwick, Miss Alice Balfour)

The contributions of the Balfour family to the new women's laboratory, as it turned out, were tinged by personal tragedy as well as by generosity. Francis Maitland Balfour, elected professor of animal morphology in May 1882, was killed two months later in a climbing accident in Switzerland, while negotiations for the new laboratory were in progress. His loss was a terrible blow to Cambridge and indeed to the entire scientific establishment of Britain, but it was especially devastating for the scientific women of Cambridge, whom he had fostered.[22] To honor his memory, the new women's laboratory was christened the Balfour Biological Laboratory for Women at Cambridge.

Renovations and outfitting the new laboratory took the better part of 1883 and early 1884. In part, the delay came from ensuring that the laboratory was up to the standards of modern experimental biology. Again, the reformist university science dons served the women's colleges well. Coutts Trotter, for example, offered to refloor the lab to "make it better & healthier" and also donated his collection of valuable physical apparatus. Adam Sedgwick, appointed lecturer in animal morphology upon Balfour's death, gave advice concerning the facilities for housing marine organisms, noting that "sinks are more valuable than basins because such creatures as water mussels, crayfish &c can be kept alive in a sink." Thanks to the lecturer in physiology, Walter Holbrook Gaskell, the laboratory was outfitted with the new water-driven "cutting apparatus" for cytological work—the automatic Caldwell microtome recently designed by two Gonville and Caius College zoology students who subsequently founded the Cambridge Scientific Instrument Company. Still other dons contributed "several valuable Physiological instruments, including a spectroscope, galvanometer and pendulum myograph."[23]

When the new laboratory opened its doors for the first time in the spring of 1884, it was the pride of the Cambridge women's colleges (see Figure 2). Practical work in physiological chemistry and physics was not at first provided, but the second-floor gallery was well outfitted for biological studies. There, in a central location, at the request of the students of Newnham "who had had the privilege of attending his lectures," was placed a copy in marble of the bronze bust of Frank Balfour by Adolf von Hildebrand. It would serve as a reminder of both the vibrant program of experimental biology Balfour fostered

---

in memory of Professor F. M. Balfour": [History of the Balfour Laboratory, 1914], BLR, NCA. In July 1883, for example, Girton purchased "six new and specially good Student's Microscopes by Zeiss of Jena" for natural sciences students: *Girton Review,* July 1883, p. 14.

[22] See Geison, *Foster and the Cambridge School of Physiology* (cit. n. 4), pp. 124–127. Balfour's family donated both his "very valuable scientific outfit" and his zoological library to the University Morphological Laboratory he had directed; see Shipley, "*J.*" (cit. n. 10), p. 282. His worktable, however, was donated to the new Balfour Laboratory.

[23] *Newnham College Club: Cambridge Letter,* 1884, p. 9. Coutts Trotter's brother William donated a cupboard with glass doors, and Trinity College also contributed several cupboards; see Walter Holbrook Gaskell to Eleanor Sidgwick, 2 Dec. 1883, BL: P&C; and BLR, NCA. John Newport Langley also donated funds toward purchasing equipment; see Account Ledger Sheet, Balfour Laboratory, NCA. The professor of chemistry, George D. Liveing, a longtime supporter of women's higher education at Cambridge, donated the spectroscope; see Bathurst Fund, 1883, NCA.

The microtome was a tremendous aid to experimental biology, greatly reducing the time it took to prepare a continuous series of sections of an organism. Arthur Shipley recalled, for example, that in the practical zoology class he took at Cambridge in 1880, "we used to spend our time cutting Amphioxus, embedded in cocoa-butter, into little slices under the supervision of Mr. Balfour and of Mr. Adam Sedgwick. There were then no means of sticking the sections in series on a slide, so the thin slices had to be mounted, a few at a time, on a separate microscopic slide, and the labour was considerable": Shipley, "*J.*" (cit. n. 10), pp. 280, 282. See also "Caldwell's Automatic Microtome," *Quarterly Journal of Microscopical Science,* N.S., 1884, *24*:648–654.

Figure 2. *Exterior view of the Balfour Biological Laboratory for Women, Downing Street, Cambridge. (Contemporary photograph taken by Marsha Richmond.) Formerly a Congregational chapel, the building served as the women's biological laboratory from 1884 to 1914. Newnham College leased the building to the Beit Memorial Fellowships for Medical Research in 1919 and sold it in 1928. For several years it was used by Frederick Gowland Hopkins's Department of Biochemistry and then by the Department of Geography. Today the building has reverted to religious use, serving as a church hall.*

at Cambridge and his ongoing support for women's education in the sciences (see Figure 3).[24]

The description of the new laboratory carried in the 1884 Newnham College magazine provides a vivid picture of the internal layout of the new facility:

---

[24] Clough, *Memoir of Anne Jemima Clough* (cit. n. 2), pp. 312–313. Hildebrand's original was at Balfour's University Morphological Laboratory. Facilities for instruction in chemical physiology were added in 1888, for physics in 1891, and for geology (crystallography) in 1906. The reasons why physics was not immediately included in the plan of the new laboratory were threefold: few women were studying physics at the time; those who did were welcome in the practical classes of the Cavendish Laboratory; and since practical work in physics and chemistry was not stressed in the scientific paper of the Higher Local Examination, the need for laboratory instruction was not as pressing in these areas as in biology.

Figure 3. *Interior views of the Balfour Biological Laboratory for Women, showing the second-floor gallery, where demonstrations in morphology, physiology, and botany were conducted.* Top: *The bust of Francis Maitland Balfour, placed in a prominent position overlooking the students' worktables, on which can be found a representative selection of the laboratory's equipment and instruments.* Bottom: *The window tables, at which microscopical work was carried on, and the tables for morphological and physiological work. (Reproduced by permission of the Principal and Fellows of Newnham College, Cambridge.)*

The gallery round three sides of the laboratory is now chiefly used for microscopical work, the windows having been enlarged to increase the amount of light. The furniture consists of tables fixed against the wall running nearly all round the gallery, with all the necessary fittings of gas, water, shelves, &c., a row of tables standing further back, and a line of cupboards fixed to the edge of the gallery all round. A demonstrator's room is partitioned off at one corner. The gallery is used for demonstrations in Comparative Anatomy and Physiology, to which the liberal allowance of light and space renders it admirably adapted. It is capable of accommodating in a luxurious manner eighteen students at once, and double that number could be taken in fairly comfortably.

Eight years later, after a further refurbishment, the principal of Newnham College could report that the Balfour Laboratory was "one of the best Laboratories in Cambridge after the University Laboratory."[25]

Having converted a former temple of God into a new temple of science,[26] Newnham College, along with Girton, now turned to preparing a new generation of women for pursuing biological studies at a university that, despite the loss of Balfour, was still the premier scientific institution in the country.

BIOLOGICAL INSTRUCTION AT THE BALFOUR LABORATORY

For the next thirty years, from 1884 until 1914, the Balfour Biological Laboratory formed the backbone of instruction in the life sciences for the women of Cambridge. The laboratory directors, according to the governing committee, were charged with providing instruction required by students in botany, zoology and comparative anatomy, physiology, and physics, "either by obtaining for them admission to University or College lectures and demonstrations, or by providing lectures and demonstrations themselves." This aim was accomplished through the appointment of a series of demonstrators drawn from the pool of former students who had distinguished themselves in Part 2 of the new Natural Sciences Tripos and who were carrying on independent research (see Table 1).[27]

The initial direction of the laboratory was placed in the hands of a recent "graduate" in the natural sciences. Indeed, the college was in the enviable position of having to choose between two qualified candidates, who shared the distinction of having been the first Newnham students to take first-class honors in the new "divided" tripos. Florence Elizabeth Eves (d. 1911), a student of physiology, was holder of a Bathurst studentship in 1881–1882. She was one of the first beneficiaries of the scholarship fund established in 1879 by

[25] *Newnham College Club: Cambridge Letter,* 1884, pp. 8–9; and "Principal's Report," *Newnham College Letter,* 1892, p. 8. Another contemporary, however, was not so generous. The physiologist W. B. Hardy described the building as "a queer, ugly block of a building, once a chapel. How it came by so surprising a change I never heard. Some contraction in the spiritual life of Cambridge must have thrown it, a spiritual derelict, on to the market. At any rate it became the laboratory for women science students and, as Cambridge was still stirred by the genius and the tragic death of Francis Maitland the most brilliant of the Balfour brothers, it bore his name": W. B. Hardy, "Mrs. G. P. Bidder," *Nature,* 1932, *130:*689–690.

[26] The analogy is quite literal: the pews of the former chapel were used to build the partitions and cupboards in the new laboratory: *Newnham College Club: Cambridge Letter,* 1884, p. 9.

[27] For the founding statement see Report of the Laboratory Committee, 15 Mar. 1884, BL: P&C, NCA. The physiologist W. H. Gaskell replaced Frank Balfour on the committee. The NST was divided into two parts in February 1880. Part 1 was intended to "test general knowledge of up to four subjects," while Part 2, generally taken the following year, required the candidate "to demonstrate 'special proficiency in one or more of the eight subjects,' and (for a First) 'competent knowledge of some cognate subject.' " Most students sat only Part 1; only those aspiring to an academic or specialized career in the sciences generally sat the more specialized and demanding Part 2 examination. See Roy MacLeod and Russell Moseley, "Breaking the Circle of the Sciences: The Natural Sciences Tripos and the 'Examination Revolution,' " in *Days of Judgement: Science, Examinations, and the Organization of Knowledge in Late Victorian England,* ed. MacLeod (Driffield: Studies in Education, 1982), pp. 189–212, esp. p. 202. Although there were periods in which demonstrators who had completed only Part 1 of the NST were appointed at the Balfour, it was recognized that such appointments "could hardly be anything but tentative": Report of [May 1892], BLR, NCA.

Table 1. Balfour Laboratory Staff, 1884–1914

Years	Name	Position	NST Part1 (Class)	NST Part 2 (Class)
1884–1890	Alice Johnson (N)	Director	1881 (1)	
1884–1890		Demonstrator, Animal Morphology		
1884–1888	Marion Greenwood (G)	Demonstrator, Physiology and Botany	1882 (1)	1883 (1)
1888–1899		Demonstrator, Physiology		
1890–1899		Director		
1902–1903		Demonstrator, Physiology		
1886–1887	Anna Bateson (N)	Assistant, Botany	1884 (2)	1886 (2)
1886–1887	Lilian Sheldon (N)	Assistant, Botany	1883 (2)	1884 (2)
1893–1898		Demonstrator, Animal Morphology		
1888–1890	Edith Rebecca Saunders (N)	Demonstrator, Botany	1887 (2)	1888 (1)
1890–1914		Director		
1890–1892	Laura Russell Howell (G)	Demonstrator, Animal Morphology	1888 (1)	
1890–1891	Rachel Alcock (N)	Demonstrator, Animal Morphology	1889 (2)	1890 (3)
1898–1899		Demonstrator, Animal Morphology		
1903–1904		Demonstrator, Biology		
1891–1901	Helen Gertrude Klaasen (N)	Demonstrator, Physics	—	—
1892–1896	Agnes Isabella Mary Elliot (N)	Demonstrator, Vertebrate Morphology	1889 (1)	1891 (2) Zoology
1897–1899	Elizabeth Dale (G)	Assistant, Botany	1890 (1)	1891 (2)
1897–1898	Anne Purcell Sedgwick (G)	Assistant, Physiology	1892 (2)	1893 (2)
1898–1902	Elinor Gladys Philipps (N)	Demonstrator, Animal Morphology	1894 (1)	1895 (2)
1898–1900	Florence Margaret Durham (G)	Demonstrator, Animal Morphology	1891 (2)	1892 (2)
1900–1910		Demonstrator, Physiology		
1901–1902	Sibille Ormston Ford (N)	Assistant, Animal Morphology	1898 (2)	1899 (1) Botany and Zoology
1903–1904		Assistant, Botany		
1902–1912	Igerna Brünhild Johnson Sollas (N)	Demonstrator, Animal Morphology	1899 (1)	1901 (1) Zoology
		Lecturer, Animal Biology		
1894–1914	Gertrude Lilian Elles (N)	Demonstrator, Geology	1894 (2)	1895 (1) Geology
1907–1914	Muriel Wheldale (N)	Demonstrator, Physiological Botany	1902 (1)	1904 (1) Botany
1908–1909	Mary Gladys Sykes (G)	Assistant, Botany	1905 (1)	1906 (1) Botany
1909–1910		Demonstrator, Physiology		
1910–1911		Demonstrator, Vegetable Biology		
1910–1912	Susila Anita Bonnerjee (N)	Demonstrator, Physiology	1894 (2)	—
1911–1914	Agnes Robertson (N)	Demonstrator, Systematic Botany	1901 (1)	1902 (1) Botany

NOTE.—N = Newton College; G = Girton College.

SOURCES.—*Newnham College Register, 1871–1950*, Vol. 1: *1871–1923* (Cambridge: Newnham College, [1963]); *Girton College Register, 1869–1946* (Cambridge: Girton College, 1948); Balfour Laboratory Reports, 1884–1919; and "List of Classes held at the Balfour Laboratory, Jan 1910 to 1914," Newnham College Archives, Newnham College, Cambridge.

a benefactor of Newnham College "for the purpose of promoting the study of Natural Science among women," a grant awarded annually to women of both Newnham and Girton who had "passed the Natural Sciences Tripos with credit, and who wish to carry their Studies further, independently, but under the advice of the Cambridge Teachers."[28] Alice

[28] *Newnham College Report,* Feb. 1887, NCA. The fund was established by Miss S. Bathurst, who also contributed to the Balfour Laboratory fund. Eves, educated at the North London Collegiate School and University College London before coming up to Newnham in 1878, served as a demonstrator in chemistry for Newnham College from 1881 to 1887. She was "the first woman to obtain a degree in physiology," graduating in 1881. See E. M. Tansey, " 'To Dine with Ladies Smelling of Dog'? A Brief History of Women and the Physiological Society," in *Women Physiologists: An Anniversary Celebration of Their Contributions to British Physiology,* ed. Lynn Bindman, Alison Brading, and Tilli Tansey (London/Chapel Hill, N.C.: Portland, 1993). Between 1883 and 1894, Eves published two papers in the *Journal of Physiology,* the first with J. N. Langley, on internal secretions involving sugar and starch metabolism. In 1887 she was appointed assistant mistress at Manchester

Johnson (1860–1940), Bathurst student in 1882–1883, had studied animal morphology under Balfour and was pursuing independent embryological research under Sedgwick.[29]

In the event, it was Johnson rather than Eves who was chosen to head the new Balfour Laboratory.[30] As demonstrator in physiology the committee appointed a Girton student, Marion Greenwood (1862–1932), who had gained distinction by obtaining first-class honors in both Part 1 (1882) and the more difficult Part 2 (1883) of the NST. At the time of her appointment Greenwood was engaged in postgraduate research in Foster's University Physiological Laboratory. Indeed, her work was so highly thought of that Foster's demonstrator, John Newport Langley, believed that she should not accept the position at the Balfour, telling Eleanor Sidgwick that it "would probably be best for Miss Greenwood to have as much time as possible to continue the work she is doing now." However, Greenwood decided to take the post, trusting that it would "mean no lessening of research, but only a better arranging of the day's work."[31] In addition to conducting demonstrations in physiology, Greenwood was also asked to demonstrate in botany; she continued to do so until the college was able to establish a separate position in 1889, when Edith Rebecca Saunders was appointed.[32] (See cover illustration.)

In October 1884, before the appointment of the director, Eves and Johnson submitted a joint prospectus to the Council of Newnham College outlining their views as to how best

---

High School and then at St. Leonards School in St. Andrews. She later became a worker for social reform and was head of the Women's House of the Christian Socialist Union in Hoxton; see *Newnham College Letter,* 1911.

[29] Johnson, whose father was a schoolmaster in Cambridge, studied at private schools in Cambridge and Dover before coming up to Newnham in 1878. A Cambridge Senior Local Scholar in 1878 and College Scholar in 1880, she served as demonstrator in animal morphology at the Balfour Laboratory between 1884 and 1890. Between 1882 and 1886, she published several scientific papers in the *Proceedings of the Royal Society* (1884) and the *Quarterly Journal of Microscopical Science* (1883–1886). After resigning as director of the Balfour Laboratory, she served as private secretary to the principal, Eleanor Sidgwick, until 1903. Becoming closely involved, like the Sidgwicks, with the Society for Psychical Research, she held a number of offices, including editor, and also belonged to the Society on Census of Hallucinations; see *Newnham College Register, 1871–1950,* Vol. 1: *1871–1923* (Cambridge: Newnham College, [1963]), p. 60.

[30] Johnson's selection may have been influenced by the poor recommendation that Eves received from university staff members in physiological botany, including the reader in botany, Sydney Howard Vines, and his demonstrator, neither of whom thought that Eves was "up to teaching efficiently" nor had a "good opinion of her work": Francis Darwin to Eleanor Sidgwick, n.d. [1884], BL: P&C, NCA. Francis Darwin, lecturer in botany, had married the Newnham lecturer in history and English literature and council member Ellen Wordsworth Crofts in 1883.

[31] J. N. Langley to Eleanor Sidgwick, 13 June 1884; and Marion Greenwood to Eleanor Sidgwick, 2 July 1884, BL: P&C, NCA. Greenwood, who was born into a Nonconformist family in Hull, Yorkshire, and whose father was a shipping agent, came up to Girton in 1879 at the age of seventeen after earning the Brown Scholarship from the Girls' Grammar School at Bradford. Bathurst student in 1883–1884 and winner of the Gamble Prize in 1888, Greenwood was "one of the first women to pursue independent research in Cambridge": *Girton College Register, 1869–1946* (cit. n. 10), p. 636. See her obituaries in the *Times* (27 Sept. 1932); the *Girton Review,* 1833, pp. 3–5; and *Nature,* 1932, *130*:689–690.

[32] Report for 1888, BLR, NCA. Saunders, the daughter of a Brighton hotel keeper, was educated at Handsworth Ladies College before entering Newnham College. She took a second class in Part 1 of the NST (1887) and a first class in Part 2 (1888). A Bathurst student in 1888–1889, Saunders served as demonstrator in botany at the Balfour Laboratory from 1889 until 1899, then as director until it closed in 1914. She was director of science studies at Girton College (1904–1914) and at Newnham College (1918–1925). In 1895 Saunders began research in plant hybridism with William Bateson that gained her wide recognition in the new field of Mendelian genetics. Her later study of floral morphology, synthesized in a two-volume monograph (1937–1939), was also well known and influential. Among the first women to gain admission to London scientific societies, she was elected fellow of the Linnean Society of London in 1905, the Royal Horticultural Society (1925–1945; Banksian Medallist, 1906), the British Association for the Advancement of Science (president of section K [botany] in 1920), and the Genetical Society (president, 1936). See *Newnham College Register,* Vol. 1 (cit. n. 29), p. 7; and Mary R. S. Creese, "Edith Rebecca Saunders," *Dictionary of National Biography, Missing Persons,* ed. C. S Nicholls (Oxford: Oxford Univ. Press, 1993), pp. 584–585.

to organize the instruction at the Balfour Laboratory.[33] This document is insightful on several fronts: it provides a detailed picture of the aims of the new laboratory from the standpoint of its prospective scientific directors, reveals fundamental aspects of life in one of the first British university biology laboratories, and illuminates the status of women's higher education in the sciences in Britain in the mid 1880s.

One of Eves and Johnson's major recommendations was that the college should discourage students from coming up to Newnham before taking the elementary paper in Group E (the natural sciences paper) of the Higher Local Examination. This would encourage the girls' secondary schools to prepare their students better in the sciences. The focus of the college should be to provide university-level scientific instruction. From the beginning, then, the Balfour Laboratory aimed to offer the same caliber of work in the sciences for women as that provided to Cambridge undergraduates reading for the NST.

Recognizing that initially the college was nonetheless committed to preparing beginning students in the natural sciences, Eves and Johnson proposed that three courses of lectures be offered at the laboratory in the first year: a course on physics in the Michaelmas term, "with occasional experiments or demonstrations during the lecture, at the discretion of the lecturer; a course on chemistry, the nonmetals, in the Lent term taught by Pattison Muir"; and a course on biology in the May term, "three lectures a week with demonstrations, which would generally not last more than two hours, after each." Biology, they emphasized, even more than chemistry or physics, should be "studied experimentally." If students could not spend as much time as the prospectus suggested, they should "come *only to the* demonstrations, *not* only to the lectures." This stipulation reflected Eves and Johnson's belief that biology could not be taught "merely by means of books & lectures," for "not only is it difficult to attain to any adequate conception of an animal or plant in that way, but it encourages the method of beginning to look at natural facts from the wrong end."[34]

The views of Eves and Johnson do not just provide insight into the pedagogical precepts followed in the new laboratory. Their emphasis on hands-on experience with representative forms of the plant and animal kingdoms echoes the wider reform movement in the sciences at the time. Specifically, it reflects the precepts of the program of instruction first articulated by Huxley in the early 1870s and introduced to Cambridge by Foster. These were disseminated in Huxley and Henry Newell Martin's influential textbook, *A Course of Elementary Instruction in Practical Biology* (1875), derived from the South Kensington courses for teachers. As Huxley there proclaimed, "the road to a sound and thorough knowledge of Zoology and Botany lay through Morphology and Physiology; and that, as in the case of all other physical sciences, so in these, sound and thorough knowledge was only to be obtained by practical work in the laboratory." The aim was "not to make botanists, nor zoologists, nor anatomists, of the members of the class, but to give them a practical insight into the structure and activities of living things, in such a way as to enable them to observe for themselves the relations and connections of the various forms of life, and to follow from actual examples the characteristics and increasing complexity of different plans of structure."[35]

The "type system" that Huxley developed in his text served the university laboratories well. It allowed students to examine, over the course of a term, representative organisms

---

[33] The proposal is not dated but is in an envelope postmarked 19 Oct. 1884, BL: P&C, NCA.

[34] Quotations are from the prospectus by Eves and Johnson.

[35] T. H. Huxley and H. N. Martin, *A Course of Elementary Instruction in Practical Biology* (London: Macmillan, 1875; rpt., 1882), p. v; and Edwin Ray Lankester, "Instruction to Science Teachers at South Kensington," *Nature,* 1871, 4:361–364, on p. 363 (I thank Adrian Desmond for drawing this article to my attention).

chosen for embodying the most characteristic features of a group or class. Practical classes gave students a firsthand acquaintance with the organisms discussed in the lectures. For women, moreover, the importance of this plan of study took on an added dimension. They were informed that it stressed the "habits of mental discipline" that had application in everyday life as well as in the sciences: "The advantages of this system are by no means inconsiderable. The hand and eye are educated as well as the brain. A methodical training of these induces a methodical habit of mind; and this is important, because few things are of more consequence in the study of Science than a clear and systematic arrangement of one's ideas and the power of expressing them in a lucid and incisive manner."[36] Hence, analytical as well as practical skills were stressed. For women, many of whom would never become practicing scientists, it was this facet of biological training that provided an added incentive to study.

Day-to-Day Work in the Laboratory

Following the plan developed in the university laboratories by Foster and Balfour, then, the Balfour Laboratory offered demonstrations in various subjects immediately following university lectures. It was the job of the demonstrators, according to Foster (referring to Huxley's course), "to make each member of the class see for himself or herself, so far as was possible, the actual thing of which the master had spoken." Thereafter, students followed up what had been demonstrated by independent work, making their own dissections or performing appropriate experiments.[37]

The practical work offered in the laboratory depended on the sequence of university lectures in the academic year. In elementary morphology, demonstrations were given three times "a week through three terms and the Elementary biology 3 demonstrations a week during the Lent & part of the Easter terms." In physiology, the instructor demonstrated for an elementary class that attended Foster's lectures three times a week. Elementary biology, which was initially taught by the demonstrator of morphology but, like the university course, later shared with the botanical demonstrator, was offered three mornings a week for one-and-a-half terms. Advanced courses, such as invertebrate morphology and osteology, were also sometimes offered at the Balfour, since "the teaching provided by the University in this part of the subject & open to women is inadequate." For the advanced courses, however, demonstrators tried to arrange to use the appropriate university laboratories, since the Balfour Laboratory did not have the specialized equipment and instruments these required.[38]

[36] "The Natural Sciences Tripos," *Girton Review*, Dec. 1882, p. 6. On Huxley's "type system" see Geison, *Foster and the Cambridge School of Physiology* (cit. n. 4), pp. 134, 139. This plan of study is outlined in the letter from Michael Foster and F. M. Balfour to the Museums and Lecture-Rooms Syndicate, [Michaelmas term 1881], CUR 39.36: Professor of Animal Morphology, Cambridge University Archives, MSS Room, CUL. The American Mary Ann Willcox, who read natural sciences at Newnham between 1880 and 1883 and was professor of biology at Wellesley College from 1883 to 1910, recalled that it was "under [Foster's] guidance that Huxley's *Elementary Biology* was used in our little stone-floored laboratory at Newnham": Willcox, "Sidgwicks in Residence" (cit. n. 13), p. 14.

[37] Michael Foster, "Thomas Henry Huxley," *Proceedings of the Royal Society of London*, 1896, *58*:xlvi–lxvi on p. lx. See also Gooday, " 'Nature' in the Laboratory" (cit. n. 7). According to Mary Ann Willcox, Huxley called Foster the "Archangel": Willcox, "Sidgwicks in Residence," p. 14. Huxley admitted a woman to his first course, made up of nineteen school science teachers, and she received one of two microscopes awarded for achievement; see Lankester, "Instruction to Science Teachers" (cit. n. 35), p. 363. See also Greenwood to Eleanor Sidgwick, 8 May 1888, BL: P&C, NCA.

[38] Eleanor Sidgwick to Lilian Sheldon, [May 1890], BL: P&C; Eleanor Sidgwick to Greenwood, [June 1884], BL: P&C; Report of [May 1892], BLR, NCA (quotation); and Edith Rebecca Saunders, "Mrs. G. P. Bidder (Marion Greenwood)," *Newnham College Letter*, 1932, pp. 61–66, on p. 63 (lack of specialized equipment).

The number of students who worked in the Balfour Laboratory in any given year fluc-tuated greatly. Indeed, the college had a difficult time predicting how many new students intended to read for the NST, gauging this by noting how many enrolled in elementary biology. Although precise annual figures are lacking prior to 1900, it appears that through-out the 1880s there were on average around forty students working in all subjects at the Balfour Laboratory in any given year, including both elementary and advanced classes. In 1890, for example, there were fifty-two in all: nine students in elementary physiology and seven in revision (i.e., preparation for the NST), and five in advanced physiology; thirteen in elementary and two in advanced botany; and seven in elementary morphology (zoology) and nine in revision. Only about fourteen, on average, prepared each year to sit the NST.[39] After 1896, when lectures in morphology, physics, and geology were added, the number of students working in the laboratory increased substantially, with an average of around sixty per year.

At the Balfour, demonstrators had to spend considerable time preparing for instruction, although this varied by subject. In botany, for example, it took an hour or less to prepare the material needed for demonstrations. In physiology and morphology, however, dem-onstrators might spend up to twelve hours a week preparing the required fresh specimens. In addition, staff members also had to care for the mounted specimens. A personal account of the work of demonstrators was given by Edith Rebecca Saunders in her obituary of the former director of the Balfour Laboratory, Marion Greenwood:

> It is difficult in these days [1932], when students go daily from one well equipped laboratory to another for different subjects, to appreciate the amount of work involved in teaching two subjects under the conditions which then existed. The courses had to be organized from the beginning. All equipment had to be bought or borrowed, the strictest economy having to be exercised in every way. Reagents had to be purchased, apparatus set, microscopic sections prepared, with only a young untrained boy to act as general laboratory attendant. Had Miss Greenwood not been endowed with a splendid constitution she could hardly have accomplished all that was required and have carried out research work at the same time.[40]

Demonstrators, moreover, were also generally expected to guide the study of students outside of the laboratory. For advanced students preparing for Part 2 of the NST, they directed the "choice of original papers to read, the discussion of any points raised in those papers which may not be clear, and the discussion and explanation of lecture notes." This included such help as "revision dissection" or "revision physiology" with "special theo-retical work" offered outside of class time. For students in elementary classes, there was the "superintendence of the working through of experiments which students have seen done in demonstration." In elementary biology, it was recognized that students who were inexperienced were often slow and in need of special attention. Demonstrators in these cases were called upon to give "that sort of a review of a whole course with emphasis of the most important points in it" that was known as "coaching."[41]

---

[39] Attendance figures have been drawn from the annual reports and data included in the "Principal's Report" published annually in the *Newnham College Letter*. Not until 1910 are annual figures readily available; see "List of Classes held at the Balfour Laboratory, Jan 1910 to [Easter Term 1914]," NCA. Data for 1890 come from Greenwood to Eleanor Sidgwick, 30 Nov. 1890, BL: P&C, NCA. For the average NST figures see MacLeod and Moseley, "Fathers and Daughters" (cit. n. 3), pp. 326–327.

[40] Saunders, "Mrs. G. P. Bidder" (cit. n. 38), p. 63.

[41] Greenwood to Eleanor Sidgwick, 8 May 1888, BL: P&C; Report of [May 1892], BLR; and Alice Johnson to Eleanor Sidgwick, 6 Feb. 1888, BL: P&C, NCA.

THE BALFOUR LABORATORY AND THE SCIENTIFIC SUBCULTURE
OF CAMBRIDGE WOMEN

The handicaps that women faced as marginal students at Cambridge could in part be alleviated through creating a scientific culture that mirrored the one enjoyed by the male undergraduates. There were many facets of a scientific education that occurred outside of the lecture room and laboratory. Because they could not participate fully in the scientific community of Cambridge, women experienced a number of distinct disadvantages compared with men who were studying the sciences.

It was generally recognized, for example, that women coming up to Cambridge to read natural sciences frequently required extra tutoring owing to the deficient scientific instruction provided by the girls' secondary schools. However, this was not significantly more than what was required in the university laboratories, for many of the men who came up to Cambridge also required coaching.[42] Yet the situation of women science students was less favorable than that of the men because they did not have access to the same opportunities to remedy their educational deficiencies, including the Cambridge Natural Sciences Club, the informal "departmental" and formal college tutorial system, and prizes and fellowships that supported research students. Women responded to this situation by constructing their own intellectual and social network in the sciences.

The Natural Sciences Clubs

Women were denied membership in the various university scientific societies that were central to undergraduates and senior members alike. In such forums as the Cambridge Philosophical Society and the student-run Cambridge Natural Sciences Club, both students and dons had the opportunity to hear discussions of the latest scientific research and to present the results of their own current studies prior to publication, often in a journal edited by the department head.[43]

At a time when the social segregation of the sexes was unquestioned in the university, it is not surprising that the women's colleges established their own natural sciences clubs, Newnham College in 1883 and Girton in 1884. Membership in the women's Natural Sciences Clubs was open to students who were reading for the NST; at Newnham, those preparing for honors in Group E (natural sciences) of the Higher Local Examination were also eligible. Club meetings, at which students read and discussed papers they had written, were generally held every two weeks. Meetings were well attended and were deemed an integral part of the course of studies. Newnham women were informed, for example, that "this is the only Club which considers its work sufficiently important to justify meetings being held in study hours."[44]

It is indicative of the standing of these organizations in college life that after 1903 there were two annual joint meetings of the Girton and Newnham Natural Sciences Clubs, which

---

[42] See Michael Foster's and George Liveing's comments to the Degrees for Women Syndicate, 10 Nov. 1896, Synd. II.2, MSS Room, CUL. Foster commented in 1885 that the poor performance of many students in the NST was due to their having "come up to the University, not only wholly ignorant of even the elementary facts and ideas of chemistry and physics, but with their minds unprepared by any adequate discipline for the reception of the truths of experimental science": MacLeod and Moseley, " 'Naturals' and Victorian Cambridge" (cit. n. 13), p. 184.

[43] Sonia Uyterhoeven, "Student Culture at Late Victorian Cambridge: The Natural Sciences Club," paper presented at the annual meeting of the History of Science Society, New Orleans, 1994.

[44] *Newnham College Letter,* 1883, p. 14; 1885, p. 12.

served to enhance the significance of the proceedings. These meetings served as forums for discussion of new discoveries—such as Marie Curie's "brilliant discovery" of radium—as well as novel theories and controversies; the results of original research conducted by advanced students or instructors were also presented. In the 1880s and 1890s topics discussed at meetings included "Symbiosis" (Emily Elizabeth Wood, 1884), "The Influence of Climate on Vegetation" (Anna Bateson, 1886), "Pasteur's Treatment of Hydrophobia" (Clara Eisdell Vavasseru, 1887), "Dr Gaskell's Theory of the 'Origin of the Vertebrate from a Crustacean-like Ancestor'" (Elizabeth Dale, 1890), "Diptheria" (Marion Greenwood, 1894), "The Struggle for Existence among Plants" (Edith Saunders, 1894), and "Recent Work upon the Nucleus in Plants" (Ethel Sargant, 1898).[45]

The science clubs, then, were a central feature of the women's subculture of science at Cambridge. In addition to enabling women to keep abreast of the latest scientific developments, they promoted the research ethic. Students reading the natural sciences were able to hear Part 2 students presenting their current investigations, and sometimes this sparked their own interest in pursuing postgraduate research.

The Tutorial System

The lack of complete access to university lecturers was a more acute difficulty for advanced than for beginning students. For third- and fourth-year students reading for Parts 1 and 2 of the NST, it was largely through working with demonstrators in the university laboratories that, in addition to developing experimental skills, they received guidance concerning the scientific literature to read to prepare for examinations set by Cambridge dons. By the mid 1880s, with overcrowding in the laboratories and lectures, these women were frequently excluded from university practical classes, to the detriment of their programs of study. In 1884, for example, the Newnham student Florence Maberley Buxton was preparing for Part 2 of the NST when she wrote to Eleanor Sidgwick requesting urgent assistance. Frustrated by the lack of guidance she was receiving from the head of physiological botany, Sydney Vines, she lamented:

> I am so far in a worse position than the men in that they *can* go to the laboratory & so meet Mr. Vines & the demonstrators, while Mr. Vines himself said that none of us, even those going to the Lectures, who had done the Course before were to go to the Practical work, that being the only time when there is a chance of seeing any one to ask about difficulties or to obtain advice. I do not want teaching, but some direction in reading. If Mr. Vines would mention in his lectures what papers &c to read on the Subject one would have something to go upon—but as it is, one has to face such a mass of undigested literature in various magazines & periodicals nearly all being in German, that one does not know where to begin or what to choose. Mr. Gardiner promised to send me a list of some papers as soon as he had time to think about it. I should be very grateful if the College would give me some help next term from Miss [Florence] Eves, who would be thoroughly capable of giving me what I want if she knew beforehand that it wd. be required.[46]

The Balfour Laboratory, when it opened several months later, helped to remedy this situation. The laboratory gave women students access to qualified, salaried demonstrators

---

[45] The *Girton Review* regularly reported the Natural Sciences Club's activities each term.

[46] F. M. Buxton to Eleanor Sidgwick, 25 Nov. [1884], BL: P&C, NCA. Vines responded to this complaint in a letter of 2 Dec. 1884 (BL: P&C), stating that Buxton should have attended his lectures, even though she had already attended them twice before, because he always introduced new material based on recent literature. As for practical work, "It was clearly impossible for me to have all the students at the practical class, on account of the limited accomodation [sic], and in view of the large numbers of new students."

who were highly capable of guiding them in their studies. Often such assistance was given informally outside of the laboratory. Marion Greenwood, for example, volunteered time in the evenings to coach students who were having difficulties.[47] Even so, advanced students lacked the kind of nurturing that comes only from close and frequent contact with instructors.

Toby Appel has drawn attention to the role that women's physiological laboratories played in the United States in "mentoring women students in the family-like setting of the women's college."[48] Certainly the Balfour Laboratory served a similar function for the women natural science students at Cambridge, providing a place where they received the guidance needed to achieve their goals. Most natural science students aimed to teach in the secondary schools, but increasingly Part 2 students hoped to pursue their research interests either at the university or in one of the few positions open to women in government agencies or private industry.

Evidence of a feminine laboratory culture at Cambridge is rare, particularly as the women strove to make their pedagogy comparable to that of the university. Yet we can catch a glimpse of what obviously was sometimes an informal atmosphere in the Balfour Laboratory in an anecdote recounted by Edith Saunders in her obituary of Marion Greenwood: "It was her practice, and also mine, for many years to lunch at the Balfour Laboratory—bread and butter, chocolate and milk sufficed. . . . After lunch the short interval before an afternoon class was spent in battledore and shuttle-cock, a game with queer hazards played around a stove and among tables and chairs. Another recreation occasionally indulged in was whist—bridge was still in the future." Eating in the laboratory—"which called forth," Saunders noted, "a fatherly remonstrance from a University member of the Laboratory Committee who prophesied ill consequences"—let alone playing games there, suggests an atmosphere that was comfortable and even "homey."[49]

RESEARCH IN THE BALFOUR LABORATORY

At the Balfour, in contrast to American physiological laboratories, mentoring was carried on as much through the example of the demonstrator's own research accomplishments as through other means. Colleges were proud of the publication records of their staff members and research fellows and students, which were highlighted in the college magazines. Although the Balfour Laboratory primarily served as an undergraduate teaching facility, research students were also accommodated. It was customary for Bathurst students and other research fellows to be given worktables there.[50] However, the number of such re-

[47] Greenwood to Eleanor Sidgwick, 8 May 1888, BL: P&C, NCA. See also M.B.T., "In Memoriam: Mrs. G. P. Bidder," *Girton Review,* 1933, pp. 3–5.

[48] Appel, "Physiology in American Women's Colleges" (cit. n. 5), p. 28.

[49] Saunders, "Mrs. G. P. Bidder" (cit. n. 38), p. 65.

[50] Greenwood to Eleanor Sidgwick, 11 Aug. 1890, BL: P&C, NCA. The research interests of Newnham women were regularly published in the *Newnham College Letter* beginning in 1908; Girton students and alumnae learned about current research topics in the *Girton Review* in the description of the Natural Sciences Club. On the role of mentoring see also Dyhouse, *No Distinction of Sex?* (cit. n. 1). In addition to the Balfour demonstrators, Eleanor Sidgwick herself provided an important role model. A student at Newnham in 1881, Helena Powell, recalled the impression made on students when they learned of Mrs. Sidgwick's scientific attainments: "When we learnt that Mrs. Sidgwick had done original work in physics; that she helped Lord Rayleigh to correct Mathematical Tripos papers; that Mr. [Arthur] Balfour discussed philosophical problems with her on equal terms; and set beside that the great difficulty she found in making conversation with a drawing-room of students . . . [or in delivering] a speech when occasion called for it, . . . Mrs. Sidgwick corrected one's standard of values": quoted in Ethel Sidgwick, *Mrs. Henry Sidgwick* (cit. n. 17), p. 77.

search students was never great. The most who ever worked there at one time was in 1911, when there were four.

There are reasons for the small number of research students working at the Balfour. It was preferable for such students to work with specialized equipment, alongside other postgraduate students, under the direction of distinguished research scientists in the university laboratories. Beginning in the early 1880s, a number of women enjoyed this opportunity. Alice Johnson, for example, carried on embryological research in the University Morphological Laboratory, first under Frank Balfour and then under Adam Sedgwick. Likewise, Lilian Sheldon, appointed demonstrator in morphology after Johnson's resignation in 1890, studied the development of *Peripatus* (a contender for the ancestor of vertebrates) under Sedgwick's guidance, assisting with his monograph on the genus. Florence Eves and Marion Greenwood carried on their physiological research in Foster's University Physiological Laboratory. And Anna Bateson and Dorothea F. M. Pertz worked with Francis Darwin in the Botanical Laboratory.[51] Yet ultimately this privilege depended on the tolerance of the professor who headed the laboratory.

Even within laboratories that were ostensibly open to women research workers, an undercurrent of prejudice could sometimes turn into hostility and make working there problematic. Despite serving as mentor to several Balfour demonstrators, Adam Sedgwick, for example, was known for his hot temper, sometimes directed at women students.[52] In Foster's laboratory, too, there could be an air of intolerance, depending on the individuals involved. The physiologist William Bate Hardy, who shared a bench with Greenwood in the 1880s and 1890s, referred in his obituary of her to the prejudices against women research students commonly held by Cambridge men:

> At that time women were rare in scientific laboratories and their presence by no means generally acceptable—indeed, that is too mild a phrase. Those whose memories go back so far will recollect how unacceptability not infrequently flamed into hostility. The woman student was rather expected to be eccentric in dress and manner; she was still unplaced, so far as the male in possession was concerned. Miss Greenwood, it so happened, was not only a woman of quite unusual intellectual distinction but she had also great personal charm and a great gift of comradeship. Science by no means absorbed all her interest which covered a wide knowledge of literature.

Greenwood earned an international reputation for her studies of the digestive processes in protozoa and lower invertebrates, published between 1884 and 1896. Her marriage to the

---

[51] Women frequently published the results of their research in journals edited by the heads of laboratories. Johnson and Sheldon, along with Agnes Isabella Mary Elliot and Igerna Brünhild Johnson Sollas, published papers in the *Quarterly Journal of Microscopical Science*, coedited by Sedgwick. Sheldon's study of *Peripatus* also appeared in *Studies from the Morphological Laboratory at Cambridge*, which Sedgwick edited. Those working in the Physiological Laboratory often published their results in the *Journal of Physiology*, edited first by Foster and then by J. N. Langley. Papers by women were sometimes published jointly with male mentors or were communicated by them to scientific societies, especially the Royal Society of London or the Cambridge Philosophical Society, and published in the respective transactions or proceedings of those organizations: see *Newnham College Letter*, 1908, pp. 49–56, 1909, pp. 58–59. On the "gendered politics of collaboration in scientific work" illustrated by women research students at the Cavendish Laboratory see Gould, "Women and the Culture of University Physics" (cit. n. 3), pp. 137–140. Gould stresses collaboration rather than confrontation, arguing that at the Cavendish women "fashioned spaces for themselves instead of fighting tooth and nail to widen a prescribed sphere" (p. 149).

[52] Arthur Shipley, who was a demonstrator in Sedgwick's laboratory in the early 1880s, related an incident in which a woman student attempted to speak to Sedgwick when he was engaged in a tedious dissection. "After a third or fourth interruption Adam, whose temper was in those days always within easy call, said, 'Oh, tell Miss "Chose" to go to the Devil.' I was not best pleased a moment later to hear Brocket [head laboratory attendant], who is always a diplomatist, saying to the lady, 'Mr. Sedgwick, Miss, thinks you'd better see Mr. Shipley' ": Shipley, "*J.*" (cit. n. 10), p. 280.

Cambridge zoologist George Parker Bidder in 1899, however, effectively brought her research career to an end.[53]

Other women research students apparently found it harder than Greenwood to fit into the male-dominated environment of the university research laboratory. Lilian Sheldon, for example, despite achieving a creditable second-class standing in Part 2 of the NST zoology paper, received but a lukewarm recommendation in 1884 from the university lecturer in invertebrate morphology, Walter Frank Raphael Weldon, when he was asked to support her candidacy for a Bathurst studentship. Weldon paid Sheldon a backhanded compliment when he informed Eleanor Sidgwick of his perception of her abilities:

> I see no reason to doubt that Miss Sheldon would make good use of any opportunity which might be afforded her of doing original work.
> So far as I can judge, Miss Sheldon's ability is quite equal to that of many men who, immediately on taking a degree, have come here and done work of a useful kind.
> I have very little experience in such matters, but it seems to me that the kind of work usually put into the hands of a student, while it is useful and worth doing, requires rather a certain amount of diligence, than any great power of original thought—these things, I am sure Miss Sheldon possesses: as to the rest, I am not fit to offer an opinion, but I certainly do not believe that her degree disproves her possession of such qualities.

Over the course of the next thirteen years Sheldon, who received the Bathurst award despite this reference, produced a number of publications that can be judged to be more than simply "of a useful kind."[54]

---

[53] W. B. Hardy, "Mrs. G. P. Bidder," *Nature,* 1932, *130:*689–690. As Hardy attested, "Miss Greenwood made solid contributions to science" (p. 689). The German physiologist Max Verworn described her work as "very interesting" in his popular text *General Physiology: An Outline of the Science of Life,* trans. Frederic S. Lee (London: Macmillan, 1899), pp. 152–154. At the Fourth International Physiological Congress, held in Cambridge in 1898, Greenwood was one of seven British women delegates in attendance, and the only one from Cambridge. An anecdote illustrating the reputation of Greenwood's research is recorded in an obituary in the *Times* (27 Sept. 1932): "The story is told that when [the Nobel Prize–winning microbiologist Elie] Metchnikoff visited Cambridge his host, Sir Michael Foster, asked whether there was anyone whom he would specially like to meet. The great man replied that there was a young man of the name of M. Greenwood, whose papers had much impressed him. Sir Michael promised with alacrity that Metchnikoff should have 'M. Greenwood' for his neighbour at luncheon." According to Anna Bidder, Greenwood's daughter, the *Times* got this story wrong. Foster's answer was rather: "You shall see him tonight at the soirée given for him." When asked why her mother gave up scientific research after marriage to her father, Anna Bidder remarked that her mother was a perfectionist in all things, and that if she could not devote her whole attention to her studies, it was better to give them up altogether (interview with Marsha L. Richmond, July 1993). Greenwood did, however, keep up with the current literature in physiology after her marriage. She was responsible for the information on physiology included in the textbook *Domestic Economy in Theory and Practice* (Cambridge: Cambridge Univ. Press, 1901), coauthored with Florence Baddeley.

[54] W. F. R. Weldon to Eleanor Sidgwick, 20 July 1884, BL: P&C, NCA. J. N. Langley, however, in a letter to Sidgwick, [July 1884], also gave his opinion that the other candidate was "somewhat better than Miss Sheldon both in Comp. Anat & in Physiol.": BL: P&C, NCA. Lilian Sheldon took a second class in Parts 1 (1883) and 2 (1884) of the NST. Sheldon is an example of someone who did not have sufficient private means to undertake a scientific career without a salaried position that could fully support her. Between 1885 and 1896, she published six articles and also contributed a description of the nemertines to the *Cambridge Natural History.* Sheldon was asked to become director of the Balfour Laboratory upon Johnson's resignation in 1890 since she had "done a good deal of original work under the direction of Mr Sedgwick—publishing several papers—& Mr Sedgwick thinks very well of her": Report of 10 May 1890, BLR, NCA. However, Sheldon declined; she had primary responsibility for her aging parents and stated that under such circumstances "I could never feel that it was a permanent settlement, as I should know that at any time I might be called upon to give it up." In 1892, when the morphological demonstrator Laura R. Howell resigned following her marriage to the zoologist Sydney F. Harmer, Sheldon was able to accept an invitation to return to Cambridge and take over these duties. In 1898 she left Cambridge for Devonshire to keep house for her brother, with whom she carried on archaeological investigations published in the *Devon Association of Science, Literature, and Art.*

Still, some women found their research interests well accommodated at the Balfour Laboratory. After 1900, many of the women who were members of William Bateson's school of genetics conducted their breeding experiments at the Balfour. Muriel Wheldale (later Onslow) (1880–1932), who became demonstrator in physiological botany in 1907, carried out much of her work on the genetics and chemical physiology of plant pigmentation in the Balfour Laboratory before moving to Frederick Gowland Hopkins's new biochemistry laboratory after World War I.[55] She and other Bateson collaborators may have worked in the Balfour because Bateson, even after his 1908 appointment as professor of biology, did not have adequate university accommodations for his genetical breeding experiments.[56]

It was never the aim of the women's colleges to produce brilliant research scientists or "original geniuses," despite the fact that some pointed to this shortfall as proof of the intellectual inferiority of women. In 1887, for example, George John Romanes, one of the leading Darwinians of the day, argued that "the disabilities under which women have laboured with regard to education, social opinion, and so forth, have certainly not been sufficient to explain this general dearth among them of the products of creative genius." As a counter to this argument, Emily Davies pointed out that genius would always be rare and that indeed it required "little encouragement from educational opportunities." It was, rather, "the women of ordinary endowment" whom she believed received the greatest benefit from higher education. Eleanor Sidgwick expressed a similar opinion in 1897:

> Some have said that women have not the same capacity as men for original work. . . . But even if very few women should prove capable of the greater and more important part of the work of discovery and research—fewer even than the very few men—I have good hope that women will do excellent work in the subordinate fields of science and learning, will do well much laborious work that needs to be done, though it is not very brilliant or striking, and will in particular prove excellent assistants.[57]

After 1900, Sidgwick's somewhat limited view of women serving as mere scientific "assistants" to men began to appear outmoded.[58] A new generation of women who had

---

[55] See Marjory Stephenson, "Muriel Wheldale Onslow (1880–1932)," *Biochemical Journal,* 1932, 26:915–916; and Creese, "British Women" (cit. n. 4), p. 284. The biochemist Dorothy Needham, who began research in Cambridge in 1920, recalled that "women research workers were very rare in Cambridge at that time—indeed I think they were accepted only in the Department of F. G. Hopkins (later Sir Gowland) who had the Chair of Biochemistry. He welcomed them, and there were perhaps nine of us, about the same number as the men": Dorothy Needham, "Women in Cambridge Biochemistry," in *Women Scientists: The Road to Liberation,* ed. Derek Richter (London: Macmillan, 1982), pp. 158–163, on p. 159.

[56] This may have influenced Bateson to leave Cambridge in 1910 to take up the directorship of the John Innes Horticultural Institute. See Beatrice Bateson, *William Bateson, F.R.S., Naturalist: His Essays and Addresses, Together with a Short Account of His Life* (Cambridge: Cambridge Univ. Press, 1928). On the women who were members of Bateson's School of Genetics see Marsha L. Richmond, "The Entry of Women into Academic Science: William Bateson's Cambridge School of Genetics," in preparation.

[57] George John Romanes, "Mental Differences between Men and Women" (1887), in *The Education Papers: Women's Quest for Equality in Britain, 1850–1912,* ed. Dale Spender (New York: Routledge & Kegan Paul, 1987), pp. 10–31, on p. 11; Emily Davies, cited in "The Women's Congress," *Girton Review,* July 1899, p. 1; and Eleanor Sidgwick, *The University Education of Women* (Cambridge: Macmillan & Bowes, 1897), p. 14. See also Flavia Alaya, "Victorian Science and the 'Genius' of Woman," *Journal of the History of Ideas,* 1977, *38:* 261–280.

[58] On problems women faced in getting credit for collaboration see Gould, "Women and the Culture of Physics" (cit. n. 3), pp. 137–140. Someone who agreed with Sidgwick, however, was Florence Buchanan, who held that such positions were more readily available to women than independent research posts. See Florence Buchanan, "Work for Women in the Biological Sciences," in *Women in Professions: Being the Professional Section of the International Congress of Women* (London: Unwin, 1900), pp. 174–179.

gained the opportunity to pursue original research did not want to settle for this subordinate status. They recognized that to be able to carry on independent research, women needed the same kind of support available to men.

Several individuals were in the forefront of the campaign to endow more research fellowships. In 1899 Florence Margaret Durham (1869–1949), demonstrator in physiology at the Balfour and a member of William Bateson's school of genetics, urged that "more effort should be made by Girton and Newnham to encourage advanced and research work, and thus to show to the world that women mean to do serious work and have higher aims in view than mere success in examination." The true aim of college study was not to place well in the tripos but to form "a secure foundation for serious work in the future." The botanist Ethel Sargant expressed the same opinion in 1901, when she launched an appeal for Girton alumni to support new research fellowships for women. Sargant noted that the movement for the higher education of women had entered a new phase. The "bricks and mortar" of the women's colleges were in place, and a second generation of students were entering their gates. It was now time to prove that women were worthy of sharing in "the great inheritance" of Cambridge and Oxford universities by taking "the first steps in a career of independent research." Women, she believed, could distinguish "themselves in research as they have already distinguished themselves in the class lists" and contribute "original work of real value" were they only given "a year's work under skilled guidance followed by one or two more of independent research." The lack of such fellowships, however, continued for many years to be a major impediment to women entering academics.[59]

The Balfour Laboratory, then, assumed an importance for the women biology students at Cambridge well beyond its role as a teaching laboratory. It became the nucleus for the scientific subculture of Cambridge women, providing a nurturing environment for both beginning and advanced students. Most important, it also created a number of teaching positions for women who would otherwise not have been placed at a British university.[60] Soon, however, it began to serve yet another function: as the venue for lectures as well as laboratory instruction in biology.

THE EXCLUSION OF WOMEN FROM UNIVERSITY LABORATORIES AND LECTURES

When the Balfour Laboratory opened in 1884, it took its place as a college teaching laboratory comparable to those of Trinity, St John's, and Gonville and Caius colleges. For the previous ten years, women had been welcome in both the lectures and the practical classes of all but a few science professors. To be sure, they were sometimes made to feel "dreadfully uncomfortable" in the science lectures, with the segregation of the sexes an unwritten rule. In Foster's lectures, for example, women were required to sit "in a gallery

---

[59] Florence Margaret Durham to the editor, 10 Apr. 1899, *Girton Review*, 1899, p. 8; and Ethel Sargant, "The Inheritance of a University," *ibid.*, 1901, pp. 17–21. Durham, who was William Bateson's sister-in-law, took a second class in Parts 1 (1891) and 2 (1892) of the NST. She taught biology at Royal Holloway College and the Froebel Institute in London from 1893 to 1899 before returning to Cambridge to serve on the staff of the Balfour Laboratory. When Bateson left Cambridge in 1910, Durham joined him at the Innes Institute to continue work on plant physiology. Serving as a hospital pathologist during World War I, she joined the staff of the Medical Research Council in 1917, retiring in 1930. See *Girton College Register* (cit. n. 10), p. 49. On the continuing shortage of fellowships for women see Fernanda Perrone, "Women Academics in England, 1870–1930," *History of Universities*, 1993, *12*:339–367; Dyhouse, *No Distinction of Sex?* (cit. n. 1), p. 141; and Dorothy Needham, "Women in Cambridge Biochemistry" (cit. n. 55), p. 161.

[60] Eschbach, *Higher Education of Women* (cit. n. 1).

where they had their microscopes at a low window, and the demonstrator went up to help them in their seclusion." In the lectures of the anatomist G. M. Humphry, "they sat modestly in the back in a little row by themselves." Embarrassing incidents were not infrequent; in Humphry's class, for example, "when a specimen of the human brain was passed round for inspection, they became aware that the undergraduates in front had all turned round to see whether this would discompose the ladies."[61] By the mid 1880s, however, with the growing crisis over science accommodation, the admission of women to lectures began for the first time to be threatened. Only shortly after it opened, the Balfour Laboratory had to assume a more prominent role in science instruction than did the men's laboratories. In effect, it became the locus rather than the periphery for women's biological instruction.

The first warning sign of the new challenge to women's science education came in 1885. Despite the completion of a new university botanical laboratory in that year, the reader in botany, Sydney Vines, informed Eleanor Sidgwick in February that, having assessed the situation, it was apparent to him

> that there is not more accomodation [*sic*] than is necessary for the men. Under these circumstances we regret to have to say that it will be impossible to have a class of women students any longer. . . . I hope that you will not find much difficulty in providing practical teaching for your students, at least for those doing elementary work. As regards the advanced students, those, namely, who are preparing for the second part of the Natural Sciences Tripos, we are willing, in case of necessity, to do what we can, for a time at any rate, in the way of providing demonstrations for them.

This new circumstance was mentioned in passing in the annual report of the Museums and Lecture-Room Syndicate for 1886, which stated that the elementary practical class previously open to women had been closed to them "owing to want of room."[62] However, here as in future incidents, it seems clear that a "hidden agenda" on the part of the instructor was a subtext to the official pronouncements. Vines in fact believed women should have separate laboratory facilities; the opening of the Balfour Laboratory provided the opportunity for him to exclude them from his practical classes.[63]

This new development put extra pressure on the Balfour Laboratory. The laboratory became more crowded, "chiefly owing to the fact that practical instruction in botany has been carried on there instead of in Mr Vines laboratory." There was also a need for a separate demonstrator in botany, but no funds were immediately available for the purpose. As an expediency, the college provided Marion Greenwood, demonstrator in both physiology and botany, with assistants: Lilian Sheldon and the botanist Anna Bateson (1863–1928). Only in 1888 was the college able to establish a separate demonstratorship in

---

[61] Stephen, *Emily Davies and Girton College* (cit. n. 2), p. 287.

[62] S. H. Vines to Eleanor Sidgwick, 26 Feb. 1885, BL: P&C, NCA. The botanical staff reported that the class in elementary botany had doubled in size since 1881 and that it currently consisted of forty-seven men in a classroom designed to hold eighteen: Report to the Chairman of the Special Board for Biology and Geology, 26 Feb. 1886, CUR, New Museums, Vol. 2, no. 237, MSS Room, CUL. Vines's colleague Francis Darwin had prepared Eleanor Sidgwick for this, writing that "I am afraid you are right and that Mr Vines does intend to try and get rid of the women from his laboratory": [1884], BL: P&C, NCA.

[63] See S. H. Vines to the Degrees for Women Syndicate, 2 Nov. 1896, in which he stated: "The relief given by this move [the opening of the Balfour Laboratory] to my department, as doubtless also to the other biological departments, as also the advantage gained in the way of comfort by the women-students, and finally the institution of one or more demonstratorships for women, have convinced me of the great importance of providing distinct laboratory-accomodation [*sic*] for women-students." Synd. II.12, MSS Room, CUL.

botany, to which Edith Saunders was appointed.[64] Saunders offered practical classes in elementary and advanced botany; she and the morphology assistant also helped with the classes in elementary biology.

Vines's decision set a precedent. Women soon found themselves excluded from practical classes in elementary biology, physiology, and zoology as well as botany. The Balfour Laboratory was nonetheless able to respond to the situation without much difficulty. However, the women's colleges were pushed to the brink in 1897, when a further development threatened the very existence of women's education in the biological sciences.

In November 1897 Adam Sedgwick, head of the School of Morphology, informed Newnham and Girton colleges that henceforth women were unwelcome in the lectures in elementary morphology and biology. This announcement presented the colleges with the most difficult challenge they had yet faced, that of providing all the instruction in two core areas for students reading for the NST. To understand how Sedgwick felt able to take this step, we need to look at the general milieu at Cambridge in the wake of the defeat in May 1897 of the 1896 memorial (petition) requesting degrees for women.

The 1896–1897 Degrees for Women Movement

In 1896, prompted by the presentation of three memorials requesting that the university grant degrees to women, Cambridge undertook an extensive reexamination of its commitment to the higher education of women. In the fifteen years since women had been formally admitted to the tripos examinations, the climate of Cambridge had changed considerably. The early "experiment" of providing a university education to women was no longer a novelty. The annual tripos results more than proved that women could not only compete on an intellectual par with men but sometimes even surpass them.[65] In the natural sciences, women were increasingly obtaining first-class honors in Part 1 and even Part 2 of the NST, and a few were gaining recognition for their research publications in leading scientific periodicals. These accomplishments, however, served to harden the opposition to women among certain factions of the university community rather than to assuage their concerns.[66]

One of the central issues driving the 1896 degrees for women campaign related to

---

[64] Report of 1886 (crowding due to practical instruction in botany); Report of 19 Nov. 1887 (Sheldon and Bateson); and Report for 1888, BLR, NCA (Saunders). Sheldon was nearing the end of her Bathurst studentship. Anna Bateson was the daughter of William Henry Bateson, master of St John's College, and the sister of both William Bateson and the Newnham historian Mary Bateson. Her mother served on the Council of Newnham College. When her position at the Balfour finished, Anna Bateson served as an assistant to Francis Darwin, publishing two papers with him in 1887 and 1888, a paper with her brother in 1891, and two other independent papers. Disappointed by her seconds in both parts of the NST (1884, 1886), she believed that she was precluded from following an academic career; see William Bateson Correspondence, Add. MS 8634, MSS Room, CUL. She left Cambridge in 1890, becoming owner of a nursery in Bashley, Hampshire. See *Newnham College Register*, Vol. 1 (cit. n. 29), p. 71; and Alan Cock, "Anna Bateson of Bashley: Britain's First Professional Woman Gardener," *Hampshire*, May 1979, pp. 59–62.

[65] In 1890 Philippa Garrett Fawcett, daughter of the Cambridge political economist Henry Fawcett and the prominent feminist Millicent Fawcett, placed above the senior wrangler (the highest in first-class honors) in Part 1 of the Mathematical Tripos. For a description of the jubilant celebrations this prompted during the conferring of degrees see Ethel Sidgwick, *Mrs. Henry Sidgwick* (cit. n. 17), p. 113.

[66] Opposition to similar petitions at Oxford in 1896 and 1919, according to Annie Rogers, was based on the "dislike and fear" of the presence of women in the university. See Susan J. Leonardi, *Dangerous by Degrees: Women at Oxford and the Somerville College Novelists* (New Brunswick, N.J.: Rutgers Univ. Press, 1989), p. 19. On the 1896 degrees for women movement see McWilliams-Tullberg, *Women at Cambridge* (cit. n. 3); and Eschbach, *Higher Education of Women* (cit. n. 1), Ch. 6.

women in the sciences. The women's colleges argued that women who had passed the NST examinations and yet left Cambridge with a certificate rather than a degree were at a disadvantage now that all other British universities, with the exception of Oxford, offered degrees for women. As Eleanor Sidgwick explained in a speech at University College, Liverpool, in May 1896, the reality of the present-day situation was that most of the students who attended university, both men and women, did so with a view toward preparing for a profession in which they could support themselves. Only two professions, teaching and medicine, were generally open to women, and they had many more opportunities as teachers than as physicians. An 1887 survey conducted by Newnham College found that "77 per cent. of the former students (women) of Cambridge and Oxford had engaged in teaching as a regular profession after leaving college"; by 1894 more than half of the former students of Newnham College were engaged in teaching. Yet the job market in teaching had become highly competitive, and Cambridge women were finding themselves passed over in favor of those who held university degrees rather than certificates.[67]

For the heads of the women's colleges, however, these considerations paled next to the more immediate concern of providing adequate scientific instruction for their students. The head of Girton College, Elizabeth Walsh, pointed out to the Degrees for Women Syndicate the "drawbacks connected with the present insecurity of tenure as regards admission to lectures and laboratories." Eleanor Sidgwick was even more direct, informing the syndicate that while women students had at present little to complain of with respect to university privileges, these were nonetheless accorded on sufferance:

> At any moment, they may be crowded out of lecture rooms or laboratories. At any moment, professors or lecturers may be appointed who may decline to have anything to do with them; and the fact that they have hitherto met with so much kindness and encouragement as they have, does not secure the future. Already the students of Newnham and Girton have, from want of room, been excluded from the laboratory courses in Elementary Biology, Physiology, Zoology and Botany for the First Part of the Natural Sciences Tripos. It was quite right that this should happen, and in these large elementary classes . . . it is, on the whole, no disadvantage for the women to have separate laboratories. But the loss to women's education would be very serious indeed were Newnham and Girton students unable to obtain admission to the University laboratories for advanced work, . . . or were they, in any department, to be deprived of the opportunities they now enjoy of attending the lectures and classes of University teachers and of being taught by them as undergraduates are.

Although the Balfour Laboratory could provide the necessary practical work in elementary subjects, it was not equipped with the specialized instruments needed by those preparing for Part 2 of the NST. Indeed, this problem was addressed by Edith Saunders, who informed the syndicate of a case in which two women had been excluded from the practical work accompanying Francis Darwin's 1891 lectures on plant physiology. Darwin had informed them that since he had eight men to fit into a room that accommodated only seven, he "was therefore obliged to exclude the women students from the demonstrations, since he 'must consider first members of the University strictly so-called.' " Saunders added that "the course was one which required special apparatus, & which therefore could

---

[67] Eleanor Sidgwick, *University Education of Women* (cit. n. 57), pp. 14–15. On the perception of universities as primarily teacher-training institutions for women see Dyhouse, *No Distinction of Sex?* (cit. n. 1), pp. 18–22. On the disadvantages of certificates compared with the status of degree titles see MacLeod and Moseley, "Fathers and Daughters" (cit. n. 3), pp. 330–331.

not be organised at the Balfour laboratory."[68] In effect, then, by the 1890s women science students at Cambridge were gravely hindered in their pursuit of both elementary and advanced work in the life sciences.

In part, as Eleanor Sidgwick acknowledged, the declining position of women in the university reflected the wider crisis in the institution itself. Given the battles that the science instructors had been waging with the university for the past two decades over their need for expanded facilities, it is perhaps not surprising to find that the major opposition to the movement to grant degrees to women came from the science faculty. It was precisely this circumstance to which Sydney Vines alluded in opposing the granting of degrees to women. Having left Cambridge in 1888 to become Sheridian Professor of Botany at Oxford, Vines nonetheless responded to the circular sent by the syndicate to science lecturers inquiring about the effect the admission of women had had on their lectures and laboratory classes. Granting degrees to women, Vines pointed out, would require the university to construct new laboratory facilities, entailing a massive outlay of funds.[69]

This argument carried a good deal of weight. However, other science instructors expressed more personal concerns about admitting women to their classes. Although a few noted that their classes had been improved by the addition of "eager and enthusiastic" women students, others complained that the presence of women resulted in a number of inconveniences and often real impediments. Those who had to lecture on subjects touching on sexual reproduction found it personally embarrassing to discuss the topic in front of a mixed-sex audience. Some believed that the presence of women students deterred men from discussing any difficulties they might have with the subject matter. Several laboratory instructors stated that women often required more elementary demonstrations than did the men, making it more difficult to organize practical work. Many acknowledged practical difficulties in providing separate lavatory accommodations. Several dons referred to the custom of gender segregation, with women generally being seated either in a gallery or in the front rows of a lecture room. When this segregation was carried over to the laboratories, they noted, it prompted competitiveness between men and women: those who had better access to specimens or demonstrations clearly benefited from the experience. In a rare glimpse of a less intellectualized level of gendered discourse, one respondent commented flippantly: "Then finally there is always a certain and obvious difficulty if the girl is either a very pretty or a very ugly one, or has red hair or no hair at all."[70]

---

[68] "Report of the Degrees for Women Syndicate," Appendix 5: "Replies to Inquiries Addressed to the Heads of Girton College and Newnham College," *Cambridge University Reporter,* 1897, pp. 614–617; and Edith R. Saunders to the Degrees for Women Syndicate, 23 Nov. 1896, Synd.II.16, MSS Room, CUL. The women in question were Elizabeth Dale (b. 1868) of Girton and Frances Gregory Whitting (1868–1914) of Newnham. Francis Darwin nonetheless supported the admission of women to science lectures, responding favorably to the 1896 questionnaire regarding women in lectures and laboratories: Synd.II.12, MSS Room, CUL.

[69] Vines to the Degrees for Women Syndicate, 2 Nov. 1896, Synd.II.12, MSS Room, The physiologist W. B. Hardy was a secretary of a group of dons who actively opposed the memorial: McWilliams-Tullberg, "Women and Degrees at Cambridge University" (cit. n. 3), pp. 297–298.

[70] The author was the botanist Walter Gardiner; the remark about the custom of gender separation was published in the *Cambridge University Reporter,* 1897, p. 600. For the printed circular and the responses by the "Biology and Geology" instructors, see Synd.II.3; for a summary of the comments see Synd.II.2, MSS Room, CUL. That there existed in Cambridge a level of "unofficial" discourse on these matters apart from the intellectualized tone of the printed reports was pointed out by William Bateson in the university discussion of the syndicate's report. Bateson, who strongly favored conferring degrees on women, confronted opponents with the comment: "We are not solely dependent on official statements. We understand our opponents' views. We do not look solely to what we hear in this room. We do not look solely to the fly-sheets and to formal and official documents. Do we close our ears when we leave this room, and do we pretend not to know what we hear in Hall and in other places where we meet? We know very well that there is quite a different feeling in the matter

The final "Report of the Degrees for Women Syndicate," published in February 1897, demurred from including in the memorial to be voted on in May "any proposal for regulating the attendance of women at Lectures and Laboratories," noting that science "Professors and Lecturers have shewn themselves willing to give to Women Students such facilities for study as they could." When the votes of members of the university (those holding Cambridge M.A.'s) were counted on 21 May, the opposition carried by a margin of 1,713 to 662.[71]

The defeat was obviously a tremendous blow to the movement for the education of women, but it especially impacted on biological studies. It gave Adam Sedgwick institutional support for his decision to exclude women for the first time from lectures as well as laboratories in zoology (see Figure 4). Sedgwick's opposition to the degrees for women movement had been made public in an 18 May 1897 letter to the *Standard,* published just days before the vote. Drawing attention to the recent mock vote in which undergraduates expressed their opposition to the degree proposal, Sedgwick warned that were the grace to carry, "the glorious career of this University as a producer of great men will receive a most serious check." Indeed, Sedgwick had long been opposed to the granting of degrees to women. Ten years earlier, when Cambridge first considered the matter, Sedgwick had threatened that "he would turn all the women out of his laboratory and not allow them to attend lectures" if they did not stop agitating for degrees.[72]

Bolstered by the defeat, Sedgwick notified the heads of the women's colleges at the beginning of the Michaelmas term of 1897 that women would no longer be welcome in the lectures or the laboratory of elementary morphology. As Eleanor Sidgwick lamented to Alexander MacAlister, professor of medicine and a longtime supporter of women's education, "Mr. Sedgwick's wish no longer to have our students at his lectures is the last straw making it necessary for us to have a lecture-room of our own." The ever-resourceful Mrs. Sidgwick met this challenge head-on. The college appointed a new lecturer in morphology, Elinor Gladys Philipps (1872–1965), added a new lecture room onto the laboratory, completed by the Lent term of 1899, and provided all the instruction in morphology and biology.[73]

---

altogether, and that what our opponents mean is opposition to women being here altogether. [Hear, hear.]" Bateson then urged those who said " 'hear, hear' " to be good enough "to put that expression of opinion into print and send it through the Senate . . . because then we shall know what we have got to deal with." See *Cambridge University Reporter,* 1897, p. 798.

[71] *Cambridge University Reporter,* 1897, p. 589. McWilliams-Tullberg interprets the defeat as a reflection of the "fierce resistance to women's enfranchisement in prewar Britain" because graduates of Cambridge and Oxford were given a vote in university affairs: McWilliams-Tullberg, "Women and Degrees at Cambridge" (cit. n. 3), p. 117. The disappointment of the defeat led many women, including Emily Davies, to turn from supporting higher education as the primary means of improving the position of women in British society to advocating women's suffrage. See Sheila Fletcher, *Feminists and Bureaucrats: A Study in the Development of Girls' Education in the Nineteenth Century* (Cambridge: Cambridge Univ. Press, 1980), pp. 181–182; and Susan Kingsley Kent, *Sex and Suffrage in Britain, 1860–1914* (Princeton, N.J.: Princeton Univ. Press, 1987), pp. 197–219.

[72] Adam Sedgwick to the editor, *Standard,* 19 May 1897 (I thank Jill Allbrooke, Information Officer at the British Library, for providing me with a copy of this letter). See also McWilliams-Tullberg, *Women at Cambridge* (cit. n. 3), p. 135. Sedgwick's 1887 statement is quoted in Gould, "Women and the Culture of Physics" (cit. n. 3), p. 148. In an ominous sign of things to come, the *Times* carried an editorial on 22 May lauding the outcome: "Cambridge has hitherto treated her women students with indulgence, some may think excessive indulgence." But now "concessions must cease if it is to maintain its ancient tradition as a seat of learning . . . governed in the interests of men." Quoted in Eschbach, *Higher Education of Women* (cit. n. 1), p. 32.

[73] Eleanor Sidgwick to Alexander MacAlister, 12 Nov. 1897, BL: P&C, NCA. According to Edith Saunders, the new addition was made possible by another donation from Eleanor Sidgwick; see Saunders, [History of the Balfour Laboratory, 1914], BLR, NCA. Another wing was added ten years later, providing accommodations for chemical physiology and histology and a second room for lectures and coaching: Report of 15 May 1909; and

Figure 4. *Adam Sedgwick. (Photograph by A. G. Dew Smith. By permission of the Balfour and Newton Library, Department of Zoology, University of Cambridge.)*

THE CHANGE IN THE "MIND OF THE UNIVERSITY"

By the outbreak of World War I, however, a fundamental change in the "mind of the University," to recall Henry Sidgwick's prediction after the defeat of 1897, began to occur.[74] The place of women in the scientific establishment of the university began to be

---

Report of 8 May 1911, BLR, NCA. Elinor Philipps, who had taken a first class in Part 1 of the NST (1894) and a second in Part 2 (1895), was called to Cambridge from her position as assistant mistress at Bradford Girls' Grammar School. On the staff of the Balfour from 1898 to 1901, she then traveled to Japan, where she was on the staff of the Japan Women's University, Tokyo, until 1941. See *Newnham College Register,* Vol. 1 (cit. n. 29), p. 9.

[74] It was Henry Sidgwick's view, after the defeat, that the university was not yet ready to grant women full membership and that nothing the supporters could do would be likely to alter this. "We have only to wait," he stated, "for a fundamental change in the mind of the University; and we shall wait": "Discussion of a Report," *Cambridge University Reporter,* 1897, p. 754.

accepted. In 1906, the Balfour Laboratory committee learned that Adam Sedgwick was "willing for the present to readmit the students of Newnham and Girton to his course of lectures on elementary Zoology." Elected to the vacant professorship in zoology and comparative anatomy in 1907, Sedgwick left Cambridge in 1909 to become professor of zoology at the newly constituted Imperial College of Science and Technology. His successor, John Stanley Gardiner, was more sympathetic to women. Gardiner not only agreed to admit "women completely both for lectures & practical work to the University Laboratory," but also appointed the first woman, Kathleen Haddon (1888–1961), as assistant demonstrator in the University Morphological Laboratory.[75] From this time forward, the position of women in the university began to improve.

This "progress" may have been prompted as much by economic expediency as by changes in mentality. By 1910, the university had constructed facilities large enough to accommodate women as well as men. In 1912 the Balfour Laboratory committee informed Newnham College that "the Committee believe that it is a good thing for our students to work at the University Laboratory, & gladly availed themselves of Professor Gardiner's offer." By 1914, with enrollment declining owing to wartime mobilization, the professors of botany and physiology followed Gardiner's lead, and it was decided to close the Balfour Laboratory "in consequence of the admission of women students to these University courses."[76] The committee recommended that the building should be maintained for university women or men engaged in biological research. As Edith Saunders later recalled: "The full admission of the students of Newnham & Girton Colleges to all University courses relieved the need for organising separate courses for the women students in a separate laboratory and all teaching work at the laboratory came to an end in 1914 after it had served the needs of the Science students of Newnham and Girton for 30 years."[77]

CONCLUSION

The Balfour Biological Laboratory for Women was born out of the need to provide a facility in which women could have hands-on experience with living organisms and learn the practical techniques associated with the new experimental biology. As such, it was crucial to the success of women sitting the Natural Sciences Tripos at Cambridge. At first it merely complemented the university courses and practical classes; later, after women began to be excluded from those lectures and laboratories, it assumed the more critical

---

[75] Report of [1901]; and Report of 12 May 1906, BLR, NCA. Gardiner had married the former natural sciences research fellow at Newnham, Edith Gertrude Willcock (1879–1953), in 1909. Kathleen Haddon, the daughter of Alfred Haddon, professor of zoology at Dublin University, took first-class honors in Part 1 of the NST (1910) and a second in Part 2 (1911): *Newnham College Register,* Vol. 1, p. 224. The readmission of women to university lectures was not entirely advantageous, however, for it reduced the number of lectureships available to women. Upon Sedgwick's 1906 decision, Agnes Isabella Mary Elliot (1863–1946) was dismissed after three years' teaching; and in 1912 the more senior zoologist, Igerna Brünhild Johnson Sollas (1863–1946), staff lecturer and demonstrator in zoology since 1903, was left "without any teaching": Report of 7 May 1912, BLR, NCA.

[76] Report of 2 May 1912; and Report of 1914, BLR, NCA. An economic motive also influenced opening the university to women at Oxford, wartime inflation having greatly reduced the value of the university's endowment. See Adams, *Somerville for Women* (cit. n. 2), pp. 149–150. In 1912 J. N. Langley, Foster's successor since 1903, agreed for the first time to accept women in the histological part of both the elementary and intermediate courses in physiology; in 1914 the professor of botany, Albert Charles Seward, gave women permission to attend all botanical demonstrations, as did the new professor of biochemistry, Frederick Gowland Hopkins.

[77] Saunders, [History of the Balfour Laboratory, 1914], BLR, NCA. Before selling the building in 1928, Newnham College leased it in 1919 to the Beit Memorial Fellowships for Medical Research for the use of Hopkins's biochemistry department, and for many years it was home to the Department of Geography. Today, ironically, the building has again reverted to religious use, serving as a church hall.

pedagogical role of providing the entire range of biological instruction for the women of Cambridge.

Paradoxically, the exclusion of women from the lectures in elementary morphology, as well as from the laboratories of zoology, botany, and physiology, in many respects strengthened the edifice of women's scientific instruction at Cambridge. Forced to rely upon their own graduates to staff their biological laboratory and provide instruction in elementary subjects, the women's colleges created academic science positions that would otherwise have been unavailable to women. Indeed, few of the women who served as staff members of the Balfour Laboratory enjoyed subsequent academic careers in the sciences.[78]

Eleanor Sidgwick, surveying the women's colleges' achievements in science in the decades prior to 1915, did not see the necessity for Newnham and Girton to erect a separate women's biological laboratory as in any sense extraordinary, "for many of the men also received their practical instruction in college laboratories which have since, for the most part, been done away with and their work absorbed in a more centralised system." Her view may well have been shared by other pioneers who were intimately involved in the early movement to extend university education to women. Certainly the participation of women in university courses and laboratories was accepted by the 1920s. As the embryo ecologist G. Evelyn Hutchinson, a student at Cambridge in the early 1920s, recalled: "Senior undergraduates, at least in the biological sciences, regarded the women students as equally as much a part of the place as themselves." Yet Hutchinson also recognized the disadvantages the dons of Newnham and Girton colleges faced, for they "took very little part in teaching outside college supervision, though they were not completely excluded from university classrooms."[79]

Even though, by 1914, women were assured places in the university lectures and laboratories, this privilege was not without cost. Before 1948 they entered these facilities on terms unequal to those of undergraduates, precluded from full participation in the Cambridge scientific community, yet without the support of the Balfour Laboratory, which for thirty years had formed the core of the women's scientific subculture. Moreover, with the loss of the academic positions the Balfour's demise entailed, women academic scientists' access to an established career ladder was diminished.

The barriers that women in the life sciences faced at Cambridge in the late nineteenth and early twentieth centuries reveal much, despite the specialized context, about the general situation faced by women in science. Women were restricted from full access to university lectures and laboratories and excluded from scientific societies, which marginalized them not only from instruction but also from mainstream science. Women who wished to pursue postgraduate research were hindered should they fail to gain a male mentor, permission to work in a university laboratory, or one of the few fellowships available to women. In short, women suffered by not having as well developed a scientific support system as did the men—including the college fellowships, studentships, scholarships, prizes, demonstrator-

---

[78] See MacLeod and Moseley, "Fathers and Daughters" (cit. n. 3), p. 330, Table 3. Of the one-time Balfour demonstrators and lecturers, only three continued to hold academic appointments: Elinor Philipps, Edith Saunders, and Muriel Wheldale Onslow. Philipps, however, taught in Japan, and Saunders held college appointments. Wheldale was appointed a university lecturer in Hopkins's new Biochemical Department at Cambridge (1927), "one of the first women to receive such an appointment": Creese, "British Women" (cit. n. 4), p. 184.

[79] Eleanor Sidgwick, address delivered at the memorial to the longtime Newnham chemical lecturer Ida Freund, 7 Aug. 1915, NCA; and G. Evelyn Hutchinson, *The Kindly Fruits of the Earth: Recollections of an Embryo Ecologist* (New Haven, Conn.: Yale Univ. Press, 1979), p. 75. The less favorable position of women dons was also noted by another science student of the time. See Enid M. R. Russell-Smith, "The Art of Theorising," in *A Newnham Anthology,* ed. Phillips (cit. n. 13), pp. 155–158, on p. 157.

ships, and lectureships that could enable them to pursue careers in science. Without these, few were able to remain in science for more than a few years.

Nonetheless, during the first three decades of women's education at Cambridge, there arose an active and distinguished group of women scientists who, despite obstacles connected with their gender, were able to carry on work "of a useful kind" or sometimes even tinged with "genius." Barred from full participation in the wider scientific community available to men, women established their own scientific subculture at the university. Under such conditions few could attain scientific prominence. The history of the Balfour Biological Laboratory for Women both shows the workings of an early teaching laboratory and provides a rare glimpse into this rich subculture.

For historians, this episode may stand as a warning against a conceptual trap analogous to the late nineteenth-century mentality regarding genius. Perhaps we should approach the accomplishments of women scientists in terms other than our traditional analysis of research programs, research schools, professionalization, publication records, and other signs of "normal" academic achievement.[80] As we have seen, these categories do not always adequately capture the full range of women's experiences in the sciences.

---

[80] See Pnina G. Abir-Am and Dorinda Outram, "Introduction," in *Uneasy Careers and Intimate Lives,* ed. Abir-Am and Outram (New Brunswick, N.J.: Rutgers Univ. Press, 1987), p. 2. They argue that studies of women in science indicate that "substantial modification needs to be made to the usual picture of the history of the structures of modern scientific organization as largely determined by a gradual process of professionalization," which assumes "paid posts for the full-time pursuit of science" and "the certification of scientific competence through formal examination by recognized groups or institutional bodies."

Race and Gender: The Role of Analogy in Science

By Nancy Leys Stepan

METAPHOR OCCUPIES a central place in literary theory, but the role of metaphors, and of the analogies they mediate, in scientific theory is still debated.[1] One reason for the controversy over metaphor, analogy, and models in science is the intellectually privileged status that science has traditionally enjoyed as the repository of nonmetaphorical, empirical, politically neutral, universal knowledge. During the scientific revolution of the seventeenth century, metaphor became associated with the imagination, poetic fancy, subjective figures, and even untruthfulness and was contrasted with truthful, unadorned, objective knowledge—that is, with science itself.[2]

In the twentieth century logical positivists also distinguished between scientific and metaphoric language.[3] When scientists insisted that analogies or models based on analogies were important to their thinking, philosophers of science tended to dismiss their claims that metaphors had an *essential* place in scientific utterances. The French theoretical physicist Pierre Duhem was well known for his criticism of the contention that metaphor and analogies were important to *explanation* in science. In his view, the aim of science was to reduce all theory to mathematical statements; models could aid the process of scientific discovery, but once they had served their function, analogies could be discarded as extrinsic to science, and the theories made to stand without them.[4]

One result of the dichotomy established between science and metaphor was that obviously metaphoric or analogical science could only be treated as

---

[1] A metaphor is a figure of speech in which a name or descriptive term is transferred to some object that is different from, but analogous to, that to which it is properly applicable. According to Max Black, "every metaphor may be said to mediate an analogy or structural correspondence": see Black, "More About Metaphor," in *Metaphor and Thought*, ed. Andrew Ortony (Cambridge: Cambridge Univ. Press, 1979), pp. 19–43, on p. 31. In this article, I have used the terms *metaphor* and *analogy* interchangeably.

[2] G. Lakoff and M. Johnson, *Metaphors We Live By* (Chicago/London: Univ. Chicago Press, 1980), p. 191. Scientists' attacks on metaphor as extrinsic and harmful to science predate the Scientific Revolution.

[3] See A. J. Ayer, *Language, Truth and Logic* (New York: Dover, 1952), p. 13.

[4] On Duhem, see Carl H. Hempel, *Aspects of Scientific Explanation and Other Essays in the Philosophy of Science* (New York: Free Press, 1965), pp. 433–477. Hempel agrees with Duhem's view that "all references to analogies or analogical models can be dispensed with in the systematic statement of scientific explanations" (p. 440).

"prescientific" or "pseudoscientific" and therefore dismissable.[5] Because science has been identified with truthfulness and empirical reality, the metaphorical nature of much modern science tended to go unrecognized. And because it went unrecognized, as Colin Turbayne has pointed out, it has been easy to mistake the model in science "for the thing modeled"—to think, to take his example, that nature *was* mechanical, rather than to think it was, metaphorically, seen as mechanical.[6]

More recently, however, as the attention of historians and philosophers of science has moved away from logical reconstructions of science toward more "naturalistic" views of science in culture, the role of metaphor, analogies, and models in science has begun to be acknowledged.[7] In a recent volume on metaphor, Thomas S. Kuhn claims that analogies are fundamental to science; and Richard Boyd argues that they are "irreplaceable parts of the linguistic machinery of a scientific theory," since cases exist in which there are metaphors used by scientists to express theoretical claims "for which no adequate literal paraphrase is known."[8] Some philosophers of science are now prepared to assert that metaphors and analogies are not just psychological aids to scientific discovery, or heuristic devices, but constituent elements of scientific theory.[9] We seem about to move full circle, from considering metaphors mere embellishments or poetic fictions to considering them essential to scientific thought itself.

Although the role of metaphor and analogy in science is now recognized, a critical theory of scientific metaphor is only just being elaborated. The purpose of this article is to contribute to the development of such a theory by using a particular analogy in the history of the life sciences to explore a series of related questions concerning the cultural sources of scientific analogies, their role in scientific reasoning, their normative consequences, and the process by which they change.

RACE AND GENDER: A POWERFUL SCIENTIFIC ANALOGY

The analogy examined is the one linking race to gender, an analogy that occupied a strategic place in scientific theorizing about human variation in the nineteenth and twentieth centuries.

---

[5] For this point see Jamie Kassler, "Music as a Model in Early Science," *History of Science.* 1982. 20:103–139.

[6] Colin M. Turbayne, *The Myth of Metaphor* (Columbia: Univ. South Carolina Press, 1970), p. 24.

[7] General works on metaphor and science include Philip Wheelwright, *Metaphor and Reality* (Bloomington: Indiana Univ. Press, 1962); Max Black, *Models and Metaphor* (Ithaca, N.Y.: Cornell Univ. Press, 1962); Mary Hesse, *Models and Analogies in Science* (Notre Dame, Ind.: Univ. Notre Dame Press, 1966); Richard Olson, ed., *Science as Metaphor* (Belmont, Calif.: Wadsworth, 1971); W. M. Leatherdale, *The Role of Analogy, Model and Metaphor in Science* (Amsterdam: North-Holland, 1974); Ortony, ed., *Metaphor and Thought* (cit. n. 1); and Roger S. Jones, *Physics as Metaphor* (Minneapolis: Univ. Minnesota Press, 1982). Warren A. Shibles, *Metaphor: An Annotated Guide and History* (Whitewater, Wisconsin: Language Press, 1971), gives an extensive introduction and guide to the general problem of metaphor, language, and reality.

[8] Thomas S. Kuhn, "Metaphor in Science," in *Metaphor and Thought*, ed. Ortony, pp. 409–419, on p. 414; and Richard Boyd, "Metaphor and Theory Change: What Is 'Metaphor' a Metaphor For?" *ibid.*, pp. 356–408, on p. 360.

[9] For a defense of the centrality of analogy to science see N. R. Campbell, "What Is a Theory?" in *Readings in the Philosophy of Science*, ed. Baruch A. Brody (Englewood Cliffs, N.J.: Prentice-Hall, 1970), pp. 252–267. Shibles, in *Metaphor*, p. 3, also argues that each school of science "is based on a number of basic metaphors which are then expanded into various universes of discourse."

As has been well documented, from the late Enlightenment on students of human variation singled out racial differences as crucial aspects of reality, and an extensive discourse on racial inequality began to be elaborated.[10] In the nineteenth century, as attention turned increasingly to sexual and gender differences as well, gender was found to be remarkably analogous to race, such that the scientist could use racial difference to explain gender difference, and vice versa.[11]

Thus it was claimed that women's low brain weights and deficient brain structures were analogous to those of lower races, and their inferior intellectualities explained on this basis.[12] Woman, it was observed, shared with Negroes a narrow, childlike, and delicate skull, so different from the more robust and rounded heads characteristic of males of "superior" races. Similarly, women of higher races tended to have slightly protruding jaws, analogous to, if not as exaggerated as, the apelike, jutting jaws of lower races.[13] Women and lower races were called innately impulsive, emotional, imitative rather than original, and incapable of the abstract reasoning found in white men.[14] Evolutionary biology provided yet further analogies. Woman was in evolutionary terms the "conservative element" to the man's "progressive," preserving the more "primitive" traits found in lower races, while the males of higher races led the way in new biological and cultural directions.[15]

Thus when Carl Vogt, one of the leading German students of race in the middle of the nineteenth century, claimed that the female skull approached in many respects that of the infant, and in still further respects that of lower races, whereas the mature male of many lower races resembled in his "pendulous" belly a Caucasian woman who had had many children, and in his thin calves and flat thighs the ape, he was merely stating what had become almost a cliché of the science of human difference.[16]

So fundamental was the analogy between race and gender that the major modes of interpretation of racial traits were invariably evoked to explain sexual traits. For instance, just as scientists spoke of races as distinct "species," incapable of crossing to produce viable "hybrids," scientists analyzing male-female differences sometimes spoke of females as forming a distinct "species," individual members of which were in danger of degenerating into psychosexual hybrids

[10] See Nancy Stepan, *The Idea of Race in Science: Great Britain, 1800–1960* (London: Macmillan, 1982), esp. Ch. 1.

[11] No systematic history of the race-gender analogy exists. The analogy has been remarked on, and many examples from the anthropometric, medical, and embryological sciences provided, in Stephen Jay Gould, *The Mismeasure of Man* (New York: W. W. Norton, 1981), and in John S. Haller and Robin S. Haller, *The Physician and Sexuality in Victorian America* (Urbana: Univ. Illinois Press, 1974).

[12] Haller and Haller, *The Physician and Sexuality*, pp. 48–49, 54. Among the several craniometric articles cited by the Hallers, see esp. J. McGrigor Allan, "On the Real Differences in the Minds of Men and Women," *Journal of the Anthropological Society of London*, 1869, 7:cxcv–ccviii, on p. cciv; and John Cleland, "An Inquiry into the Variations of the Human Skull," *Philosophical Transactions, Royal Society*, 1870, 89:117–174.

[13] Havelock Ellis, *Man and Woman: A Study of Secondary Sexual Characters* (1894; 6th ed, London: A. & C. Black, 1926), pp. 106–107.

[14] Herbert Spencer, "The Comparative Psychology of Man," *Popular Science Monthly*, 1875–1876, 8:257–269.

[15] Ellis, *Man and Woman* (cit. n. 13), p. 491.

[16] Carl Vogt, *Lectures on Man: His Place in Creation, and in the History of the Earth* (London: Longman, Green, & Roberts, 1864), p. 81.

when they tried to cross the boundaries proper to their sex.[17] Darwin's theory of sexual selection was applied to both racial and sexual difference, as was the neo-Lamarckian theory of the American Edward D. Cope.[18] A last, confirmatory example of the analogous place of gender and race in scientific theorizing is taken from the history of hormone biology. Early in the twentieth century the anatomist and student of race Sir Arthur Keith interpreted racial differences in the human species as a function of pathological disturbances of the newly discovered "internal secretions" or hormones. At about the same time, the apostle of sexual frankness and well-known student of sexual variation Havelock Ellis used internal secretions to explain the small, but to him vital, differences in the physical and psychosexual makeup of men and women.[19]

In short, lower races represented the "female" type of the human species, and females the "lower race" of gender. As the example from Vogt indicates, however, the analogies concerned more than race and gender. Through an intertwined and overlapping series of analogies, involving often quite complex comparisons, identifications, cross-references, and evoked associations, a variety of "differences"—physical and psychical, class and national—were brought together in a biosocial science of human variation. By analogy with the so-called lower races, women, the sexually deviate, the criminal, the urban poor, and the insane were in one way or another constructed as biological "races apart" whose differences from the white male, and likenesses to each other, "explained" their different and lower position in the social hierarchy.[20]

It is not the aim of this article to provide a systematic history of the biosocial science of racial and sexual difference based on analogy. The aim is rather to use the race-gender analogy to analyze the nature of analogical reasoning in science itself. When and how did the analogy appear in science? From what did it derive its scientific authority? How did the analogy shape research? What did it mean when a scientist claimed that the mature male of many lower races resembled a mature Caucasian female who had had many children? No simple theory of resemblance or substitution explains such an analogy. How did the analogy help construct the very similarities and differences supposedly "discovered" by scientists in nature? What theories of analogy and metaphor can be most effectively applied in the critical study of science?

THE CULTURAL SOURCES OF SCIENTIFIC METAPHOR

How particular metaphors or analogies in science are related to the social production of science, why certain analogies are selected and not others, and why

[17] James Weir, "The Effect of Female Suffrage on Posterity," *American Naturalist*, 1895, 29:198–215.

[18] Charles Darwin, *The Descent of Man, and Selection in Relation to Sex* (London: John Murray, 1871), Vol. II, Chs. 17–20; Edward C. Cope, "The Developmental Significance of Human Physiognomy, *Amer. Nat.*, 1883, 17:618–627.

[19] Arthur Keith, "Presidential Address: On Certain Factors in the Evolution of Human Races," *Journal of the Royal Anthropological Institute*, 1916, 64:10–33; Ellis, *Man and Woman* (cit. n. 13), p. xii.

[20] See Nancy Stepan, "Biological Degeneration: Races and Proper Places," in *Degeneration: The Dark Side of Progress*, ed. J. Edward Chamberlin and Sander L. Gilman (New York: Columbia Univ. Press, 1985), pp. 97–120, esp. pp. 112–113. For an extended exploration of how various stereotypes of difference intertwined with each other, see Sander L. Gilman, *Difference and Pathology: Stereotypes of Sexuality, Race, and Madness* (Ithaca, N.Y.: Cornell Univ. Press, 1985).

certain analogies are accepted by the scientific community are all issues that need investigation.

In literature, according to Warren Shibles, striking metaphors just come, "like rain."[21] In science, however, metaphors and analogies are not arbitrary, nor merely personal. Not just any metaphors will do. In fact, it is their lack of perceived "arbitrariness" that makes particular metaphors or analogies acceptable as science.

As Stephen Toulmin recently pointed out, the constraints on the choice of metaphors and analogies in science are varied. The nature of the objects being studied (e.g., organic versus nonorganic), the social (e.g., class) structure of the scientific community studying them, and the history of the discipline or field concerned all play their part in the emergence of certain analogies rather than others and in their "success" or failure.[22] Sometimes the metaphors are strikingly new, whereas at other times they extend existing metaphors in the culture in new directions.

In the case of the scientific study of human difference, the analogies used by scientists in the late eighteenth century, when human variation began to be studied systematically, were products of long-standing, long-familiar, culturally endorsed metaphors. Human variation and difference were not experienced "as they really are, out there in nature," but by and through a metaphorical system that structured the experience and understanding of difference and that in essence created the objects of difference. The metaphorical system provided the "lenses" through which people experienced and "saw" the differences between classes, races, and sexes, between civilized man and the savage, between rich and poor, between the child and the adult. As Sander Gilman says in his book *Seeing the Insane,* "We do not see the world, rather we are taught by representations of the world about us to conceive of it in a culturally acceptable manner."[23]

The origin of many of the "root metaphors" of human difference are obscure. G. Lakoff and M. Johnson suggest that the basic values of a culture are usually compatible with "the metaphorical structure of the most fundamental concepts in the culture."[24] Not surprisingly, the social groups represented metaphorically as "other" and "inferior" in Western culture were socially "disenfranchised" in a variety of ways, the causes of their disenfranchisement varying from group to group and from period to period. Already in ancient Greece, Aristotle likened women to the slave on the grounds of their "natural" inferiority. Winthrop Jordan has shown that by the early Middle Ages a binary opposition between blackness and whiteness was well established in which blackness was identified with baseness, sin, the devil, and ugliness, and whiteness with virtue, purity, holiness, and beauty.[25] Over time, black people themselves were compared to apes, and their childishness, savageness, bestiality, sexuality, and lack of intellectual capacity stressed. The "Ethiopian," the "African," and especially the "Hottentot"

---

[21] Shibles, *Metaphor* (cit. n. 7), p. 15.

[22] Stephen Toulmin, "The Construal of Reality: Criticism in Modern and Postmodern Science," *Critical Inquiry,* 1982, 9:93–111, esp. pp. 100–103.

[23] Sander L. Gilman, *Seeing the Insane* (New York: John Wiley, 1982), p. xi.

[24] Lakoff and Johnson, *Metaphors We Live By* (cit. n. 2), p. 22. The idea of root metaphors is Stephen Pepper's in *World Hypothesis* (Berkeley/Los Angeles: Univ. California Press, 1966), p. 91.

[25] Winthrop D. Jordan, *White over Black: American Attitudes toward the Negro, 1550–1812* (New York: Norton, 1977), p. 7.

were made to stand for all that the white male was not; they provided a rich analogical source for the understanding and representation of other "inferiorities." In his study of the representation of insanity in Western culture, for instance, Gilman shows how the metaphor of blackness could be borrowed to explicate the madman, and vice versa. In similar analogical fashion, the laboring poor were represented as the "savages" of Europe, and the criminal as a "Negro."

When scientists in the nineteenth century, then, proposed an analogy between racial and sexual differences, or between racial and class differences, and began to generate new data on the basis of such analogies, their interpretations of human difference and similarity were widely accepted, partly because of their fundamental congruence with cultural expectations. In this particular science, the metaphors and analogies were not strikingly new but old, if unexamined and diffuse. The scientists' contribution was to elevate hitherto unconsciously held analogies into self-conscious theory, to extend the meanings attached to the analogies, to expand their range via new observations and comparisons, and to give them precision through specialized vocabularies and new technologies. Another result was that the analogies became "naturalized" in the language of science, and their metaphorical nature disguised.

In the scientific elaboration of these familiar analogies, the study of race led the way, in part because the differences between blacks and whites seemed so "obvious," in part because the abolition movement gave political urgency to the issue of racial difference and social inequality. From the study of race came the association between inferiority and the ape. The facial angle, a measure of hierarchy in nature obtained by comparing the protrusion of the jaws in apes and man, was widely used in analogical science once it was shown that by this measure Negroes appeared to be closer to apes than the white race.[26] Established as signs of inferiority, the facial angle and blackness could then be extended analogically to explain other inferior groups and races. For instance, Francis Galton, Darwin's cousin and the founder of eugenics and statistics in Britain, used the Negro and the apish jaw to explicate the Irish: "Visitors to Ireland after the potato famine," he commented, "generally remarked that the Irish type of face seemed to have become more prognathous, that is, more like the negro in the protrusion of the lower jaw."[27]

Especially significant for the analogical science of human difference and similarity were the systematic study and measurement of the human skull. The importance of the skull to students of human difference lay in the fact that it housed the brain, differences in whose shape and size were presumed to correlate with equally presumed differences in intelligence and social behavior. It was measurements of the skull, brain weights, and brain convolutions that gave apparent precision to the analogies between anthropoid apes, lower races, women, criminal types, lower classes, and the child. It was race scientists who provided the new technologies of measurement—the callipers, cephalometers, craniometers, craniophores, craniostats, and parietal goniometers.[28] The low facial angles at-

---

[26] Stepan, *The Idea of Race in Science*, pp. 6–10.

[27] Francis Galton, "Hereditary Improvement," *Fraser's Magazine*, 1873, 7:116–130.

[28] These instruments and measurements are described in detail in Paul Topinard, *Anthropology* (London: Chapman & Hall, 1878), Pt. II, Chs. 1–4.

tributed by scientists starting in the 1840s and 1850s to women, criminals, idiots, and the degenerate, and the corresponding low brain weights, protruding jaws, and incompletely developed frontal centers where the higher intellectual faculties were presumed to be located, were all taken from racial science. By 1870 Paul Topinard, the leading French anthropologist after the death of Paul Broca, could call on data on sexual and racial variations from literally hundreds of skulls and brains, collected by numerous scientists over decades, in order to draw the conclusion that Caucasian women were indeed more prognathous or apelike in their jaws than white men, and even the largest women's brains, from the "English or Scotch" race, made them like the African male.[29] Once "woman" had been shown to be indeed analogous to lower races by the new science of anthropometry and had become, in essence, a racialized category, the traits and qualities special to woman could in turn be used in an analogical understanding of lower races. The analogies now had the weight of empirical reality and scientific theory. The similarities between a Negro and a white woman, or between a criminal and a Negro, were realities of nature, somehow "in" the individuals studied.

METAPHORIC INTERACTIONS

We have seen that metaphors and analogies played an important part in the science of human difference in the nineteenth century. The question is, what part? I want to suggest that the metaphors functioned as the science itself—that without them the science did not exist. In short, metaphors and analogies can be constituent elements of science.

It is here that I would like to introduce, as some other historians of science have done, Max Black's "interaction" theory of metaphor, because it seems that the metaphors discussed in this essay, and the analogies they mediated, functioned like interaction metaphors, and that thinking about them in these terms clarifies their role in science.[30]

By interaction metaphors, Black means metaphors that join together and bring into cognitive and emotional relation with each other two different things, or systems of things, not normally so joined. Black follows I. A. Richards in opposing the "substitution" theory of metaphor, in which it is supposed that the metaphor is telling us indirectly something factual about the two subjects—that the metaphor is a *literal comparison,* or is capable of a literal translation in prose. Richards proposed instead that "when we use a metaphor, we have two thoughts of different things active together and supported by a single word or phrase, whose meaning is the resultant of their interaction." Applying the interaction theory to the metaphor "The poor are the negroes of Europe," Black paraphrases Richards to claim that "our thoughts about the European poor and American negroes are 'active together' and 'interact' to produce a meaning that is a resultant of that interaction."[31] In such a view, the metaphor cannot be simply

[29] *Ibid.,* p. 311.

[30] Black, *Models and Metaphor* (cit. n. 7), esp. Chs. 3 and 13. See also Mary Hesse, *Models and Analogies in Science* (cit. n. 7); Hesse, "The Explanatory Function of Metaphor," in *Logic, Methodology and Philosophy of Science,* ed. Y. Bar-Hillel (Amsterdam: North-Holland, 1965), pp. 249–259; and Boyd, "Metaphor and Theory Change" (cit. n. 8).

[31] Black, *Models and Metaphor,* p. 38, quoting I. A. Richards, *Philosophy of Rhetoric* (Oxford: Oxford Univ. Press, 1938), p. 93.

reduced to literal comparisons or "like" statements without loss of meaning or cognitive content, because meaning is a product of the interaction between the two parts of a metaphor.

How do these "new meanings" come about? Here Black adds to Richards by suggesting that in an interaction metaphor, a "system of associated common-places" that strictly speaking belong only to one side of the metaphor are applied to the other. And he adds that what makes the metaphor effective "is not that the commonplaces shall be true, but that they should be readily and freely evoked."[32] Or as Mary Hesse puts it in *Models and Analogies in Science*, these implications "are not private, but are largely common to a given language community and are presupposed by speakers who intend to be understood."[33] Thus in the example given, the "poor of Europe" are seen in terms strictly applicable only to the "Negro" and vice versa. As a consequence, the poor are seen like a "race apart," savages in the midst of European civilization. Conversely, the "Negro" is seen as shiftless, idle, given to drink, part of the social remnant bound to be left behind in the march toward progress. Both the ideas of "savagery" and of "shiftlessness" belong to familiar systems of implications that the metaphor itself brings into play.

Black's point is that by their interactions and evoked associations both parts of a metaphor are changed. Each part is seen as more like the other in some characteristic way. Black was primarily interested in ordinary metaphors of a culture and in their commonplace associations. But instead of commonplace associations, a metaphor may evoke more specially constructed systems of implications. Scientists are in the business of constructing exactly such systems of implications, through their empirical investigations into nature and through their introduction into discourse of specialized vocabularies and technologies.[34] It may be, indeed, that what makes an analogy suitable for scientific purposes is its ability to be suggestive of new systems of implications, new hypotheses, and therefore new observations.[35]

In the case of the nineteenth-century analogical science of human difference, for instance, the system of implications evoked by the analogy linking lower races and women was not just a generalized one concerning social inferiority, but the more precise and specialized one developed by years of anthropometric, medical, and biological research. When "woman" and "lower races" were analogically and routinely joined in the anthropological, biological, and medical literature of the 1860s and 1870s, the metaphoric interactions involved a complex system of implications about similarity and difference, often involving highly technical language (for example, in one set of measurements of the body in different races cited by Paul Topinard in 1878 the comparisons included measures in each race of their height from the ground to the acromion, the epicondyle, the styloid process of the radius, the great trochanter, and the internal malleolus). The systems of implications evoked by the analogy included questions of com-

---

[32] Black, *Models and Metaphor*, p. 4.
[33] Hesse, *Models and Analogies in Science* (cit. n. 7), pp. 159–160.
[34] See Turbayne, *Myth of Metaphor* (cit. n. 6), p. 19, on this point.
[35] Black himself believed scientific metaphors belonged to the pretheoretical stage of a discipline. Here I have followed Boyd, who argues in "Metaphor and Theory Change" (cit. n. 8), p. 357, that metaphors can play a role in the development of theories in relatively mature sciences. Some philosophers would reserve the term "model" for extended, systematic metaphors in science.

parative health and disease (blacks and women were believed to show greater degrees of insanity and neurasthenia than white men, especially under conditions of freedom), of sexual behavior (females of "lower races" and lower-class women of "higher races," especially prostitutes, were believed to show similar kinds of bestiality and sexual promiscuity, as well as similar signs of pathology and degeneracy such as deformed skulls and teeth), and of "childish" character- istics, both physical and moral.[36]

As already noted, one of the most important systems of implications about human groups developed by scientists in the nineteenth century on the basis of analogical reasoning concerned head shapes and brain sizes. It was assumed that blacks, women, the lower classes, and criminals shared low brain weights or skull capacities. Paul Broca, the founder of the Société d'Anthropologie de Paris in 1859, asserted: "In general, the brain is larger in mature adults than in the elderly, in men than in women, in eminent men than in men of mediocre talent, in superior races than in inferior races. . . . Other things being equal, there is a remarkable relationship between the development of intelligence and the volume of the brain."[37]

Such a specialized system of implications based on the similarities between brains and skulls appeared for the first time in the phrenological literature of the 1830s. Although analogies between women and blackness had been drawn be- fore, woman's place in nature and her bio-psychological differences from men had been discussed by scientists mainly in terms of reproductive function and sexuality, and the most important analogies concerned black females (the "sign" of sexuality) and lower-class or "degenerate" white women. Since males of all races had no wombs, no systematic, apparently scientifically validated grounds of comparison between males of "lower" races and women of "higher" races existed.

Starting in the 1820s, however, the phrenologists began to focus on differences in the shape of the skull of individuals and groups, in the belief that the skull was a sign faithfully reflecting the various organs of mind housed in the brain, and that differences in brain organs explained differences in human behavior. And it is in the phrenological literature, for almost the first time, that we find women and lower races compared directly on the basis of their skull formations. In their "organology," the phrenologists paid special attention to the organ of "philopro- genitiveness," or the faculty causing "love of offspring," which was believed to be more highly developed in women than men, as was apparent from their more highly developed upper part of the occiput. The same prominence, according to Franz Joseph Gall, was found in monkeys and was particularly well developed, he believed, in male and female Negroes.[38]

By the 1840s and 1850s the science of phrenology was on the wane, since the organs of the brain claimed by the phrenologists did not seem to correspond with the details of brain anatomy as described by neurophysiologists. But although the

---

[36] For an example of the analogous diseases and sexuality of "lower" races and "lower" women, see Eugene S. Talbot, *Degeneracy: Its Causes, Signs, and Results* (London: Walter Cott, 1898), pp. 18, 319–323.

[37] Paul Broca, "Sur le volume et la forme du cerveau suivant les individus et suivant les races," *Bulletin de la Société d'Anthropologie Paris*, 1861, 2:304.

[38] Franz Joseph Gall, "The Propensity to Philoprogenitiveness," *Phrenological Journal*, 1824– 1825, 2:20–33.

specific conclusions of the phrenologists concerning the anatomical structure and functions of the brain were rejected, the principle that differences in individual and group function were products of differences in the shape and size of the head was not. This principle underlay the claim that some measure, whether of cranial capacity, the facial angle, the brain volume, or brain weight, would be found that would provide a true indicator of innate capacity, and that by such a measure women and lower races would be shown to occupy analogous places in the scale of nature (the "scale" itself of course being a metaphorical construct).

By the 1850s the measurement of women's skulls was becoming an established part of craniometry and the science of gender joined analogically to race. Vogt's *Lectures on Man* included a long discussion of the various measures available of the skulls of men and women of different races. His data showed that women's smaller brains were analogous to the brains of lower races, the small size explaining both groups' intellectual inferiority. (Vogt also concluded that within Europe the intelligentsia and upper classes had the largest heads, and peasants the smallest.)[39] Broca shared Vogt's interest; he too believed it was the smaller brains of women and "lower" races, compared with men of "higher" races, that caused their lesser intellectual capacity and therefore their social inferiority.[40]

One novel conclusion to result from scientists' investigations into the different skull capacities of males and females of different races was that the gap in head size between men and women had apparently widened over historic time, being largest in the "civilized" races such as the European, and smallest in the most savage races.[41] The growing difference between the sexes from the prehistoric period to the present was attributed to evolutionary, selective pressures, which were believed to be greater in the white races than the dark and greater in men than women. Paradoxically, therefore, the civilized European woman was less like the civilized European man than the savage man was like the savage woman. The "discovery" that the male and female bodies and brains in the lower races were very alike allowed scientists to draw direct comparisons between a black male and a white female. The male could be taken as representative of both sexes of his race and the black female could be virtually ignored in the analogical science of intelligence, if not sexuality.

Because interactive metaphors bring together a *system* of implications, other features previously associated with only one subject in the metaphor are brought to bear on the other. As the analogy between women and race gained ground in science, therefore, women were found to share other points of similarity with lower races. A good example is prognathism. Prognathism was a measure of the protrusion of the jaw and of inferiority. As women and lower races became analogically joined, data on the "prognathism" of females were collected and women of "advanced" races implicated in this sign of inferiority. Havelock Ellis, for instance, in the late nineteenth-century bible of male-female differences *Man and Woman*, mentioned the European woman's slightly protruding jaw as a trait, not of high evolution, but of the lower races, although he added that in white women the trait, unlike in the lower races, was "distinctly charming."[42]

[39] Vogt, *Lectures on Man* (cit. n. 16), p. 88. Vogt was quoting Broca's data.
[40] Gould, *Mismeasure of Man* (cit. n. 11), p. 103.
[41] Broca's work on the cranial capacities of skulls taken from three cemeteries in Paris was the most important source for this conclusion. See his "Sur la capacité des cranes parisiens des divers époques," *Bull. Soc. Anthr. Paris,* 1862, *3*:102–116.
[42] Ellis, *Man and Woman* (cit. n. 13), pp. 106–107.

Another set of implications brought to bear on women by analogy with lower races concerned dolichocephaly and brachycephaly, or longheadedness and roundheadedness. Africans were on the whole more longheaded than Europeans and so dolichocephaly was generally interpreted as signifying inferiority. Ellis not surprisingly found that on the whole women, criminals, the degenerate, the insane, and prehistoric races tended to share with dark races the more narrow, dolichocephalic heads representing an earlier (and by implication, more primitive) stage of brain development.[43]

ANALOGY AND THE CREATION OF NEW KNOWLEDGE

In the metaphors and analogies joining women and the lower races, the scientist was led to "see" points of similarity that before had gone unnoticed. Women became more "like" Negroes, as the statistics on brain weights and body shapes showed. The question is, what kind of "likeness" was involved?

Here again the interaction theory of metaphor is illuminating. As Black says, the notion of similarity is ambiguous. Or as Stanley Fish puts it, "Similarity is not something one finds but something one must establish."[44] Metaphors are not meant to be taken literally but they do imply some structural similarity between the two things joined by the metaphor, a similarity that may be new to the readers of the metaphoric or analogical text, but that they are culturally capable of grasping.

However, there is nothing obviously similar about a white woman of England and an African man, or between a "criminal type" and a "savage." (If it seems to us as though there is, that is because the metaphor has become so woven into our cultural and linguistic system as to have lost its obviously metaphorical quality and to seem a part of "nature.") Rather it is the metaphor that permits us to see similarities that the metaphor itself helps constitute.[45] The metaphor, Black suggests, "selects, emphasizes, suppresses and organizes features" of reality, thereby allowing us to see new connections between the two subjects of the metaphor, to pay attention to details hitherto unnoticed, to emphasize aspects of human experience otherwise treated as unimportant, to make new features into "signs" signifying inferiority.[46] It was the metaphor joining lower races and women, for instance, that gave significance to the supposed differences between the shape of women's jaws and those of men.

Metaphors, then, through their capacity to construct similarities, create new knowledge. The full range of similarities brought into play by a metaphor or analogy is not immediately known or necessarily immediately predictable. The metaphor, therefore, allows for "discovery" and can yield new information through empirical research. Without the metaphor linking women and race, for example, many of the data on women's bodies (length of limbs, width of pelvis, shape of skull, weight or structure of brain) would have lost their significance as signs of inferiority and would not have been gathered, recorded, and interpreted

[43] Alexander Sutherland, "Woman's Brain," *Nineteenth Century*, 1900, 47:802–810; and Ellis, *Man and Woman*, p. 98. Ellis was on the whole, however, cautious about the conclusions that could be drawn from skull capacities and brain weights.

[44] Stanley Fish, "Working on the Chain Gang: Interpretation in the Law and Literary Criticism," in *The Politics of Interpretation*, ed. W. J. T. Mitchell (Chicago: Univ. Chicago Press, 1983), p. 277.

[45] Max Black, as cited in Ortony, *Metaphor and Thought* (cit. n. 1), p. 5.

[46] Black, *Models and Metaphor* (cit. n. 7), p. 44.

in the way they were. In fact, without the analogies concerning the "differences" and similarities among human groups, much of the vast enterprise of anthropology, criminology, and gender science would not have existed. The analogy guided research, generated new hypotheses, and helped disseminate new, usually technical vocabularies. The analogy helped constitute the objects of inquiry into human variation—races of all kinds (Slavic, Mediterranean, Scottish, Irish, yellow, black, white, and red), as well as other social groups, such as "the child" and "the madman." The analogy defined what was problematic about these social groups, what aspects of them needed further investigation, and which kinds of measurements and what data would be significant for scientific inquiry.

The metaphor, in short, served as a program of research. Here the analogy comes close to the idea of a scientific "paradigm" as elaborated by Kuhn in *The Structure of Scientific Revolutions;* indeed Kuhn himself sometimes writes of paradigms as though they are extended metaphors and has proposed that "the same interactive, similarity-creating process which Black has isolated in the functioning of metaphor is vital also in the function of models in science."[47]

The ability of an analogy in science to create new kinds of knowledge is seen clearly in the way the analogy organizes the scientists' understanding of causality. Hesse suggests that a scientific metaphor, by joining two distinct subjects, implies more than mere structural likeness. In the case of the science of human difference, the analogies implied a similar *cause* of the similarities between races and women and of the differences between both groups and white males. To the phrenologists, the cause of the large organs of philoprogenitiveness in monkeys, Negroes, and women was an innate brain structure. To the evolutionists, sexual and racial differences were the product of slow, adaptive changes involving variation and selection, the results being the smaller brains and lower capacities of the lower races and women, and the higher intelligence and evolutionarily advanced traits in the males of higher races. Barry Barnes suggests we call the kind of "redescription" involved in a metaphor or analogy of the kind being discussed here an "explanation," because it forces the reader to "understand" one aspect of reality in terms of another.[48]

ANALOGY AND THE SUPPRESSION OF KNOWLEDGE

Especially important to the functioning of interactive metaphors in science is their ability to neglect or even suppress information about human experience of the world that does not fit the similarity implied by the metaphor. In their "similarity-creating" capacity, metaphors involve the scientist in a selection of those aspects of reality that are compatible with the metaphor. This selection process is often quite unconscious. Stephen Jay Gould is especially telling about the ways in which anatomists and anthropologists unself-consciously searched for and selected measures that would prove the desired scales of human superiority and inferiority and how the difficulties in achieving the desired results were surmounted.

Gould has subjected Paul Broca's work on human differences to particularly

---

[47] Thomas S. Kuhn, *The Structure of Scientific Revolutions* (Chicago: Univ. Chicago Press, 1962; 2nd ed., 1973), esp. Ch. 4; and Kuhn, "Metaphor in Science" (cit. n. 8), p. 415.

[48] Barry Barnes, *Scientific Knowledge and Sociological Theory* (London: Routledge & Kegan Paul, 1974), p. 49.

thorough scrutiny because Broca was highly regarded in scientific circles and was exemplary in the accuracy of his measurements. Gould shows that it is not Broca's measurements per se that can be faulted, but rather the ways in which he unconsciously manipulated them to produce the very similarities already "contained" in the analogical science of human variation. To arrive at the conclusion of women's inferiority in brain weights, for example, meant failing to make any correction for women's smaller body weights, even though other scientists of the period were well aware that women's smaller brain weights were at least in part a function of their smaller body sizes. Broca was also able to "save" the scale of ability based on head size by leaving out some awkward cases of large-brained but savage heads from his calculations, and by somehow accounting for the occasional small-brained "geniuses" from higher races in his collection.[49]

Since there are no "given" points of measurement and comparison in nature (as Gould says, literally thousands of different kinds of measurements can theoretically be made of the human body), scientists had to make certain choices in their studies of human difference. We are not surprised to find that scientists selected just those points of comparison that would show lower races and women to be nearer to each other and to other "lower" groups, such as the anthropoid apes or the child, than were white men. The maneuvers this involved were sometimes comical. Broca, for instance, tried the measure of the ratio of the radius to the humerus, reasoning that a high ratio was apish, but when the scale he desired did not come out, he abandoned it. According to Gould, he even almost abandoned the most time-honored measure of human difference and inferiority, namely, brain weights, because yellow people did well on it. He managed to deal with this apparent exception to the "general rule of nature" that lower races had small heads by the same kind of specious argumentation he had used with small-brained geniuses. Broca claimed that the scale of brain weights did not work as well at the upper end as at the lower end, so that although small brain weights invariably indicated inferiority, large brain weights did not necessarily in and of themselves indicate superiority![50]

Since most scientists did recognize that the brain weights of women were in fact heavier in proportion to their body weights than men, giving women an apparent comparative advantage over men, not surprisingly they searched for other measures. The French scientist Léonce Pierre Manouvrier used an index relating brain weight to thigh bone weight, an index that gave the desired results and was in confirmation with the analogies, but that even at the time was considered by one scientist "ingenious and fantastic but divorced from common sense."[51] Even more absurd when viewed from the distance of time was the study mentioned by Ellis by two Italians who used the "prehensile" (i.e., apish) character of the human toe to compare human groups and found it was greater in normal white women than in white men, and also marked in criminals, prostitutes, idiots, and of course lower races.[52]

One test of the social power (if not the scientific fruitfulness) of an analogy in science seems in fact to be the degree to which information can be ignored, or

[49] Gould, *Mismeasure of Man* (cit. n. 11), pp. 73–112. For another example see Stephen Jay Gould, "Morton's Ranking of Race by Cranial Capacity," *Science*, 1978, *200*:503–509.
[50] Gould, *Mismeasure of Man*, pp. 85–96.
[51] Sutherland, "Woman's Brain" (cit. n. 43), p. 805.
[52] Ellis, *Man and Woman* (cit. n. 13), p. 53.

interpretation strained, without the analogy losing the assent of the relevant scientific community. On abstract grounds, one would expect an analogy of the kind being discussed here, which required rather obvious distortions of perception to maintain (at least to our late twentieth-century eyes), to have been abandoned by scientists fairly quickly. Since, however, interactive metaphors and analogies direct the investigators' attention to some aspects of reality and not others, the metaphors and analogies can generate a considerable amount of new information about the world that confirms metaphoric expectations and direct attention away from those aspects of reality that challenge those expectations. Given the widespread assent to the cultural presuppositions underlying the analogy between race and gender, the analogy was able to endure in science for a long time.

For instance, by directing attention to exactly those points of similarity and difference that would bring women and lower races closer to apes, or to each other, the race-gender metaphor generated data, many of them new, which "fit" the metaphor and the associated implications carried by it. Other aspects of reality and human experience that were incompatible with the metaphor tended to be ignored or not "seen." Thus for decades the Negro's similarity to apes on the basis of the shape of his jaw was asserted, while the white man's similarity to apes on the basis of his thin lips was ignored.

When contrary evidence could not be ignored, it was often reinterpreted to express the fundamental valuations implicit in the metaphor. Gould provides us with the example of neoteny, or the retention in the adult of childish features such as a small face and hairlessness. A central feature of the analogical science of inferiority was that adult women and lower races were more childlike in their bodies and minds than white males. But Gould shows that by the early twentieth century it was realized that neoteny was a positive feature of the evolutionary process. "At least one scientist, Havelock Ellis, did bow to the clear implication and admit the superiority of women, even though he wriggled out of a similar confession for blacks." As late as the 1920s the Dutch scientist Louis Bolk, on the other hand, managed to save the basic valuation of white equals superior, blacks and women equal inferior by "rethinking" the data and discovering after all that blacks departed more than whites from the most favorable traits of childhood.[53]

To reiterate, because a metaphor or analogy does not directly present a preexisting nature but instead helps "construct" that nature, the metaphor generates data that conform to it, and accommodates data that are in apparent contradiction to it, so that nature is seen via the metaphor and the metaphor becomes part of the logic of science itself.[54]

CHANGING METAPHORS

Turbayne, in his book *The Myth of Metaphor*, proposes as a major critical task of the philosopher or historian of science the detection of metaphor in science. Detection is necessary because as metaphors in science become familiar or com-

---

[53] Gould, *Mismeasure of Man*, pp. 120–121.

[54] Terence Hawkes, *Metaphor* (London: Methuen, 1972), p. 88, suggests that metaphors "will retrench or corroborate as much as they expand our vision," thus stressing the normative, consensus-building aspects of metaphor.

they were far from being at the periphery of the biological and human sciences in the nineteenth and early twentieth centuries. I believe other studies will show that what was true for the analogical science of human difference may well be true also for other metaphors and analogies in science.

My intention has also been to suggest that a theory of metaphor is as critical to science as it is to the humanities. We need a critical theory of metaphor in science in order to expose the metaphors by which we learn to view the world scientifically, not because these metaphors are necessarily "wrong," but because they are so powerful.

"Women's Work" in Science, 1880–1910

By Margaret W. Rossiter

THE PRACTICE OF SCIENCE, it has often been asserted, was always open to both sexes—or, to use sociological terms, was "universalistic" or "sex-blind"—but in fact separate labor markets have long existed for men and women in the sciences.[1] Such markets seem to have emerged in the United States in the 1880s and 1890s, when women first began to seek scientific employment in significant numbers, and they were firmly established in several fields by 1910.[2] Although the practice of sex segregation was usually justified with the essentially conservative rhetoric that women had "special skills" or "unique talents" for certain fields or kinds of work, the phenomenon seems to have been basically an economic one, originating in and sustained by three forces: the rise of a supply of women seeking employment in science, including the first female college graduates; strong resistance to their entering traditional kinds of scientific employment, for example, university teaching or government employment; and the changing structure of scientific work in the 1880s and after, which provided new roles and fields for these entering women. As a result, women were incorporated into the world of scientific employment but segregated within it, as the prevailing stereotypes of appropriate sexual roles interacted with expanding scientific research work and changing research strategies between 1880 and 1910.

When the movement to give women a higher education had begun to take hold in the United States in the 1870s and 1880s, little thought had been given to the eventual careers that such graduates might take up. Because of the prevailing notion of "separate spheres" for the two sexes, most women were assumed to be seeking personal fulfillment and to be planning to become better wives and mothers. Advocates of their studying science saw it as offering a rigorous and satisfying intellectual experience to women who led essentially "aimless lives." Even such accomplished scientists as entomologist Mary Murtfeldt of St. Louis, Missouri, astronomer Maria Mitchell of Vassar College, ornithologist Graceanna Lewis of Philadelphia, and physicist Edward C. Pickering of the Massachusetts Institute of Technology expected that the women would participate in science only as amateurs. There were still so few women scientists

I wish to thank Sally Gregory Kohlstedt for many useful criticisms over a number of years, Jeffrey Escoffier for several helpful theoretical discussions, and the History and Philosophy of Science Program of the National Science Foundation for financial assistance (grants SOC 77-22159 and SOC 79-07562).

[1] The tendency to discuss idealized, stereotypical versions of social institutions (i.e., scientific "norms") rather than actual behavior seems to have been typical of the conservative social thought of the 1950s. See Douglas T. Miller and Marion Nowak, *The Fifties, The Way We Really Were* (Garden City, N.Y.: Doubleday & Co., 1977), Ch. 8.

[2] Although this discussion is limited to the United States, similar phenomena may have occurred elsewhere. The national differences are worth exploring.

in the United States in the 1870s, hardly any of whom were employed, that there was barely a hint yet of sexual stereotyping or "women's work" in science: all fields were presumed to be open to women, who were assumed to be equally adept at all of them.[3]

As the numbers of college women increased in the 1880s, however, expectations began to rise on all sides that their training should lead somewhere. If it had been intolerable in the 1870s that so many uneducated women should be sitting home idle, how much more of a waste that unmarried college graduates should lack useful and respectable work. Accordingly the advocates of higher education for women began in the 1880s to talk hesitantly of improved job prospects as well as personal fulfillment for college graduates.[4] In her 1882 article "Scientific Study and Work for Women," Mary Whitney, Mitchell's student and successor in astronomy at Vassar, went beyond the standard view that science would help to develop a woman's mind and would "introduce a more definite purpose into her life" to proclaim that it also laid the basis for "useful, and I hope, in the future, remunerative labor" for them. Unfortunately this was not yet a real possibility for most women, since as Whitney put it, "we cannot say the present offers many examples." Some women had been successful as physicians and others as professors, but she also thought there would soon be opportunities for women in those other fields for which they were particularly well suited and in which some women (no names were given) had already been active, like practical chemistry, architecture, dentistry, and agriculture. To her credit Whitney resisted the temptation to minimize the problems this pioneer generation would face. She warned them that they would have difficulty in finding any kind of employment, since even at the women's colleges "the chances are largely in favor of the man." But after urging young women to prepare themselves for such careers, Whitney then showed her own ambivalence by concluding that in any case scientific training would make them excellent mothers, which was after all "the highest profession the world has to offer."[5]

Fortunately for the women seeking to enter science in the 1880s and after, at least three forces were shaping scientific work at that time which would provide new roles and opportunities for them: the rise of "big science" with large budgets that could support staffs of assistants at a few research centers; a new concern for the nation's growing social problems, which created the need for several new hybrid or "service" professions (or semi-professions) designed to solve them; and the need for new faculty and other personnel at the coeducational land-grant agricultural colleges. Since some of these new jobs could be seen as offering women a chance to use their special

[3] Mary E. Murtfeldt, "Woman and Science," *Moore's Rural New Yorker,* Jan. 4, 1873, 27:19; Maria Mitchell, "Address of the President," *Papers Read at the Third Congress of Women, Syracuse, October 1875* (Chicago: Fergus Printing Co., 1875), pp. 1–7; Graceanna Lewis, "Science For Women," *ibid.*, pp. 63–73; [Edward C. Pickering], "Education," *Atlantic Monthly,* 1874, 33:760–764. Maria Mitchell, "The Need for Women in Science," *Papers Read at the Fourth Congress of Women, Philadelphia, 1876* (Washington, D.C.: Todd Bros., 1877), p. 9, suggested that women students would make good assistants in observatories and museums.

[4] This theme is discussed in Daniel T. Rodgers, *The Work Ethic in Industrial America, 1850–1920* (Chicago: University of Chicago Press, 1978), Ch. 7 ("Idle Womanhood"), and Roberta Frankfort, *Collegiate Woman, Domesticity and Career in Turn-of-the-Century America* (New York: New York University Press, 1977), Ch. 6. See also Jane M. Bancroft, Ph.D., "Occupations and Professions for College-bred Women," *Education,* 1885, 5:486–495; Charles F. Thwing, "What Becomes of College Women?" *North American Review,* 1895, 161:546–553; Alice M. Gordon, "The After-Careers in University-Educated Women," *Nineteenth Century,* 1895, 37:955–960; Frances M. Abbott, "Three Decades of [Vassar] College Women," *Popular Science Monthly,* 1904, 65:350–359.

[5] Mary W. Whitney, "Scientific Study and Work for Women," *Education,* 1882, 3:58–69, on pp. 58, 66, 68 and 69.

"feminine" skills in ways that did not threaten men directly and even enhanced their dominant role, the 1880s and 1890s saw much explicit channeling of the newly available women into certain "appropriate" callings.

Advocates of such "women's work" had no trouble developing a rationale for separate kinds of jobs for women. They had merely to urge women to capitalize on the relative welcome that they were already receiving in the marketplace for two kinds of jobs: those that were so low-paying or low-ranking that competent men would not take them, which often required great docility or painstaking attention to detail, and those that involved social service, as working in the home or with women or children, which were also often poorly paid. The literature on women's work glorified these positions, considered them very suitable for women in science, and advocated more of them.[6] This message was quickly communicated to an eager audience by a new social mechanism, the middle-class magazines, whose contributors seized upon any work found suitable for such women and with evident relief advised others to enter it as well. Like the more explicit vocational guidance of the twentieth century, these magazines needed only a hint of a success story to unleash a torrent of articles (many written by women) extolling the new opportunities awaiting women in the newest area of women's work.[7]

WOMEN IN ASTRONOMY

Events were moving so rapidly in the early 1880s that even as Mary Whitney had been writing her transitional and ambivalent article in 1881–1882, the first kinds of "women's work" in science were appearing. Although a few women (including Maria Mitchell herself) had previously worked at home as "computers" for others' astronomical projects, women's work in astronomy was just entering a new phase. The change apparently grew out of a fortunate but not unusual set of circumstances at the Harvard College Observatory in 1881. In that year, the story goes, Edward Pickering, the advocate of advanced study for women cited above, who had recently been elected director of the Observatory and was its first astrophysicist, became so exasperated with his male assistant's inefficiency that he declared even his maid could do a better job of copying and computing. He promptly put Williamina P. Fleming (a Scottish immigrant, then twenty-four years old, a public-school graduate, a mother, and separated from her husband) to the test, and she did so well that he kept her on for the next thirty years. She not only became one of the best known astronomers of her generation, but she also showed such good executive ability and "energy, perseverance and loyalty," as one obituary put it, that Pickering put her in charge of hiring a whole staff of other women assistants, whom he paid the modest sum of twenty-five to thirty-five cents per hour to classify photographs of stellar spectra. Between 1885 and 1900 she had twenty such assistants, including the college graduates Antonia C. Maury, Vassar 1887; Henrietta P. Leavitt, Radcliffe 1892; and Annie Jump Cannon, Wellesley 1884, who came after living at home in Delaware for a decade. Several of them made such prodigious contributions not only to the Observatory's main project in those years, the

---

[6] These rationalizations are more "conservative" than those used in the same years by the advocates of women's admission to the new graduate schools, who employed the more liberal rhetoric that women were fully equal to men and that it was "only fair" that they be accepted and awarded degrees. See Margaret W. Rossiter, "Women Scientists in America to 1940" (forthcoming), Ch. 2, "Doctorates and Fellowships."

[7] E.g., Mrs. Emily Crawford, "Journalism as a Profession for Women," *Contemporary Review*, 1893, *64*:362–371 and Margaret E. Sangster, "Editorship as a Profession for Women," *Forum*, 1895/6, *20*:445–455.

Figure 1. The Harvard stellar classification team, known as "Pickering's harem." Photograph courtesy Harvard College Observatory.

Henry Draper Star Catalog of stellar spectra, but also to other areas of astrophysics that they became highly regarded in their own right.[8]

Within a few years the fame of this novel employment practice had begun to spread, and women astronomers had become the subject of several favorable magazine articles.[9] In addition, Fleming openly propagandized for the Harvard arrangement in an address on "A Field for Woman's Work in Astronomy" at the World Columbian Exposition (World's Fair) in Chicago in 1893. Like the other contemporary accounts, she praised Pickering's progressive attitude in hiring the women and talked of their many contributions, but she also moved beyond this into some sex stereotyping of the skills involved. Thus Fleming was on safe ground when she urged other observatory directors to hire female assistants, since such women, "if granted similar opportunities would undoubtedly devote themselves to the work with the same untiring zeal." But she was on more precarious ground when she tried to determine what this case meant about the comparative skills and abilities of the two sexes, and in fact it is not clear

[8] The best accounts of the women astronomers at Harvard (and a few from elsewhere) are Bessie Z. Jones and Lyle Boyd, *The Harvard College Observatory: The First Four Directorships, 1839–1919* (Cambridge: Harvard University Press, 1971), Ch. 11, "A Field For Women," with full bibliographic notes, and Pamela E. Mack, "Women in Astronomy in the United States, 1875–1920" (B.A. honors thesis, Harvard University, 1977). See also entries for Cannon, Fleming, and Leavitt in *Notable American Women* (Cambridge: Harvard University Press, 1971, 3 vols.) to be referred to hereafter as *NAW*; Annie J. Cannon, "Williamina Paton Fleming," *Astrophysical Journal*, 1911, *34*:314–317, and Helen Buss Mitchell, "Henrietta Swan Leavitt and Cepheid Variables," *Physics Teacher*, 1976, *14*:162–167. (Two of the women were also handicapped: Leavitt was "extremely deaf" and Cannon less so.)

[9] Women astronomers received an inordinate amount of attention in popular journals in these years, e.g., E. LaGrange, "Women in Astronomy," *Pop. Sci. Monthly*, 1885/6, *28*:534–537; Esther Singleton, "Women as Astronomers," *The Chautauquan*, 1891, *14*:209–212 and 340–342; Helen Leah Reed, "Women's Work at the Harvard Observatory," *New England Magazine*, 1892, N.S. *6*:165–176; Herman S. Davis, "Women Astronomers," *Popular Astronomy*, 1898, *6*:129–138, 211–220 and 220–228. The last is in part a review of Alphonse Rebière's *Les Femmes dans la science* (Paris: Nony & cie., 1897; 2nd ed.), as is Dorothea Klumpke, "La Femme dans l'astronomie," *Astronomie*, 1899, *13*:162–170 and 206–215. See also Edward S. Holden, "On the Choice of a Profession, II. Science," *Cosmopolitan Magazine*, 1898, *24*:543–549, which hypothesizes that women will never find happiness in science; and Anne P. McKenney, "What Women Have Done for Astronomy in the United States," *Pop. Astron.*, 1904, *12*:171–182, which contains errors and in part copies from Fleming (n. 10 below).

what she did mean in her rather confused conclusion: "While we cannot maintain that in everything woman is man's equal, yet in many things her patience, perseverance and method make her his superior. Therefore, let us hope that in astronomy, which now affords a large field for woman's work and skills, she may, as has been the case in several other sciences, at least prove herself his equal."[10] Apparently she agreed with the prevailing idea that women were generally inferior to men, but she felt that by overachieving in the bottom ranks, far outstripping what persons at their level were expected to do, they might prove themselves "equal" to men with far greater opportunities.

This pattern of segregation and sex typing proved popular and spread to most other major observatories in the United States in the 1890s and after. Of the twenty women astronomers listed in the third edition of the *American Men of Science* (1921), eight worked as assistants at major observatories. (Seven others taught at women's colleges, as had three who were retired. Fleming, who died in 1911, was listed in an earlier edition.) Yet even this sizeable proportion of the nation's women astronomers only skims the surface of the phenomenon, for Pamela Mack has collected a list of 164 women, mostly high school graduates, who worked at various observatories for a year or more between 1875 and 1920. (Of the few college graduates in her list, those trained at Vassar by Mitchell, Whitney, and later Caroline Furness were particularly sought after by observatory directors.) Almost every large observatory hired these women, as Mack found 24 at the Dudley Observatory in Albany, New York, 12 at Yerkes Observatory in Wisconsin, 12 at Mt. Wilson in southern California, 6 at the U.S. Naval Observatory in Washington, D.C., and smaller numbers at most of the others, such as Columbia, Allegheny, Lick, and Yale, where talented doctorate Margaretta Palmer was especially notable.[11]

Such a wide acceptance of female assistants in astronomy in these years would seem to have been the result of something more pervasive than Pickering's personality and his practice at Harvard. More likely it grew out of certain competitive forces within the field of astronomy itself. Although still a small field, astronomy was growing very rapidly in the 1880s and 1890s, when several new observatories were built and the whole new field of astrophysics was just appearing. This rapid expansion created two problems for the older observatories: maintaining a large staff of good assistants would be very difficult, since the more experienced male assistants might be offered at any moment better positions elsewhere, and competing in observational astronomy would be very difficult, since the newer observatories often had larger telescopes and were located in areas with better viewing conditions. Thus, as Mack has explained it, most of the newer observatories used their advantages to concentrate on the traditional observational astronomy and offered men from the older institutions exciting new opportunities in this field. Although these new observatories also hired some women, they restricted them to the tedious and laborious computations women had long done for male astronomers. For the women the better opportunities in this period of change were not in observational astronomy at all but in the newer specialties, where they got some of the work, although not the actual jobs, that the more mobile young men had left behind. They fared much better at Harvard, for example, than they did out West, because in order to compete with his new rivals, Pickering moved away from observational astronomy and into another

---

[10]Mrs. W. Fleming, "A Field for Woman's Work in Astronomy," *Astronomy and Astrophysics*, 1893, *12*:688–689.
[11]Pamela Mack, "Women in Astronomy," Ch. 4.

specialty, the new field of photographic astrophysics. His adaption of this more advanced technology of cameras and spectroscopes had great implications for women in science, since it required a different labor force: Pickering needed fewer observers (men's work) and many more assistants (women's work) to classify as cheaply as possible the thousands of photographic plates his equipment was generating.[12]

If Pickering and some other observatory directors were progressive in greatly expanding women's employment in astronomy in the 1880s and 1890s, they were not so far ahead of their time as to promote them for important or even outstanding work. Not only did the women have no chance for advancement, they rarely received a raise—at least at Harvard—even after years of devoted service. Because they were not promoted to other more administrative or managerial duties when their scientific work was good, as happened to the more talented men, the female assistants were forced (or expected) to make a whole career out of a job that should have been just a stepping stone to more challenging and prestigious roles. In a sense, science benefited from this practice, since the women, having no alternatives, remained on the job for decades and completed many major projects: women carried out the massive Henry Draper Star Catalog, for example, when Pickering himself became too involved in administration and fund raising to do much astronomy. Knowing their "place" and having few if any options, the women graciously (and for the most part gladly[13]) accepted what was offered to them and stayed as long as they were wanted. Thus the new "women's work" in astronomy trapped many into low-level jobs; this apparently happened to Fleming, who from accounts of her strong executive abilities, might have made a very good director of one of the many new observatories opening up in these decades. Indeed the only female directors of American observatories to date have been those at the women's colleges and at the small Maria Mitchell Observatory on Nantucket Island, a feminist memorial and outpost in the Atlantic Ocean.[14]

One should also note that some arguments were not deployed to restrict women's place in astronomy in the 1890s. For the most part the arguments remained economic—the new technology of spectrophotography yielded hundreds of prints of the stellar spectra that required much cheap and intelligent labor to sort out. Women were willing to do this and did it very well in the late nineteenth century, apparently far better than men did for the same price. Apparently no one went a step further and

---

[12] *Ibid.*; Deborah Warner, "Women Astronomers," *Natural History*, May 1979, 88:12–26, pp. 12–14. The term "proletariatinization" has been applied to this phenomenon of downgrading the job and then allowing it to be feminized; Renate Bridenthal and Claudia Koonz, "Beyond *Kinder, Küche, Kirche*: Weimar Women in Politics and Work," in Bernice A. Carroll, ed., *Liberating Women's History, Theoretical and Critical Essays* (Urbana: University of Illinois Press, 1976), p. 318. A related form of the same phenomenon started even earlier in the museum world (late 1860s in Boston), as women were given low-paying staff positions to help cope with the big expansion in collections brought back by ever more numerous expeditions to far-off places. Yet these positions were not acclaimed "women's work" and received far less publicity than did the women astronomers. See Sally Gregory Kohlstedt, "The Nineteenth-Century Amateur Tradition: The Case of the Boston Society of Natural History," in Gerald Holton and William A. Blanpled, eds., *Science and its Public* (Dordrecht; D. Reidel Publishing Co., 1976), pp. 183–184 and Ralph Dexter, "Guess Who's Not Coming to Dinner: Frederic Ward Putnam and the Support of Women in Anthropology," *History of Anthropology Newsletter*, 1978, 5(2):5–6.

[13] Although most of the women astronomical assistants rejoiced in being allowed to do such interesting work, Antonia C. Maury reportedly rankled at the close supervision and lack of recognition accorded her work. She sought jobs elsewhere and gave several short series of lectures, but apparently always returned to the Harvard College Observatory. (Jones and Boyd, *Harvard College Observatory*, pp. 395–400.)

[14] Thomas E. Drake, *A Scientific Outpost: The First Half Century of The Nantucket Maria Mitchell Association* (Nantucket, Mass.: The Nantucket Maria Mitchell Association, 1968).

used the second standard argument for sex stereotyping of occupations, one commonly used to justify it in libraries and public schools: the claim that the subject matter (here, the stars or the heavens) was somehow inherently feminine or that the tools or facilities of that science (telescopes or observatories) were really homes which women could tend better or more naturally than men. Thus the sex stereotyping in astronomy did not take on the additional rhetorical and psychological trappings that other areas of women's work did at the time, for example, child psychology, home economics, social work, and librarianship. Not only did low pay feminize those fields, but generations of persons believed and advised others that women belonged in them, because there was something uniquely feminine about the subject matter.

SCIENTIFIC EMPLOYMENT IN THE FEDERAL GOVERNMENT

Interestingly, sex typing and sex segregation did not occur in all the fields and areas that it might have. The development of scientific employment within the federal government in the 1880s and 1890s provides examples that help clarify the relatively rare circumstances necessary for the emergence and spread of "women's work" within a science. First, the field had to have a shortage of available men, since there was a general reluctance to appoint women scientists to any job for which men were in good supply. Yet in practice the situation was more complex than this, since the actual supply of suitable workers depended on the job description, which was often in flux in these years. Many scientific jobs could be upgraded (masculinized) or downgraded (feminized) over time or as the budget required, thus manipulating the kind of workers desired. Second, and closely related to the first, the employer had to want to hire women, for the appointment process required him—officially under federal Civil Service rules, unofficially elsewhere—to predetermine which sex he wished to appoint. Generally only those employers with strong feelings or economic incentives would request a woman. Otherwise inertia and prevailing stereotypes meant that most appointments would go to men. Third, women had to be alerted that this was work for which they would be hired. This receptivity was usually communicated by participants like Fleming or by enthusiastic journalists who described the job's duties almost exclusively in terms of prevailing sexual stereotypes. Anthropology could, for example, be described as a field in which women could make unique contributions by studying women and children better than men could. This publicity was a kind of "market signal" to both potential workers and future employers as to what type of person would be hired.[15] The result of these processes was that what was accepted as men's work and what as women's was oftentimes not particularly logical or even consistent, but rather the result of a series of employer preferences and economic incentives. In general, however, the women got the less powerful, less prestigious, and lower-paying jobs. The rest were reserved for the men.

The prime example of sex segregation in the federal government occurred in the field of botany, or, more precisely, the new specialty of plant pathology, in the 1880s. One sign that the signal had by this time already been passed to the public that botany was a "feminine" science was an article reprinted in *Science* in 1887, "Is Botany a Suitable Study for Young Men?" Although none of its four arguments (mental discipline, outdoor exercise, practicality, and lifelong happiness) seems very

---

[15] Michael Spence, *Market Signaling: Information Transfer in Hiring and Related Screening Processes* (Cambridge: Harvard University Press, 1974). I thank Jeffrey Escoffier for this reference.

sex-linked today, popularizers had already propagandized botany's suitability for young ladies so effectively that some persons thought this protest necessary.[16] Perhaps this feminine image was one reason why several rapidly growing divisions of the U.S. Department of Agriculture, which would soon unite to form its Bureau of Plant Industry (BPI), made it a regular practice to hire women as "scientific assistants." Erwin Frink Smith, the USDA's pathologist-in-charge, who hired Effie Southworth in 1887, was quite proud of this practice and continued it until his retirement in the 1920s. Over the years he hired more than twenty women assistants, including such talented ones as Nellie Brown, Clara Hasse, Charlotte Elliott, Agnes Quirk, Della Watkins, and Mary Bryan, who earned modest fame for their outstanding work on such agricultural problems as crown galls, citrus canker disease, and corn and chestnut blight.[17]

Smith's motives for this unusual employment practice are unclear. His 650-page biography gives no clues as to what they might have been. His reasons may simply have been economic: Smith and others at the BPI may simply have taken advantage of certain highly discriminatory restrictions on women taking Civil Service examinations at the time. Although sources on this point differ, before 1919 women were presumably prohibited from taking the examinations for the standard entry positions of "junior botanist" or "junior plant pathologist": even those women with masters degrees in botany could take only the exam for the lower category of "scientific assistant." Under these conditions a shrewd laboratory director could easily hire highly qualified personnel at bargain rates. For their part the women might have been very glad to work at projects suited to their skills—and indeed several praised Smith for this—rather than at the more tedious ones to which the Civil Service rules limited them. Then too, as for women astronomers, any training given them would be a good investment, since they would not be promoted to an administrative position or leave for a better position in a state university or an experiment station, both of which institutions were also expanding rapidly in these same years. In addition to these economic motives for preferring women assistants, there may also have been a psychological one: Smith and the other male bosses at the USDA (as perhaps did Pickering at the Harvard Observatory) may have liked the "harem effect" of being surrounded by a bevy of female subordinates, competent but less threatening than an equal number of bright young men. Employing women was in any case an effective way to limit turnover and competition while maintaining a good staff in a period of great opportunity and rapid growth. The Bureau of Plant Industry was, however, the only agency of the federal government to become so highly feminized before World War I.[18]

Yet if there were compelling economic and psychological advantages to hiring

[16] J. F. A. Adams, M.D., "Is Botany a Suitable Study for Young Men?" *Science*, 1887, 9:117–118 and Emmanuel D. Rudolph, "How It Developed That Botany Was the Science Thought Most Suitable for Victorian Young Ladies," *Children's Literature*, 1973, 2:92–97. I thank Sally Kohlstedt and Ravenna Helson, respectively, for these references.

[17] Andrew Denny Rodgers III, *Erwin Frink Smith: A Story of North American Plant Pathology*, Memoirs of the American Philosophical Society, 31 (Philadelphia, 1952), passim, esp. pp. 174, 211, 379–380, 432, 481–482, 651. See also obituaries, e.g., of Nellie Brown, *Washington Evening Star*, Sept. 15, 1956 (I thank Paul Lentz of the USDA for this); of Della E. Watkins, *Washington Evening Star*, Apr. 12, 1977; Alice L. Robert and John G. Moseman, "Charlotte Elliott, 1883–1974," *Phytopathology*, 1976, 66:237; and Gladys Baker, "Women in the United States Department of Agriculture," *Agricultural History*, 1976, 50:190–201.

[18] Margaret W. Rossiter, "Women Scientists in the United States Before 1920," *American Scientist*, 1974, 62:316; data based on first three editions of *American Men of Science (AMS)*.

women subordinates in government agencies, one would expect that the practice would have become as widespread as the employment of women astronomical assistants. Since the economic potential for sex stereotyping was there, perhaps in all fields, it is worth noting those federal agencies where it did not take place. Several other agencies (besides the Naval Observatory, mentioned above) appointed one or more women to their staffs—usually on a temporary basis before 1900—but such appointments were not perceived as precedents leading to the reservation of places for other women and thus to the feminization of a field or role. In two cases, however, feminization almost occurred. Three women clerk-copyists, apparently the first in the federal government, were appointed to the Patent Office in the 1850s; they included Clara Barton, later the founder of the American Red Cross. Their appointments might have led to a series of women patent examiners doing such detailed, painstaking, and indoor work—the type often perceived as feminine—but did not.[19]

The other case of potential feminization failed because of the active resistance of some key men in the 1880s and after. At the Bureau of American Ethnology, Director John Wesley Powell financed the expeditions of at least two women anthropologists, Erminnie Smith and Matilda Stevenson, in the 1880s. Since there were also a great many other women anthropologists around Washington at the time, including Alice Fletcher and patron Phoebe Apperson Hearst, one suspects that female fieldwork was on the verge of becoming a regular feature of the Bureau's projects, and might even have justified a separate women's division within it. In fact, one can surmise from the all-male Anthropological Society of Washington's stern refusal to admit women members in 1885 that the feminization of anthropology was an all too likely possibility in the 1880s. The women were numerous enough to form their own Women's Anthropological Society of America in response. Then in 1901 Franz Boas, just embarking on his plan to professionalize American anthropology, wrote Phoebe Hearst via archaeologist Zelia Nuttall that the way to upgrade the field was to train "a small number of young men."[20]

WOMEN IN HIGHER EDUCATION

Meanwhile at least three other kinds of "women's work" were developing at coeducational colleges and universities in the 1890s.[21] They demonstrate two kinds of occupational sex segregation: a hierarchical form as in astronomy, where women were employed as assistants to other higher-ranking persons, and a territorial kind, where women did all the work in a specific, highly sex-typed, field or location. Although one can list a series of historical "firsts" to document the slow trickle of

[19] Biography of an Ideal: A History of the Federal Civil Service (Washington, D.C.: Government Printing Office, 1973), pp. 161–162; Barton in NAW.

[20] Smith, Stevenson, Fletcher, and Hearst in NAW; Organization and Historical Sketch of the Women's Anthropological Society of America (Washington, D.C.: The Society, 1889); "Women in Science, Mrs. Nuttall Believes That a Promising Field in Archaeology is Open to Them," New York Tribune, June 30, 1900 (account of recent meeting of American Association for the Advancement of Science), clipping in Alice Fletcher Papers, National Anthropological Archives, Smithsonian Institution, Washington, D.C.; Franz Boas to Zelia Nuttall, May 18, 1901, in Phoebe Apperson Hearst Papers, Bancroft Library, University of California, Berkeley. See also George W. Stocking, "Franz Boas and the Founding of the American Anthropological Association," American Anthropologist, 1960, 62:1–17.

[21] Although most of the women scientists in the U.S. employed before 1910 taught at the women's colleges, and many persons must have thought this was clearly one kind of "women's work," few said so, since many men taught there as well, often in the highest ranking positions. (Rossiter, "Women Scientists . . . 1920," p. 318).

women onto the faculties of coeducational schools from the 1850s on, the opposition to even these "exceptions" was usually intense. Maria Mitchell had been aware as early as the 1870s that the coeducational schools were not hiring many women and had mentioned it in an address to the Association for the Advancement of Women (AAW) in 1875. She recounted the tale of a relatively liberal president of a coeducational college who said that he would hire a woman scientist if she was as good as Mary Somerville, the renowned British mathematician. Mitchell pointed out that he was creating a double standard and requiring more of the women than he was of the men, for, as she put it, "If he applied the same standard to his choice of gentlemen professors, his chairs must be vacant today."[22]

By the 1890s the topic was no longer humorous, and other women were calling attention to the small proportion of women in and their systematic exclusion from faculties of coeducational schools. In what was probably the first of that genre later known as "reports on the status of women," Octavia Bates reported to the AAW in 1891 that though women were now attending coeducational colleges and universities in large numbers, few of the faculty at these institutions were women.[23] The idea was surfacing that the proportion of women in the student body should bear some correlation to their representation on the faculty. In a way, this "share-of-the-market" argument did lead to the reserving of a certain percentage of faculty jobs for women, but in a way that was very typical of the separate and subordinate world of women's work in the 1890s. Those women who were allowed on the faculty found themselves restricted to the segregated fields of home economics and hygiene and to the subfaculty position of dean of women. That they were lucky to get even this much at the coeducational schools is clear from the experience at Cornell University in these years.

The opposition to having women on the faculty at Cornell University was so intense in the 1890s and after as to politicize greatly the few appointments of women that were made. Women's status and rank were deliberately lowered, thus setting precedents and limits for future appointments at that school. Cornell had opened in 1867 and after much hesitation admitted its first women students in 1872, but did not allow any women on the faculty until the late 1890s. Even then two women appointees were allowed only in the bottom ranks. When Liberty Hyde Bailey, Cornell's beloved and energetic professor of horticulture and later dean of its college of agriculture, urged the appointment of entomologist Anna Botsford Comstock as assistant professor of nature study, the trustees insisted that she be only a lecturer. Nor would they allow Agnes Claypole to rank higher than a mere assistant despite her doctorate in zoology. It was thus to be expected that the appointment of the first women full professors at Cornell in 1911 would call forth a pitched battle. Faculty arguments against their promotion included those that the university would lose status by appointing women to professorships, that the women did not have families to support and did not need the money (a common fallacy, but the women's pay always lagged anyhow), that these women were not as well trained as most men (probably true in this particular case), and that there was no need to bring the women into competition with the men. The debate was finally settled in favor of the women,

[22]Maria Mitchell, "Address of the President," *Papers Read at the Third Congress of Women*, pp. 4–5.
[23]Octavia W. Bates, "Women in Colleges," *Papers Read Before the Association for the Advancement of Women, 19th Woman's Congress, Grand Rapids, Michigan, October 1891* (Syracuse: C.W. Bardeen, 1892), pp. 19–21.

but only because they were in the new department of home economics, or as the historian of Cornell has put it: "After a long and acrimonious argument, the faculty voted (18 October 1911) that 'while not favoring in general the appointment of women to professorships, it would interpose no objection to their appointment in the Department of Home Economics.'"[24] The women had inched forward but were still kept at a safe distance. There were no women full professors in Cornell's College of Arts and Sciences until 1960 and not even an assistant professor until 1947.[25]

The increasing numbers and percentage of women students at coeducational institutions in the 1890s played a major role in staking out an area of "women's work" for the women faculty, but in a way that now seems to have hurt them as much as it helped. When there had been only a few female students, they had been tolerated without any great difficulty, but as their numbers grew, their presence had become so visible as to disturb the men and create much pressure for segregation, both in the curriculum with special women's subjects (the humanities plus home economics) and in student housing with special women's dormitories. Soon this duplication would require new personnel whose status was unclear, to take care of the women's "special problems." And who was better suited to worry about these matters than the young woman on the faculty, who would never be promoted anyway? She knew the school and was of high moral character; even though single, she had always taken a special interest in the women students and might even be thought to have the appropriate "maternal instincts" for the job. In a flash a woman chemist could become a home economist, or an assistant professor a dean of women, almost whether she wanted to or not.

HOME ECONOMICS AS WOMEN'S WORK

Probably the largest area of scientific "women's work" in academia was the whole field of "home economics." Several factors contributed to its rise in the 1890s and its rapid institutionalization as an academic field for women after 1910. Although no history of it has been written, the subject seems in brief to have been the product of two long-range trends which merged in the 1890s and after: one of nutrition research, which was creating a large supply of new information, and the other of popularization, which fed a strong and increasing demand for practical advice. The feminization of the field, which had become pronounced by 1900, was the result of men's aversion and possibly their inability to advise women on domestic matters and their willingness to let the women do it instead. Few men tried to enter the field in the 1890s and after. Had men chosen to take over the field and make it into a profession like medicine or religion or even the law, where men commonly advise female patients or clients, it is hard to see how the women could have stopped them.

The field of nutrition research grew out of the nineteenth-century sciences of analytical chemistry and biochemistry, which received great impetus from the works of the great German chemist Justus Liebig, published in the 1840s and after. He and his many German and American followers greatly stimulated the scientific study of

[24] Morris Bishop, *A History of Cornell* (Ithaca: Cornell University Press, 1962), pp. 337–338, 379–381, and 388–389; Flora Rose, "Forty Years of Home Economics at Cornell University," in *A Growing College: Home Economics at Cornell University* (Ithaca: Cornell University Press, 1969), pp. 19 and 34–38. Comstock is in *NAW*.

[25] Charlotte Williams Conable, *Women at Cornell: The Myth of Equal Education* (Ithaca: Cornell University Press, 1977), pp. 126–30.

foods and human metabolism. By the 1880s workers at the young American experiment stations, especially W. O. Atwater at Connecticut, were studying these subjects, and by 1910 other scientists, like Graham Lusk, Russell Chittenden, Lafayette B. Mendel, and Francis Benedict, were working on them in medical schools, universities, and private institutes. There were no women in this classic research tradition until 1909, when Mary Swartz Rose earned her doctorate at Yale and started her own research program at Teachers College, Columbia University.[26]

There were many women, however, in the second long-term root of home economics: the tradition of "advice literature" for the family and home. The classic work in the 1840s and several decades thereafter was the *Treatise on Domestic Economy for the Use of Young Ladies at Home and at School* (1841) by non-scientist Catherine Beecher. This sort of literature sought to popularize and circulate among persons who would later be called consumers the latest scientific advice on how best to run their lives, their homes, and their families. Here chemistry, bacteriology, and psychology all in due course made their contributions to a primitive form of the activity later partially institutionalized and subsidized by the federal government in the Department of Agriculture's extension service. One particularly receptive audience or market for this literature was the female one of wives and mothers whose needs and willingness to listen created a demand for a series of lay female advisors.[27]

Perhaps it was inevitable that the researchers or experts would be dissatisfied with much of the advice that these popularizers circulated and would want to have some hand in upgrading it. But in addition two other non-scientific factors began to bring the researchers and the popularizers closer together in the 1890s and to create the need for new well-trained hybrids or home economists who could better fit the women's needs. These new forces were the massive immigration to eastern cities and the rise of the agricultural college. The cities seemed to many reformers of the time to require numerous social services—including libraries, schools, settlement houses, hospitals, and social welfare agencies—to train, "Americanize," and generally homogenize and upgrade these unwashed hordes into respectable middle-class citizens. For many reasons, women, especially the new college graduates, seemed best able to take on this overwhelming social task. They were presumed to have the female's traditional interest in the home (which could now be extended to include the neighborhood and even the whole city), to be more venturesome than their stay-at-home sisters who had not gone to college, and to be available and willing to work at low pay on these herculean social problems. Thus, like the schoolteachers, social workers, librarians, and settlement house workers, the women home economists could act as missionaries trying to save society and its victims—through better nutrition and home life. There were enough diseases and other public health problems, especially among city children, who accounted for half of all deaths in the 1890s, to create a real demand at least in the eyes of the middle-class reformers, if not among the immigrants themselves, for new and better methods of hygiene and diet. It would be easier

---

[26] Rose in *NAW*; Elmer V. McCollum, *A History of Nutrition: The Sequence Ideas in Nutrition Investigations* (Boston: Houghton Mifflin Co., 1957); Margaret W. Rossiter, *The Emergence of Agricultural Science: Justus Liebig and the Americans, 1840–1880* (New Haven, Conn.: Yale University Press, 1975); Edward C. Kirkland, "'Scientific Eating:' New Englanders Prepare and Promote a Reform, 1873–1907," *Proceedings of the Massachusetts Historical Society,* 1974, 86:28–52.

[27] Catherine Beecher, *Treatise on Domestic Economy for the Use of Young Ladies at Home and at School* (Boston: Marsh, Capen, Lyon and Webb, 1841); Kathryn Kish Sklar, *Catherine Beecher: A Study in American Domesticity* (New Haven: Yale University Press, 1973), Ch. 11; Joan N. Burstyn, "Catherine Beecher and the Education of American Women," *New England Quarterly,* 1974, 47:386–403.

for such self-appointed ministers to the unfortunate, however, if they had some authority or expertise other than that they thought they knew best. A scientific background and thus a claim to the role of expert gave these women some authority to tell others how to live. In time master's degrees would be required of them and of other women in similar sub-professional roles, but in the 1890s women college graduates who "meant well" felt equal to the task.[28]

Besides this large but unorganized and perhaps manufactured urban demand for home economics, there was also a rural demand for home economics, one that in the long run created more jobs for women at the university level. As a result of the Morrill Land-Grant Act of 1862 many colleges had been created in rural areas of the country to teach the "agricultural and mechanic arts" to local youths of both sexes. Since agriculture was still at that time as much an art as a science, the content of the new colleges' curriculum was somewhat problematical. On the one hand there was much anti-elitist sentiment that such colleges should offer practical instruction; on the other hand its offerings had to be more rigorous or worthwhile than what their students could pick up at home on the farm. Some fields, like economic entomology or soil chemistry, straddled this problem successfully and were seen as both practical and scientific at the same time.

But what were the young women at such schools to study? Most chose to enroll for courses on cookery, sewing, and the household arts. What sort of faculty was best for this? It apparently had to be female—the association with women's work was too strong for any man to teach such subjects in late nineteenth-century America. Otherwise one can imagine that the agricultural colleges could have imported male chefs (perhaps from Europe) to teach cookery if they had wished. There is no evidence men ever contested the issue. Then too what teaching methods were suitable for teaching such domestic arts? Could one lecture on cookery and assign readings on sewing or were demonstrations enough? The fact that the students often had little or no scientific background (and would often, as the immigrants did, balk at having to learn chemistry just to cook) limited the amount of rigor that could be maintained. One could easily go too far, however, in pleasing the students and their parents and end up scorned by the rest of the college faculty and the university administration for not having a German doctorate and for teaching a subject that lacked intellectual rigor. Thus there was pressure on the early faculties of domestic science or home economics at the land-grant colleges to upgrade their curriculum, to make it seem scientific and demanding, and to hire women, preferably with Ph.D.s, to teach the subject as rigorously as the traffic would bear. The women were also expected to do research in some aspect of the field and upgrade it nearly to the level of a regular academic science. Treading the narrow path between practicality and prestige has long continued to be the dominant theme in the history of the field.[29]

The founder of home economics was Ellen Swallow Richards, who earned a B.A. from Vassar in 1870 and a B.S. from Massachusetts Institute of Technology in 1873.

[28] Dee Garrison, "The Tender Technicians: The Feminization of Public Librarianship, 1876–1905," *Journal of Social History*, 1972/3, 6:131; Roy Lubove, *The Professional Altruist: The Emergence of Social Work as a Career, 1880–1930* (Cambridge: Harvard University Press, 1965); Mrs. Frances Fisher Wood, "The Scientific Training of Mothers," *Papers Read Before the Association for the Advancement of Women, 18th Woman's Congress, Toronto, Canada, October 1890* (Fall River, Mass.: J. H. Franklin & Co., 1891), pp. 39–48.

[29] Eugene W. Hilgard, "Progress in Agriculture by Education and Government Aid," *Atlantic Monthly*, 1882, 49:531–541 and 651–661. Alfred C. True, *A History of Agricultural Education in the U.S., 1785–1925* (Washington, D.C.: Government Printing Office, 1929), pp. 267–272.

By all accounts her character and leadership touched her contemporaries deeply. Between 1880 and 1910 she almost single-handedly created the field—she propagandized for it, ran demonstration projects, raised money, performed many chemical analyses, wrote several handbooks, trained and inspired her co-workers, and organized its main activities and professional associations. In a sense she had been preparing for this role all her life. As a student at MIT she had learned how to make a place for herself and other women by capitalizing on woman's traditional role, or as she put it in 1871: "Perhaps the fact that I am not a Radical or a believer in the all powerful ballot for women to right her wrongs and that I do not scorn womanly duties, but claim it as a privilege to clean up and sort of supervise the room and sew things, etc., is winning me stronger allies than anything else."[30] After marrying Robert Richards, an engineering professor at MIT (who proposed, appropriately enough, in the chemistry laboratory), she volunteered her services and about $1,000 annually to the "Woman's Laboratory" there, which she induced philanthropic Bostonians to support from 1876 until 1883. Her curriculum was not initially sex-typed; her students were mostly schoolteachers whose normal school training had lacked laboratory work and who now wished to do analytical chemistry and learn mineralogy, one of her own specialties. By 1880, however, Richards was stressing the value of chemistry to the homemaker, asserting in an address to the AAW that "laboratory work, rightly carried out, makes women better housekeepers, better cooks, better wives, and mothers more fitted to care for the versatile American youth. . . ." This change in emphasis was probably the result of both the increasing interest in what would later be called pure food and drug issues and her own precarious position at MIT.[31]

After MIT began to admit women directly in 1878, the need for the separate Women's Laboratory lessened; it was closed in 1883, and for a year Richards lacked a position. But when in 1884 MIT set up a new laboratory to study sanitation, apparently the first of its kind in the nation, she was appointed an instructor in sanitary chemistry, a position she held until her death twenty-seven years later in 1911. There she helped MIT professors associated with the laboratory analyze the state's water samples and developed her interests in the composition of food and groceries, safe drinking water, and low cost diets for the poor. She prepared many popular works for their use and in 1889 helped several college women in Boston start the New England Kitchen, where they prepared nutritious soups for the city's poor. This whole experience convinced Richards that the field of nutrition education, as it is now called, offered great opportunities to women college graduates who should, she felt, be using their trained minds and special talents to understand and help solve

[30] Caroline L. Hunt, *The Life of Ellen H. Richards* (Boston: Whitcomb and Barrows, 1912), p. 91.

[31] The best short biography of Richards is that by Janet Wilson James in *NAW*; tributes by contemporaries include Marion Talbot, "Mrs. Richards' Relation to the Association of Collegiate Alumnae," *Journal of the Association of Collegiate Alumnae*, 1912, 5:302–304, and Isabel Bevier, "Mrs. Richards' Relations to the Home Economics Movement," MS in the Isabel Bevier Papers, University of Illinois Archives, Urbana. A provocative recent biography is Robert Clarke, *Ellen Swallow: The Woman Who Founded Ecology* (Chicago: Follett Publishing Co., 1973). The early days of the Woman's Laboratory are discussed in Ellen Richards to Edward Atkinson, July 1878, included as an appendix to his Report to the Committee on Subscriptions, 1882 (mimeographed) in MIT Archives, Cambridge, Mass. (I thank Helen Slotkin and Deborah Cozort for bringing this to my attention.) Richards's 1880 address on "Woman's Work in the Laboratory" to the Association for the Advancement of Women is reported in Mary Clemmer, "Women in the Laboratory," *Pacific Rural Press*, Jan. 8, 1881, p. 22. [Ellen Richards] to Flora _____, July 29, 1883, MIT Historical Collections, describes her unease at being without a job—one of the few moments when Richards seemed to falter.

the social problems around them. In 1890 she presented a paper on "The Relation of College Women to Progress in Domestic Science" to the Association of Collegiate Alumnae, of which she had been a founder (Marion Talbot later remarked that she was also "like an elder sister" to the other early members). There Richards stressed how challenging efficient housework could be and how much it could be improved by the application of scientific principles. She thought the new subject of domestic science should be taught at all the women's colleges. It would help college women not only to lead more efficient home lives but also to bring them into touch with pressing local social problems. When few liberal arts colleges seemed to follow this lead, she taught domestic science herself to the more vocationally oriented women students at Simmons College in Boston.[32]

By 1893 Ellen Richards was (like everyone else) at the World's Columbian Exposition in Chicago, the greatest showpiece of women's achievements since the Centennial Exposition in Philadelphia in 1876. There she ran a Rumford Kitchen, offering nutritious, scientifically cooked lunches to visitors for thirty-two cents. Six years later Richards and Melvil Dewey, Director of the New York State Library, author of the Dewey Decimal System of book classification and advocate of women's work in librarianship, called the first of the ten Lake Placid Conferences on Home Economics. These annual meetings (1899–1908) brought together the diverse elements within the movement—the urban cooking school leaders, the public school supervisors, and the increasingly strong contingent of faculty members from the agricultural and teachers colleges—to discuss the field and its problems, especially its terminology and objectives, and to formulate model curricula. At its tenth conference in 1908 the group formed the American Home Economics Association and elected Ellen Richards its first president.[33]

Yet even before her death in 1911, the movement was already moving beyond Richards's vision to a deliberately more academic phase, as it saw its future tied less to urban cooking schools and demonstration kitchens and more to the growing agricultural colleges of the Midwest and West. By 1911 many of these colleges had already formed programs and even departments of home economics, and others were eager to follow. Some of the great leaders in the field in the next few decades were getting their start around 1910, like Isabel Bevier at the University of Illinois, Mary Swartz Rose at Teachers College, Columbia University, Agnes Fay Morgan at the University of California, Abby Marlatt at the University of Wisconsin, and Flora Rose and Martha Van Rensselaer at Cornell University. Yet the very success of this kind of "women's work" on major campuses only increased sexual segregation for future generations. Rather than leading to the acceptance of women in other kinds of scientific work, as the pioneers had expected, their success in "women's work" merely consigned their

[32]Ellen Richards, *The Relation of College Women to Progress in Domestic Science,* Publications of the Association of Collegiate Alumnae, Ser. 2, No. 27, 1890. (Some of her notes for this talk are in the Ellen Richards Papers, Sophie Smith Collection, Smith College Library.) Besides the Richards Collections at MIT and Smith, the best collection of her letters is her correspondence with Edward Atkinson, 1889–1900, in his papers at the Massachusetts Historical Society, Boston. See also Sally Gregory Kohlstedt, "Single-Sex Education and Leadership: The Early Years of Simmons College" (in press).

[33]Emma Seifrit Weigley, "It Might Have Been Euthenics: The Lake Placid Conferences and the Home Economics Movement," *American Quarterly,* 1974, 26:79–96, analyzes the conferences' published proceedings. Isabel Bevier and Susannah Usher, *The Home Economics Movement* (Boston: Whitcomb & Barrows, 1912) describes the roots of the movement. Bevier expanded this in her *Home Economics in Education* (Philadelphia: J. B. Lippincott Co., 1924). Rumford Kitchens were named for the American Benjamin Thompson, Count Rumford, who had run soup kitchens for the Electorate of Bavaria.

followers to segregated employment. The movement snowballed: since women were finding such relatively good opportunities in this field, many persons (even women scientists in other fields) urged ambitious young women interested in science to head for home economics. Fortunately for them the field expanded greatly in the 1920s and 1930s; it was the only one in which a woman could hope to be a full professor, a department chairman, or even a dean at a coeducational university.[34]

OTHER ACADEMIC POSITIONS

Another new and highly feminized field, which looked for a while as if it might attract women physiologists the way home economics absorbed women chemists, was that of hygiene or hygiene and physical education. Started by a number of male physicians in the late nineteenth century, this field sought to understand the scientific bases of both personal and public health, especially in relation to physical exercise. About the same time, several of the early women's colleges and coeducational universities employed women doctors both to teach hygiene and to be the college physician, since administrators were anxious to minimize the outbreak of physical ailments suspected at the time to accompany mental exertion in females. Some women doctors apparently found this position a congenial one, as did for example the sisters Dr. Clelia Mosher, appointed at Stanford University in 1893, and Dr. Eliza Mosher, hired by the University of Michigan in 1896. Another, Dr. Lillian Welsh of Goucher College (appointed in 1894), not only taught hygiene for thirty years but also gradually developed the only full department of physiology and hygiene at a woman's college into a strong pre-medical program famed for the number of its graduates who went on to the nearby Johns Hopkins Medical School. The subject of hygiene, however, did not flourish like home economics but floundered; by the 1920s it had been replaced in American colleges and universities by the more popular subject of physical education. By that time specialists in physical education were trained in the normal schools and accredited by their own professional societies, though they often taught the subject of health as well in the public schools. Thus though hygiene might have followed the path of home economics, with many universities hiring women with doctorates in physiology to do research and train high school teachers of physical education, the field's subsequent history developed quite differently.[35]

---

[34] Bevier, Marlatt, and Rose in *NAW*; Lita Bane, *The Story of Isabel Bevier* (Peoria, Ill.: Charles A. Bennet Co., Inc., 1955); Richard G. Morris, *Fields of Rich Toil: The Development of the University of Illinois College of Agriculture* (Urbana: University of Illinois Press, 1970), pp. 176–204; Eugene Davenport, "Home Economics at Illinois," *Journal of Home Economics*, 1921, *13*:337–341; Ida Hyde, "The Beginnings of the Science as a Development of Intellectual Personality," unpublished manuscript (1910) in Ida Hyde Papers, University of Kansas Archives, Spencer Research Library, Lawrence, Kansas, esp. pp. 8, 9, 12.

[35] E. Mosher and Welsh in *NAW*; C. Mosher in *AMS*, 2nd ed. (1910), and Kathryn Allamong Jacob, "Clelia Duel Mosher," *Johns Hopkins Magazine*, June 1979, pp. 8–16; William W. Guth, "Impressions of Lillian Welsh," *Goucher Alumnae Quarterly,* Dec. 1924, pp. 1–10, and Florence Sabin, "Dr. Lillian Welsh," *ibid.*, Feb. 1931, pp. 3–5 and "Dr. Mary Sherwood," *ibid.*, July 1935, pp. 9–13; *A Tribute to Lillian Welsh* (Baltimore: Goucher College, 1938); see also Thomas A. Storey, *The Status of Hygiene Programs in Institutions of Higher Education in the United States,* Stanford University Publications, University Series, Medical Sciences, Ser. 2, No. 1 (Stanford, 1927); James F. Rogers, *Instruction in Hygiene in Institutions of Higher Education,* U.S. Office of Education, Bulletin No. 7 (1936); Willystine Goodsell, *The Education of Women, Its Social Background and Its Problems* (New York: The Macmillian Co., 1923), Ch. 9; Dorothy S. Ainsworth, *The History of Physical Education in the United States* (New York: A. S. Barnes and Co., 1942); Arthur Weston, *The Making of American Physical Education* (New York: Appleton-Century-Crofts, 1962); Deobold B. Van Dalen et al., *A World History of Physical Education* (Englewood Cliffs, N. J.: Prentice-Hall, Inc., 1953), Chs. 23 and 24.

A third kind of academic women's work arising at the coeducational colleges in the 1890s was that of the dean of women. As the number of women students increased, and especially as they were required to live in dormitories on campus, the need arose for some sort of supervision. Although the duties of the office were not at first clear, male administrators considered almost all faculty women capable of the job. Some of the first office holders were scientists or physicians: Marion Talbot, assistant professor of sanitary science at the new University of Chicago, was appointed dean of women in 1892; Eliza Mosher, M.D., became a dean of women as well as professor of hygiene at the University of Michigan in 1896; psychologist Margaret Washburn was appointed Warden of Sage College (the women's dormitory) at Cornell in 1900; Mary Bidwell Breed, assistant professor of chemistry at Indiana University, was appointed their dean of women in 1901; Lucy Sprague (later Mitchell), who was later active in the child study movement, replaced a woman physician as dean of women at the University of California in 1906; and Fanny Cook Gates, formerly professor of physics at Goucher and Grinnell Colleges, became the dean of women at the University of Illinois in 1916. Despite these early appointments, it soon became clear that not all women were suited to such a position of moral authority over the students, for personality and temperament were of paramount importance. Thus the research-oriented Margaret Washburn, who disliked her job at Cornell intensely, was only too glad to give it up after two years, and Fanny Gates left Illinois after two years when gossip hinted that she may have been addicted to drugs. The president of the University of Illinois vowed never to make the mistake of hiring a Ph.D. in science for that job again.[36]

Another problem facing these early deans of women was the declining status of their position. Although the first such deans tended to have full faculty status and taught a few classes in their specialty, the position was in constant danger of being downgraded to a staff or administrative appointment. This had apparently happened by 1911, when a report on the duties and status of deans of women at 55 colleges and universities across the nation revealed that, though the deans were usually required to have doctorates, they were only barely tolerated on the institutions' often nearly all-male faculty.[37]

[36] Talbot, E. Mosher, and Washburn in *NAW*; Breed in Thomas D. Clark, *Indiana University: Midwestern Pioneer* (Bloomington: Indiana University Press, 1973), Vol. II, pp. 26–27 (which incorrectly reports her Bryn Mawr doctorate as from Heidelberg); I wish to thank Dolores Lahrman of the Indiana University Archives for copies of Breed's correspondence with President Joseph Swain about her title and the faculty status of the dean in 1901; Lucy Sprague Mitchell, *Two Lives: The Story of Wesley Clair Mitchell and Myself* (New York: Simon and Schuster, 1953), pp. 192–193; Gates in Mary Louise Filbey, "The Early History of the Deans of Women, University of Illinois, 1897–1923," typescript, 1969, p. 90, in Filbey Family Papers, University of Illinois Archives, Urbana; a few letters from Gates's earlier happier days are in the Edwin G. Conklin Papers, Special Collections, Princeton University Library, and the Ernest Rutherford Papers, Cambridge University Library (on microfilm at Office for the History of Science and Technology, University of California, Berkeley; I thank J. L. Heilbron for bringing them to my attention). The early assumption that women professors would be very solicitous about the personal needs of their female students is evident (humorously so) in Mrs. Caroline A. Soule, "A Collegiate Education for Women, and the Necessity of a Woman-Professor in the Mixed College," *Papers and Letters Presented at the First Woman's Congress of the Association for the Advancement of Women, New York, October 1873* (New York: Mrs. William Ballard, Printer, 1874), pp. 60–67.

[37] Gertrude S. Martin, "The Position of the Dean of Women," *J. Assoc. Collegiate Alumnae*, 1911, Ser. 4, 4:65–77; see also Marion Talbot, *More Than Lore: Reminiscences of Marion Talbot* (Chicago: University of Chicago Press, 1936), esp. Ch. 9; Lulu Holmes, *A History of the Position of Dean of Women in a Selected Group of Co-Educational Colleges and Universities in the United States,* Teachers College, Columbia University, Contributions to Education, No. 767 (New York: 1939); Sarah M. Sturtevant et al., *Trends in Student Personnel Work, ibid.,* No. 787 (1940) and Jane Louise Jones, *A Personnel Study of Women Deans in Colleges and Universities, ibid.,* No. 326 (1928), plus several others in the same series.

Yet attitudes were beginning to change somewhat by 1910. The revival of the women's rights movement (which led eventually to the ratification of the suffrage amendment in 1920) emboldened some women scientists to speak out, among them Marion Talbot, who as professor of household science and dean of women at the University of Chicago, was an archetype of women's work. She and others began to realize around 1910 that sex-typed employment had not proven the opening wedge to broader opportunities that it might have. It had brought them jobs in science and academia, but it was now clear that many of these were only marginal or subordinate positions, easily downgraded and rarely accorded professional recognition (e.g., a star in the *American Men of Science*).[38] Even the jobs in home economics, which were some of the best positions for women outside the women's colleges, were both separate and unequal (deliberately so, as in the Cornell faculty's vote), and thus had not brought women that much closer to the final and now for the first time visible goal of full equality. "Women's work" no longer seemed as progressive a step as it had in the 1880s and 1890s. Unfortunately, however, such segregation suited so many other needs and constituencies so well that later generations would find it difficult to move beyond it into the mainstream of scientific employment.

Thus the women's experience in science in these years promises to add a new dimension to our knowledge of the development of scientific employment, especially its professionalization, in the United States in the period 1880–1910. What had formerly been a fairly nonhierarchical collection of independent investigators had become, in some fields at least, highly bureaucratized "big science" with all the gradations in status and role that this implied: henceforth, some persons would have to be "hired hands" on projects directed by others. Government science was also expanding rapidly, as were several applied or service-oriented fields that academics scorned and tried to keep at a distance. It was a tense time of much jockeying for position and status, as some jobs and roles were downgraded, others created, and still others expanded and promoted. The presence of women created new opportunities for the more liberal "empire-builders" of the time, but it worried other more vulnerable men, whose scientific standing it seemed to threaten. Thus a segregated, low-status, almost invisible kind of "women's work" offered a harmonious way to incorporate the newcomers into the scientific labor force in ways that divided the ever-expanding labor, but withheld the ever more precious recognition. A pattern had been set for the twentieth century.

[38] E.g., Ellen Hayes, "Women in Scientific Research," *Science*, 1910, *32*:864–866; and Marion Talbot, "Eminence of Women in Science," *ibid.*, p. 866. I thank John Burke for these items.

Physiology in American Women's Colleges

The Rise and Decline of a Female Subculture

By Toby A. Appel

W OMEN HAVE FACED IMMENSE OBSTACLES to full participation in all fields of science, but some disciplines have provided a more favorable environment for them than others. Each discipline developed its own institutional structure and professional culture, which created unique constraints or opportunities for its women members. In the period before 1940, bacteriology and psychology attracted sizable numbers of women in part because they offered nonacademic areas of "women's work"—in public health departments for bacteriologists and in public schools for psychologists. Moreover, in these fields it was not difficult for women to become members of the main disciplinary societies, the American Society of Bacteriologists and the American Psychological Association. Both societies provided a modest affirmation of their female members by electing a woman as president before 1930. In other fields, such as botany and ornithology, women could attain high recognition through their research without having a paid position. Creditable work in observational fields such as these could still be carried out in the early twentieth century in a domestic setting or in the context of an amateur tradition.[1]

Physiology had few of these advantages. Unlike botany, physiology had little con-

Much of the research at women's colleges and several of the interviews of women physiologists were undertaken in common with Louise H. Marshall. I thank Robert Friedel, Gerald Geison, Peggy Kidwell, Louise Marshall, Ellen More, Edward Morman, Margaret Rossiter, Jean Soderlund, Deborah Warner, and the anonymous *Isis* referees for their helpful comments on drafts of this essay, and the college archives staffs, especially Elaine Trehub of Mount Holyoke, for providing research materials. Research for this article was funded in part by a Travel to Collections Grant from the National Endowment for the Humanities and by the Visiting Professorships for Women Program of the National Science Foundation.

[1] On women in various scientific disciplines see Margaret W. Rossiter, *Women Scientists in America: Strategies and Struggles to 1940* (Baltimore: Johns Hopkins Univ. Press, 1982), pp. 229–230, 238–243, 245–246, 282; Marianne Gosztonyi Ainley, "Field Work and Family: North American Women Ornithologists, 1900–1950," in *Uneasy Careers and Intimate Lives: Women in Science, 1789–1979,* ed. Pnina Abir-Am and Dorinda Outram (New Brunswick, N.J.: Rutgers Univ. Press, 1987), pp. 60–76; Nancy G. Slack, "Nineteenth-Century American Women Botanists: Wives, Widows, and Work," *ibid.,* pp. 77–103; and James H. Capshew and Alejandra C. Laszlo, " 'We Would Not Take No for an Answer': Women Psychologists and Gender Politics during World War II," *Journal of Social Issues,* 1986, *42:*157–180.

305

nection with traditional feminine values. In contrast to psychology and bacteriology, the discipline provided few opportunities for the employment of women. Although women might readily obtain training in physiology in medical school departments after about 1920, these departments rarely hired women even as research associates. Furthermore, there were no significant service roles for physiologists outside academia. Unlike ornithology, physiological research required laboratory space, instrumentation, and staff assistance unavailable outside a supportive institutional environment. Because of its restrictive membership policy, even women trained in the best departments of physiology found it difficult to qualify as members of the American Physiological Society (APS). Moreover, no woman served as president or even as a member of the Council of the APS until the 1970s.[2]

Yet physiology was unique in generating a women's subculture located in women's liberal arts colleges. Whereas almost all other departments of physiology were situated in medical schools, in the early twentieth century several elite women's colleges developed full-fledged departments of physiology, staffed by women with Ph.D.'s, that offered undergraduate majors and even master's degrees. Often not taught in men's liberal arts colleges, physiology as biomedical science was almost always part of the curriculum in women's colleges. In the first half of the twentieth century women physiologists taught at Barnard, Bryn Mawr, Connecticut, Douglass, Flora Stone Mather (Western Reserve), Goucher, Hood, Hunter, Mount Holyoke, North Carolina, Rockford, Simmons, Skidmore, Smith, Vassar, Wellesley, Wheaton, and Winthrop colleges. Even in a school such as Florida State College for Women, where in the early 1930s every science department except home economics was headed by a man and no woman in these departments held a rank higher than assistant professor, physiology—not a department, but listed separately in the catalogue— was the province of a woman with the rank of associate professor. While other sciences in women's colleges might be taught by either a man or a woman, physiology was almost always taught there by a woman. In this context, physiology was perceived as a women's domain in a way that astronomy, chemistry, mathematics, physics, zoology, and even botany and psychology were not.[3]

Physiology in the women's colleges developed as a women's subculture largely because of the circumstances under which the subject entered the curriculum, beginning in the 1860s. The dominant Victorian discourse of physiology differed radically from that of the twentieth century. Physiology, understood in its chief cultural meaning as personal hygiene, became part of the women's college curriculum because it was perceived to serve the practical needs of Victorian women students. In the late nineteenth century a new meaning of physiology as a biomedical discipline emerged and found a home in reformed medical schools. In the early twentieth century physiology in women's colleges was transformed into an experimental science

[2] Bodil Schmidt-Nielsen has been the only woman to serve as president of the APS in one hundred years. See Toby A. Appel, Marie M. Cassidy, and M. Elizabeth Tidball, ''Women in Physiology,'' in *History of the American Physiological Society: The First Century, 1887–1987*, ed. John R. Brobeck, Orr E. Reynolds, and Appel (Bethesda, Md.: American Physiological Society, 1987), pp. 381–390, esp. pp. 384–385.

[3] Florida State College for Women, *Bulletin* (includes *Catalogue, 1933–1934*) (Tallahassee, Fla., June 1934), esp. pp. 185–186. The professor of physiology was Viola Graham. The prevalence of physiology in the curricula of women's colleges has been previously noted by Mabel Newcomer, *A Century of Higher Education for Women* (New York: Harper, 1959), pp. 28, 78–79, 82–83, 101, 102. The list of schools is drawn from analysis of the careers of women members of the APS plus the schools represented in the informal conferences on physiology in the women's colleges (discussed below).

similar to that taught in medical schools. It not only remained in the curriculum but expanded because it met the vocational and personal needs of another generation of women in the 1920s and 1930s.

This article focuses on teaching and research at five colleges in which physiology attained the status of a major field: Mount Holyoke, Vassar, Smith, Wellesley, and Goucher. It contrasts two meanings of or discourses on "physiology"—physiology as hygiene and physiology as biomedical science—and it contrasts two generations of teachers of physiology in women's colleges—the female college physicians who taught physiology as hygiene and a later generation of women academics who promoted physiology as a laboratory science. The physicians were outspoken pioneers whose careers were lived out primarily in a world of women and for whom science and research were secondary to advocating reforms in personal and public hygiene. The later generation of Ph.D.'s in physiology were unmarried women scientists who attempted to bridge two worlds—that of the discipline dominated by men and that of mentoring women students in the family-like setting of the women's college. The significance of female role models for encouraging women to enter science is now widely appreciated. This essay provides a case study of how such mentoring functioned within a particular discipline. Limited research facilities as well as the barriers raised by the structure of the discipline prevented these women physiologists from achieving more than a marginal status in physiology. Yet they creatively used the resources of the women's college to fashion satisfying research programs (some of which involved issues relating to women) that enabled them to maintain ties to sympathetic male physiologists while at the same time enriching the personal lives and careers of their female students.

PHYSIOLOGY AS HYGIENE ENTERS THE NINETEENTH-CENTURY WOMEN'S COLLEGE

For most of the nineteenth century, "physiology" was nearly synonymous with personal hygiene and health reform. Physiology encompassed an understanding of the anatomy and functioning of the human body in the service of health and well-being. In popular usage there was little demarcation between descriptive anatomy and function and prescriptive hygienic rules. Physiology thus constructed became the key to learning how best to live.[4]

Physiology as hygiene was a discourse in which women could participate as both contributors and audience. In 1837, fifty years before the disciplinary society, the American Physiological Society, was founded, another short-lived society of the same name was formed as a popular self-help organization of men and women that emphasized diet reform. At midcentury women in a number of cities joined Ladies' Physiological Institutes, among the earliest of women's organizations, so that they might acquire "a knowledge of the Human System, the Laws of Life and Health, and the means of relieving sickness and suffering." And late in the century the Wom-

---

[4] On the health reform movements and popular meanings of physiology see James Whorton, *Crusaders for Fitness: The History of American Health Reformers* (Princeton, N.J.: Princeton Univ. Press, 1982); Harvey Green, *Fit for America: Health, Fitness, Sport, and American Society* (Baltimore: Johns Hopkins Univ. Press, 1986); and Regina Markell Morantz-Sanchez, *Sympathy and Science: Women Physicians in American Medicine* (New York: Oxford Univ. Press, 1985), pp. 28–46.

en's Christian Temperance Union advocated the teaching of physiology in high schools using approved texts that argued that alcohol was a poison.[5]

In the antebellum period health reform was predominantly advanced in a religious context as a moral duty. A healthy body was deemed a prerequisite for salvation. Thus Mary Lyon, who shared the religious fervor of her era, included physiology as part of the first year's course of study at Mount Holyoke Seminary (South Hadley, Massachusetts) when it opened in 1837. In 1844 she called on a friend, the clergyman-naturalist Edward Hitchcock of nearby Amherst College, to present a special series of lectures on physiology with the aid of his $700 male manikin, a complex layered apparatus revealing the internal organs. When a woman physician joined the Mount Holyoke faculty in the 1860s to tend to the health of the students, she took over the teaching of physiology, and a female manikin was specially made for the seminary.[6]

In the latter half of the nineteenth century hygienic reform was advocated in a predominantly secular context as a preventative to the myriad disorders prevalent in an increasingly industrial, urban, and technological age. Women, especially, were consumers of hygienic knowledge because women's bodies were widely regarded as frail and subject to a host of physical and nervous ailments.[7] Defenders of the new women's colleges faced the additional task of combating arguments, couched in the language of "physiology," that women's expenditure of mental energy in studying posed a danger to fragile reproductive organs.

In reaction to the nineteenth-century women's movement, which pressed for education and a wider sphere of activity than the domestic, some men constructed arguments drawing on the new post-Darwinian authority of science. Edward H. Clarke's much-debated book *Sex in Education* (1873) represented one line of justification for a restricted role for women, loosely based on the doctrine of conservation of energy. Human beings were assumed to possess a fixed store of vital energy that was easily exhausted by the needs of bodily growth and maintenance and by physical and mental activity, including the demands of the stressful modern environment on the nervous system. Clarke, a former professor at Harvard Medical School, claimed that women should not be educated with men or even in the same manner as men because their reproductive systems would suffer irreparable harm if, during menstruation, they diverted their vital energies to study rather than to the building of their reproductive

---

[5] Whorton, *Crusaders for Fitness*, pp. 62–91; Martha H. Verbrugge, "The Social Meaning of Personal Health: The Ladies' Physiological Institute of Boston and Vicinity in the 1850s," in *Health Care in America: Essays in Social History*, ed. Susan Reverby and David Rosner (Philadelphia: Temple Univ. Press, 1979), pp. 45–66, esp. p. 48; Verbrugge, *Able-Bodied Womanhood: Personal Hygiene and Social Change in Nineteenth-Century Boston* (New York: Oxford Univ. Press, 1988), esp. pp. 81–96; and Philip J. Pauly, "The Struggle for Ignorance about Alcohol: American Physiologists, Wilbur Olin Atwater, and the Women's Christian Temperance Union," *Bulletin of the History of Medicine*, 1990, 64:366–392.

[6] Green, *Fit for America* (cit. n. 4), pp. 3–29; Verbrugge, *Able-Bodied Womanhood*, pp. 4, 28–30; Mount Holyoke College, *First Annual Catalogue . . . 1837*, pp. 8–9; Abby Howe Turner, "Episodes in the History of Physiology at Mount Holyoke," *Mount Holyoke Alumnae Quarterly*, May 1940, 24(1):5–8, esp. pp. 5–6; and Sydney R. MacLean, "Mary Lyon," in *Notable American Women, 1607–1950: A Biographical Dictionary*, 3 vols., ed. Edward T. James, Janet Wilson James, and Paul S. Boyer (Cambridge, Mass.: Harvard Univ. Press, Belknap, 1971), Vol. 2, pp. 443–447. On the founding and early history of the elite women's colleges see Helen Lefkowitz Horowitz, *Alma Mater: Design and Experience in Women's Colleges from Their Nineteenth-Century Beginnings to the 1930s* (Boston: Beacon, 1986). Mount Holyoke began to award the baccalaureate degree in 1889.

[7] Verbrugge, *Able-Bodied Womanhood*, pp. 97–138; and Green, *Fit for America*, pp. 103–105.

organs. "Identical education of the two sexes," Clarke warned, "is a crime before God and humanity, that physiology protests against, and experience weeps over."[8]

Even those women who contested any connection between mental effort and the reproductive system believed the latter to be exceedingly fragile and deemed menstruation, in particular, a time for great caution. In her critique of *Sex in Education*, Dr. Alida C. Avery, Vassar's first physician and professor of physiology, assured Clarke that "at the beginning of every collegiate year the students are carefully instructed regarding the precautions which are periodically necessary for them." Her students were "positively forbidden to take gymnastics at all during the first two days of their period," and they were strongly advised not to dance, run up and down stairs, or do anything that might cause physical shock to the body. Dr. Lilian Welsh, who taught physiology and hygiene at Goucher for thirty years, recalled that when she arrived there in 1894 "we were still in the midst of the discussion precipitated by Dr. Clarke's book entitled 'Sex in Education.' The reproductive organs of women were looked upon as the source of most of their ills and the function of menstruation as a monthly recurrent disabling period."[9]

Anxious about the effect of the college regime on women's health, those establishing women's colleges hired women physicians to care for students, to examine them in conjunction with programs of physical exercise, and to teach a required course in physiology. Women physicians were perceived to be more sympathetic to women's health problems than male physicians and better able to provide care according to cultural standards of feminine modesty. Typically given the title of "professor of physiology and hygiene," college physicians taught courses that mixed descriptive information on the structure and function of the human body and practical prescriptions for personal and public hygiene. Courses were illustrated by means of female manikins, a skeleton, models of organs, charts, and sometimes dissections of lower animals.[10]

For example, at Vassar College (Poughkeepsie, New York), which was organized

---

[8] E. H. Clarke, *Sex in Education; or, A Fair Chance for the Girls* (1873; rpt. New York: Arno, 1972), p. 127. On the menstruation controversy see Rosalind Rosenberg, *Beyond Separate Spheres: Intellectual Roots of Modern Feminism* (New Haven, Conn.: Yale Univ. Press, 1982), pp. 1–27; Carroll Smith-Rosenberg and Charles Rosenberg, "The Female Animal: Medical and Biological Views of Woman and Her Role in Nineteenth-Century America," in *Women and Health in America*, ed. Judith Walzer Leavitt (Madison: Univ. Wisconsin Press, 1984), pp. 12–27; and Vern Bullough and Martha Voght, "Women, Menstruation, and Nineteenth-Century Medicine," *ibid.*, pp. 28–37. On science and "the woman question" see Cynthia Eagle Russett, *Sexual Science: The Victorian Construction of Womanhood* (Cambridge, Mass.: Harvard Univ. Press, 1989), esp. pp. 104–129, 160–164.

[9] See Avery's comments in Julia Ward Howe, ed., *Sex and Education: A Reply to Dr. E. H. Clarke's "Sex and Education"* (1874; rpt. New York: Arno, 1972), pp. 191–195, quotation on pp. 192–193; see also Howe's views, pp. 17–19. Lilian Welsh, *Reminiscences of Thirty Years in Baltimore* (Baltimore: Norman, Remington, 1925), p. 119; see also Welsh, *Fifty Years of Women's Education in the United States* (Baltimore: Goucher College, ca. 1923), pp. 10–11.

[10] See Morantz-Sanchez, *Sympathy and Science* (cit. n. 4), pp. 56–63; Regina Markell Morantz, "The 'Connecting Link': The Case for the Woman Doctor in Nineteenth-Century America," in *Sickness and Health in America: Readings in the History of Medicine and Public Health*, ed. Judith Walzer Leavitt and Ronald L. Numbers, 2nd ed. (Madison: Univ. Wisconsin Press, 1987), pp. 161–172. For course descriptions see, e.g., Mount Holyoke College, *Forty-ninth Annual Catalogue . . . 1885–6*, p. 23. Although physiology as hygiene played a particularly prominent role in the curricula of early women's colleges, it was also often present in men's liberal arts colleges, where it was likely to be associated with physical education. Amherst and Williams, for example, had required courses in physiology in the late nineteenth and early twentieth centuries. But these courses were not transformed into research-oriented biomedical science courses, as they were in women's colleges. I consulted catalogues of Amherst, Williams, Haverford, and Bowdoin colleges for the period 1900–1940.

by departments from its opening in 1865, the Department of Physiology and Hygiene was under the charge of Alida C. Avery, M.D. In her first annual report to the president of Vassar, in June 1866, she noted, "Since the Christmas holidays a part of one evening each week has been devoted to familiar lectures on matters pertaining to health, and the physiological 'conduct of life,' the influence of which has been most gratifying for those who labor here for the hand-in-hand advancement of mental and physical culture among women." Topics included sleep, exercise, constipation, mental hygiene, "tight lacing," "the philosophy of bathing," and, of course, menstruation. Physiology was likewise taught at Wellesley College (Wellesley, Massachusetts) by the "lady physician" from the opening of the school in 1875. At Smith College (Northampton, Massachusetts), which also opened in 1875, outside male lecturers taught physiology and hygiene until a female college physician was hired in 1888. When departments were formed in 1894, physiology was made a department under the college physician. Goucher (Baltimore, Maryland), too, when it opened in 1888, hired a woman physician to teach physiology and hygiene.[11]

Physiology and hygiene courses varied greatly according to the interests and eccentricities of those teaching them. The most influential of the college physician physiologists, because of their strong personalities and long tenure, were Elizabeth Burr Thelberg (1860–1935) at Vassar and Lilian Welsh (1858–1938) at Goucher. Thelberg, known affectionately as "Dr. T.," began a more than forty-year career at Vassar in 1887 (Figure 1). She considered it a supreme duty to prepare students for their future roles as wives, mothers, and active members of the community. Freshmen were given a series of lectures on personal hygiene that included discussion of constipation and menstruation. To upperclassmen she gave special lectures on "reproduction, maternity, and the care and feeding of young children." She informed them of the dangers of venereal diseases and "criminal abortion" and of the promise of eugenics, and firmly insisted that they nurse their babies. "These lectures are not easy to give," she wrote. "I feel as if they were at once the most important and hardest part of my year's work." Even her elective course in human physiology remained more practical than scientific. It too gave primacy to the reproductive system, illustrated by models and microscopic slides.[12]

The forceful Lilian Welsh, one of the pioneer women physicians in Baltimore, took over the teaching of physiology at Goucher from two short-term predecessors in 1894. One graduate who went on to a scientific career recalled that Welsh's "dark piercing eyes . . . twinkled with delight as she briskly bade us in her bass voice to live by common sense and put aside many of the common feminine habits of dress and eating which were bound to ruin our health." But her course gave greatest emphasis to preventative medicine and the importance of a well-run municipal health department.[13]

[11] Alida C. Avery, "Report of the Professor of Physiology and Hygiene for the 1st College Year 1865–66," 25 June 1866, Special Collections, Vassar College, Poughkeepsie, New York (hereafter **VC-SC**); *The First Wellesley Announcement, December 1874*, p. 3; Margery N. Sly, Smith College Archivist, to Toby A. Appel, 31 Aug. 1987; and Anna Heubeck Knipp and Thaddeus P. Thomas, *The History of Goucher College* (Baltimore: Goucher College, 1938), pp. 32–33.

[12] Elizabeth Burr Thelberg, "Instruction of College Students in Regard to Reproduction and Maternity," *New York Medical Journal*, 1912, *95*:1269–1270; see also Toby A. Appel, interview of Louise H. Marshall, San Francisco, Calif., 4 May 1984. Thelberg's courses were also described in the series of annual reports of the Department of Physiology and Hygiene to the President of Vassar College, VC-SC.

[13] Florence B. Seibert, *Pebbles on the Hill of a Scientist* (St. Petersburg, Fla.: Privately printed, 1968),

Figure 1. *Elizabeth B. Thelberg, M.D., professor of physiology and Vassar College physician from 1887 to 1890 and from 1892 to 1930. (Courtesy of Special Collections, Vassar College Libraries.)*

The nineteenth-century female physicians, although they esteemed science, were not researchers. They were, rather, practitioners who lived in a world of women and social reform. Welsh, a graduate of Woman's Medical College of Pennsylvania, formed a long-term relationship resembling a marriage with the physician Mary Sher-

p. 8. On Welsh see Welsh, *Reminiscences of Thirty Years in Baltimore* (cit. n. 9); *A Tribute to Lilian Welsh* (Baltimore: Goucher College, 1938) (a 44-page pamphlet of speeches given at her funeral); Florence Sabin, "Dr. Lilian Welsh," *Goucher Alumnae Quarterly*, 1931, *9*(2):3–5; William W. Guth, "Impressions of Lilian Welsh," *ibid.*, 1924, *4*(1):3–10; and Genevieve Miller, "Lilian Welsh," in *Notable American Women*, ed. James *et al.* (cit. n. 6), Vol. 3, pp. 567–568.

wood, with whom she participated in numerous educational and social welfare projects in Baltimore. An ardent feminist, active in the suffrage movement, Welsh recalled in her autobiography, "Always in my lectures in hygiene I was obliged to
point out the essential need of the ballot as a tool for securing conditions in the
community favorable to health." Thelberg, a graduate of Woman's Medical College
of the New York Infirmary, was passionately interested in public health, eugenics,
and medical and hospital care for the unfortunate. She was a leader in the American
Women's Hospitals Committee of the Medical Women's National Association and
served as president of the association in 1927/1928.[14]

In the late nineteenth century a different meaning of physiology, as a biomedical
discipline, emerged. This physiology came to be characterized by highly technical
experimental research on the physical and chemical basis of vital functions in animals; its results could be recorded graphically or in tables. Physiologists experimented on animals with the goal of understanding human physiology. The classic
multipurpose instrument of physiology was the kymograph, a revolving drum covered with smoked paper on which a leveraged stylus recorded minute mechanical
movements of tissue to produce a permanent graphic record, for example, of blood
pressure or muscle contraction. To further this new understanding of physiology, the
American Physiological Society was founded in 1887 as an elite disciplinary society.
Almost all its twenty-eight charter members had received part of their training in
Europe or were students of those trained in Europe. The part-time clinicians who
taught physiology as a lecture course in medical schools were not admitted into the
APS. As part of the reform of medical education that transformed medical schools
from proprietary institutions to endowed centers for teaching and research, departments of physiology gradually came under the control of full-time salaried academics, members of the APS, who established both research and teaching laboratories.[15]

Physiology as biomedical science was an elite and manly discourse based on surgical manipulations of live animals, a discourse from which women were initially
excluded. The APS had no women members until Ida Hyde, who had obtained a
German doctorate in 1893, was elected in 1902—and then only after a specific discussion of whether to admit women; no other woman was elected until 1913. Women
were rather to be found in the front ranks of those opposed to the new experimental
science, namely, those who organized the antivivisection campaigns of the 1880s
and 1890s.[16]

[14] Welsh, *Reminiscences of Thirty Years in Baltimore*, p. 156; and Esther Pohl Lovejoy, "In Memoriam: Dr. Elizabeth Burr Thelberg," *Women in Medicine*, July 1935, *49*:19. Clelia Duel Mosher,
who wrote on menstruation and gathered material for a survey of women's sexual beliefs and habits,
was of this generation of women physician-reformers. In 1910 she became physician to women students
at Stanford University and taught personal hygiene. Though she listed her field as "physiology" in the
1921 edition of *American Men of Science*, she would not have been considered for membership by the
APS. See Rosenberg, *Beyond Separate Spheres* (cit. n. 8), pp. 179–186, 196–197; and Carl N. Degler,
"What Ought to Be and What Was: Women's Sexuality in the Nineteenth Century," *American Historical Review*, 1974, *79*:1467–1490.

[15] On the emergence of physiology as a discipline see Gerald Geison, ed., *Physiology in the American
Context, 1850–1940* (Bethesda, Md.: American Physiological Society, 1987). My essay in that volume
discusses shifting meanings of physiology in the context of the founding of the APS: Toby A. Appel,
"Biological and Medical Societies and the Founding of the American Physiological Society," *ibid.*, pp.
155–176. See also W. Bruce Fye, *The Development of American Physiology: Scientific Medicine in the
Nineteenth Century* (Baltimore: Johns Hopkins Univ. Press, 1987); and Brobeck *et al.*, eds., *History of
the American Physiological Society* (cit. n. 2).

[16] Toby A. Appel, "First Quarter Century, 1887–1912," in *History of the American Physiological*

Bryn Mawr College, founded in 1884, offers an instructive contrast to both the hygienic and biomedical models of physiology. Unlike in other women's colleges, physiology did not enter the Bryn Mawr curriculum by way of the college physician, but was established on a variant biological model. In this minority understanding, based on biology at Johns Hopkins before the opening of the medical school, physiology was to serve as the counterpart to morphology in a department of biology. Bryn Mawr president Martha Carey Thomas hired up-and-coming men in biology, most of whom left after a few years for major university positions. Physiology was first taught by Frederic Schiller Lee, who in 1891 became professor of physiology at Columbia University, and then by Jacques Loeb, who went on to Chicago, Berkeley, and finally the Rockefeller Institute for Medical Research. Although a few educators, like the zoologist Charles Otis Whitman, promoted the biological model of physiology and sought to establish the field in a university environment, physiology, like biochemistry, ultimately aligned itself with medical schools and medical problems.[17]

THE TRANSFORMATION TO PHYSIOLOGY AS BIOMEDICAL SCIENCE IN THE WOMEN'S COLLEGES

That physiology remained in the curriculum of women's colleges in the twentieth century and was transformed by the 1920s in conformity with the scientific discipline can be attributed to several interrelated circumstances. First, the term *physiology* gradually became identified in the broader culture with experimental biomedical science removed from immediate practical goals. Earlier courses in "physiology" created a niche for the later experimental courses in the same way that, as Robert Kohler has shown, "medical chemistry" in nineteenth-century medical schools provided a niche for biochemistry.[18] Second, physiology as a laboratory science served the needs of a broad spectrum of twentieth-century women students. Knowledge of the functioning of the human body was widely regarded as beneficial to women in their roles as wives, mothers, and teachers. The colleges could thus count on a large attendance for introductory courses in physiology. Third, courses in physiology served new vocational needs of women in the 1920s and 1930s. It is only because physiology benefited larger constituencies that it could also accommodate the needs of those few

Society, ed. Brobeck *et al.*, pp. 31–52, esp. p. 33. Hyde's difficulties in obtaining a doctorate in physiology are described in Ida H. Hyde, "Before Women Were Human Beings: Adventures of an American Fellow in German Universities of the '90s," *Journal of the American Association of University Women*, 1928, *31*:226–236. On women among the antivivisectionists see James Turner, *Reckoning with the Beast: Animals, Pain, and Humanity in the Victorian Mind* (Baltimore: Johns Hopkins Univ. Press, 1980), esp. pp. 92–95, 114–119.

[17] Philip J. Pauly, *Controlling Life: Jacques Loeb and the Engineering Ideal in Biology* (New York: Oxford Univ. Press, 1987), pp. 55–64; and Lucy Fisher West, Bryn Mawr College Archivist, to Appel, 21 Nov. 1985. Two women taught physiology briefly at Bryn Mawr: Florence Peebles and Anna Baker Yates. On biological versus medical styles of physiology see Jane Maienschein, "Physiology, Biology, and the Advent of Physiological Morphology," in *Physiology in the American Context*, ed. Geison (cit. n. 15), pp. 177–193; Alejandra C. Laszlo, "Physiology of the Future: Institutional Styles at Columbia and Harvard," *ibid.*, pp. 67–96; and Pauly, "The Appearance of Academic Biology in Late Nineteenth-Century America," *Journal of the History of Biology*, 1984, *17*:369–397. On biochemistry see Robert E. Kohler, *From Medical Chemistry to Biochemistry: The Making of a Biomedical Discipline* (Cambridge: Cambridge Univ. Press, 1982).

[18] Kohler, *From Medical Chemistry to Biochemistry*, pp. 158–168.

women who desired to earn a Ph.D. Finally, whereas men trained in medical school departments of physiology could expect to find positions in medical schools, women with equivalent training, barred from such positions, were available to teach in women's colleges.

As the narrower disciplinary definition of physiology gained acceptance in the twentieth century, the older association of physiology and hygiene became outmoded. At the same time, the elite women's colleges were placing increasing emphasis on science, research, and publication. No longer was it enough for women faculty members to be good teachers. They were expected to obtain doctoral-level training and carry out a modest research program. The presence of practical physiology and hygiene courses in the women's college curriculum created a niche for introducing a more scientific style of physiology.

The transformation came about in one of two ways. In some schools—Mount Holyoke, Smith, and Wellesley—women in a zoology department who were seeking to expand its offerings managed to take over physiology from an overworked college physician. This maneuver, most easily accomplished when there was frequent physician turnover, usually entailed the separation of physiology from hygiene; the college physician continued to give required lectures in personal hygiene and sometimes, as at Mount Holyoke, offered a course in public health as well.[19] At Vassar and Goucher, on the other hand, college physicians themselves presided over the transformation of their department.

By 1910, at Wellesley, Smith, and Mount Holyoke, physiology had entered the zoology department. The chair of Wellesley's Department of Zoology, Mary Alice Wilcox, an invertebrate zoologist who had obtained some training in physiology from Michael Foster at Cambridge and C.-E. Brown-Séquard in Paris, secured the required physiology and hygiene course for her department in the 1880s and by 1887 added an advanced laboratory course. The department began to refer to itself as the Department of Zoology and Physiology (Figure 2). In 1906 Wellesley's newly formed Department of Hygiene and Physical Training reclaimed personal hygiene and began a rival course in physiology, ceded to the Department of Zoology and Physiology only after more than a decade of wrangling. At Smith Inez Whipple Wilder, a member of the zoology department and wife of the chairman, Harris Wilder, took over physiology from the college physician in 1906, leaving the physician with the Department of Hygiene.[20]

At about the same time, in 1904, physiology came under the purview of the zoology department at Mount Holyoke. The dynamic zoologist Cornelia Clapp, through her graduate study at the University of Chicago and summers at the Marine Biological Laboratory at Woods Hole, Massachusetts, came under the influence of Charles Otis Whitman. Like Whitman, she regarded physiology as a biological science co-

---

[19] Hygiene, in so far as it was developed as a science, became a loose synonym for public health or preventative medicine. Courses in public health were usually offered in elite women's colleges but, except at Goucher, not in the same department as physiology. On hygiene see Judith Walzer Leavitt, "Public Health and Preventative Medicine," in *The Education of American Physicians: Historical Essays*, ed. Ronald L. Numbers (Berkeley: Univ. California Press, 1980), pp. 250–272.

[20] On zoology at Wellesley see Marian Hubbard, "The Plight of Our Zoology Department," *Wellesley Alumnae Magazine*, 1928, *12*(3):122–130, esp. p. 125; *Wellesley College Calendar, 1892–93*, pp. 51–52; *Wellesley College Calendar, 1915–16*, pp. 154–156; and Wilma Slaight, Wellesley College Archivist, to Appel, 15 Apr. 1985. On Smith see Sly to Appel, 31 Aug. 1987; and Myra M. Sampson Oral History, 20 and 23 Mar. 1971, transcript, p. 14, Smith College Archives, Northampton, Massachusetts.

Figure 2. *Physiology class taught by Caroline Woodman (upper right), instructor in zoology at Wellesley College, 1893/1894. By the 1890s physiology and hygiene had been transferred from the college physician to the zoology department. The department taught a required physiology and hygiene course and an elective laboratory course, which is probably the course here illustrated. Note the diagram of the female pelvis on the right. (Courtesy of Wellesley College Archives; photograph by Partridge.)*

ordinate with morphology. According to Abby Turner, "Dr. Clapp saw the future of this new biological science and she coveted a share in it for Mount Holyoke,— real courses in comparative physiology with the laboratory approach. She saw that such teaching was incompatible with the interrupted life of the college physician. She therefore arranged to have zoology adopt physiology temporarily, while the physician kept the instruction in personal hygiene and in public health." Clapp was able to accomplish this "adoption" because Mount Holyoke had an especially rapid turnover of physicians.[21]

When Clapp acquired physiology, she invited her former student Abby Howe Turner (1875–1957) back to Mount Holyoke to help with the teaching. Since obtaining her bachelor's degree in 1896, Turner had taken additional training in zoology and physiology at the University of Pennsylvania and at the University of Chicago. In 1909

[21] Turner, "Episodes in Physiology at Mount Holyoke" (cit. n. 6), p. 6. On Clapp see Charlotte Haywood, "Cornelia Maria Clapp," in *Notable American Women*, ed. James *et al.* (cit. n. 6), Vol. 3, pp. 336–338; and "Cornelia Maria Clapp, March 17, 1849–December 31, 1934: An Adventure in Teaching," *Mount Holyoke Alum. Quart.*, 1935, *19*(1):1–9.

she began a two-year leave of absence at Harvard Medical School, working with the physiologist William T. Porter, founder of the *American Journal of Physiology*. By 1913 Turner had become a full professor at Mount Holyoke and virtual head of a section of the renamed Department of Zoology and Physiology. Instead of focusing on comparative physiology, as Clapp had envisioned, her course followed the medical model and became oriented to human physiology.[22]

At Goucher and Vassar, longtime college physicians Lilian Welsh and Elizabeth Thelberg were not about to cede physiology to biology or zoology departments (which in those schools happened to be headed by men). Instead, they themselves enlarged the curriculum and eventually hired academic physiologists to succeed them. Thelberg hired an assistant physician for her Department of Physiology and Hygiene, added laboratory experiments to her advanced course in physiology, and expanded the department's offerings to include courses in municipal sanitation and child hygiene. But academic physiologists did not invade the department until after World War I.[23]

On the initiative of Lilian Welsh, Goucher became the first women's college to acquire a faculty member with a doctorate in physiology. Welsh had hoped to undertake research on problems in exercise physiology that might be studied on women in the gymnasium, but she admitted that with her teaching and community activities her contributions "were nil." In 1910 she taught two courses, a required physiology and hygiene course and an elective course for which she had just introduced laboratory work. When in the following year she was authorized to hire another staff member, she deliberately sought a "physiologist" who could devote full time to the department and conduct research. She recalled, "I felt this was my opportunity to secure the services of a physiologist, but two great difficulties stood in the way. First, women physiologists were very rare and, second, Dr. [E. A.] Noble, then president of the College, could not see his way clear to offer any adequate salary. It was rather a hopeless quest, successful finally only because while women physiologists were scarce, positions open to them were still scarcer."[24]

Welsh found her physiologist in Jessie Luella King (1881–1956), who was to be associated with the Department of Physiology and Hygiene at Goucher College from 1911 until her retirement in 1947. King appears to have been the first woman to receive a doctorate in physiology through an American medical school department of physiology—Cornell Medical College (Ithaca, New York), in 1911. Despite her unique credentials, King had difficulty finding a position. In the year before she was to receive her degree she sent letters of inquiry to a number of women's colleges. Though Welsh was able to offer only a small salary and unpromising teaching and research conditions, King agreed to come to Baltimore for an interview on her way home to Indiana after Cornell's commencement. As Welsh recalled, "She came and

[22] Abby H. Turner, "Biographical Data Given for Directory at Centenary Celebration Mount Holyoke College—1937," Faculty Files, Mount Holyoke College Archives, South Hadley, Massachusetts (hereafter **MHC-A**); D. Elizabeth Williams, "Abby Howe Turner," *Mount Holyoke Alum. Quart.*, 1940, 24(2):61–62; Turner, "Episodes in Physiology at Mount Holyoke," p. 6; and Turner, *Laboratory Directions for the Course in General Physiology (Zoology B-I and B-II), Mount Holyoke College* (privately printed, 1914).

[23] See the annual reports of the Department of Physiology and Hygiene, VC-SC.

[24] Welsh, *Reminiscences of Thirty Years in Baltimore* (cit. n. 9), pp. 123–124, 130; and *Twenty-Second Annual Program of the Woman's College of Baltimore, 1910*, pp. 56–57.

seemed rather contemptuous. . . . I had no idea I would ever see her again." But in August King wrote to say that she would accept the position.[25]

In appearance and personality, Welsh and King made a striking contrast. Where Welsh was brilliant, outgoing, and sometimes acerbic, King was recalled as a gentle and intellectual Quaker lady. Florence Seibert, a graduate of Goucher in 1918, described her as "slim and quiet to the extent of being mouse-like, and she had all the understanding kindness that was needed by us on our rebounds from Dr. Welsh's onslaughts." King had little interest in "hygiene" and left the required course in physiology and hygiene to Welsh and her successor as college physician. Instead, she took over the elective course and turned it into a strictly scientific laboratory course on the organ systems of the human body—the physiology of muscles, nerves, special sense organs, circulation, and respiration.[26]

Vocational opportunities for college graduates in health fields coupled with limited opportunities for women doctorates in physiology encouraged the rapid expansion of physiology in women's colleges after World War I. In all the colleges World War I marked a watershed in the fortunes of physiology in the curriculum. The war, with its call on women for emergency work, followed by the influenza epidemic of 1918, in which more than half a million Americans died, pointed up the need for more women in health fields, especially in public health. By the 1920s graduates of women's colleges were expected to work outside the home at least until they married. Employment bureaus and vocational guidebooks enthusiastically steered graduates to the many new areas of work that were opening up.[27]

Among the new fields of work especially intended for women were a number of health-related areas other than medicine proper. Public health nursing, medical and psychiatric social work, dietetics, occupational therapy, physical therapy, medical laboratory technical work (performing routine medical tests), and laboratory research assistance were all women's work for which a course in physiology would serve as useful preparation. Marian Hubbard, chairman of the Department of Zoology and Physiology at Wellesley, recalled in 1928: "In the early years of the college, teaching, medicine and nursing were the only fields open to women who were trained in zoology. Today a number of opportunities beckon. Conspicuous among these are openings for technicians, for assistants in research, and for workers in the new field of public health."[28]

[25] Welsh, *Reminiscences of Thirty Years in Baltimore*, pp. 130–131. King was trained in a preclinical department in Ithaca that served as a component of Cornell Medical School in New York City. Margaret Rossiter has noted that four (unidentified) women received doctorates in physiology before 1900, two from Bryn Mawr and two from the University of Chicago. Both schools would have taught physiology in a biological mode. None of the women seems to have pursued the discipline; see Rossiter, *Women Scientists in America* (cit. n. 1), p. 36. On King see Bessie L. Moses, "Dr. Jessie King," *Goucher Alum. Quart.*, 1947, 25(4):16–17; and Florence B. Seibert, "Jessie Luella King, 1881–1956," *ibid.*, 1956, 34(3):6–7.

[26] Seibert, *Pebbles on the Hill of a Scientist* (cit. n. 13), p. 8; and *Goucher College Register for 1913*, pp. 53–54. King's career will be discussed further below.

[27] Rossiter, *Women Scientists in America* (cit. n. 1), pp. 263–265; Rossiter, "Vocational Guidance for Women," paper presented at the American Historical Association meetings, Los Angeles, 30 Dec. 1981; and Barbara Miller Solomon, *In the Company of Educated Women: A History of Women and Higher Education in America* (New Haven, Conn.: Yale Univ. Press, 1985), pp. 115–140.

[28] Hubbard, "Plight of Our Zoology Department" (cit. n. 20), p. 128. On career choices of students in zoology or allied fields see Hubbard, "Handicaps and Achievements," *Wellesley Alum. Mag.*, 1929, 13(4):206–208. The Vassar department noted in 1931 that "demand for sound knowledge of human physiology is . . . increasing in several fields of activity open to liberal arts students; in public health work, dietetics and nutrition and child welfare, for example": "Annual Report, Department of Physi-

Paradoxically, an undergraduate course in physiology was not regarded as useful preparation for medicine. Medical schools actively discouraged undergraduate schools from teaching physiology to premedical students because they felt that physiology was their domain. Zoology, rather than physiology, was required of premedical students.[29] Since men prepared for careers in medicine rather than in the lower-paying and more subordinate fields dominated by women, men's colleges had little need of academic human physiology. Moreover, men trained in physiology at medical schools were usually not available to teach at liberal arts colleges. Male physiologists could expect to find positions in medical schools (or in medical practice, since they were more likely to have M.D.'s than were women physiologists). Women physiologists like Jessie King had few alternatives to teaching in a women's college.

The appreciation of the vocational benefits of physiology coincided with the opening up of more Ph.D. programs in medical school departments to women. The 1920s and 1930s were a heyday for women seeking advanced degrees in physiology.[30] Women could obtain Ph.D.'s in the same elite schools as the future male leaders in the field (Chicago and Harvard were especially favored), but they could not obtain jobs in medical school departments. Thus there were excellently trained women available to teach, at least for a while, in the women's colleges.

Departments of physiology expanded enormously after World War I. In the 1920s the curriculum blossomed, majors were established, and master's programs were initiated. Academic physiologists took over from the remaining college physicians who were teaching physiology. At Goucher the offerings expanded from two year-long courses in 1915 to nineteen semester courses in 1917/1918. That year the department acquired the first of a succession of "assistants"—recent college graduates who helped with the laboratories and quizzes. By 1920/1921 the department had three regular teaching staff members and two assistants and was offering a major.[31] Jessie King, made a full professor in 1919, became chairman when Lilian Welsh retired in 1924.

The role of physiology in Wellesley's war emergency activities convinced Marian Hubbard that physiology ought to be expanded on a par with zoology so that majors might be offered. In 1919 the Department of Zoology and Physiology hired the first of a series of biomedical scientists. In the next few years advanced courses were set up, and the department acquired its first master's student in physiology, Eleanor

---

ology, May 1931,'' VC-SC. For an example of career advice concerning health-related fields see Elizabeth Kemper Adams, *Women Professional Workers* (Chautauqua, N.Y.: Chautauqua Press, 1921), Ch. 5, ''Health Services Other than Medicine,'' pp. 83–102.

[29] Goucher's catalogue explicitly stated that the courses of the department were not premedical; see Goucher College, *Catalogue for 1923–1924*, pp. 71–73. In a 1953 report on premedical education in liberal arts colleges the authors specifically admonished, ''All premedical students should avoid courses which are previews of medical school work.'' They specifically mentioned histology, medical bacteriology, and human physiology. See Aura E. Sevringhaus, Harry J. Carman, and William E. Cadbury, *Preparation for Medical Education in the Liberal Arts College: The Report of the Subcommittee on Preprofessional Education of the Survey on Medical Education* (New York: McGraw Hill, 1953), p. 93.

[30] The percentage of members elected to APS by five-year intervals who were female reached a high point in the 1920s and 1930s. It peaked at 12.2 percent in 1920–1924 and declined to a low of 4.9 percent in 1950–1954. As of 1980–1984, the percentage of those elected to regular membership who were women was only 11.3, the same as in 1930–1934. See Appel *et al.*, ''Women in Physiology'' (cit. n. 2), pp. 382–384.

[31] Goucher College, *Announcements for 1917–1918*, pp. 70–72; and Goucher College, *Catalogue for 1920–1921*, pp. 79–81.

Dewey Mason, a recent graduate of Mount Holyoke who would go on to earn her doctorate at Harvard Medical School.[32]

At Mount Holyoke and Smith, women well along in their careers returned to graduate school with support from their institutions so that they could obtain credentials needed to manage research programs in physiology. Myra Melissa Sampson (1887–1984), who taught physiology at Smith from 1909 to the 1950s, obtained a Ph.D. at the University of Michigan in zoology; with the aid of additional instruction from the biochemist Lafayette B. Mendel at Yale, she began a productive research program in experimental nutrition. Her department, which she chaired after 1929, though officially titled zoology, was in effect a department of zoology and physiology, since Sampson developed a separate introductory course and a separate track for physiology. When Cornelia Clapp of Mount Holyoke retired in 1916, two strong women vied for power in the Department of Zoology and Physiology: Abby Turner and the zoologist Ann Haven Morgan. As President Mary Woolley preferred to give each professor her own domain, she split the department in 1922, creating an independent department of physiology under Turner.[33] In 1924 the department moved into expanded quarters in the new Clapp Hall, which it shared with botany and zoology. That same year Turner, at the age of nearly fifty, took leave to take a doctorate in physiology at the Harvard School of Public Health.

Vassar's Department of Physiology and Hygiene also became more academic in the 1920s. In 1919 the department, still under the charge of college physician Elizabeth Thelberg, acquired its first faculty member with a Ph.D. Vassar, like Wellesley, experienced considerable turnover in personnel in the 1920s, but among those hired was one woman who was to stay until 1961. Ruth Conklin (1895–1988), who arrived in 1924, was a student of Turner at Mount Holyoke and had since obtained a master's degree in physiology at the University of Rochester. After teaching for three years at Vassar she took a leave of absence in order to follow her mentor to the Harvard School of Public Health, where she obtained a doctorate in physiology in 1930 for work with Turner's advisor, Cecil K. Drinker.[34]

Physiology at Vassar took a path somewhat different from that at Mount Holyoke, for it became linked with the movement to create a new interdisciplinary program in "euthenics." Inspired by Ellen Swallow Richards, who coined the term, the Blodgett family donated funds to build a large Gothic stone building on the edge of campus for euthenics, "the application of the arts and sciences to the betterment of human living."[35] The splitting of physiology from hygiene (which remained with the

[32] Wellesley College, "Report of the Department of Zoology and Physiology for the Year 1917–1918," "Report of the Department of Zoology and Physiology for 1918–1919," and "Report of the Department of Zoology and Physiology for 1919–1920," 1DB/1899–1966, President's Office, Academic Departments: Zoology and Physiology Annual Reports (1914–43), Wellesley College Archives, Wellesley, Massachusetts (hereafter **WC-A**). Mason spent her career teaching physiology at Women's Christian College, Madras, India. Basic biographical information on women was obtained from various editions of *American Men of Science*.

[33] Myra M. Sampson, "Zoology in Smith College," *Smith Alumnae Quarterly*, 1934, 26(1):1–10; and *Smith College Bulletin, 1935–36*, pp. 177–182. On Sampson see "Myra Melissa Sampson," *Smith Alum. Quart.*, Summer 1984, pp. 67–68; and Sampson Oral History (cit. n. 20). On the division of Mount Holyoke's Department of Zoology and Physiology see Turner, "Episodes in Physiology at Mount Holyoke" (cit. n. 6), p. 7.

[34] The first female faculty member with a Ph.D. was Grace Medes, a biochemist who later had a distinguished career at the Lankenau Hospital Research Institute in Philadelphia. On Conklin see Faculty Files, VC-SC; and Elbert Tokay, "In Memoriam: Ruth E. Conklin," *Vassar Quarterly*, 1988, 84(4):3.

[35] Quotation from the plaque on Blodgett Hall. Euthenics, which functioned as an interdisciplinary

college physician) coincided with Thelberg's retirement and with the move of the Department of Physiology into a large wing of the new Blodgett Hall in 1929. Previously physiology and hygiene had been housed in cramped quarters at the center of campus with the departments of botany, zoology, and geology. After the relocation, physiology and zoology were more separated at Vassar than on other campuses. Physiology, nominally under the charge of Ruth Wheeler, a biomedical scientist who also headed the euthenics program, was in fact under the control of Conklin. Conklin recalled in an interview, "We were kind of horrified to have so much hygiene taught before us and we tried to get away from it as much as possible." She modeled her own introductory course in physiology on the Mount Holyoke course.[36]

PHYSIOLOGY COURSES AND THEIR AUDIENCES

By the 1930s physiology at the women's colleges had reached its fullest expression in departments offering majors, opportunities for undergraduate research, and master's degrees. In effect, these departments were small-scale versions of medical school departments. Each of the colleges had from two to five regular faculty teaching physiology and two or three assistants. Although small, the departments flourished because women scientists designed the curriculum to appeal to a number of different constituencies.[37]

Most important for the continued existence of the department was a large introductory course open to freshmen and attractive to those who had no intention of majoring in science. Strong science requirements—two years in two different departments at Mount Holyoke and Vassar—contributed to large enrollments in these courses. Vassar's course reached a hundred students in the 1930s. Myra Sampson's introductory physiology course at Smith had eighty students in that decade. Although Mount Holyoke's course drew only sixty in the 1930s, later, under the inspired teaching of Charlotte Haywood (1897–1971), the numbers reached as high as two hundred. Goucher's introductory course, which by the 1930s had become experimentally oriented, was required of all students. The printed laboratory manuals of Turner, Haywood, and Conklin went through numerous editions.[38]

What did students learn? The Mount Holyoke introductory course, which was the model for the Vassar course, was organized by organ system and function. Laboratory work began with the chloroforming and dissection of a cat as a means of teaching students human anatomy. Students performed chemical tests on food nu-

---

major, sponsored no courses of its own during the school term except for a seminar. See Marshall interview (cit. n. 12); and Horowitz, *Alma Mater* (cit. n. 6), pp. 295–302.

[36] Elizabeth B. Thelberg, "Report of the Professor of Physiology, 1927–1928," 17 May 1928, VC-SC; Thelberg, "Annual Report 1928–1929 to Henry N. MacCracken, President," 7 May 1929, VC-SC; and Toby A. Appel, interview (with Louise Marshall) of Ruth Conklin, Poughkeepsie, N.Y., 14 Nov. 1986. On Wheeler see Elizabeth Neige Todhunter, "Ruth Wheeler," in *Notable American Women*, ed. James *et al.* (cit. n. 6), Vol. 3, pp. 576–577.

[37] General statements are based on the annual catalogues of the five colleges considered, annual reports of the department for Wellesley, Vassar, and Mount Holyoke, and articles in alumnae magazines cited elsewhere in these footnotes.

[38] Conklin interview (cit. n. 36); and Toby A. Appel, interview (with Louise Marshall) of Jane McCarrell, Sykesville, Md., 2 Jan. 1986. On enrollments see Sampson, "Zoology in Smith College" (cit. n. 33), p. 4; "Report of the Department of Physiology, 1932–33," MHC-A; and Curtis G. Smith, "In Memoriam: Charlotte Haywood," *Mount Holyoke Alum. Quart.*, 1971, 55(1):18. See notes 22 and 39 on laboratory manuals.

Figure 3. Physiology experiment, Wellesley College, 1950s. Respiratory mechanics are measured with a pneumograph, a closed rubber tube fastened about the thorax and attached to a tambour, and recorded graphically by means of the kymograph. (Courtesy of Wellesley College Archives; photograph by Olive Sawyer.)

trients, digestive enzymes, familiar foods, blood, and urine. After resting, standing, and exercise they measured their vital capacity, counted their inhalations or pulses, and recorded their blood pressure. With the aid of the kymograph they produced graphic recordings of the movements of a laboratory partner's chest in respiration and her arterial pulse under various experimental conditions (Figure 3). They recorded heart movements and the results of mechanical or electrical stimulation of muscles and nerves in frogs or turtles. Vivisection experiments were performed only on these animals, though a variety of animal models were dissected. There was no single text or specific reading assignments. Students were encouraged to obtain background information or to prepare themselves to discuss specified topics from a variety of recommended sources, including standard medical physiology texts.[39]

[39] I have examined Abby H. Turner, *Laboratory Directions for the Course in General Physiology (Zoology B-I and B-II), Mount Holyoke College* (1914), 2nd ed. (Privately printed, 1916); Turner and Charlotte Haywood, *Laboratory Experiments: Elementary Survey Course, Physiology 101-102, Mount Holyoke College* (South Hadley, Mass.: Mount Holyoke College, 1944); and [Ruth Conklin], *Vassar College Physiology 105: Laboratory Directions*, 5th ed. (Privately printed, 1956). I am grateful to Elaine D. Trehub, College History and Archives Librarian at Mount Holyoke, for providing me with a copy of Turner's manual. In the 1950s students in the course at Vassar dissected a rat rather than a cat.

It is difficult to assess the social content of these courses. Laboratory work emphasized the generation of scientific knowledge rather than application of this knowledge to daily living. Yet in some areas, especially in nutrition, students did acquire practical knowledge. Students determined the nutrients in common foods, examined their personal diets, and designed a nutritious daily menu. Laboratory manuals sometimes directed students to information on social issues. Turner stated in the section on reproduction in her 1916 manual that "references to discussions of related hygienic, social, and moral topics, and the literature on heredity and eugenics will be given to those who wish them." Haywood's manual in 1944 told students to know what was meant by venereal disease, birth control, masturbation, and continence.[40]

Physiology was justified for nonscientists because it provided useful knowledge for personal and family living and because it introduced students to the experimental method in science. Abby Turner reported in 1932 that the introductory course at Mount Holyoke was intended for the "average student" who might take it to fulfill a science requirement "and whom we hope to make glad of her choice, because she finds [the] subject matter useful as a basis for intelligent personal hygiene for herself and, later, for her family, and because she comes to know by personal participation how scientific work is done." At each commencement, she claimed, alumnae would tell her of the value of physiology in their daily lives, "perhaps none more frequently than the mothers." In a period in which students enjoyed the new freedom of dating, a course in physiology provided basic information on reproductive physiology. Ruth Conklin, professor of physiology at Vassar, was often told that the "main drawing card" of introductory physiology was the several weeks devoted to the reproductive system toward the end of the year.[41]

The remainder of the physiology curriculum was directed to majors in physiology and such related fields as zoology, chemistry, psychology, and, at Vassar, euthenics. Several colleges offered an advanced survey of physiology, with science prerequisites, as an alternative to the more elementary survey. At the next level courses treated the various organ systems separately; there might be offerings in circulation and respiration, the nervous system, or endocrinology, for example. All departments included courses in nutrition and metabolism taught as experimental biomedical sciences. At the most advanced level there were independent study courses and seminars for discussion of recent journal articles. At both Vassar and Mount Holyoke master's students also met prominent biomedical scientists, male and female, through departmental lectures.[42]

Physiology majors numbered no more than six to eight a year, and sometimes there were as few as two or three. Majors filled out their program in other sciences: courses in zoology and advanced courses in chemistry were generally required, and physics was also urged. Knowledge of foreign languages, especially German, was

---

[40] Turner, *Laboratory Directions*, p. 107; and Turner and Haywood, *Laboratory Experiments*, pp. 75–79.

[41] "Report of the Department of Physiology, 1931–1932," MHC-A; "Report of the Department of Physiology, 1937-1938," MHC-A; and Conklin interview (cit. n. 36). On women's college culture in the 1920s and 1930s see Solomon, *Company of Educated Women* (cit. n. 27), pp. 157–171.

[42] In 1938, e.g., Mount Holyoke hosted A. Baird Hastings, a prominent biochemist at Harvard Medical School; Eugene Landis, soon to become chairman of the department of physiology at Harvard Medical School; and Marion Fay, professor of biochemistry at Woman's Medical College of Pennsylvania; see "Report of the Department of Physiology, 1938–1939," MHC-A.

essential for reading the journal literature. Turner summed up the career choices of majors: "Our major students find their places after leaving college in teaching, in further study, in nutritional work, in various responsible nursing positions, in laboratory assistance of a high grade, and in other fields." Turner's students were especially well equipped to find jobs as technicians.[43]

As they evolved at the various colleges, the departments acquired somewhat different orientations. Goucher's was the only department to include bacteriology or public health; Jessie King introduced courses in bacteriology that eventually became as important in the curriculum as physiology. (In the other women's colleges bacteriology was taught in the botany or hygiene department.) At Vassar and Smith nutrition played an especially strong role. More biologically oriented than the other departments, Mount Holyoke's was the only one to include "general physiology" as well as "chemical physiology" (biochemistry), which in other schools was taught in the chemistry department.

The master's programs, begun at Mount Holyoke, Vassar, Wellesley, and Smith in the 1920s, were a curious blend of job and fellowship. Recent graduates, hired as assistants, taught laboratories and quiz sections of the survey course, took advanced courses, and carried out a thesis project; if all went well, they would receive a master's degree in two years. (Goucher, although it had no master's program, had similar short-term assistant positions.) Professors at the various colleges recommended their students for one another's assistant slots. Interviews were scarcely necessary, since the women knew each other well and trusted each other's choices. Usually a department could accommodate only two assistants at a time. Through this arrangement a young college graduate might make a modest living while pursuing an advanced degree in a sheltered and friendly environment. At the end of her appointment she would be encouraged to go on for her doctorate in a medical school department. With the experience thus acquired, she would be exceptionally well prepared and a likely candidate for fellowship support.[44]

The physiology subculture at women's colleges attained a sufficient identity that women faculty organized three day-long conferences of students and teachers to discuss special problems of teaching physiology to women. The first was held at Mount Holyoke in 1922, the second at Vassar in 1932, and the third at Wellesley in 1936. Delegates attended from such schools as Wellesley, Goucher, Mount Holyoke, Vassar, Smith, Wheaton, Simmons, Connecticut College, and Bryn Mawr (the one male representative). Faculty members presented discussion papers on teaching the introductory course, requirements for the major, senior comprehensive examinations, interdepartmental cooperation in major field elections, individual research problems for advanced students, and the need for the undergraduate college to support research. Esther Greisheimer, a former Wellesley faculty member who had become a professor of physiology at Woman's Medical College of Pennsylvania, spoke at the Wellesley conference on "How should we advise young women in regard to phys-

---

[43] See, e.g., "Report of the Department of Physiology, 1932–33," "Report of the Department of Physiology, 1931–1932," and "Report of the Department of Physiology, 1934–35," MHC-A.

[44] Sarah Chapman, a graduate of Goucher recommended by King for an assistant position at Vassar, recalled that she received three offers of graduate fellowships after her master's training in 1935; she thought it unlikely that she would have found support immediately after undergraduate school. Chapman went to Yale, but owing to John Fulton's preoccupation with other students, her marriage, and the impossibility of continuing her research with primates elsewhere she did not complete her degree. Toby A. Appel, interview of Sarah Chapman Finan, Washington, D.C., 8 Dec. 1988.

iology as a life work?" These conferences, Turner noted, were especially welcome because the national disciplinary societies entirely ignored problems of teaching.[45]

INTEGRATING RESEARCH AND MENTORING

The biomedical scientists who taught physiology in women's colleges, unlike their college physician predecessors, straddled two worlds, that of the male-dominated discipline and that of their female students and colleagues. They derived considerable pleasure from their research and the contacts it afforded with sympathetic men in the discipline, but they also had a strong sense of responsibility to act as mentors and role models for their students. The most successful of the women were able to integrate research and teaching creatively so that they functioned in a synergistic manner. How, then, did the female subculture of physiology in women's colleges relate to the wider world of the male discipline?

Taken as a whole, the discipline of physiology was inhospitable to women. Physiology departments in medical schools trained women but hired very few of them. Of the forty-nine women who were members of the American Physiological Society in 1940 (of 690 total members), only eleven were associated in any capacity with medical school physiology departments. Aside from two at Woman's Medical College of Pennsylvania, only one other woman had achieved the rank of associate professor.[46] The APS also posed a barrier to women through its restrictive membership policy. To be nominated for membership by the APS Council, one must have conducted and published original research in physiology and had to be continuing to do research. In all but a handful of cases, a Ph.D. or M.D. was also tacitly required. Moreover, the Council tended to define "physiology" narrowly, especially in the 1930s. One might be rejected for membership if one's publications were not physiological enough or if it was thought that one was more appropriately a candidate for some other disciplinary society. As women Ph.D.'s were unlikely to get positions in physiology departments, they often took lower-level research positions in related fields and therefore could not qualify for membership. King, with credentials and publications, became the third female member of the APS in 1914. Turner first applied for membership in 1915 and again in 1916. Although she was a full professor of physiology and had published three papers, she was rejected, presumably because she did not have a degree and was not continuing to publish. It was only after she obtained her Ph.D. in 1926 and published additional papers that she was elected in 1928. Haywood and Conklin succeeded in becoming members, but Sampson was rejected in 1930 and again in 1931, probably because her research dealt with nutri-

---

[45] "Report of Physiology, 1931–32," VC-SC; Annual Report of the Department of Zoology and Physiology, Wellesley College, 1935–36, WC-A; "Report of the Department of Physiology: 1935–1936," MHC-A; and "Report of the Department of Physiology for the Year 1940–1941," MHC-A. Greisheimer's tantalizing paper has unfortunately not been found.

[46] Frances Hellebrandt's case was anomalous, because she taught physical education students rather than medical students in a department, at the University of Wisconsin, that was unusual in being university-wide: Frances A. Hellebrandt to Bernard Schermetzler (copy to Toby A. Appel), 20 Sept. 1985. Data on women in the APS are based on the *Yearbook of the Federation of American Societies for Experimental Biology, 1939–40* (which includes the American Physiological Society), and on biographical information from various editions of *American Men of Science*.

tion; she instead joined the American Institute of Nutrition, a related biomedical disciplinary society.[47]

It might be noted that it was far less difficult to publish in the APS journal and to present papers at its meetings than to become a member of the society. Any member had the right to present a paper each year or introduce a nonmember to present a paper. Thus physiologists commonly gave papers to the APS before they were elected to membership. After Turner and Conklin were elected they were able to place their undergraduate students on the program and have their abstracts printed in the society's journal. The journal, which was financed independently of society dues (membership did not include a subscription), published all papers within its defined scope; there was no peer review until the 1930s.[48] King, Turner, and Conklin published most of their articles in the *American Journal of Physiology*; many of Haywood's pieces appeared there also.

Before World War II the American Physiological Society was strictly a research organization, providing no forum whatever for discussion of teaching at either the medical school or the undergraduate level; nor did the society take an interest in problems of the profession. Women could become members and participate in the scientific programs, but they were otherwise ignored.

Although most male physiologists paid little attention to the women's college subculture, a few were highly supportive, among them Lafayette B. Mendel of Yale, Anton J. Carlson of Chicago, William Henry Howell of Johns Hopkins, and George W. Corner of Rochester. Two male physiologists especially stand out as benefactors of physiology in women's colleges—Francis Gano Benedict (1870–1957), head of the Carnegie Institution's Boston Nutrition Laboratory, and Cecil K. Drinker (1887–1956), chairman of the Department of Physiology of the Harvard School of Public Health. An expert in calorimetry and respiratory gas analysis, Benedict developed instrumentation to measure basal metabolism. His investigations, which involved measurements in humans of different ages, sexes, races, and conditions, provided a research program in which women in women's colleges could readily participate (Figure 4). In 1926/1927 Florence Gustafson, a master's candidate at Wellesley, worked with Benedict on her thesis project, "Seasonal Variation in Basal Metabolism in College Girls." Benedict provided the equipment for her to carry out an extended series of measurements on twenty Wellesley students and coauthored the published report. That same year Benedict gave a lecture at Wellesley on "Recent Advances in the Study of Metabolism and the Contribution that Women's Colleges Can Make to Research."[49] Also in the 1920s, Benedict aided Abby Turner with her doctoral

[47] See *History of the American Physiological Society*, ed. Brobeck *et al.* (cit. n. 2), pp. 21–22, 32–35, 66–67. Information on elections is found in the Council Minutes of the American Physiological Society, American Physiological Society Archives, Bethesda, Maryland. On the election of women to membership and on their role in the society see Appel *et al.*, "Women in Physiology" (cit. n. 2).

[48] Geison has shown that authors of articles in the APS journal represented a far broader spectrum of researchers than members of the society; see Gerald L. Geison, "International Relations and Domestic Elites in American Physiology, 1900–1940," in *Physiology in the American Context*, ed. Geison (cit. n. 15), pp. 115–154. Policies of the society's meetings and publications are discussed in chapters by Toby A. Appel in *History of the American Physiological Society*, ed. Brobeck *et al.*, pp. 97–130, esp. pp. 67–78.

[49] "Annual Report of the Department of Zoology and Physiology, 1925–26," " Annual Report of the Department of Zoology and Physiology, 1926–27," "Annual Report of the Department of Zoology and Physiology, 1927–28," WC-A; and Florence L. Gustafson and Francis G. Benedict, "The Seasonal Variation in Basal Metabolism," *American Journal of Physiology*, 1928, 86:43–58. These experiments were also mentioned in Hubbard, "Plight of Our Zoology Department" (cit. n. 20), pp. 124, 129.

Figure 4. *Francis Gano Benedict, Hazeltine Stedman (assistant professor of physiology at Mount Holyoke), and two Mount Holyoke students with respiratory apparatus, circa 1925. Benedict, a male physiologist who was especially supportive of physiology in women's colleges, coauthored articles with Stedman and with the professor of physiology, Abby Turner. (Courtesy of Mount Holyoke College Library/Archives.)*

research on the circulatory physiology of women at the nearby Harvard School of Public Health. In 1935 he and Turner coauthored a paper that compared basal metabolism and nitrogen excretion of foreign-born Oriental students at Mount Holyoke with American-born students.[50]

Cecil K. Drinker's department at the Harvard School of Public Health was closely allied with, but far more supportive of women than, Walter B. Cannon's physiology department at Harvard Medical School. Drinker, whose wife Katherine was a graduate of Woman's Medical College of Pennsylvania and a member of the APS, was probably the most sympathetic of the male mentors of women in physiology. He supervised the dissertations of no fewer than six women members of the APS, all

Another example of work in this area by women physiologists is Mary Elizabeth Collett and Roberta Hafkesbring, "Day to Day Variations in Basal Metabolism of Women," *Amer. J. Physiol.*, 1924, *70*:73–83. Collett later taught at Flora Stone Mather College and Hafkesbring at Woman's Medical College of Pennsylvania.

[50] Abby H. Turner and Francis G. Benedict, "Basal Metabolism and Urinary Nitrogen Excretion of Oriental Women," *Amer. J. Physiol.*, 1935, *113*:291–295. Benedict had earlier asked Turner to collaborate on a project correlating mental activity of college women with basal metabolism: F. G. Benedict to Abby H. Turner, 10 Aug. 1926, MHC-A.

of whom had been associated with women's colleges: Ann Stone Minot (Ph.D., 1923), Turner (1926), Conklin (1930), Madeleine Field Crawford (1932), Florence Haynes (1933), and Jane McCarrell (1940). After they completed their degrees he helped them to find jobs, saw that they were treated with appropriate deference, advised them on further research, and sponsored them as members of the APS.[51]

Central to women's ability to act as mentors in physiology were their research programs. In a liberal arts college with limited facilities and support staff and large teaching loads, women could not hope to rival the achievements of men in medical schools. From the perspective of most of the leaders of the discipline their research was marginal, yet for the women themselves research provided great satisfaction. The most successful of the women adopted creative strategies to make the most of the opportunities available in the women's college setting.[52] Jessie King at Goucher, Abby Turner and Charlotte Haywood at Mount Holyoke, Ruth Conklin at Vassar, and Myra Sampson at Smith were all able to establish research programs that allowed them to interact as colleagues with male researchers, provided topics for student research projects, and enabled them to coauthor published papers with master's students or exceptional undergraduates. Only at Wellesley was the program in physiology a relative failure in the 1930s—the woman in charge, Ada Roberta Hall, published little and belonged to none of the biomedical disciplinary societies. Physiology at Wellesley stagnated until the 1940s and 1950s, when two much more productive and better connected women, Louise Palmer Wilson and Virginia Fiske, took over.[53]

One strategy, adopted by King and Turner, was to choose research projects involving as subjects what women's colleges had in abundance—so-called healthy young women. Although much of physiology involved animal research, some areas—respiration, circulation, metabolism, and to some extent reproduction—were amenable to research using human subjects. Both King and Turner had to abandon animal experimentation begun in a medical school setting to begin new research programs more appropriate to the women's college setting.

King's and Turner's research dealt with the controversial subject of sex differences. While women social scientists of the same era—such as Helen Thompson

---

[51] Conklin interview (cit. n. 36); McCarrell interview (cit. n. 38); Toby A. Appel, telephone interview of Madeleine Field Crawford, Needham, Mass., 4 Jan. 1985; R. H. Kampmeier, "Ann Stone Minot (1894–1980): Clinical Chemist and Teacher," *Clinical Chemistry*, 1986, *32*:1602–1609; and "Report of the Department of Physiology, 1932–1933," p. 7, MHC-A. On Drinker see "Katherine Rotan Drinker, M.D., 1889–1956, and Cecil Kent Drinker, M.D., 1887–1956," *A.M.A. Archives of Industrial Health*, 1957, *15*:74–75.

[52] Some women doctorates, especially at less elite women's colleges, were unable or unwilling to adjust and gave up research upon taking a teaching position. Jane McCarrell, after a number of years of research at the Harvard School of Public Health and Massachusetts General Hospital, became professor and chairman of the biology department of Hood College in 1946 and published no further. Similarly, Madeleine Field Crawford, after several years working with Drinker at the Harvard School of Public Health, spent the latter part of her career teaching biology at Pine Manor Junior College, where she did no research. Even King's successor, Phoebe Crittenden, until then a productive researcher, ceased publication after arriving at Goucher: McCarrell interview; Crawford interview; and Phoebe Crittenden to Louise Marshall, 13 Jan. 1987.

[53] Sampson's research was in the area of experimental nutrition, especially the effects of vitamin A on metabolism, and used the typical animal model, the albino rat. She was proud of the theses she supervised and of the women who went on to graduate school and medical school. But as she had fewer research and personal ties to the discipline of physiology, she was not as significant a mentor of physiologists as King, Turner, Haywood, and Conklin. Hall worked on problems of endocrinology and metabolism in the albino rat but published nothing from her arrival at Wellesley in 1930 until 1942. See Ada Hall Faculty Biographical File, WC-A; and Louise P. Wilson, "Ada R. Hall," *Wellesley Alum. Mag.*, May 1964, *48*(4):239–240.

Woolley, Leta Hollingworth, and Jessie Tate—combined professional scientific research on sex differences with a deliberate social agenda, King, Turner, and, later, Charlotte Haywood chose this area of research primarily because it was readily adaptable to the women's college. They seemed more interested in generating publishable physiological data than in actively promoting social reform based upon their work. Physiologists in women's colleges furthered women's causes far more by their mentoring activities than through the content of their research.[54]

At Cornell King had carried out her doctoral research under Sutherland Simpson in the area of neuromuscular physiology. The experiments involved anesthetizing and suspending a sheep, laying bare the cerebral cortex, and electrically stimulating the brain in an effort to localize the areas controlling the movement of each of the legs. A project made possible by the resources of a medical school department, it could not be pursued at Goucher College. King published her dissertation results and then had to find a new area of research. Welsh, the feminist college physician, suggested that King investigate the physiology of menstruation in healthy women. King measured pulse rate, diastolic and systolic pressures, and temperature of a group of college students and instructors twice a day through the course of several menstrual cycles. For this research, as well as for much of her other work, King took advantage of Goucher's location in Baltimore to obtain advice and instruments (in this case a modern sphygmomanometer) from men at Johns Hopkins Medical School. Although she cited the physician Mary Putnam Jacobi, whose Boylston Prize essay of 1876 had used similar physiological measurements to dispute the damaging conclusions of Edward H. Clarke, King's own conclusion was understated. She claimed that her results, "as far as they go, seem to indicate that there has been a tendency to over-emphasize the inefficiency of women during the menstrual period."[55] But King was more interested in science as an end in itself than in the social debate over menstruation. Her later papers cited none of the contemporary work on menstruation by the more outspoken social scientists.

In the physiology of the female reproductive cycle King found a research program that she could exploit throughout her career. Another early paper, suggested by the physiologist William Henry Howell of Johns Hopkins, extended the recent popular work on the knee-jerk response to women during various phases of the menstrual cycle. In a later paper King showed that menstrual cycles were far more variable than was commonly thought and that the twenty-eight-day cycle should not be considered "normal." Here again, King did not discuss the broader implications of her work.[56]

[54] On research on sex differences by women social scientists see Rosenberg, *Beyond Separate Spheres* (cit. n. 8).

[55] For King's dissertation results see Sutherland Simpson and Jessie L. King, "Localization of the Motor Area in Sheep," *Quarterly Journal of Experimental Physiology*, 1911, 4:53–65; and King, "Localization of the Motor Area in the Sheep's Brain by the Histological Method," *Journal of Comparative Neurology*, 1911, 21:311–321. For King's research on the menstrual cycle see King, "Concerning the Periodic Cardiovascular and Temperature Variations in Women," *Amer. J. Physiol.*, 1914, 34:202–219, on p. 218; she cites Mary Putnam Jacobi, *The Question of Rest for Women during Menstruation: The Boylston Prize Essay of Harvard University for 1876* (New York, 1877).

[56] Jessie L. King, "Possible Periodic Variations in the Extent of the Knee-Jerk in Women," *Amer. J. Physiol.*, 1918–1919, 47:404–409; King, "Menstrual Records and Vaginal Smears in a Selected Group of Normal Women," *Contributions to Embryology* (Carnegie Institution of Washington, Publication no. 363), 1928, no. 95, pp. 79–94; and King, "Menstrual Intervals," *American Journal of Obstetrics and Gynecology*, 1933, 25:583–587.

With the help of the anatomist George W. Corner, whom she met at Hopkins, King also undertook animal research on reproductive physiology. She spent 1927 in Corner's laboratory at the University of Rochester learning the latest embryological and endocrinological techniques. She then carried out research on the estrous cycle in guinea pigs, rats, and, following Corner, in uteri of sows obtained from a local abattoir. Through Corner and contacts at Johns Hopkins, she was able to get funding from the National Research Council's Committee for Research on Problems of Sex and use of the facilities of the Carnegie Institution's Department of Embryology at Johns Hopkins. At her retirement, Corner hailed her as one of the pioneers in research in problems of human reproduction.[57]

King's research not only gained her some modest recognition but also provided problems for student research. For her best honors students she recruited Hopkins professors as collaborators. One student, featured in an alumnae magazine article on the honors program in 1933, studied the effect of injections of adrenal cortex extract on the growth of normal rats and rats from which the adrenal glands had been removed, a project derivative from King's own research. In addition to the benefits of working with King, the article emphasized the excitement of meeting eminent researchers at Hopkins. One man did the adrenalectomies for the project, and another provided the hormone extract made in his laboratory and "also follows the progress of her experiment with the greatest interest, comparing what she is doing with his results with dogs, giving her helpful criticism and advice." In 1934 King had a colleague at Hopkins, the anatomist Franklin Snyder, introduce another honors student when she presented her research before a meeting of the Society for Experimental Biology and Medicine. The following year the paper, coauthored by King, appeared in the *American Journal of Physiology*.[58]

King's students, a number of whom went on to careers in the medical sciences, recalled her readiness to serve as a mentor. Florence Seibert, a graduate of Goucher and a noted tuberculosis researcher, recalled, "No effort was too great on the behalf of any student, and every one of us felt the untiring sympathy and concern for the future that this teacher possessed." When alumnae visited the department, King "used such occasions to bring them in contact with young students as examples of the opportunities ahead."[59]

Abby Turner's first publications, based on work done with William T. Porter at Harvard Medical School in 1909/1910, dealt with nervous control of respiration in the rabbit, cat, and dog and required complex surgical procedures. Turner published nothing more for ten years. Then, while in graduate school at the Harvard School of Public Health, Turner deliberately selected a research project that could be continued at Mount Holyoke. Her dissertation used twenty-five "healthy young women"

[57] Jessie L. King, "Observations on the Activity and Working Power of the Uterine Muscle of the Non-Pregnant Sow," *Amer. J. Physiol.*, 1927, *81*:725–737; and King, "Oxygen Usage of Uterine Muscle of the Sow," *ibid.*, 1931–1932, *99*:631–637. On Corner's use of sow uteri as research material see Adele E. Clarke, "Research Materials and Reproductive Science in the United States, 1910–1940," in *Physiology in the American Context*, ed. Geison (cit. n. 15), pp. 323–350, esp. pp. 329–331; for Corner's praise see Moses, "Dr. Jessie King" (cit. n. 25).

[58] Elizabeth Nitchie, " 'A Contribution to Knowledge . . . ,' " *Goucher Alum. Quart.*, 1933, *11*(2):23–27, on p. 26; Edith C. Hoopes, "Prolonged Pregnancy in Albino Rat Following Injection of Pregnancy Urine Extract," *Proceedings of the Society for Experimental Biology and Medicine*, 1934, *31*:1115–1117; and Hoopes and Jessie L. King, "Prolongation of Pregnancy in the Rat by the Injection of Human Pregnancy Urine Extract," *Amer. J. Physiol.*, 1935, *111*:507–514.

[59] Seibert, "Jessie Luella King" (cit. n. 25).

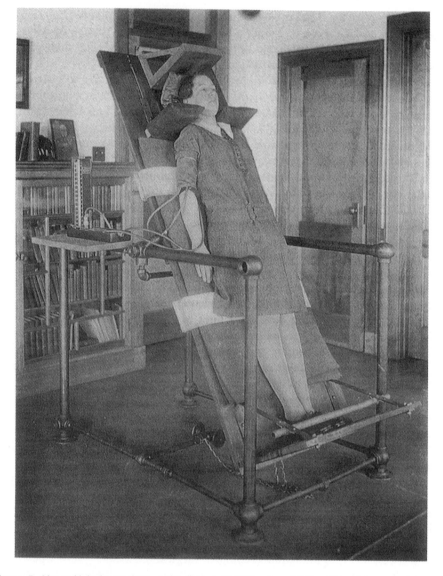

Figure 5. *Mount Holyoke student on the Turner tilt-table, circa 1930s. Abby Turner, professor of physiology, used this apparatus in her teaching and research to measure circulatory efficiency. (Courtesy of Mount Holyoke College Library/Archives.)*

as subjects in an effort to investigate the concept of circulatory efficiency. From various measures of pulse and respiratory gases she calculated the circulatory reaction to postural changes—reclining, sitting, and standing. Later papers investigated the circulatory reaction to prolonged standing or to being tilted at various angles on what became known as the Turner tilt-table (Figure 5). These experiments could even be incorporated into Turner's advanced courses, leading generations of college women to measure the strain on each other's circulatory systems of standing perfectly still against a wall. Yet other Turner papers set standards, adopted by the medical profession, for "vital capacity" (maximum inhalation followed by maximum exhalation) in college women. Turner's research obtained for women physiological data that were previously available only for men and at the same time provided research problems for several master's theses.[60]

A second research strategy for women in women's colleges was to pursue work in the area of general physiology, taking advantage of the facilities available during summers to those who rented tables at the Marine Biological Laboratory (MBL) in Woods Hole. General physiology—nonmedical physiology—investigated physiological processes common to a wide range of animals, using as subjects lower vertebrates and invertebrates that could be maintained at a women's college. The MBL, which Turner and Haywood visited nearly every summer, provided a unique opportunity for women physiologists to interact with the more biologically oriented male physiologists on a relatively equal footing. Master's students at Mount Holyoke were encouraged to take the physiology course at Woods Hole during the summer between their two years of study, not only for the opportunity to work with marine animals, but also for the "stimulus from the association with many investigators unequalled by any university or college."[61] Research facilities at Woods Hole thus supplemented those at Mount Holyoke; experiments begun during the summer could be completed and written up during the academic year.

Turner and Haywood both developed lines of research that they pursued at the MBL. After spending 1932 at the Copenhagen laboratory of Nobel laureate August Krogh, one of Drinker's mentors and another supportive advisor of women, Turner began a series of investigations on blood serum (colloid osmotic pressure, protein content, etc.) in fishes and turtles.[62] Haywood, a graduate of Mount Holyoke, took her doctorate under the general physiologist Merkel Jacobs in the Medical School of the University of Pennsylvania in 1927. After three years of teaching at Vassar she was recalled to her alma mater in 1930, eventually taking over the department

[60] For Turner's early publications see William T. Porter and Abby H. Turner, "On the Crossing of the Respiratory Impulse at the Level of the Phrenic Nuclei," *Amer. J. Physiol.*, 1911–1912, *29*:xxxi; Turner, "Remarks on the Origin of the Phrenic Nerve in the Rabbit, Cat, and Dog," *ibid.*, 1913, *32*:65–69; Porter and Turner, "Direct and Crossed Respiration upon Stimulation of the Phrenic, the Sciatic, and the Brachial Nerves," *ibid.*, pp. 95–106; and Porter and Turner, "Further Evidence of a Vasotonic and a Vasoreflex Mechanism," *ibid.*, 1915–1916, *39*:236–238. For work based on her dissertation see Turner, "The Circulatory Minute Volumes of Healthy Young Women in Reclining, Sitting, and Standing Positions," *ibid.*, 1927, *80*:601–630. For the later work see Turner, "Vital Capacity in College Women, I: Standards for Normal Vital Capacity in College Women; II: A Study of Students with High and Low Vital Capacity," *Archives of Internal Medicine*, 1930, *46*:930–945. On Turner's research see transcript of a lecture on Turner by Curtis Smith, Mount Holyoke College, 30 Sept. 1986, MHC-A. The work of a master's student and of an undergraduate (who went on to become a member of the APS) is incorporated in Turner, M. Isabel Newton, and Florence W. Haynes, "The Circulatory Reaction to Gravity in Healthy Young Women: Evidence Regarding Its Precision and Its Instability," *Amer. J. Physiol.*, 1930, *94*:507–520.

[61] "Report of the Department of Physiology, 1938–1939," MHC-A. The atmosphere of the Marine Biological Laboratory is discussed in Philip J. Pauly, "Summer Resort and Scientific Discipline: Woods Hole and the Structure of American Biology, 1882–1925," in *The American Development of Biology*, ed. Ronald Rainger, Keith R. Benson, and Jane Maienschein (Philadelphia: Univ. Pennsylvania Press, 1988), pp. 121–150.

[62] See, e.g., Mildred L. Campbell and Abby H. Turner, "Serum Protein Measurements in the Lower Vertebrates . . . ," *Biological Bulletin*, 1937, *73*:504–526. Turner is mentioned in the list of foreign students of Krogh in E. Snorrason, "Schack August Steenberg Krogh," *Dictionary of Scientific Biography*, 18 vols. (New York: Charles Scribner's Sons, 1976–1990), Vol. 7, pp. 501–504. In Copenhagen Turner coauthored papers with Krogh and Eugene M. Landis, who later became chairman of the Department of Physiology at Harvard Medical School.

from Turner. She worked with developing sea urchins, frogs, and fishes, measuring physiological effects of increased pressure of gases, especially of carbon dioxide; a later set of papers, inspired by a year's leave spent at her graduate alma mater, dealt with the permeability of the liver in marine vertebrates to various substances. This work gave rise to a steady stream of related projects that could be undertaken by advanced undergraduates and master's students. After Turner retired in 1940, Haywood also took up a line of research in respiratory physiology, using Mount Holyoke students as subjects.[63]

Mount Holyoke's was by far the most successful of the departments in sending its undergraduates on to pursue doctorates in physiology.[64] No fewer than eleven graduates of Mount Holyoke before 1940 received doctorates in physiology; most of these became members of the American Physiological Society. Mount Holyoke's record was in part due to the advantage of having two especially competent teachers and researchers, Turner and Haywood, in the department. By all accounts, Charlotte Haywood was one of the most inspiring teachers of physiology in this century. She brought more women into physiology than any other single mentor.

Ruth Conklin's research at Vassar, while less obviously suited to a women's college, also creatively linked collegial interactions with male physiologists and mentoring relations with female master's students. In graduate school at the Harvard School of Public Health under Cecil Drinker, Conklin had worked on the lymphatic system, Drinker's area of specialty, in the frog. At Drinker's suggestion, Conklin set to work on the lymphatic system of the snake on her return to Vassar. Three master's students, all of whom went on to graduate school, wrote theses on some aspect of the lymphatic system of the snake. Two of these theses were published with Conklin as coauthor. Conklin recalled that the master's students stimulated her research and were actually necessary to its continuation, since it took two people to do the delicate manipulation required. Like Turner, Conklin was able to reinvigorate her research through leaves of absence spent at major laboratories—an additional year at her graduate alma mater and a semester in 1938 with August Krogh in Copenhagen.[65] Though Vassar had as many physiology majors (or more) each year as Mount Holyoke, fewer went on to graduate school. Vassar was far more successful in turning out physiologists through the master's degree (which attracted graduates of other schools) than through the baccalaureate.

[63] For examples of coauthored papers in these areas see Charlotte Haywood, Thelma O. Stevens, Helen M. TeWinkel, and Margaret Schott, "The Relative Effects of Increased Carbon Dioxide Tensions and Decreased Oxygen Tensions upon the Heart Frequency of Young Trout," *Journal of Cellular and Comparative Physiology*, 1934–1935, 5:509–518; and Haywood and Margaret E. Bloete, "Respiratory Responses of Healthy Young Women to Carbon Dioxide Inhalation," *Journal of Applied Physiology*, 1969, 27:32–35. On Haywood see Elizabeth Carlsen Gerst, "Retiring Faculty Members, Charlotte Haywood '19," *Mount Holyoke Alum. Quart.*, 1962, 46(2):92; and Smith, "In Memoriam: Charlotte Haywood" (cit. n. 38).

[64] Pamela Mack and Miriam Levin are preparing a monograph on Mount Holyoke's outstanding role in training women in the sciences.

[65] Sarah W. Chapman and Ruth E. Conklin, "The Lymphatic System of the Snake," *Journal of Morphology*, 1935, 58:385–417; and Lois A. Gillilan and Conklin, "Removal of Foreign Substances by the Lymphatics of the Snake Lung," *Amer. J. Physiol.*, 1938, 123:598–607. Jane McCarrell's 1934 thesis, "Studies on the Lymphatic System of the Snake," appears not to have been published. See also Conklin interview (cit. n. 36). A later line of research, also productive of master's theses and coauthored publications, dealt with regulation of blood pressure and circulatory and respiratory responses to postural changes in rabbits, rats, turtles, and snakes. Conklin described her 1938 leave in Copenhagen in "Glimpses of the Life of a Physiologist," *Radcliffe Quarterly*, 1939, 23(1):19–20.

It is no accident that M. Elizabeth Tidball, who in the 1970s wrote now-classic articles on women faculty as role models for women achievers and on the importance of the women's colleges in educating future scientists, majored in physiology at Mount Holyoke. With Vera Kistiakowsky, Tidball collected data on the undergraduate origins of female holders of the doctorate that documented the major role that women's colleges, both private and public, had played in sending women to graduate school. She acquired the inspiration for these much-cited studies from her personal ideal of a female role model—Charlotte Haywood, her major field advisor, whose introductory course awakened her interest, whose undergraduate independent study gave her a sense of empathy with her professor, and whose advice and encouragement sent her on her way to graduate school.[66]

DECLINE OF THE FEMALE SUBCULTURE

After 1940 women's colleges, especially Mount Holyoke, continued to turn out women who went on to earn doctorates in physiology. Among the 1940s and 1950s graduates and master's degree students were several who, like Elizabeth Tidball, eventually acquired mainstream faculty positions in medical school departments of physiology.[67] But physiology in women's colleges gradually lost some of its distinctiveness as a subculture of the discipline. As Margaret Rossiter has noted, after 1940 administrators in women's colleges began to hire more married women and, in an effort to increase prestige and move away from the image of a faculty of spinsters, more men. Ruth Conklin and Charlotte Haywood were eventually replaced by men whom they had hired. Haywood was said to have been greatly disappointed that she had trained no female disciple to continue the line of succession that could be traced through Turner and Cornelia Clapp back to Mary Lyon.[68]

After a century of success in women's colleges, departments and majors in physiology came to an end in the period between 1957 and 1966. In one college after another, usually at the urging of the administration, departments of botany, zoology, physiology, and bacteriology were combined to form departments of the biological sciences. In these unified departments physiology played at most a minor role. The demise of the physiology major coincided with the rise of molecular and cellular biology, which broke down boundaries of earlier disciplines and made separate departments of botany and zoology seem old-fashioned, and with the rise of federal funding for biology, which encouraged departmental mergers. It coincided also with

[66] Toby A. Appel, interview of M. Elizabeth Tidball, Washington, D.C., 27 Nov. 1985; M. Elizabeth Tidball, "Perspective on Academic Women and Affirmative Action," *Educational Record*, 1973, *54*:130–135; Tidball and Vera Kistiakowsky, "Baccalaureate Origins of American Scientists and Scholars," *Science*, 1976, *193*:646–652; and Rossiter, *Women Scientists in America* (cit. n. 1), pp. 144–145. Tidball served as chair of the APS Task Force on Women in Physiology in the 1970s; see Appel *et al.*, "Women in Physiology" (cit. n. 2), pp. 385–387.

[67] Another example is Jean McE. Marshall (M.A., Mount Holyoke, 1946), professor of physiology at Brown University, who in 1985 became the first woman to be appointed to the powerful APS Publications Committee since its inception in 1898.

[68] Margaret Rossiter, "Majors, Money, and Men at the Women's Colleges, 1945–65," paper presented at the History of Science Society annual meeting, Cincinnati, Ohio, 28 Dec. 1988. Myra Sampson of Smith, who carefully monitored the replacement of women by men at Smith, was an outspoken critic of this trend; see B. Elizabeth Horner, "Myra Melissa Sampson," *Smith Alum. Quart.*, Summer 1984, pp. 67–68. On Haywood's regrets see Smith, "In Memoriam: Charlotte Haywood" (cit. n. 38); see also Tokay, "In Memoriam: Ruth E. Conklin" (cit. n. 34). The first male faculty member in the departments under discussion was hired at Vassar in 1939.

the end of the era of classical or organ-system physiology symbolized by the kymograph. Finally, it coincided with the opening of a wider range of employment opportunities for women physiologists and their students.

CONCLUSION

This essay has examined a women's subculture within a broader discipline. In the Victorian era physiology, understood in its then-dominant meaning as hygiene, entered the women's colleges and was likely also to have been found in modified form in other types of colleges. Toward the turn of the century physiology began to be redefined as a male-oriented experimental biomedical science. Physiology in the five women's colleges under discussion was transformed in light of that new meaning, but in such a manner as to preserve a female network with an identity of its own.

In the twentieth century the dominant culture of physiology was centered in medical school departments. In these departments men taught physiology as a service to (mostly) male medical students, they carried out research programs that often required elaborate equipment and large animals for use as research subjects, and they trained the most promising medical students as well as male Ph.D. candidates for research careers in medical schools. Male physiologists, more often than not, were initially recruited to physiology as medical students.[69]

In the female subculture, in contrast, women physiologists in women's colleges taught a simplified medical physiology to provide knowledge and skills for women's work in the home and for gender-based occupations in the marketplace. Women adapted their research programs to the facilities available in the women's colleges (and sometimes to women's issues) and to the needs of research training of women students. A few exceptional undergraduates and master's students were encouraged to continue on to graduate school in the dominant culture, where opportunities for their employment were scarce and the dropout rate was high. Women physiologists were more likely to encounter physiology as undergraduates and to obtain a Ph.D. alone, rather than an M.D. or both degrees, as was common among men. Indeed, some women saw physiology as a way to pursue a medically related career while bypassing the considerable obstacles to becoming a woman physician.[70] The notion of a female subculture is useful because it sets the focus on the multiple roles of physiology within the women's college setting, not just on research programs and the training of those students who eventually earned doctorates and were elected members of the APS.

To what extent the patterns discussed here apply to physiology outside these five women's colleges and to disciplines other than physiology must be a subject of further research. It is likely that, across the spectrum of women's colleges, physiology provided preparation for women's work in biomedical occupations other than medicine. Possibly the same considerations that led to the expansion of physiology in the women's colleges in the 1920s held true for other biomedical sciences such as bacteriology, public health, nutrition (as experimental science), and biochemistry. One would like to know more about the location and clientele of physiology courses

---

[69] For a sampling of careers of leading male physiologists see the chapters on APS presidents in *History of the American Physiological Society*, ed. Brobeck *et al.* (cit. n. 2).

[70] Conklin interview (cit. n. 36); Chapman Finan interview (cit. n. 44); and Crawford interview (cit. n. 51).

at large state universities and about physiology in its relationship to home economics or physical education for women. These fields, as well as academic programs in nursing, appear to have provided settings outside of women's colleges in which women physiologists could teach other women.[71]

Any understanding of careers of women in physiology in the twentieth century must take into account the significant role of women's colleges in recruiting women into the discipline. Of ninety-four women elected to membership in the American Physiological Society through 1950, more than half spent part of their education or career in a women's liberal arts college.[72] Many more alumnae of women's colleges contributed to the *American Journal of Physiology* or listed themselves as physiologists in *American Men of Science*.[73] That women were attracted to physiology despite the barriers to their participation is in large part due to the supportive network of women physiologists at women's colleges. The demise of this subculture has meant the loss of a special route for women into a biomedical discipline in which, compared to other sciences, there are still very few female role models in leadership positions.[74]

[71] Brooklyn College, e.g., provided employment in the 1940s for at least three female members of the APS in the Department of Health and Physical Education (Women).

[72] If one includes Woman's Medical College of Pennsylvania, the figure increases to more than 60 percent.

[73] Gerald Geison has compiled information on women contributing to the journal; Margaret Rossiter has provided me with entries of women "physiologists" in *American Men of Science*.

[74] As of 1986, there was still only one woman chairing a department of physiology in a medical school. Earlier in the century, Woman's Medical College of Pennsylvania was the sole medical school in which a woman chaired the department of physiology; see Appel *et al.*, "Women in Physiology" (cit. n. 2), p. 390.

Nettie Maria Stevens (1861–1912). Photograph taken in 1904 (courtesy Carnegie Institution of Washington).

Nettie M. Stevens and the Discovery of Sex Determination by Chromosomes

By Stephen G. Brush

I

D URING THE FIRST DECADE of the twentieth century it was established that the sex of almost all many-celled biological organisms is determined at the moment of fertilization by the combination of two kinds of microscopic entities, the X and Y chromosomes.[1] This discovery was the culmination of more than two thousand years of speculation and experiment on how an animal, plant, or human becomes male or female; at the same time it provided an important confirmation for the recently revived Mendelian genetics that was to become a central part of modern biology.

According to most biologists and historians who have written on the subject, the crucial step in the discovery of chromosomal sex determination was taken in 1905 by Nettie M. Stevens (1861–1912) and Edmund B. Wilson (1856–1939). But the scientific and chronological relation between their contributions has rarely been specified, and the role of Stevens, who died in 1912 before she could attain a reputation comparable to that of Wilson, has sometimes been forgotten. In fact neither Stevens nor Wilson is now given adequate recognition by writers of texts and popular works on biology; most of the credit for the establishment of modern genetics usually goes to Thomas Hunt Morgan (1866–1945), who would not accept the chromosome theory until several years after the work of Stevens and Wilson had been published.

This article could not have been written without the generous assistance of the staff of the Carnegie Institution of Washington and Mrs. Gertrude Reed (Bryn Mawr College), who made available the basic documents. For detailed criticisms and suggestions I am indebted to Garland Allen, Lindley Darden, Raymond Doetsch, Scott Gilbert, Donna Haraway, Jane Oppenheimer, and Philip Pauly. I fear they have been only partly successful in overcoming my ignorance of biology and its history, and they are certainly not responsible for any errors that remain. This research has been supported by a grant from the History and Philosophy of Science Program of the National Science Foundation.

[1]This conclusion was not accepted by all biologists in subsequent decades; during the 1920s and 1930s there was strong support for a quantitative "balance" theory of sex determination, with no sharp turning point in development. See C. B. Bridges, "Triploid Intersexes in *Drosophila melanogaster*," *Science*, 1921, *54*:252–254; "The Origin of Variations in Sexual and Sex-Limited Characters," *American Naturalist*, 1922, *56*:51–63; "Sex in Relation to Chromosomes and Genes," *Am. Nat.*, 1925, *59*:127–137; "The Genetics of Sex in Drosophila," in *Sex and Internal Secretions*, ed. E. Allen (Baltimore: Williams and Wilkins, 1932), pp. 55–93. R. Goldschmidt, "The Determination of Sex," *Nature*, 1921, *107*:780–784. W. E. Castle, "The Quantitative Theory of Sex and the Genetic Character of Haploid Males," *Proceedings of the National Academy of Sciences*, 1930, *16*:783–791. G. E. Allen, "Opposition to the Mendelian-Chromosome Theory: The Physiological and Developmental Genetics of Richard Goldschmidt," *Journal of the History of Biology*, 1974, 7:49–92.

In this article I do not attempt a definitive assessment of Stevens' contribution to the chromosome theory of sex determination; that would require much more knowledge of the technical details of her work and that of others than I possess. My purpose is primarily to call attention to some documents (held at the Carnegie Institution of Washington) that illuminate her relations with Wilson and Morgan, and perhaps to persuade historians of biology to investigate her work more thoroughly.[2]

II

Nettie Maria Stevens was born July 7, 1861, in Cavendish, Vermont. Her father was a carpenter; she was the second of three children, the elder of two daughters. She attended Westfield State Normal School in Massachusetts to prepare for a teaching career. Her interest in science may have been aroused during summer courses at Martha's Vineyard in 1890 and 1891. She saved enough money from her teaching jobs to go to Stanford University, where she enrolled in 1896 and received a B.A. degree in 1899.[3] The following year she obtained her M.A. from Stanford and entered Bryn Mawr College as a doctoral student.

One might think that going to a small women's college would be fatal to the career of an aspiring scientist at that time, but in this case the opposite was true. Two of the leading American biologists, E. B. Wilson and T. H. Morgan, taught successively in the biology department at Bryn Mawr. Both were on the editorial board of the *Journal of Experimental Zoology* and otherwise influential in the scientific establishment. Though Wilson had gone to Columbia in 1891, he retained close ties with Morgan, who stayed at Bryn Mawr until 1904. As a student and later a colleague of Morgan's, Stevens was able to keep in touch with current research in the fast-moving fields of genetics, cytology, and embryology. As a promising woman scientist who had already published nine papers by the time she completed her Ph.D. in 1903,[4] she was eligible for special fellowships and prizes. The Bryn Mawr President's European Fellowship enabled her to study at the Naples Zoological Station and at the University of Würzburg, where she worked with the German biologist Theodor Boveri. Then she received an award of $1,000 from the "Association for Maintaining the American Woman's Table at the Zoölogical Station at Naples and for Promoting Scientific Research among Women," offered for the best paper written by a woman on a scientific subject.[5] A postdoctoral research assistantship from the Carnegie

[2]The most comprehensive accounts of the history of this subject are G. E. Allen, "Thomas Hunt Morgan and the Problem of Sex Determination, 1903–1910," *Proceedings of the American Philosophical Society,* 1966, *110*:48–57, and S. Gilbert, "Sex Determination and the Embryological Origins of the Gene Theory," Masters thesis (Johns Hopkins University, 1975). Gilbert says "Wilson was the major spokesman and Nettie Stevens the major source of evidence" for chromosomal determination of sex (p. 42).

[3]I am indebted to Rosamund Bacon of Stanford, California, for detailed information on Nettie Stevens' early life. The article by Hans Ris is the best published biographical source so far but contains some inaccuracies: see *Notable American Women,* ed. E. T. James, *et al.* (Cambridge, Mass.: Harvard University Press, 1971–1972), Vol. III, pp. 372–373. See also *The Bryn Mawr Alumnae Quarterly,* June 1912, pp. 124–126; *Stanford Alumnus,* Feb. 1913; "Life" inserted at the end of N. M. Stevens, "Further Studies on the Ciliate Infusoria, Licnophora and Boveria." Ph.D. dissertation (Bryn Mawr College, 1903). A comprehensive account of Stevens' life and work is being prepared by Marilyn Ogilvie at the University of Oklahoma.

[4]*Bryn Mawr Alum. Q.,* June 1912, p. 125; these papers dealt primarily with regeneration in different organisms. I have not found a complete bibliography of Stevens' publications; a fairly extensive list can be found in the card catalogue of the Bryn Mawr College Library.

[5]N. M. Stevens, "A Study of the Germ Cells of Aphis Rosae and Aphis Oenotherae," *Journal of Experimental Zoology,* 1905, 2:313–333 (dated Dec. 20, 1904; published in the Aug. 1905 issue).

Institution of Washington allowed her to continue research at Bryn Mawr, unburdened by teaching duties during the crucial years 1904–1905, when her research on the sex chromosomes was being done.

III

During the previous decade Boveri, W. S. Sutton, Wilson, and others had found strong evidence for the individuality of chromosomes as identifiable parts of cells (re-identifiable after division) and suggested that each chromosome may be responsible for a definite part of the hereditary endowment of an organism. Chromosomes could be duplicated during the process of reproduction but otherwise remained unaffected by their environment; thus they seemed to behave like August Weismann's "germ plasm," which was supposed to be isolated from the rest of the organism and excluded the possibility that acquired characteristics could be inherited. In the first edition (1896) of his influential book *The Cell in Development and Inheritance,* Wilson collected evidence that the seat of heredity is in the chromosomes in the cell nucleus. This treatise helped prepare American biologists to receive Mendel's theory when it was independently rediscovered and advocated four years later by Hugo de Vries, Carl Correns, and Erich Tschermak von Seysenegg[6] and organized support for the Sutton-Boveri chromosome hypothesis.

But Wilson, Morgan, and most other biologists were not yet ready to accept the idea that sex is completely determined by chromosomes at the moment of fertilization. Instead, many believed that environmental factors such as nutrition play a major role in determining the sex of the developing embryo.[7] Wilson wrote, in the 1900 edition of *The Cell,* that "sex as such is not inherited. What is inherited is the capacity to develop into either male or female, the actual result being determined by the combined effect of conditions external to the primordial germ-cell."[8] In 1903 Morgan concluded an extensive review of the subject with the paragraph:

> Our general conclusion is that while recent theories have done good service in directing attention to the early determination of sex in the egg, those of them which have attempted to connect this conclusion with the assumption of the separation of male from female primordia in the germ-cells have failed to establish their point of view. The egg, as far as sex is concerned, appears to be in a sort of balanced state, and the conditions to which it is exposed, even when it is not fully formed, may determine which sex it will produce. It may be a futile attempt to try to discover any one influence that has a deciding influence for all kinds of eggs. Here, as elsewhere in organic nature, different stimuli may determine in different species which of the possibilities that exist shall become realized.[9]

In 1902 C. E. McClung suggested that the "accessory chromosome," discovered earlier by Hermann Henking, plays an important part in sex determination. His

[6] H. J. Muller, "Edmund B. Wilson—An Appreciation," *Am. Nat.,* 1943, *77*:5–37, 142–172; see pp. 29 ff.

[7] Allen, "Thomas Hunt Morgan," p. 49; Gilbert, "Sex Determination," pp. 5–8. According to W. E. Castle, "The last forty years have seen the rise, culmination, and at least incipient decline of a plausible but fundamentally erroneous idea about sex,—the idea that it is subject to control through the environment of the developing organism. The latest manifestation of this idea is found in Schenk's theory of sex-control in man through regulation of the mother." See Castle, "The Heredity of Sex," *Bulletin of the Museum of Comparative Zoology, Harvard College,* 1903, *40*:189–218, p. 190. He cites L. Schenk's theory presented at the International Zoological Congress in Berlin, Aug. 1901.

[8] E. B. Wilson, *The Cell in Development and Inheritance* (2nd ed., New York: Macmillan, 1900), p. 145. Wilson allowed this statement to remain in reprints of the book as late as 1919 (copy in University of Maryland Library).

[9] T. H. Morgan, "Recent Theories in Regard to the Determination of Sex," *Popular Science Monthly,* 1903, *64*:97–116.

hypothesis was that this chromosome is "the bearer of those qualities which pertain to the male organism."[10] This hypothesis turned out to be wrong, but it attracted the attention of other workers such as William Bateson and Miss E. R. Saunders at Cambridge University and W. E. Castle at Harvard.[11] There is some disagreement among later writers as to how much credit McClung should receive in the discovery of chromosomal sex determination, but there is no doubt that his hypothesis was extremely important in stimulating work by others.

Stevens was quick to see the possibilities opened up by McClung's hypothesis in conjunction with the Mendelian theory of heredity. Morgan's letter to the Carnegie Institution, November 19, 1903, in support of her application for a grant, makes it clear that she was the one who wanted to investigate sex determination by chromosomes, whereas he, the senior (though younger) partner of the team, was still under the influence of environmental theories (see Appendix). Yet this difference of opinion did not prevent them from collaborating. In fact, given a situation of mutual respect, the disagreement was fortunate for Stevens in the long run, since it meant that she published her results under her own name alone. If Morgan's name had been on her 1905 publication, the scientific world would undoubtedly have given most of the credit to him. (See Sec. V, below.)

Both Morgan and Wilson were enthusiastic in recommending Stevens to the Carnegie Institution. Morgan wrote that "of the graduate students that I have had during the last twelve years I have had no one that was as capable and independent in research work as Miss Stevens . . ." (see Appendix). Wilson stated: "I know Miss Stevens' work well, and it is of a very independent and admirable character from every point of view. I consider her not only the best of the women investigators, but one whose work will hold its own with that of any of the men of the same degree of advancement."[12]

Stevens' initial letter of application to the Carnegie Institution, dated July 19, 1903, indicated her situation quite clearly: she needed money to live on, and "College positions for women in Biology this year seem, however, to be very few." She wanted to pursue the "histological side of the problems in heredity connected with Mendel's Law" (the complete letter is given in the Appendix).

The grant was awarded, and Stevens proceeded with her research, which involved detailed examination of the chromosomes of several insects and comparison with the sex of the progeny. In a paper on the germ cells of aphids completed at the end of 1904, she failed to find McClung's extra chromosome, but the direction of her research was clear: a review of the current state of research had convinced her that "the evidence is overwhelmingly on the side of the view that sex is determined in the egg; but to the question how sex is determined in the egg, no thoroughly convincing answer has yet been given."[13]

[10] C. E. McClung, "The Accessory Chromosome—Sex Determinant?" *Biological Bulletin,* 1902, *3*:43–84, p. 72. McClung's subsequent work was not very fruitful; see letter from E. B. Wilson to C. D. Walcott, Oct. 24, 1903, in Wilson file at Carnegie Institution of Washington.

[11] W. Bateson and Miss E. R. Saunders, "Experimental Studies in the Physiology of Heredity," *Reports to the Evolution Committee of the Royal Society,* Report I (London: Harrison & Sons, 1902), pp. 138–139. Castle, "The Heredity of Sex."

[12] E. B. Wilson to C. D. Walcott, Nov. 27, 1903; letter in Stevens file at Carnegie Institution of Washington, quoted by permission.

[13] Stevens, "A Study of the Germ Cells," p. 328. Morgan remarked that "Miss Stevens denied at first the presence of an unpaired sex chromosome in the spermatogenesis, but later corrected this error. She failed to note, at first, that the male had fewer chromosomes than the female, but later recognized this difference." T. H. Morgan, "The Scientific Work of Miss N. M. Stevens," *Science,* 1912, *36*:468–470, p. 469.

Stevens was more fortunate with *Tenebrio molitor,* the common mealworm; males are produced by spermatozoa containing one chromosome that is clearly much smaller than the corresponding chromosome in the spermatozoa that produce females. On May 23, 1905, she submitted to the Carnegie Institution a manuscript on "Studies in Spermatogenesis" for publication in the Carnegie monograph series. It was sent on May 29 to Wilson, as a member of the institution's advisory committee, for his opinion. He returned it on June 13 with the brief statement: "It is in every way a most admirable piece of work which is worthy of publication by any learned society, and I do not hesitate to recommend it to you for publication by the Institution."[14]

Stevens' monograph on spermatogenesis was published in September 1905. After describing her experiments with *Tenebrio,* she concluded:

> Since the somatic cells of the female contain 20 large chromosomes, while those of the male contain 19 large ones and 1 small one, this seems to be a clear case of sex-determination, not by an accessory chromosome, but by a definite difference in the character of the elements of one pair of chromosomes of the spermatocytes of the first order, the spermatozoa which contain the small chromosome determining the male sex, while those that contain 10 chromosomes of equal size determine the female sex. This result suggests that there may be in many cases some intrinsic difference affecting sex, in the character of the chromatin of one-half of the spermatozoa, though it may not usually be indicated by such an external difference in form or size of the chromosomes as in Tenebrio.[15]

IV

It is generally stated that E. B. Wilson obtained the same results as Stevens, at the same time. Roughly speaking, this is true, but the statement must be qualified in three ways. First, Wilson happened to choose a species in which the male has one less chromosome than the female, whereas Stevens investigated the much more common case in which the male has a small chromosome (Y) corresponding to the large chromosome (X) in the female. It could later be argued that the cases are "the same in principle,"[16] but in the context of early-twentieth-century biology the Wilson case (X,O) looks like a simple reversal of McClung's hypothesis and does not bring out clearly the dominant-recessive feature that distinguishes the modern (X,Y) theory.

Second, Wilson probably did not arrive at his conclusion on sex determination until after he had seen Stevens' results. This is perhaps the most important point established by the documents at the Carnegie Institution. Morgan, in his obituary of Wilson, implied that Wilson discovered the crucial difference between the chromosome numbers in *Anasa tristis* (22 in the female, 21 in the male) before Stevens submitted her paper:

> The question is sometimes asked as to the priority of Stevens' and Wilson's papers. Stevens' paper was handed in on May 15, 1905[17] and printed in September of that year. In

[14]E. B. Wilson to R. S. Woodward, June 13, 1905; letter in Stevens file at the Carnegie Institution, quoted by permission.

[15]N. M. Stevens, *Studies in Spermatogenesis with Especial Reference to the "Accessory Chromosome"* (Washington, D.C.: Carnegie Institution of Washington, Publication No. 36, September 1905), p. 13. The paper is dated May 15, 1905, but Stevens waited a week before sending it to the Carnegie Institution in the hope of getting the official name of one of the species she had studied; see the letter of transmittal from Stevens to R. S. Woodward, May 23, 1905, in the Stevens file at the Carnegie Institution.

[16]T. H. Morgan, "Biographical Memoir of Edmund Beecher Wilson 1856–1939," *Biographical Memoirs of the National Academy of Sciences,* 1940, *21*:315–342, p. 333.

[17]It was actually sent on May 23 (see above, n. 15), but this does not affect the priority question.

Wilson's paper "Studies on Chromosomes" I (dated May 5, 1905; published August 1905) he says in a footnote: "The discovery, referred to in a preceding footnote, that the spermatogonial number of Anasa is 21 instead of 22, again goes far to set aside the difficulties [of McClung's hypothesis] here urged. Since this paper was sent to press I have also learned that Dr. N. M. Stevens (by whose kind permission I am able to refer to her results) has independently discovered in a beetle, Tenebrio, a pair of unequal chromosomes that are somewhat similar to the idiochromosomes in Hemiptera and undergo a corresponding distribution to the spermatozoa. She was able to determine, further, the significant fact that the small chromosome is present in the somatic cells of the male only, while in those of the female it is represented by a larger chromosome. These very interesting discoveries, now in course of publication, afford, I think, a strong support to the suggestion made above; and when considered in connection with the comparison I have drawn between the idiochromosomes and the accessory show that McClung's hypothesis may, in the end, prove to be well founded."[18]

But Morgan overlooked the fact that the "preceding footnote" in Wilson's paper, mentioned in the above quotation, was not in the original paper as submitted on May 5; it begins "Since this paper was sent to press I have determined beyond the possibility of doubt, I think, that the number of spermatogonial chromosomes in Anasa tristis is 21, not 22. . . ."[19] Wilson emphasizes in this footnote the special care he took to verify this result, which disagreed with the results of other biologists; it is hardly likely (and he does not claim) that he established the result in the period of less than four weeks between May 5 and the time he received Stevens' paper from the Carnegie Institution for review.

It is true that Wilson's paper "Studies on Chromosomes" (part I) was published before Stevens'—in fact it appeared in the same issue of the *Journal of Experimental Zoology* (August 1905) as Stevens' paper on aphids that had been submitted to this quarterly journal in December 1904.[20] But in view of Wilson's position on the editorial board, the fact that his paper was published with a relatively short time lag (three months, compared to eight months for Stevens') perhaps should not be used to establish his priority.

Nevertheless, one may still reject the above two arguments—one may claim that Wilson did arrive at essentially the same conclusion as Stevens, before he received her paper at the end of May 1905. Then look at his short article, dated October 3, 1905, published in *Science* on October 20 of that year. This is the article usually cited as the first report of Wilson's discovery of sex determination.[21] Although he begins by stating that there is "no doubt that a definite connection of some kind between the chromosomes and the determination of sex exists in these animals," he concludes by reverting to a semi-environmental theory:

> . . . great, if not insuperable, difficulties are encountered by any form of the assumption that these chromosomes are specifically male or female sex determinants. It is more

[18]T. H. Morgan, "Biographical Memoir," p. 333, quoting from E. B. Wilson, "Studies on Chromosomes. I. The Behavior of the Idiochromosomes in Hemiptera," *J. Exp. Zool.*, 1905, 2:371–405, p. 403.

[19]*Ibid.*, p. 399.

[20]The fact that Wilson's article was published in the August 1905 issue of the *Journal of Experimental Zoology* does not necessarily mean that it actually appeared before Stevens' monograph, which was dated September 1905. According to Alice Baxter, "In a letter from T. H. Morgan to Ross G. Harrison (September 11, 1939, Ross G. Harrison collection at Yale University, New Haven, Conn.), the former inquired as to who deserved priority. Harrison answered in a letter (September 14, 1939, Ross G. Harrison collection) saying that he had received Stevens' paper a day before Wilson's." Alice Levine Baxter, "Edmund Beecher Wilson and the Problem of Development: From the Germ Layer Theory to the Chromosome Theory of Inheritance," Ph.D. dissertation (Yale University, 1974), p. 308.

[21]E. B. Wilson, "The Chromosomes in Relation to the Determination of Sex in Insects," *Science*, 1905, 22:500–502.

probable . . . that the difference between eggs and spermatozoa is primarily due to differences of degree or intensity, rather than of kind, in the activity of the chromosome groups in the two sexes; and we may here find a clue to a general theory of sex determination that will accord with the fact observed in hemiptera . . . during the synaptic and growth periods . . . these chromosomes play a more active part in the metabolism of the cell in the female than in the male. The primary factor in the differentiation of the germ cells may, therefore, be a matter of metabolism, perhaps one of growth.[22]

Wilson in 1905 was reluctant to come down as firmly as Stevens on the side of instantaneous sex determination by the mere presence or absence of a particular type of chromosome. Perhaps his greater knowledge of the complexities of the data available at that time made him properly cautious about jumping to such a simplistic conclusion; perhaps Stevens was only making a rash generalization unjustified by the evidence. Her understanding of the dominant-recessive properties of X and Y chromosomes, based on Castle's modification of Mendelian inheritance,[23] was not the same as the modern view (in which Y is always dominant and X always recessive), so one cannot accuse Wilson of rejecting a completely correct theory.

Nevertheless, there is no doubt that as late as 1906 Stevens was still ahead of Wilson in realizing the significance of their discovery. She wrote:

Wilson[24] suggests as alternatives to the chromosome sex determinant theory according to Mendel's Law, (1) that the heterochromosomes may merely transmit sex characters, sex being determined by protoplasmic conditions external to the chromosomes; (2) That the heterochromosomes may be sex-determining factors only by virtue of difference in activity or amount of chromatin, the female sex chromosome in the male being less active.[25]

She could cite evidence against both of these alternatives, and concluded: "On the whole, the first theory, which brings the sex determination question under Mendel's Law in a modified form, seems most in accordance with the facts, and makes one hopeful that in the near future it may be possible to formulate a general theory of sex determination."[26]

V

Nettie Stevens benefited from the encouragement and support of the scientific establishment during the time she was doing her most important work, but it does not appear that she later gained a reputation or material rewards commensurate with her accomplishments. Failure to win a Nobel Prize (if one thinks the discovery deserved it) can be explained by the fact that the significance of sex chromosomes in genetics was not generally appreciated before her death; Morgan had to wait until 1933 for his

[22] *Ibid.*, pp. 501–502.

[23] Castle, "The Heredity of Sex"; Allen, "Thomas Hunt Morgan," p. 49. Castle's theory assumed both male- and female-producing eggs as well as male- and female-producing sperm, with selective fertilization.

[24] E. B. Wilson, "Studies on Chromosomes. III. The Sexual Differences of the Chromosome-Groups in Hemiptera, with some Considerations on the Determination and Inheritance of Sex," *J. Exp. Zool.*, 1906, *3*:1–40, submitted Dec. 8, 1905.

[25] N. M. Stevens, *Studies in Spermatogenesis. Part II. A Comparative Study of the Heterochromosomes in certain Species of Coleoptera, Hemiptera and Lepidoptera, with Especial Reference to Sex Determination* (Washington, D.C.: Carnegie Institution of Washington, Publication No. 36, Part II, October 1906), p. 55. To balance the impression of Wilson's position given by this quotation, I quote his own statement ("Studies on Chromosomes," p. 26): "The observations here brought forward, together with those of Stevens on Tenebrio, establish the predestination (in a descriptive sense) of two classes of spermatozoa, equal in number, as male-producing and female-producing forms . . . it is evident that a substantial basis now exists for . . . the Mendelian interpretation of sex-production worked out by Castle."

[26] Stevens, *Studies in Spermatogenesis, Part II*, p. 56.

trip to Stockholm, and Wilson never did make it.[27] More serious is the fact that most modern textbooks, if they mention Stevens at all, give the impression that she worked with or following Wilson.[28] Because of Wilson's more substantial contributions in other areas, he tends to be given most of the credit for this discovery, as a result of the operation of the "Matthew effect" noted by sociologist Robert Merton.[29] ("Unto every one that hath shall be given, and he shall have abundance; but from him that hath not shall be taken away even that which he hath"—Matthew XXV: 29). The most extreme example of this effect is the occasional ascription of the discovery of chromosomal sex determination to T. H. Morgan, simply because he is considered the most important American geneticist in the first half of the twentieth century and hence the only one mentioned in superficial accounts.[30]

Those who seek outstanding female scientists to inspire the next generation of talented women to follow scientific careers seem to have overlooked Nettie Stevens. This may be partly because most books that do mention her work identify her only as "N. M. Stevens," giving no hint of gender.[31]

Even Bryn Mawr College was somewhat tardy in recognizing the achievements of its most eminent woman biologist, as the *Stanford Alumnus* pointed out in its obituary (February 1913). Her highest position there was Associate in Experimental Morphology. The Trustees of Bryn Mawr finally created for her a research professorship, but she was never able to occupy it; shortly afterwards she died of carcinoma of the breast, on May 4, 1912, at Johns Hopkins Hospital in Baltimore.

APPENDIX. STEVENS' APPLICATION FOR A CARNEGIE RESEARCH GRANT AND MORGAN'S LETTER OF RECOMMENDATION[32]

Mt. View, California
July 19, 1903

Sec'y of the
   Carnegie Inst.,
     Washington D. C.

Dear Sir—President Thomas of Bryn Mawr College advised me some time ago to apply for one of the Carnegie fellowships and continue research work instead of teaching next year, but considering the condition of my finances, I thought it better to

---

[27] One indicator of current reputations is the length of articles in the *Dictionary of Scientific Biography:* Wilson's is 14 pages long, Morgan's is 12 pages (both by G. E. Allen), but Stevens has no article at all.

[28] Of 11 authors who mention the names of both discoverers, 7 say "Stevens and Wilson" and 4 say "Wilson and Stevens." This is perhaps about what one might expect from random choice with a preference for alphabetical order. But 6 authors give credit *only* to Wilson: Carl Correns (1913), L. Doncaster (1914), Conway Zirkle (1959), Isaac Asimov (1960), M. J. Sirks and C. Zirkle (1964), E. J. Gardner (1972). One author, A. M. Winchester, gives all the credit to Stevens in one book (1966) and all the credit to Wilson in another (1972). (Complete references will be supplied on request.)

[29] R. K. Merton, "The Matthew Effect in Science," *Science,* 1968, *159*:56–63.

[30] Ruth Moore, *The Coil of Life: The Story of the Great Discoveries in the Life Sciences* (New York: Knopf, 1961), p. 217. G. R. Taylor, *The Science of Life: A Picture History of Biology* (New York: McGraw-Hill, 1963), p. 322. J. J. Fried, *The Mystery of Heredity* (New York: Day, 1971), p. 41. J. H. Otto and A. Towle, *Modern Biology* (New York: Holt, Rinehart & Winston, 1973), pp. 152–153.

[31] T. H. Morgan *et al.* (1915), C. B. Bridges (1939), F. A. E. Crew (1946), B. Dawes (1952), A. H. Sturtevant (1951), E. A. Carlson (1966), L. Levine (1969), U. Mittwoch (1973). (Complete references will be supplied on request.)

[32] Both letters are in the Nettie Stevens file at the Carnegie Institution of Washington; published by permission of the Carnegie Institution.

try for a position to teach. College positions for women in Biology this year seem, however, to be very few; and I should like to know whether there are still any fellowships to be awarded this year. I have been doing research work for four years and should prefer to continue to do that instead of teaching if there were no money question involved, but I am dependent on my own exertions for a living, and have used nearly all that I had saved while teaching before I began my college-work seven years ago.

I am especially interested in the histological side of the problems in heredity connected with Mendel's Law, and know that there is need of a great deal of painstaking work along that line.

So far my research work has been on two new species of Protozoa, problems in regeneration, ovogenesis and spermatogenesis of Sagitta and Planaria lugubris.

I heard rumors before I left Mass. of a school for research work, in which the members were to receive salaries and give their time to investigation. That is exactly what I should like, an opportunity to devote my time to research work, and freedom from anxiety over the money question.

Yours truly

N. M. Stevens

References:—
    Prof. T. H. Morgan, Woods Hole
    Prof. C. O. Whitman,   ″   ″

• • •

Bryn Mawr Penna Nov 19 1903

To the Carnegie Institution of Washington:

Dear Sirs:

At my suggestion Miss N M Stevens has made an application to the Carnegie Institution for a research Assistantship and I beg to urge as strongly as possible her appointment. Of the graduate students that I have had during the last twelve years I have had no one that was as capable and independent in research work as Miss Stevens and now that she has her degree she is devoting all of her time to research. <work> [deleted] Miss Stevens has not only the training but she has also the natural talent that is I believe much harder to find. She has an independent and original mind and does thoroughly whatever she undertakes. I fear to say more lest it may appear that I am overstating her case.

I have begun a piece of research work in collaboration with Miss Stevens that will take a year or more to bring to completion provided we can work together which will be impossible after January unless Miss Stevens obtains the appointment she seeks. The question of the factors that determine the sex of the egg is one that is now coming rapidly to the front and bids fair to give results of far reaching importance not only of theoretical but of practical interest as well. Our first problem wil[l] be to examine the conditions in the aphids where it appears possible to cause the appearance of the males and females (as contrasted with parthenogenetic forms) by changing the food. I am carrying out the experimental side of the work and Miss Stevens is examining at the same time the internal changes in the egg (the origin of the eggs, the number of polar bodies formed, the number of chromosomes present in the males, females, and parthenogenetic forms, together with the question of the reduction divisions).

At the same time Miss Stevens proposes to examine in another more favorable form the so-called accessory chromosome in eggs and sperm-cells and its possible relation to sex.

I trust that the importance of the questions involved, as well as Miss Stevens special aptitude for the work, will prevail upon the Carnegie Institution to consider her application favorably. It is also of the greatest importance to me to have some one working with me on this problem and I know of no one who is so well suited to carry out work of this sort as Miss Stevens.

> Respectfully yours
> T. H. Morgan
> Professor of Biology
> Bryn Mawr

Marcella O'Grady Boveri (1863–1950)

Her Three Careers in Biology

By Margaret R. Wright

ABSTRACT

The career of Marcella O'Grady Boveri (1863–1950), a nineteenth-century Catholic woman educated in biology at MIT and Bryn Mawr, is discussed both in the biological context of the times and with regard to the position of women in science. The thesis is that her life pattern differed strikingly from that of other woman biologists of her generation and that the character of her contributions to biology varied with that pattern. Perhaps it is in consequence that the significance of her considerable achievement has been hidden. Boveri's circumstances led her to collaboration rather than independence in research: she worked with skill and interest, but without formal recognition, on her husband's theoretically important and already established research program in Germany (1900–1915). She thought it a privilege to do so. Earlier (1889–1896), at Vassar College, and later (1927–1943), at the newly established Albertus Magnus College, she was an innovator who introduced and developed new curricula in biology, a stimulating and influential teacher, a mentor, and a role model. In addition, she did much to promote international communication, as exemplified both by her English translation of Theodor Boveri's prescient theory of cancer and by her influence in bringing important scientists to the United States.

I regret that I can no longer give my thanks directly to those friends who helped me so greatly in gathering information—the late Dorothea Rudnick and the late Sister Thomas Aquin Kelly, past archivist at Albertus Magnus College. It is my pleasure to thank Evalyn A. Clark, professor emerita of history at Vassar, for her interest and confidence in encouraging me in my new enterprise. I also appreciate permission to quote from the collections at Albertus Magnus, at Bryn Mawr, and at Vassar and the generous help of Nancy MacKechnie, curator of rare books and manuscripts at Vassar.

The reader may wonder whether I knew Marcella O'Grady Boveri personally; the answer is a definite "Yes." I was a graduate student in zoology at Yale and needed an undergraduate physics course to fill a hole in my undergraduate record. At the time Yale's physics department would not let a woman take an undergraduate course, so in 1942 I took it at Albertus Magnus College. The course was given by Boveri's friend and former student Edna Carter, who had just retired as professor of physics at Vassar. They invited me to dinner; I invited them to dinner; I walked Boveri home from seminar; she asked me to lecture to her class; friendships developed; reminiscences occurred. I remember those times with pleasure and consider myself lucky to have known my primary source as well as I did.

D ESPITE THE WEALTH OF INFORMATION uncovered during the past two decades
by research on the experience of American women in science, the accomplishments
of Marcella O'Grady Boveri (1863–1950) remain, with one exception, unrecognized in
the literature.[1] I hope to provide a biographical sketch that shows some of the unusual
circumstances and opportunities that arose in her life and her accommodation to them.

Analytical accounts of nineteenth-century women in biology suggest that Boveri's position was unique among women of her generation.[2] Even her education was unusual for
a Catholic woman: she went to public school in Boston and then to the Massachusetts
Institute of Technology, where she concentrated in biology. Like many other single
women, Boveri began her professional life in science as a teacher in a women's college,
Vassar. Unlike many others, however, after seven highly successful years and a sabbatical
year of research at the University of Würzburg, she resigned to marry. Her husband,
Theodor Boveri, was the director of a research laboratory, and she soon became his collaborator.

Boveri's career as a collaborator differed in a variety of ways from those of many women
who worked jointly with husbands. For some, collaboration was an entry to science, an
initial step that might lead to independent work in the same or a related field. Some women
with little or no scientific background contributed other skills—illustration, for example—
that complemented the partner's expertise and enhanced the quality of the research. Some
partners were equally well established in the same field or closely related fields of interest
before their marriage. Finally, in any of these categories the partners may or may not have
shared equally in the theoretical as well as the practical aspects of the science. Some
spouses were simply helpmates.[3]

Boveri did not fit in any of these categories: her collaboration was not her entry into
science; she had no special skills to complement those of her husband; she was not as well
established as he. Furthermore, in contrast to the descriptive biology that had prevailed,
especially in the nineteenth century, their work was experimental. Their purpose was to
test, by experiment, Theodor Boveri's hypothesis about the role of the chromosomes in
heredity. From the beginning of their collaboration Marcella Boveri was not merely a
helpmate, but a scientific partner interested in the analysis and the theoretical significance
of their experiments. Like many other women, however, she received no formal credit for
the collaborative work. Even the report on their last experiments, published after Theodor
Boveri's death by Marcella Boveri herself, carried his name only.

---

[1] The exception is Victor McKusick, "Marcella O'Grady Boveri (1865–1950) and the Chromosome Theory
of Cancer," *Journal of Medical Genetics,* 1985, 22:431–440. McKusick praises Marcella Boveri for recognizing
the importance of acquainting the English-speaking world with her husband's hypothesis on the origin of cancer
and making it available in translation: Theodor Boveri, *The Origin of Malignant Tumors,* trans. Marcella Boveri
(Baltimore: Williams & Wilkins, 1929).

[2] Margaret W. Rossiter, *Women Scientists in America: Struggles and Strategies to 1940* (Baltimore: Johns
Hopkins Univ. Press, 1982) (hereafter cited as **Rossiter, *Women Scientists in America: Struggles and Strategies
to 1940;*** pagination for citations is from the 1984 paperback). See also its successor: Rossiter, *Women Scientists
in America: Before Affirmative Action, 1940–1972* (Baltimore: Johns Hopkins Univ. Press, 1995).

[3] Important resources include Pnina G. Abir-Am and Dorinda Outram, eds., *Uneasy Careers and Intimate
Lives: Women in Science, 1789–1979* (New Brunswick, N.J.: Rutgers Univ. Press, 1987); and Helena M. Pycior,
Nancy G. Slack, and Abir-Am, eds., *Creative Couples in the Sciences* (New Brunswick, N.J.: Rutgers Univ.
Press, 1996). Specific examples of collaborative relationships are investigated in Marilyn B. Ogilvie, "Marital
Collaboration: An Approach to Science," in *Uneasy Careers and Intimate Lives,* ed. Abir-Am and Outram, pp.
106–109; Slack, "Nineteenth-Century American Women Botanists: Wives, Widows, and Work," *ibid.,* pp. 92–
100; Slack, "Botanical and Ecological Couples: A Continuum of Relationships," in *Creative Couples in the
Sciences,* ed. Pycior *et al.,* pp. 237–244; Janet Bell Garber, "John and Elizabeth Gould: Ornithologists and
Scientific Illustrators, 1829–1841," *ibid.,* pp. 86–97; and Pamela M. Henson, "The Comstocks of Cornell: A
Marriage of Interests," *ibid.,* pp. 112–125.

In the last chapter of her life, Boveri returned to the United States to found a science department at the newly established Albertus Magnus College in New Haven, Connecticut, where she would teach biology until her retirement at the age of eighty. Her dedication to serving science never weakened, and her own wide-ranging experience and vast acquaintance with scientists on both sides of the Atlantic made her an influential force in promoting international communication.

GROWING UP IN BOSTON

Marcella Imelda was born to Thomas and Anne O'Grady on 7 October 1863, in Boston; it was her good fortune to join an affluent and professionally oriented family that believed in education for women as well as men. Her father and brother were architects. Unlike her older sister, who at eighteen chose to become a nun, Marcella attended Boston Girls' High School and then entered MIT. She was granted the Bachelor of Science degree in General Studies in 1885, the first MIT woman to graduate with a concentration in biology.[4]

Surely this educational path was unusual for a young Catholic woman of the nineteenth century. Few went to college; indeed, there were no Catholic women's colleges before 1900. Why MIT? Perhaps it was her brother's influence: Thomas O'Grady, Jr., completed his degree in architecture at MIT in 1881. And why science? The O'Grady family apparently had a bent for the arts—and at the end of the nineteenth century a Catholic woman scientist was "an extreme rarity."[5]

Perhaps the intellectual climate of Boston stimulated her interest. Boston had a long history of popularizing science, thanks to organizations like Daniel Webster's Society for Diffusion of Useful Knowledge, the Lowell Institute, and the Boston Society of Natural History.[6] Lectures, a library, and extensive botanical, zoological, and geological museum collections were open to the public. In the 1870s, when Marcella O'Grady would have been a student, the public schools were being pressed to increase science instruction, and this led to the first summer schools of natural history intended especially for teachers. The Women's Educational Association of Boston gave enthusiastic support to these developments. Was O'Grady importantly influenced by Boston's progressive attitude toward science? There seems no way to be sure.

O'Grady's biological studies and thesis at MIT were directed by William Thompson Sedgwick. In her last year there she met a visiting investigator, Edmund Beecher Wilson. Sedgwick and Wilson were friends, having been students together at the Sheffield Scientific

---

[4] There is confusion in the literature about Marcella O'Grady's birthdate. The evidence for 1863 is from her curriculum vitae, in the Archives of the Bryn Mawr College Library, Bryn Mawr, Pennsylvania. The claim that she was the first MIT woman to graduate with a concentration in biology was verified by Elisabeth Kaplan, assistant archivist at the Institute Archives and Special Collections, The Libraries, MIT, Cambridge, Massachusetts. Ellen Swallow Richards, a Vassar graduate of 1870, was the first woman to obtain a degree from MIT— a second Bachelor of Science, in chemistry—in 1873.

[5] This comment, made by an anonymous referee, is based on his or her study of the careers of 105 women. The date of Thomas O'Grady's degree was confirmed in Elisabeth Kaplan, assistant archivist, MIT, to Margaret R. Wright, 1 Nov. 1996. Information about Catholic women's colleges comes from Edward J. Power, "Catholic Higher Education for Women in the United States" (1958), in *Higher Education for Catholic Women: An Historical Anthology,* ed. Mary J. Oates (New York: Garland, 1987), pp. 114–135, on p. 121.

[6] See Margaret W. Rossiter, "Benjamin Silliman and the Lowell Institute: The Popularization of Science in Nineteenth-Century America," *New England Quarterly,* 1971, *44*(4):602–626; Sally Gregory Kohlstedt, "The Nineteenth-Century Amateur Tradition: The Case of the Boston Society of Natural History," in *Science and Its Public: The Changing Relationship,* ed. Gerald Holton and William A. Blanpied (Boston Studies in the Philosophy of Science, 33) (Dordrecht/Boston: Reidel, 1976), pp. 173–190; and Kohlstedt, "Parlors, Primers, and Public Schooling: Education for Science in Nineteenth-Century America," *Isis,* 1990, *81:*425–445.

School at Yale and while earning their Ph.D.'s (1881) at the Johns Hopkins University. Though young, they were already recognized as research biologists of stature.[7] Sedgwick and Wilson were to become O'Grady's mentors; as Margaret Rossiter has shown, mentors played an important role in the emergence of women scientists.[8]

A PROMISING CAREER

A Taste of Teaching and Graduate Study

In the summer of 1885, thanks to (or in spite of) Wilson's qualified recommendation, O'Grady was appointed to organize and teach science in a girl's preparatory school, the Bryn Mawr School in Baltimore, which was to open in the fall. Wilson wrote: "She is faithful, industrious, earnest, and has had a good training. So far as *knowledge* goes I should not hesitate to recommend her, and she is entitled to great respect as a woman. The only trouble is her lack of experience and her mild almost shrinking modesty and lack of self-assertion." No one who knew Marcella O'Grady Boveri in later life could fail to be amused by his caution: a gentlewoman she was; a shrinking violet she certainly was not. After two successful years of teaching science at the Bryn Mawr School, she won the Fellowship in Biology for 1887–1889 for advanced study at Bryn Mawr College, awarded as an "endorsement of previous attainments." In the 1880s it was unusual for a woman to be accepted by a college or university for graduate study, even more unusual for her to be considered a candidate for the Ph.D. At Bryn Mawr College, however, Dean M. Carey Thomas gave cordial support to women's graduate education.[9]

O'Grady's main interest lay in the fields of comparative zoology and embryology, work guided at Bryn Mawr by E. B. Wilson and Frederic S. Lee. Her Ph.D. program included a special research project in fish embryology, which she pursued at Woods Hole.

The Woods Hole Experience

Women were welcome in the early days of the Marine Biological Laboratory (MBL) at Woods Hole, Massachusetts. When the first session met in its single wooden building in 1888, O'Grady was a participant in the Department of Investigation, meant to serve teachers and other students preparing for original work. Four of the seven investigators were women: Cornelia M. Clapp of Mount Holyoke, Helen Torrey Harris of Wellesley, Isabella

---

[7] O'Grady's thesis, as recorded at the MIT Archives, was entitled "The Sympathetic Nervous System of *Columba livia*." (*C. livia* is a pigeon.) For an account of the leading role of the Johns Hopkins University in the professionalization of biology see Keith R. Benson, "From Museum Research to Laboratory Research: The Transformation of Natural History into Academic Biology," in *The American Development of Biology*, ed. Ronald Rainger, Benson, and Jane Maienschein (Philadelphia: Univ. Pennsylvania Press, 1988), pp. 49–83.

[8] See Rossiter, *Women Scientists in America: Struggles and Strategies to 1940*, pp. 184–187; for reference to Sedgwick see pp. 238–240. The limits of Sedgwick's support for women in science were not apparent in O'Grady's case; he soon advocated science education to prepare women not for research, but to become technicians in public health laboratories—"woman's work." In fact, his editorials during World War I make it clear that he considered such work their patriotic duty. See, e.g., William Thompson Sedgwick, "A Welcome to Women in Public Health Work," *American Journal of Public Health*, 1918, *8:*164; and Sedgwick, "Women's New Fields of Work," *ibid.*, 1919, *9:*371–372.

[9] E. B. Wilson to M. Carey Thomas, 1 Aug. 1885, Bryn Mawr College Library; and Teresa Taylor, acting archivist, Bryn Mawr College Library, to Margaret R. Wright, 22 Jan. 1986. The majority of Ph.D. degrees awarded to women before 1889 were from Syracuse University, Boston University, and the University of Wooster. See Rossiter, *Women Scientists in America: Struggles and Strategies to 1940*, p. 32.

Mulford of Vassar, and Marcella O'Grady of Bryn Mawr. O'Grady's work in progress for the next two summers is recorded as "The Origin and Significance of Kupffer's Vesicle"; it was guided by Charles Otis Whitman, the director of the MBL.[10]

The qualities O'Grady prized at Woods Hole were direct reflections of Whitman's vision in planning the institution: it was designed to foster the professionalization of biology in this country through teaching and independent research, to be a place where all biologists— inexperienced and professional alike—could form a community through communication and cooperation. As Philip Pauly has put it: "Biology was an unusually powerful ideal, at least during the period from 1890 to 1915, because it expressed the life of a community of scientists. . . . At Woods Hole, experimentation and scientific discussion were part of a larger long-term framework of friendships and families. This unique environment crucially shaped the concept of biology."[11]

It was O'Grady's good luck to share in the excitement of these early days as she began her own professional development. She would miss the research atmosphere when she moved to Vassar from Bryn Mawr in 1889; Woods Hole provided a summertime escape from isolation. The Biological Lectures, initiated in 1890 and delivered to mixed audiences of students and professionals, became one of the features she valued most highly. In addition, Woods Hole summers offered relaxed time when she could evaluate herself, assess her capabilities, strengthen friendships, and meet new people with different experiences in biology. At Vassar, O'Grady encouraged her students' interest in research, obtaining financial help from the college so that they, too, could attend summer courses at Woods Hole. Through the 1890s more than twenty Vassar women profited from such work, among them graduating seniors, assistants who were studying for the master's degree, and instructors. O'Grady herself visited frequently as a member until she left for Germany in 1896. She never forgot her debt to Woods Hole.

Developing Biology at Vassar

The trustees of Bryn Mawr not only renewed O'Grady's appointment for 1889/1890 but voted her "Fellow by Courtesy." When Isabella Mulford resigned as teacher of botany at

---

[10] See *ibid.*, pp. 86–88, on the feminization of the summer marine laboratories in the 1870s and 1880s and their progressive defeminization in the 1890s. Information on those in the Department of Investigation comes from the first annual report (for 1888), Archives of the Marine Biological Laboratory Library, Woods Hole, Massachusetts. Jane Maienschein, *Defining Biology: Lectures from the 1890s* (Cambridge, Mass.: Harvard Univ. Press, 1986), p. 58, presents a picture showing Whitman, Sedgwick, O'Grady, and others in the laboratory. The title of O'Grady's work is taken from the annual reports of the MBL for 1889 and 1890; this information was supplied to me by Jean Monahan at the Marine Biological Laboratory Library. No copies of the work exist at the MBL, and it was apparently not published. The discovery of the neurenteric canal in teleost embryos would add to knowledge of vertebrate homologies.

[11] Philip J. Pauly, "Summer Resort and Scientific Discipline: Woods Hole and the Structure of American Biology, 1882–1925," in *American Development of Biology*, ed. Rainger *et al.* (cit. n. 7), pp. 121–150, on p. 122; see also Maienschein, *Defining Biology*, p. 50. On the history and nature of the MBL and, by way of comparison, the Naples Zoological Station see *ibid.*; Jane Maienschein, *One Hundred Years Exploring Life, 1888–1988: The Marine Biological Laboratory at Woods Hole* (Boston: Jones & Bartlett, 1989); Pauly, "Summer Resort and Scientific Discipline"; Emily A. Nunn, "The Naples Biological Station," *Science*, 1883, *1*:479–481, 507–510; Charles O. Whitman, "The Advantage of Study at the Naples Zoological Station," *ibid.*, 1883, *2*:93–97; Frank R. Lillie, *The Woods Hole Marine Biological Laboratory* (Chicago: Univ. Chicago Press, 1944); Kenneth R. Manning, *Black Apollo of Science: The Life of Ernest Everett Just* (Oxford: Oxford Univ. Press, 1983); and three articles in *Biological Bulletin*, 1985, *168* (Suppl.): *The Naples Zoological Station and the Marine Biological Laboratory: One Hundred Years of Biology*: Maienschein, "Agassiz, Hyatt, Whitman, and the Birth of the Marine Biological Laboratory," pp. 26–34; Maienschein, "First Impressions: American Biologists at Naples," pp. 187–191; and Alberto Monroy and Christiane Groeben, "The 'New' Embryology at the Zoological Station and at the Marine Biological Laboratory," pp. 35–43.

Vassar in June 1889, however, O'Grady could hardly fail to be interested: women's colleges provided the best opportunity for women to enter the scientific world at the time.[12] Leaving Bryn Mawr would mean delaying her research and postponing the Ph.D., but she made that difficult decision and, with Wilson's approval, became teacher of biology at Vassar.

O'Grady found the Vassar community well acquainted with woman scientists (Maria Mitchell, by then retired, and Mary Whitney in astronomy and Achsah Ely in mathematics had paved the way), but the college's biological offerings needed revitalizing.[13] Once again she was in the right place at the right time. As part of an administrative reorganization, the Department of Natural History was about to be divided into two separate departments: biology and geology/mineralogy. After her first year, then, O'Grady, now promoted to associate professor, became the sole member of the new Department of Biology, with freedom to design the curriculum as she saw fit.

Perhaps it was a measure of O'Grady's success that enrollments increased so markedly that in 1891/1892 she had an assistant, then two, and by 1895 an instructor was appointed: Elizabeth Bickford, whom she had known at Woods Hole in 1890. In 1893 O'Grady became a full professor, a status a woman was most unlikely to attain except at a women's college.[14]

The curriculum of the new department expanded rapidly under her insightful leadership. (See Figure 1.) The Vassar catalogue for her last year lists a total of seven courses, most of which included substantial laboratory work. Most novel, however, were two courses for seniors: "Current Biological Literature," a weekly seminar; and "Higher Biology," a lecture course. Gone were the descriptive courses that had characterized the preceding ten years. These senior courses are best appreciated by reading O'Grady's own accounts, presented in her beautifully handwritten reports to President James M. Taylor. In 1893 she wrote of "Current Biological Literature": "The aim [is] to familiarize the student with the more important lines of thought, . . . but also to give the experience and confidence accruing from the practice acquired in the preparation and presentation of papers before a class ready to discuss them. . . . It has given incentive to the recent alumnae to continue their

[12] In her second year at Bryn Mawr, O'Grady held the title "Demonstrator in Biology," which meant that she was a laboratory teaching assistant and drew a stipend in addition to her fellowship. The proffered fellowship for her third year would have continued this arrangement. Mulford came to Vassar from the Normal School in Trenton, N.J., in 1883: *Vassar Miscellany*, 1883, *13*:743. While teaching botany she earned both the B.A. and M.A. degrees, awarded by Vassar in 1886, and in 1895 Washington University (?) awarded her the Ph.D.: *Bulletin of Vassar College: Alumnae Biographical Register Issue* (Poughkeepsie, N.Y., 1939) (in fact, this source lists "Univ. Wash." as the institution that granted Mulford the Ph.D.; however, Washington University in St. Louis had a Ph.D. program with the Missouri Botanical Garden for a time in the 1890s). Mulford is not listed in *American Men of Science*. On the women's colleges as gateways to careers in science see Rossiter, *Women Scientists in America: Struggles and Strategies to 1940*, pp. 1–28.

[13] See Benson, "From Museum Research to Laboratory Research" (cit. n. 7); and Toby A. Appel, "Organizing Biology: The American Society of Naturalists and Its Affiliated Societies, 1883–1923," in *American Development of Biology*, ed. Rainger *et al.* (cit. n. 7), pp. 87–120.

[14] Bickford's background was somewhat similar to O'Grady's. She was an MIT graduate and a teacher at the Bryn Mawr School. The *Annual Catalogue of the Officers and Students of Vassar College* for 1895/1896 credits her with a doctorate, but the MIT record seems ambiguous, and the source for the obituary statement (in *Technical Review* [MIT, July 1940]) that she obtained her doctorate at the University of Freiberg is unknown. The only doctorates obtained in the natural sciences and mathematics during O'Grady's tenure at Vassar were Ph.D.'s for LeRoy C. Cooley and Charles Moulton—who were, respectively, professors of physics and chemistry—and an M.D. for the physician Elizabeth B. Thelberg, who was also professor of physiology and hygiene. No woman on the faculty had a Ph.D., but four, in addition to O'Grady and Thelberg, were the only full professors in their departments: Mary Whitney in astronomy, Abby Leach in Greek, Lucy Maynard Salmon in history, and Achsah Ely in mathematics. The title "chairman" was not used in the catalogue listings at that time.

Figure 1. *Marcella Imelda O'Grady, the young professor of biology at Vassar College in the 1890s. (Personal collection of Margaret R. Wright.)*

studies."[15] (The literature studied included papers in both French and German.) It seems likely that O'Grady's experience at Woods Hole inspired her to stimulate the exchange of ideas among her students in this way, whether or not they intended to join the professionals.

In 1894 she described "Higher Biology" as "a history of the development of the Biological Sciences with special reference to the growth of the evolution theory." Though the pros and cons of Darwinism had been discussed in the past both in courses and by visiting speakers (among them Alfred Russel Wallace), "Higher Biology" was the first formal

[15] Marcella O'Grady, annual report to President James Monroe Taylor, 1893, Special Collections, Vassar College Libraries, Poughkeepsie, New York.

course offered on the subject.[16] It doubtless caused some unease among the Baptist ministers in Vassar's administrative circles.

In a review of Henry Fairfield Osborn's book, *From the Greeks to Darwin,* O'Grady wrote: "This volume is admirably adapted to meet the needs of those students who wish to gain a general, and at the same time, a concise, connected, and unbiased conception of the growth of the Idea of Evolution."[17] It is clear that this young woman, a practicing Roman Catholic, found no irreconcilable conflict between her religion and science.

As a scientist, woman, teacher, and friend, Marcella O'Grady was a role model. She set the same high standards of quality for her students that she demanded of herself: critical thinking, industry, accuracy in observation and expression. A surprising number of those students went on to have careers, more often than not in science teaching. Many of them stayed in touch with her for the rest of her life. One of the best known was Helen D. King (Vassar, 1892), who earned her Ph.D. at Bryn Mawr and had a long career as professor of embryology at the Wistar Institute of the University of Pennsylvania. In 1932 King won an Ellen Richards Prize, awarded by the Association to Aid Women in Science—and her teacher was there to congratulate her. Another student, Edna Carter (Vassar, 1894), was encouraged by her physics professor to pursue a career, but she credited O'Grady with first stimulating her interest in science. A professor emerita at the time of Vassar's 100th anniversary celebration, Carter wrote of O'Grady's warm interest in people and her "inborn desire to be of service." She commented too that Margaret Washburn (Vassar, 1891), later a distinguished professor of psychology at Vassar, "always credited Miss O'Grady as the one who opened her eyes to the scientific method of research."[18]

Teaching, however, was not enough. O'Grady's interest in the cytological investigations being conducted in Germany was whetted by E. B. Wilson. His research in Munich and at the Zoological Station of Naples had included work with Theodor Boveri, the foremost European cytologist of the time. Wilson's monumental volume, *The Cell in Development and Inheritance,* bears the dedication: "To my friend Theodor Boveri."[19]

Wilson described some of Boveri's experiments in a lecture at Vassar; the student reviewer reported that "the work of Boveri . . . added to the scientific data upon which a theory of heredity may be founded, and Dr. Wilson thinks that . . . we are on the eve of a great discovery." And so they were. Boveri's fine cytological studies gave evidence that the chromosomes are individual units that maintain their integrity from one cell generation to the next.[20] This was a significant contribution toward defining the physical basis for the Mendelian genetics to come.

---

[16] Marcella O'Grady, annual report to President Taylor, 1894, Special Collections, Vassar College Libraries. The first professor of natural history at Vassar, Sanborn Tenney, was a student of Agassiz and a proponent of special creation; his successor until 1877, James Orton, was an ordained minister sympathetic to evolutionary theory. In fact, Orton dedicated his popular account of his explorations, *The Andes and the Amazon* (New York: Harper, 1870), to Darwin, "whose profound researches have thrown so much light upon every department of science, and whose charming 'Voyage of the Beagle' has so pleasantly associated his name with our southern continent" (p. [ix]). A student report on Wallace's visit appears in *Vassar Miscellany,* 1886, *16*(3):107–108.

[17] For O'Grady's review of Osborn's book see *Vassar Miscellany,* 1894, *24*(3):128–129.

[18] Edna Carter's recollections are in a file on Vassar's centennial celebration, Special Collections, Vassar College Libraries. Information on students' careers comes from *Bulletin of Vassar College: Alumnae Biographical Register Issue* (cit. n. 12).

[19] Edmund B. Wilson, *The Cell in Development and Inheritance* (New York: Macmillan, 1896), p. [ix].

[20] *Vassar Miscellany,* 1891, *20*(5):180–181. For the cytological studies Wilson referred to see Theodor Boveri, "Zellenstudien, II: Bei Befruchtung und Teilung des Eies von *Ascaris megalocephala,*" *Jenaischen Zeitschrift für Naturwissenschaft,* 1888, *22*:685–882.

O'Grady was tiring of secondhand science; she was restless. Accordingly, she took a sabbatical for research in 1896/1897, leaving the department in the capable hands of Elizabeth Bickford. Probably she was attracted by the fact that some German universities had, reluctantly, begun to accept foreign women (though not Germans) to study for advanced degrees; more important, Germany was the fountainhead of the "new biology."[21] Impressed by Boveri's work and strongly influenced by Wilson, she went to the University of Würzburg, where Boveri had become director of the Zoological-Zootomical Institute. In this first American chapter of O'Grady's life, now completed, the support of influential male sponsors—Wilson, Sedgwick, and Whitman—had eased her entry to the academic world of science.

A COMPROMISE

The Sabbatical Year

Marcella O'Grady was the first woman admitted to study science at the University of Würzburg.[22] One can only imagine the kind of reception an American woman biologist would have faced. She was to join an international but entirely male set of students, doctoral candidates, and established investigators. Director Boveri was said not to favor higher education for women; mercifully, she probably did not know that. With quiet skill she soon justified her mentors' faith by her intelligent, mature approach to research.

Her research problem was designed by the director to fit into the laboratory's general sphere of inquiry. From the beginning of Boveri's researches in 1885 his interest had centered on the cell, particularly the role of the chromosomes in the formation of the germ cells and the fertilization and development of the egg. O'Grady's task was to make a detailed microscopic analysis of an anomalous mitotic mechanism that occurred in some sea urchin embryos as a result of artificial hybridization. From these investigations she derived considerable expertise in cytological work and familiarity with sea urchin development, both of which would be important assets for her later collaborative work.

Fritz Baltzer, then a doctoral candidate at Würzburg, gave the following account:

> Miss O'Grady was, personally and scientifically, becoming more and more at home in Würzburg. . . . [Boveri remarked that] "The scientific world of Würzburg has been highly exercised by her question, whether she might be introduced as a guest at the meetings of the illustrious Physico-Medical Society. Overwhelming majority in favor." The excitement aroused in academic quarters by Miss O'Grady's request proved that as far as the emancipation of women

---

[21] Few women's colleges at this time paid for sabbaticals, but O'Grady was given a full year's leave at half pay. On German universities' acceptance of foreign women students see Rossiter, *Women Scientists in America: Struggles and Strategies to 1940*, p. 41.

[22] General references for this section on Marcella O'Grady Boveri's life in Germany include Richard B. Goldschmidt, *Portraits from Memory: Recollections of a Zoölogist* (Seattle: Univ. Washington Press, 1956); Fritz Baltzer, "Theodor Boveri," *Science*, 1964, *144*:809–815 (trans. by Curt Stern and Evelyn Stern from the original article in *Naturwissenschaftliche Rundschau* [Stuttgart, June 1963], Suppl. 87 of the *Verein Deutscher Biologen*); Baltzer, *Theodor Boveri: Life and Work of a Great Biologist, 1862–1915*, trans. Dorothea Rudnick (Berkeley: Univ. California Press, 1967) (hereafter cited as **Baltzer, *Theodor Boveri*, trans. Rudnick**) (originally published as Baltzer, *Theodor Boveri: Leben und Werk eines grossen Biologen* [Stuttgart: Wissenschaftliche Verlagsgesellschaft, 1962]); and Margret Boveri, *Verzweigungen: Eine Autobiographie* (Munich: Piper, 1977) (hereafter cited as **Margret Boveri, *Verzweigungen***). Quotations from all letters to follow are in the Archives of Albertus Magnus College Library, New Haven, Connecticut; letters from Marcella Boveri to her daughter, Margret Boveri, are in RG 22, Bov. III, Box I (1927–1936), and Box II (1937–1950).

was concerned the views of the university community lagged far behind those of their young American guest.[23]

From the beginning, O'Grady had the ready support of Wilhelm Conrad Röntgen, who was then the professor of physics at the university and Boveri's closest friend.[24] He was delighted with her eager interest and plucky spirit, and she soon enjoyed the Röntgens' invitations to daily afternoon coffee and family holidays in the Black Forest and in Cadenabbia, on the shore of Lake Como.

Marriage and Life in Würzburg

Boveri was not immune to O'Grady's charms. To Victoire Boveri, his Swiss sister-in-law, he wrote: "I now have an American lady zoologist in the institute, she is not really pretty, but quite attractive. I enjoy her company and must sometimes restrain myself: I think she does not care for frivolity." True, O'Grady's seriousness was very much in contrast to his quick and imaginative humor. The courtship proceeded, to the alarm of Boveri's mother, and by June 1897 they were engaged to be married. The wedding took place on 4 October 1897 at the Convent of the Good Shepherd in Troy, New York, where O'Grady's sister was living. The concerns of the Irish-Americans at their daughter marrying a German and of the bourgeois-conservative Germans at the introduction of an American into their old family were somewhat allayed by that time.[25] A consoling feature probably was their shared Catholicism and the fact that the couple would live in strongly Catholic Bavaria.

The Old World setting for a new life was itself fascinating. The Boveris lived within pleasant walking distance of the institute, which was set in a lovely parklike space on the horseshoe-shaped promenade, the "glacis," that replaced the fortifications that originally surrounded the ancient town center of Würzburg. This interesting rococo city on the River Main was graced by a monumental Romanesque cathedral and the Marienberg fortress. The episcopal residence had been in the city from the thirteenth to the eighteenth century. The university was founded in 1582, but Boveri's institute was housed in a large, well-equipped building constructed in 1889. Much to the pleasure of the man whose first ambition had been to be an artist, his research quarters looked out to the trees in the garden.

The family's country house, "Seehaus," enjoyed as a quiet retreat from university life, was at the edge of the village of Höfen, near the small, picturesque town of Bamberg where Boveri was born. Bamberg had been the capital of a powerful ecclesiastical state for some eight hundred years before the beginning of the nineteenth century, and a prince-bishop of Bamberg was said to have built the house in the early eighteenth century as a fishing retreat. Bamberg became one of Marcella Boveri's favorite spots, second only to Würzburg.

Boveri's father, a physician and art lover, had died in 1891, but his mother lived in Bamberg and spent her summers at the Höfen country home until her death in 1910. His older brother, Albert, a government councillor, lived in Munich. The two younger brothers,

---

[23] Baltzer, *Theodor Boveri*, trans. Rudnick, p. 17. At the next meeting of the Physico-Medical Society the privilege was extended to "any woman who has had a college training and has done independent scientific work": reported *in Vassar Miscellany*, Feb. 1897, *26*(5):178.

[24] Röntgen's discovery of X rays was made in 1895 at Würzburg.

[25] See Baltzer, *Theodor Boveri*, trans. Rudnick, p. 16; and Margret Boveri to Sr. Charles, Librarian, Albertus Magnus College, 4 Mar. 1959, Albertus Magnus College Library, RG 22, Bov. VII.

Walter, living in Zurich, and Robert, in Mannheim, were both executives in the international firm of Brown, Boveri & Cie. Distances were not too great for reunions and family councils to take place at Seehaus. And the Boveris often vacationed with Walter, his wife Victoire, and their children at Maloja, among the mountains and lakes in Switzerland near the Italian border. The Boveri family's acceptance of Theodor's wife was friendly, but qualified: she was an outsider, a stranger inserted in their long-established Frankish line— different, independent, American.[26]

At home in Würzburg, Boveri's university colleagues very quickly came to know his wife as a generous and charming hostess. Boveri had a small circle of intimate friends from the university faculty, Röntgen and Wilhelm Wien, another eminent physicist, among them. Hans Spemann, a future Nobel laureate who had been one of Boveri's first students, was then a docent at Würzburg and a close friend; he later held a professorship at the University of Freiberg. And Fritz Baltzer was a doctoral candidate who stayed on as docent in the institute until Boveri's death. He later held a professorship at the University of Berne. He, too, was a good friend and Boveri's biographer.

An interest in music was likely to be the path to friendship with the Boveri family. Theodor Boveri was an accomplished pianist who enjoyed the musical companionship of duets with friends and later with his daughter. "At the Boveri home . . . one received a broad education in musical literature. . . . Boveri's performance at the piano showed two characteristic qualities, namely great rhythmic precision and the ability to comprehend at first sight the essentials of a score."[27] Mrs. Boveri played no instrument herself, but she was a knowledgeable and sympathetic listener. Music was an important bond, and the home musical with friends was their favorite kind of social life.

Letters from the Röntgens after they moved to Munich show their warm relationship with the Boveris. Anna Röntgen wrote in 1902: "The times with our good friends in Würzburg were beautiful, we were always so congenial and could discuss things of common interest. Here everything is rather cold." And Röntgen, too, in 1907 wrote to Boveri: "How I miss the mental stimulation which I received so often from you." In later years Margret Boveri recalled happy evenings at "Onkel Röntgen's" "hunting lodge" in Weilheim, when "we all sat around the big round table under the large hanging lamp, and Röntgen read us a novel or an animal story." She also remembered that "during the heathcock and woodcock season he [Röntgen] got up at 3:30 in the morning and, sometimes accompanied by my mother, went through the fog of night to the little birch-branch huts to wait for the dawn and the mating calls of the birds." Her mother delighted in these opportunities to walk about in the beautiful country at the foot of the Bavarian Alps and to see and hear the birds and other small creatures along the way, even if she did not greatly appreciate the shooting.[28]

But what of Frau Boveri's scientific career? Her research at Bryn Mawr and at Woods Hole, coupled with the substantial research program at Würzburg, had brought her close to qualifying for the Ph.D. But clearly after her marriage her studies had to be curtailed,

[26] Baltzer, *Theodor Boveri,* trans. Rudnick, p. 11. Brown, Boveri & Cie was an important company that made heavy machinery, such as turbines for hydroelectric plants and motors for locomotives. On the Boveri family see Margret Boveri, *Verzweigungen,* p. 415; and (on its origins in particular) Baltzer, *Theodor Boveri,* trans. Rudnick, p. 3. The Italian name came to Franconia from Savoy in the person of Carolus Boveri in the late sixteenth century. For three hundred years all Boveri wives, except for Victoire, had been of Franconian origin.

[27] Baltzer, *Theodor Boveri,* trans. Rudnick, p. 46.

[28] Margret Boveri, "Personal Reminiscences of W. C. Röntgen," in Otto Glasser, *Wilhelm Conrad Röntgen and the Early History of the Röntgen Rays* (Springfield, Ill.: Charles C. Thomas, 1934), pp. 132–133, 153, 157.

for as the wife of a university professor who was also director of an institute and, later, rector of the university she had to assume many new and unfamiliar responsibilities. Furthermore, nothing in her experience prepared her for the everyday problems of running a household.

What led Marcella O'Grady to change course and marry? The answer, at least in part, lies in her appraisal of her own abilities and how best to use them. In Boveri she saw a brilliant, imaginative, and resourceful investigator, already at thirty-five a recognized leader. She saw also his frail health, and as she came to know him better she learned of his battles with severe rheumatism and his periods of despair. She recognized that his time and energies would be further sapped by the increasing demands of his position in the university. She was in vigorous health, modestly confident of her own abilities, happy in the life of Würzburg, but also ambitious to contribute to biology. She surely recognized the limitations faced by a woman research scientist on either side of the Atlantic—whether single or married. She was a giving person; she didn't want Boveri's talents to be wasted; that "inborn desire to be of service" noted by Edna Carter was strong in her. She decided that Boveri could accomplish much more with her help than alone and that it would be a great privilege for her to work with him.[29]

Background for Collaboration

The Zoological Station on the Bay of Naples, the first truly international marine station (founded in 1874), was designed for professionals: it was a large laboratory where biologists were supplied with materials, the best equipment, and supporting staff for their independent work at all seasons. It attracted distinguished investigators, who found it a stimulating center for communication and exchange of ideas with their peers from around the world.[30] Theodor Boveri was to rejoice in that atmosphere. He and Anton Dohrn, the founder and director of the station, were intimate friends, and warm lifelong relations developed between their families.

Boveri had been working with sea urchins for about a decade at the end of the nineteenth century, testing his ideas about the nature and significance of cellular components. It was known that an overabundance of sperm in fertilization produces some eggs with two sperm nuclei. Dispermy alters the first cleavage so that three or four cells (i.e., trefoils and tetrafoils) are produced, rather than the normal two, and the chromosomes are randomly distributed to these cells. Boveri recognized that these embryos, having an altered number of chromosomes, could offer the experimenter valuable material for research without the need for invasive techniques—ready-made experimental material. As he later remarked: "a simple thing, if only one thinks of it, to do the experiment." It was at this phase of his research that Mrs. Boveri began to collaborate.[31]

---

[29] Margret Boveri, *Verzweigungen*, p. 15. Marcella Boveri's letters confirm Margret's interpretation of her position—indeed, they may have been her source of information. And I know, from personal conversation, that she felt that privilege.

[30] The first two Americans to visit the Naples station in the early 1880s were C. O. Whitman and E. B. Wilson. For the history and character of the Naples Zoological Station see the sources cited in note 11, above.

[31] Baltzer, *Theodor Boveri*, trans. Rudnick, p. 92. To elaborate a bit: The eggs of most organisms have one nucleus, and it contains one of each kind of chromosome (i.e., the haploid set of chromosomes). The sperm nucleus is also haploid. During fertilization of the egg, these two haploid nuclei unite to form a diploid nucleus. Through a complicated and normally extremely precise process (mitosis), the fertilized egg divides to become two cells; each of those cells has a *diploid* nucleus. (This comes about by the duplication of each chromosome, followed by their separation.) Continued mitotic division creates a new diploid organism. In the Boveri's dispermic egg experiments, fertilization provided for the union of three haploid nuclei (a triploid). In these triploids the precision of mitosis was disrupted, so that the chromosomes were randomly distributed to three centers and formed the nuclei in the resulting three cells. The chance of any cell getting a normal diploid set of chromosomes was, therefore, much reduced. The resulting embryo was dubbed a trefoil.

Collaboration (1900–1915)

The Boveris' first work together was performed at a marine laboratory in Villefranche in 1900, but I want to look here at the ingenious and important experiments done at Naples in the winter and spring of 1901–1902.[32] The challenge was to test the hypothesis, advanced by Boveri in 1888, that chromosomes differ qualitatively. The Boveris produced dispermic eggs by artificial fertilization and studied the resulting embryos and larvae.

They isolated trefoils by the hundreds for microscopic examination. The embryos were composed of three areas, each derived from one of the three cells. Visible aberrations, ranging from severe distortion or absence of parts to minor effects, frequently delineated the thirds. A small number of the embryos became normal larvae.

Assuming that specific chromosomes are necessary for specific morphological characteristics to develop, their absence should result in morphological aberrations. The observed results fit that supposition: for example, one third of a larva would lack a skeletal structure, while the rest had a skeleton.

Assuming that chromosomes differ qualitatively, a totally normal larva should have received at least one of each of the qualitatively different chromosomes in each of the three cells. The question arises, What is the probability that each cell of a trefoil would have a normal set? To answer that question, Boveri—with some help from the physicist Wien—created a mechanical apparatus for predicting frequencies in an analogous system of random distribution. Actual counts of normal larvae from trefoils fit the prediction fairly well.

The interpretation of the parallel between the prediction and experimental results was "that not a definite number, but a *definite combination of chromosomes* is essential for normal development, and this means nothing less than that *the individual chromosomes must possess different qualities.*"[33] The morphological evidence was consistent with this interpretation, and experiments of different design performed in the same year verified it.

The significance of these findings is that by correlating chromosomes with the "factors" of Mendel, Boveri could point to chromosomes as the physical basis of heredity. In the same year, Walter S. Sutton at Columbia University came independently to the same conclusion by different means. Their interpretation is often cited as the Sutton-Boveri chromosome theory of inheritance.[34]

Mrs. Boveri participated in the experiment as a colleague, even if a less experienced one. Though the imaginative design of the experiment was clearly Boveri's, she shared responsibility for the execution and analysis of it. Her quick mind, her acute observational ability, and the cytological skill she had developed under his guidance must have been

---

[32] Theodor Boveri, "Ueber mehrpolige Mitosen als Mittel zur Analyse des Zelkerns," *Verhandlungen der Physicalische-medizinische Gesellschaft zu Würzburg,* 1902, *35:*67–90; an English trans. may be found in B. H. Willier and Jane Oppenheimer, *Foundations of Experimental Embryology* (New York: Prentice-Hall, 1964). Given space limitations, my description of this work is necessarily incomplete. Boveri summarized these and related experiments in "Zellstudien, VI: Die Entwicklung dispermer Seeigeleter: Ein Beitrag zur Befruchtungslehre und zur Theorie des Kerns," *Jena. Z. Naturwiss.,* 1907, *43:*1–292.

[33] Theodor Boveri, "Ueber mehrpolige Mitosen."

[34] Lack of space prevents discussion of priorities in the Sutton-Boveri hypothesis. See Ernst Mayr, *The Growth of Biological Thought* (Cambridge, Mass.: Harvard Univ. Press, Belknap, 1982), pp. 747–749; John A. Moore, "Science as a Way of Knowing—Genetics," *American Zoologist,* 1986, *26:*583–747, esp. pp. 653–664; Victor A. McKusick, "Walter S. Sutton and the Physical Basis of Mendelism," *Bulletin of the History of Medicine,* 1960, *34:*487–497; and Jane Oppenheimer, *Essays in the History of Embryology and Biology* (Cambridge, Mass.: MIT Press, 1967), pp. 60–91.

great assets. The published paper was, regrettably, authored by Boveri alone. There is no indication that she ever expected any formal recognition of her collaboration in this or any of their later work. She found her reward in the privilege of participation and, as I have heard her say, looked back on those days as the happiest of her life.

According to Baltzer, Boveri worked at Naples in the spring seasons of 1905 and 1910–1912 and in 1914, his last period there before his early death; "his wife collaborated on his projects." She prepared the posthumous publication of the last sea urchin experiments in his name, including a short preface in her own.[35]

Boveri finished the research for her "Doktorarbeit" and published it in 1903. Unfortunately, she did not complete the requirements for the Ph.D. This omission was a source of lifelong regret—regret that later made her spur her daughter on to complete her own doctorate.[36]

Visiting American Women Scientists

Because Marcella Boveri was a biologist as well as wife of the institute director, she took special pleasure in hosting visiting investigators. She was not only a gracious hostess but a knowledgeable associate, and she came to know many of the leading embryologists on both sides of the Atlantic. Her enduring interest in the education and progress of women in science must have been stirred by the arrival of American women, several with Bryn Mawr connections, at the laboratory.

Apparently Boveri had dispelled her husband's hesitancy to accept women in his laboratory, for another woman recommended from Bryn Mawr, Nettie M. Stevens, visited both Würzburg and the Naples station in 1901/1902. She was a graduate of Stanford University and received her Ph.D., working with T. H. Morgan, from Bryn Mawr in 1903. Marcella Boveri and Nettie Stevens may have enjoyed comparing experiences, for though they were about the same age their time at Bryn Mawr was more than a decade apart.[37] Two more women from Bryn Mawr joined Stevens in her second visit to Würzburg in 1908. Each was assigned a share in some correlated studies of environmental effects on development in the roundworm, *Ascaris.*

Stevens first worked with Boveri on dispermy experiments in *Ascaris,* published by Boveri with Stevens as coauthor. In the later visit she investigated and published on the

[35] Baltzer, *Theodor Boveri,* trans. Rudnick, p. 50. For the last publication see Theodor Boveri, "Zwei Fehlerquellen bei Merogonieversuchen und die Entwicklungsfähigkeit Merogonischer, partiell-merogonischer Seeigelbastarde," *Archiv für Entwicklungs-Mechanik,* 1918, *44:*417–471. See Baltzer, *Theodor Boveri,* trans. Rudnick, pp. 143–147, for the complete list of Boveri's publications and publications from the institute in his time.

[36] Marcella Boveri, "Ueber Mitosen bei einseitiger Chromosomenbindung," *Jena. Z. Naturwiss.,* 1903, *37:* 401–446; and Margret Boveri, *Verzweigungen,* p. 133.

[37] Sources for data on the American visitors are *American Men of Science* (I have used eds. 2–5) and the following analytical biographies: Marilyn R. Ogilvie and C. J. Choquette, "Nettie Maria Stevens (1861–1912): Her Life and Contribution to Cytogenetics," *Proceedings of the American Philosophical Society,* 1981, *125:* 292–311; and Ogilvie, "The 'New Look' Women and the Expansion of American Zoology: Nettie Maria Stevens (1861–1912) and Alice Middleton Boring (1883–1955)," in *The Expansion of American Biology,* ed. Keith R. Benson, Jane Maienschein, and Ronald Rainger (New Brunswick, N.J.: Rutgers Univ. Press, 1991), pp. 52–79. It is unfortunate that Boveri was irritated by both Stevens and Boring, apparently believing that they had misused the privilege of the laboratory and pilfered his ideas: Ogilvie and Choquette, "Nettie Maria Stevens," p. 298. In examining their papers I noticed two things that bear mention. The first is that Boring published a paper with a Würzburg dateline on research done at Bryn Mawr; perhaps Boveri thought the time should have been spent on research to be credited to his laboratory. Second, both Stevens and Boring properly acknowledged that Boveri suggested the problems and the methods of their research and expressed their thanks for the privilege of working with him.

effect of ultraviolet light on development in *Ascaris*.[38] Her life in research was short but productive and merits the recognition given by recent biographers.[39]

Alice M. Boring, twenty years younger than Boveri, a 1904 Bryn Mawr graduate and doctoral candidate, spent a year at Würzburg and Naples before receiving her Ph.D. at Bryn Mawr in 1910. She reported her studies of a small chromosome in *Ascaris* and temperature effects on the nuclei in *Ascaris* embryos in two publications from the laboratory.[40]

A contemporary of Boring, Mary Jane Hogue, a graduate of Goucher College in 1905, spent two years as a demonstrator at Bryn Mawr before coming to Würzburg. She worked with Boveri on a study of blastomere potentiality in *Ascaris* and was coauthor of its publication in 1909. Hogue's focus in the correlated studies was the effects of centrifugation on *Ascaris;* this work formed the basis for her dissertation for the Ph.D. from Würzburg in 1909. Hogue would hold college teaching positions until after World War I, when she turned to university positions with a greater medical orientation.[41]

Boveri was interested to learn that one of her former Vassar students, Edna Carter, had spent two years studying physics at the University of Chicago with Albert A. Michelson and Robert A. Millikan and urged her to come to Würzburg for the Ph.D. The Röntgens had left for Munich, but Wilhelm Wien welcomed Carter to his laboratory in 1904. There she joined an international group of investigators: a Finn, a Norwegian, some Russians and Germans—but no women. Happily, she quickly felt included as a member of the group. She was awarded the Ph.D. in 1906 for her studies on the energy of X rays.[42] In addition to enjoying a busy social life with her new friends in Wien's laboratory, Carter was a regular guest at Sunday dinner with the Boveris and was welcomed at Höfen during summer vacation and on excursions to the Black Forest and elsewhere. Her relationship with her former teacher developed into a lifelong friendship.

Parenthood

These early years in Germany were satisfying, intensely busy ones for Marcella Boveri. She gave birth to a daughter, Margret, on 14 August 1900, an event that sent her husband

---

[38] Theodor Boveri and N. M. Stevens, "Über die Entwicklung dispermer Ascariscier," *Zoölogischer Anzeiger,* 1904, *27:*406–417; and Stevens, "The Effect of Ultra-violet Light upon the Developing Eggs of *Ascaris megalocephala,*" *Arch. Entwick.-Mech.,* 1909, *27:*622–639. It is thought by her biographers that Stevens was probably influenced, at least indirectly, by Boveri's ideas in her subsequent researches on sex determination.

[39] In 1905 Stevens published her discovery of chromosomal sex determination in insects, providing convincing confirmation of the Sutton-Boveri theory and, therefore, of Mendelism. The priority issue between E. B. Wilson and Stevens was reexamined in Stephen G. Brush, "Nettie M. Stevens and the Discovery of Sex Determination by Chromosomes," *Isis,* 1978, *69:*163–172. See also Ogilvie and Choquette, "Nettie Maria Stevens" (cit. n. 37); and Ogilvie, "'New Look' Women" (cit. n. 37).

[40] A. M. Boring, "A Small Chromosome in *Ascaris megalocephala,*" *Archiv für Zellforschung,* 1909, *4;* and Boring, "On the Effects of Different Temperatures on the Size of the Nuclei in the Embryo of *Ascaris megalocephala,* with Remarks on the Size-Relation of the Nuclei in Univalens and Bivalens," *Arch. Entwick.-Mech.,* 1909, *28:*118–126. Boring held teaching positions at Vassar (1907–1908), the University of Maine (1909–1918), Peking Union Medical College (1918–1920), and Wellesley (1920–1923); in 1923 she returned to China as a professor at Yenching University, Peking. There she made a dramatic and productive change in her research field, moving from cytogenetics to taxonomy, and specialized in amphibia in China until 1950 (she spent part of the war years interned in Peking).

[41] Theodor Boveri and M. J. Hogue, "Über die Möglichkeit, Ascaris-Eier zur teilung in zwei gleichwertige Blastomeren zu veranlassen," *Sitz.-Ber. Phys.-Med. Ges. Würzburg,* 1909; and Hogue, "Über die Wirkung der Centralfugalkraft auf die Eier von *Ascaris megalocephala*" (dissertation), *Arch. Entwick.-Mech.,* 1910, *29:*109–145. Hogue held teaching positions at Mt. Holyoke (1911–1914) and Wellesley (1914–1918), worked in the base hospital at Fort Sill (1918–1919), and thereafter had medically oriented positions at Johns Hopkins and the University of Pennsylvania.

[42] Helen Lockwood, Barbara Swain, and Monica Healea, memorial minute for Edna Carter, read at faculty meeting, 1963, Special Collections, Vassar College Libraries.

off to Höfen a month later "to recover a bit from becoming a father."[43] Before she was a year old Margret was left, in the care of a nursemaid, with the relatives in Switzerland while her parents did their famous dispermy experiments at the Naples station. During later periods of extended investigation there she went to the International School in Naples, and the entire family especially enjoyed their association with the Dohrns.

Marcella Boveri loved Margret dearly, but she was not demonstrative, and her rationalistic and puritanical nature was more evident to her daughter than her love. They never had the warm, affectionate bonds that bound Margret to her father. It was not father, the scientist, but father, the artist, that held her. The two were temperamentally much alike and shared in creative pursuits like painting and piano performance. Mrs. Boveri, though happy to admire, was the one who demanded hours of serious practice, at which Margret, as strong-willed as her mother, often rebelled.

The matter of child-rearing brought out some differences between Boveri, the foreigner, and her Bavarian critics over the education and emancipation of women. She was a strict taskmaster; there was little compromise in Margret's daily routine; every hour had its prescribed activity and must be useful; wasting time was close to sinful. The pressure was considered excessive by Boveri's critics—there was no freedom for childish play. On the emotional level two strong wills often clashed, but as Margret wrote later: "With all her sternness. . . . I really had a lot of freedom for those days. . . . My mother wanted to make me independent as soon as possible . . . observant, and guided by laws of reason." One day, to teach her about the danger of the pond, her mother let her fall into the brook that fed it. This horrified the Bavarian domestics: the maids felt sorry for the child; the cook pronounced her mother "a godless woman."[44]

In university circles at Würzburg it was the custom to teach groups of children in private classes, and one of Boveri's pleasures was to teach them natural history. Margret knew that her mother was "not so bad," but she also saw that some other children were truly enthusiastic about her. Many years later, one of them wrote that "the hours when your mother was teaching were among the best memories of my childhood."[45]

A Period of Stress Both National and Personal

For years the Boveris resisted other universities' attempts to coax them away from Würzburg, but an especially interesting request came to Boveri in 1912. He was asked to guide the planning for and become director of the new Kaiser Wilhelm Institute for Biology in Berlin-Dahlem. It was a great honor that would also carry heavy responsibilities. Boveri was tempted, though apprehensive about the consequences for his increasingly frail health. That winter he drafted an organizational scheme appropriate for such a research institution, but after a severe bout of illness in the spring he declined any further responsibility for it. He was hospitalized in the fall of 1914. Mrs. Boveri was convinced that his worry about the war was responsible, but the cause of his death, on 15 October 1915, at the age of fifty-three, was tuberculosis.[46]

A grief-stricken widow, she did her best through the war years to nurture her daughter,

[43] Baltzer, *Theodor Boveri,* trans. Rudnick, p. 18.

[44] Margret Boveri, *Verzweigungen,* pp. 13, 15, 17. All quotations are in my free translation.

[45] *Ibid.,* p. 15.

[46] On the proposal regarding the Kaiser Wilhelm Institute for Biology see Baltzer, "Theodor Boveri" (cit. n. 22), p. 815. Information on the cause of Boveri's death is from Marcella Boveri to Margret Boveri, 23 Mar. 1934.

who at fifteen had been dealt a crushing blow by her father's death, and to guide her through her school years in Würzburg. Now a German citizen, this American-born woman felt the stress of conflicting loyalties. It is interesting that Edna Carter spoke of her time in Germany as a period of *"akademische Freiheit"*—freedom to work—implying that professors were privileged to remain detached from the politics of government. Such "freedom" and detachment would be regretted later. But Boveri's nature was to serve in some meaningful way: turning her home into a hospital, she spent long hours tending to the needs of wounded soldiers.[47]

After World War I

Boveri was generous in offering time and help to friends in trouble. Her closest friend, Anna Röntgen, died soon after the war, and every year thereafter "Onkel Röntgen" was given a family-style welcome at Christmas in memory of happier times—a salutary experience for them all. After Röntgen's death in 1923 Boveri, as executrix, settled his estate.

Boveri's sister and American friends urged her to return to the United States, but she simply was not ready to leave Germany. Her property was there; Germany had become "home"; she didn't want to uproot her daughter; she loved the countryside; and in spite of her loneliness she had friends, especially the biologists scattered around the country.

Boveri found satisfying outlets for her energy. "She was one of the chief initiators of musical life in Würzburg, having organized a Beethoven festival in 1920 and being one of the originators of the annual Mozart festivals in the baroque palace of Würzburg."[48] Margret's progress, too, was a source of pleasure: during the early 1920s she had studied Germanic and English history at the University of Würzburg, and in 1925 she began to study for a doctorate at the University of Munich.

Clearly Boveri was busy, but not with the kind of sustained and satisfying work essential to her well-being. Eventually she became restless, dispirited, tired, and unwell. Her friends prescribed a change of scene.

Boveri and her daughter visited the Dohrns in Naples over the Christmas holidays in 1926. Reinhard Dohrn was now the director of the laboratory, having succeeded his father, Anton, upon his death in 1910. Margret, from the age of eleven, had found kindred spirits in Reinhard and his wife, Tania, and they became an important source of support for her. Feeling that Margret was launched in life, Marcella Boveri, after ten years of widowhood, was free to set her own course. She sailed to the United States, supposedly for a visit.

RENEWAL

Boveri had a fine reunion with her sister and later was cordially welcomed by friends from her Bryn Mawr and Vassar days. A whole new life opened to her when she went to New Haven, Connecticut, to visit her former student, Elisabeth Woodbridge, now Mrs. Charles Morris.

Albertus Magnus College, a small liberal arts college for women founded by the Dominicans, had opened in New Haven in 1925, but it had not yet established a science department. (The 1920s saw a rapid increase in Catholic colleges for women.) The provost's wife, learning about Boveri's teaching at Vassar from Mrs. Morris, her aunt, sug-

---

[47] Undated memorandum from Margret Boveri, Albertus Magnus College Library.
[48] *Ibid.*

gested to her husband that Boveri's appointment would be of exceptional benefit to the college. Boveri took a great deal of convincing; importantly, much of the persuasion came from Ross G. Harrison at Yale, who assured her that all the resources of the Osborn Zoological Laboratory (OZL) and the Yale libraries would be available to her if she came to New Haven.[49] This assurance, implying that the local scientific community regarded her with the respect she surely felt she deserved, was of great importance in her decision. She was quick to grasp the opportunity to further science education, particularly for women, and gave her acceptance on 17 February 1927. Immediately she set about renewing her acquaintance with biology teaching in the United States by consulting her friends at Bryn Mawr, Columbia, and Mount Holyoke, as well as at Yale.

Back home in Germany, at an age (she would be sixty-four in October) when many people look toward retirement, Boveri went about preparations for teaching with the exhilaration due any prospect of high adventure: apparently all ennui had been dissipated. During the course of the next six months she spent about three months in Naples visiting the Dohrns and working in the marine laboratory, which generously provided daily plankton samples and other live sea creatures, to refresh her memory. She also stayed a month with the Spemanns in Freiberg, working as a guest in the laboratory at the university. And even though Seehaus at Höfen was as usual filled with guests during the summer, she found time for "doing zoology" there. By the time she attended the International Zoological Congress in Budapest in September 1927, she had a new sense of purpose—and a greater feeling of belonging to the scientific community than at any time since her husband's death.

Plainly she had revived in other ways as well, for in addition to her scientific work she took daily walks; went on a motor trip with her daughter and friends to Ravello and then to Switzerland; and went to the Mozart Festival in Würzburg and the Beethoven Festival in Heidelberg.

Return to Academic Life

Boveri sailed on the *Bremen* in late September. She was favored with the large, comfortable purser's cabin and the company at the captain's table of a Boveri cousin and Samuel R. Detwiler, an American zoologist whom she had met in Spemann's lab. A good sailor, she enjoyed a fine ten-day voyage and was greeted at the dock by one of her former Vassar students with a chauffeured car. After a brief visit with her sister in New Jersey, she was welcomed by Dr. Moseley (provost at Albertus Magnus) and the Morrises the following day in New Haven.

Two days later she began writing her frequent diary-like letters to her daughter; the first are filled with pleasure and anticipation mixed with a modicum of apprehension and homesickness. She found the new laboratory with

> *tables ready, no gas, no elec., no water, no chairs, no microscopes; nothing in my office. . . .* My room lovely and big with a big bath and a piazza with a fine view to the west . . . my meals in the general dining room at a table with 5 very nice teachers, but they are all so polite that when I enter they all rise and wait until I am sitting down. . . . I miss you sometimes very much.

---

[49] Dorothea Rudnick to Aikenhead, 6 Oct. 1966, Albertus Magnus College Library. On the new Catholic women's colleges see Rossiter, *Women Scientists in America: Struggles and Strategies to 1940*, p. 168.

It seems as if I must tell you about everything and hear what you think about it. . . . Homesick—went for a walk instead of dinner.[50]

But this new life was too busy and too interesting for any lasting homesickness. Largely because of Boveri's good business sense and forthright behavior, the laboratory was quickly equipped to her satisfaction. For example, an agent for microscopes who attempted to charge the college more than the agreed-on price was sent to see her:

I had no trouble. I simply said I intended to go to N.Y. to see about the microscopes on Saturday and if they didn't fill their agreement, I'd go over to the Spencer Co. and get them there. He said, "Oh no, don't do that, it must have been a mistake!" So the next day they were in my lab—Dr. Moseley said they were astonished to see how quickly and comfortably I managed it and since then they have unbounded confidence in me!

She was pleased to be invited to the weekly Journal Club at the OZL and asked to report on Spemann's latest work at a later meeting. Any worries about resuming teaching after so many years quickly abated, for she wrote: "I simply live in my work and find myself thinking of it and planning it all the time. I work out my lectures very carefully, reading a good deal in the new German publications and then speaking without notes and get along very well. So far the students are enthusiastic and inclined to work hard. . . . And I am so absorbed that I am quite happy." A friend, she reported, wrote that "I am a stoic. . . . Of course I am not. I must choose and I choose what I love, i.e., the teaching and the work in biology. It is no hardship to let the social side slip."

"Students were invariably fascinated by Madame," a later colleague, Dorothea Rudnick, wrote.

She was a charmer: babies, customs officials, dignitaries, even sophomores, fell helpless victims to her spell. Her peculiar magic was a compound of unexpected elements. She certainly created an atmosphere of peaceable grandmotherliness and had the effortless attentiveness of a born hostess (she always received a class as if they had been invited to tea). Behind the gentle charm lay harder qualities: a sharp perception, a tough and objective mind, and a very, very firm will. Even the lightest minded student was aware of dealing with a Personality. Sparing of words, she was past mistress of the perfectly placed silence and of the one light touch that says more than torrents of explanation.[51]

But life for Boveri was never all work. She and Edna Carter often met at the University Club in New York for dinner and a concert, play, or opera or for an occasional Vassar weekened together with Laura Wylie (Vassar, 1877), a professor of English and friend from the 1890s. Carter soon arranged a dinner party for her to meet the chairman of Vassar's Department of Zoology, Aaron Treadwell—he, too, had shared in the early days at Woods Hole. Boveri also went to meetings of the Foreign Policy Association whenever she could and maintained an active interest in it for the rest of her life.

E. B. Wilson was among the first to welcome her back into the American scientific community, and she greatly valued the relationship that developed with the hospitable Wilson family. These associations, both professional and social, especially with the

[50] Here and in the next few paragraphs, I quote from Marcella Boveri to Margret Boveri, 4, 9, 15, Oct. 1927, 28 Mar. 1929. The italics are my own addition.

[51] Dorothea Rudnick, "Madame Boveri and Professor Boveri," *Albertus Magnus Alumna*, 1967, *4*(1):8–11, 14, on p. 8.

Wilsons and with Ross G. Harrison and his German-born wife, were to be a great pleasure to her over the years.

Within two months of her arrival in the United States, then, Boveri had established the general pattern of the next sixteen academic years. As was certainly essential for one creating a department, she kept alert to developments in the ever-changing field of biology, though she did no further research herself. She went to professional meetings both at home and abroad: the Zoological Congress in Budapest in 1927, the Genetics Congress in Stuttgart in 1935, the Zoological Congress in Freiberg in 1936, meetings of the American Society of Zoologists in Philadelphia in 1928 and 1933 and in Atlantic City, New Jersey, in 1936, the Growth Conference at Dartmouth College in 1941, and many meetings of the Connecticut Academy of Science.[52] For years she attended nearly all of the Ph.D. thesis presentations in zoology at Yale as well as the weekly seminars. Here, indeed, was a strong, resolute, late Victorian woman keeping abreast of her field.

The Department Achieved

During her tenure Boveri gathered about her an admirable teaching department.[53] The faculty of Albertus Magnus College included both Dominican sisters and lay teachers; the college was new and its resources limited. The sisters' vow of poverty meant that they received no salaries, but available Dominican Sisters with graduate education in biology were few. The mother house in Ohio sent a young biologist, Sister Mary Urban, to Albertus Magnus as a potential teaching member, and Boveri was asked to encourage her to apply for graduate study in zoology at Yale. She was glad to smooth the way a bit as well: she extended her own recommendation to several of the senior members of the zoology faculty at a dinner party in her rooms and introduced the young nun to them that evening. Sister Mary Urban went to Yale in 1942, completed her Ph.D. in zoology, and taught at Albertus Magnus.

Down the street at the OZL there were women Ph.D. candidates who welcomed part-time positions as teaching assistants—an opportunity not then available to them at the OZL.[54] Throughout her years at Albertus Magnus, Boveri's close association with the Yale department yielded a series of these mutually beneficial teaching relationships: she got qualified assistants, and they gained an inspiring teaching experience. (See Figure 2.)

It was Boveri's good fortune to appoint two highly qualified women who found it not only feasible but interesting to combine research at Yale's OZL with teaching at Albertus Magnus. Grace E. Pickford, a British-born woman from Cambridge University with a Ph.D. in zoology from Yale, joined the department in 1935 and taught there until 1946. In anticipation of Boveri's retirement, which occurred in 1943, Dorothea Rudnick, with a Ph.D. from the University of Chicago, was appointed in 1940. She taught at Albertus Magnus for the next thirty-seven years, retiring as professor emerita in 1977. In 1945 John S. Nicholas, then chairman of zoology at Yale, wrote, to "Madame": "Albertus Magnus

---

[52] At the 1936 Atlantic City meeting Everett Just gave a lecture that Boveri called "the best talk of the entire conference. As a kind of 'barometer,' a 'reflecting pool for others' minds,' she could be trusted to give a good judgment on matters of that sort [i.e., controversial matters]": Manning, *Black Apollo of Science* (cit. n. 11), p. 293.

[53] She also introduced instruction in physics by persuading Edna Carter to teach an introductory course for two years after she retired from the Vassar faculty.

[54] In 1942, doubtless because of the decrease in available young men, discrimination ended, and I was permitted a teaching assistantship in Yale's Department of Zoology.

Figure 2. *Madame Boveri at work in the laboratory at Albertus Magnus College. She was customarily addressed as "Madame Boveri" (pronounced BoVAYree), a temptation for playful students to convert to "Madame BOvary." Fortunately, she was amused. (Photograph taken by Edna Carter in the 1930s; reproduced courtesy of the Albertus Magnus College Library.)*

has been exceptionally fortunate in having you as its initial leader in biology for you brought to this work your vast background of international acquaintance, both personal and scientific. Your successors have your attitude and have widely known scientific reputations."[55]

[55] John S. Nicholas to Marcella Boveri, 11 Oct. 1945, Albertus Magnus College Library, RG 22, Bov. III, Box IV. Both Grace Pickford and Dorothea Rudnick earned widespread recognition for their research, especially work on the endocrinology of fishes by Pickford and on experimental embryology of the chick and developmental genetics by Rudnick. In addition to her research position at the Bingham Oceanographic Laboratory at Yale, Pickford held a teaching professorship in zoology at Yale, retiring in 1970. After that she continued an active research program as scientist in residence at Hiram College, generously supported by grant aid. Rudnick was a resident guest at Yale and the honored recipient of Guggenheim and Helen Hadley Hall fellowships as well as special fellowships from the U.S. Department of Public Health.

Boveri not only inspired her students' interest in biology but took a keen interest in their further development as well. Whether a young woman needed encouragement toward graduate study or help in seeking a job, Boveri gave wise counsel and active support suited to her ability and undergraduate experience. She tried, with considerable success, to place her "girls" in positions where they could continue to grow. Of the four students I have known, all roughly the same age, one went on for the Ph.D., marriage, and an eventual professorship; one became so competent in experimental work as a research assistant that she was coauthor on one paper and published another alone before she married; and the two others, after a few years' experience as research assistants, moved to successful and rewarding careers in public school biology teaching, one of them while married.

Visiting Lecturers

Because of her friendship with many of the leading zoologists in Germany as well as her interest in their research, Boveri played a significant role in opening opportunities for them as visiting lecturers in America. Their visits helped to underline the international character of science, especially to the students in their audiences. She heard Karl von Frisch give a lecture in the summer of 1928 in Munich that impressed her greatly. His experiments on the senses and behavior of bees, so ingenious and simple, using materials no more complicated than cardboard, paint, glass dishes, and honey or sugar water, were models, particularly to present to students.[56] Consequently, when the von Frisches visited her at Höfen, she suggested that he consider a trip to America. To advance this idea she also went to Harrison to express the hope that Yale might sponsor him for a lecture series.

It took nearly two years, but Boveri persisted: in March 1930 von Frisch embarked on an extensive tour in America, beginning with three Yale lectures and ending with a lecture at Vassar in May.[57] E. B. Wilson remarked that von Frisch's talk on bees was "the finest educational lecture he had ever heard"—high praise indeed from such a seasoned scientist![58] Boveri's thoughtfulness in providing an informal trial lecture and film-showing at Albertus Magnus, which gave von Frisch a chance to try out his English before giving the Yale lectures, also offered a special benefit to her own students. She was a most competent organizer and thoughtful hostess.

To what extent Boveri may have been involved in getting Hans Spemann, a future Nobel laureate, to come to the United States on a lecture tour I do not know, but when he and his wife arrived in October 1931 she greeted them with a full weekend of entertainment before the tour began. The Spemanns went on to Rochester, Cornell, Chicago, and Harvard before returning to New Haven, where they were guests of the Harrisons while giving several lectures at Yale. Boveri's students not only heard the lectures but also met with the Spemanns at Albertus Magnus. As a colleague commented, "It was not unusual to find famous personages sitting comfortably about in the laboratory eating buns and talking with the students."[59]

---

[56] Von Frisch may also have reported on his early studies of the "language" of bees in these lectures. A popular account of his work appears in Karl von Frisch, *Bees: Their Vision, Chemical Senses, and Language* (Ithaca, N.Y.: Cornell Univ. Press, 1950).

[57] Boveri's persistence was legendary. Harrison once excused himself for neglecting something "on the plea that there is no getting away from doing anything that Mrs. B. has made up her mind to have done and that had taken up all his spare time": Marcella Boveri to Margret Boveri, undated (quoting Harrison).

[58] Wilson's remark is reported in Marcella Boveri to Margret Boveri, 1 May 1930.

[59] Rudnick, "Madame Boveri and Professor Boveri" (cit. n. 51), p. 8.

Boveri thoroughly enjoyed the Spemann visit. With evident glee she wrote to Margret: "I sat between Spemann and Harrison and had a very good time at the Lawn Club dinner ... [and again, at Morris's Newtown farm after lunch].... a little walk to get the view of a distant river and the gorgeous autumn foliage ... all invited to dine at the Harrisons, very gemütlich. A very gay time for such a hermit as your mother."[60]

Keeping the Record Straight

Every summer until the start of World War II Boveri went back to Germany. There were still some tasks left from those depressing times in the 1920s, but one project was of her own choosing: to assemble Boveri's correspondence and publish it. As she visited her European friends and talked with American biologists—including Harrison and Wilson, both of whom had actively corresponded with Boveri—she located and retrieved many letters that she stored in a leather trunk in her Würzburg apartment. It would surely have been a valuable source for historians, but the long task of collecting and editing had not been completed before everything in her apartment was destroyed in the firebombing of Würzburg in 1945. As Margret Boveri later wrote: "It was providential that it [her mother's final illness] kept her from returning to Germany and seeing what had become of the beautiful city in which she had spent the happiest years of her life."[61]

One self-imposed task during Boveri's years at Albertus Magnus was completing the translation of *The Origin of Malignant Tumors.* She was scrupulous in striving for faithful communication. As M. M. Metcalf wrote in his foreword, "The translator has sought to give the exact meaning of the original in preference to following manners of expression in some instances more customary in English."[62] As her former students still remember well: She wanted things *right!*

In a lecture to OZL's Journal Club, she pointed out that the dispermy experiments had suggested to Boveri the possibility of a causal connection between abnormal nuclear division and the growth of tumors. Over the succeeding years he attempted to test this hypothesis but was unable to find sufficient evidence to substantiate it. Nonetheless, he remained convinced; in the book he provided an account of his train of thought, concluding as follows: "If, then, I may ... send a wish along with this book, it is this: that my arguments may induce active investigators of the tumor problem to consider their work from the standpoint presented here and to ask in their future studies, whether what they find, contradicts or supports the theory I have here set forth."[63]

For whatever reason, perhaps the wartime atmosphere, Boveri's plea was apparently neglected both in Europe and in America. The translation was meant to make his principal theoretical work more widely known, though again it had no immediate effect on other investigators. Thanks to Marcella Boveri, however, his work was available for later rediscovery and appreciation—after his theory had been validated by sophisticated cytogenetic studies and the discovery of oncogenes. Ruth Sager drew the following parallel: "Boveri's

---

[60] Marcella Boveri to Margret Boveri, 18 Oct. 1931. Spemann came to Yale again in 1933 to give the Silliman Lectures, which were later published: Hans Spemann, *Embryonic Development and Induction* (New Haven, Conn.: Yale Univ. Press, 1938). A Yale Univ. Press page proof of the book is in Albertus Magnus College Library, RG 22, Bov. II.

[61] Margret Boveri to Sr. Charles, 4 Mar. 1959, 19 Nov. 1950, Albertus Magnus College Library, RG 22, Bov. VII.

[62] M. M. Metcalf, "Foreword," in Theodor Boveri, *Origin of Malignant Tumors,* trans. Marcella Boveri (cit. n. 1), pp. vii–ix, on pp. vii–viii.

[63] Theodor Boveri, *Origin of Malignant Tumors,* trans. Marcella Boveri, p. 119.

contribution to clear thinking about cancer ranks nearly with Mendel's contribution to clear thinking about genes." Eric Davidson summarizes: "Modern molecular developmental biology is firmly in the Boveri-Wilson tradition. . . . Molecular biologists are now rapidly providing the conceptual armature erected by Boveri with a body of detailed knowledge."[64]

Another opportunity to keep the historical record straight arose when Otto Glasser, a German-born and -educated biophysicist at the Cleveland Clinic Foundation who was writing a biography of Röntgen, asked for her help. Boveri wrote to her daughter:

> I made up my mind at once to come here [Cleveland] and see just what he wanted and to see what the book was like. I'd do more than that for such a good friend of ours as Uncle Röntgen was. . . . Of course I have always felt that no one knew Röntgen in his later years as we do and that it was a great pity to keep our knowledge all to ourselves; . . . I think that I have given Dr. Glasser a little conception of Röntgen's personality and of our vacations together—at least he says that he has an entirely different idea of that side now.

This account was her first gentle prod toward the fine chapter that Margret wrote, "Personal Reminiscences of W. C. Röntgen."[65]

The Threat and the Second War

The last years of Boveri's tenure at Albertus Magnus, when she was in her seventies, were the troubled years of the triumph of Nazism in Germany and the beginning of World War II. Well before this time, beginning in 1930, her letters revealed her concern about fascism in Germany. She repeatedly asked her daughter—hopefully, I think—whether the American newspaper accounts might be exaggerating its importance. She thought Hitler a "madman," and it was inconceivable to her that the German people could find him credible. (Margret had been awarded her doctorate in history and English philology from the School of Politics in Berlin in 1932.[66] She published her dissertation on "Sir Edward Grey und das Foreign Office" the following year and was soon well launched on a successful career in journalism; her mother thus had good reason to think her an exceptionally well-informed source.) Margret's replies were not reassuring.

On her last summer trip to Europe, in 1939, Boveri was in Stockholm visiting her daughter, a correspondent for the *Frankfurter Zeitung,* when the war broke out. On 22 September she sailed on a Swedish-American ship, the *Gripsholm,* that finally arrived safely in New York after fourteen days on a northern route past the Faroes, Iceland, and across the North Atlantic, with a "mixture of nationalities on board, many with tales to tell."[67]

---

[64] Ruth Sager, "Genomic Rearrangements and the Origin of Cancer: Rediscovering Boveri: The Problem of Causality," in *Chromosome Mutation and Neoplasia,* ed. James German (New York: Alan Liss, 1983), pp. 333–346 (quoted in McKusick, "Boveri and the Chromosome Theory of Cancer" [cit. n. 1], p. 439); and Eric H. Davidson, "Genome Function in Sea-Urchin Embryology: Fundamental Insights of Th. Boveri Reflected in Recent Molecular Discoveries," in *A History of Embryology,* ed. T. J. Horder, J. B. Witkovski, and C. C. Wylie (New York: Cambridge Univ. Press, 1986), pp. 397–406, on p. 403. McKusick credits Marcella Boveri for bringing her husband's work to wider attention: McKusick, "Boveri and the Chromosome Theory of Cancer."

[65] Marcella Boveri to Margret Boveri, 16 Apr. 1930; and Margret Boveri, "Personal Reminiscences of W. C. Röntgen" (cit. n. 28).

[66] Anyone familiar with Margret's autobiography knows that her life was complicated in 1928–1931 by a love affair with Everett Just—a fact, fortunately, not revealed to her mother. To my mind, the account in Ch. 5 of Kenneth Manning's *Black Apollo of Science* (cit. n. 11) emphasizes Margret's indiscretion—precipitated by the freedom from her mother's domination that began in 1927—distastefully and in disproportion to its importance.

[67] Marcella Boveri to Margret Boveri, 5 Oct. 1939.

Boveri's distress about another war in Europe, concern for her daughter's safety, and foreboding that the United States would enter the war was intense, especially in October 1940, when Margret's work brought her to New York.[68] What would be her fate if the United States joined with Britain and France? This was the disturbing personal question that worried them both while Margret was busy on assignments in New York and Washington.

Two days after the attack on Pearl Harbor on 7 December 1941, Margret was apprehended, taken to Ellis Island, and later interned at White Sulphur Springs, Virginia, as a citizen of a hostile state. She was deported in May 1942, landing at Lisbon. Beyond a telegram announcing her safe arrival there, Boveri had very little news of her daughter until after the war ended. Margret continued as correspondent for American and English affairs in Lisbon until August 1943, when Hitler prohibited the further publication of the *Frankfurter Zeitung*—a newspaper too liberal for him to tolerate. After a short interval in Madrid, Margret returned to Berlin in 1944.

In 1943, at the age of eighty, Boveri retired from her strong leadership at Albertus Magnus. Her health was failing, and the stresses of the war probably hastened the process: three years later she was taken to Rest Hill, at the Convent of the Good Shepherd in Wickatunk, New Jersey. There she was given loving care by the sisters until her death on 24 October 1950. She was buried in the cemetery of the Sisters of the Good Shepherd, where her sister had preceded her.

During those years at Wickatunk, Boveri, though physically inactive, remained interested in all that her many visitors could tell her. Sadly, Margret never could get a visa—a source of painful regret for her and a great sorrow for her mother—but her frequent long letters provided a store of pleasure. Boveri was well justified in her pride in her daughter's developing talent as a writer, for Margret was to become, according to the obituary published in the *Frankfurter Allgemeine Zeitung* on 8 July 1975, "Die grosse Dame des politischen Journalismus."[69]

PERSPECTIVE: MARCELLA BOVERI AND OTHERS

The key to Marcella O'Grady Boveri's early development, I believe, is her positive response to intellectual challenge in a supportive cultural environment, coupled with a deep-seated desire to be of service that was surely fostered by her Catholic upbringing. The first definite indication we have of her future path is her keen interest in the "new biology" introduced to her as an undergraduate at MIT by Sedgwick and Wilson. It was the challenge of problem solving that motivated her toward experimental science when the opportunity arose, first at Bryn Mawr and Woods Hole. And her teachers, leaders in the field, were her mentors. In other words, she was lucky: few women of her day had so many advantages and opportunities.[70]

---

[68] Margret had traveled by the Trans-Siberian Railroad, crossed to the California coast, and toured extensively in the States on her way.

[69] *Frankfurter Allgemeine Zeitung,* 8 July 1975, p. 15.

[70] Twenty years earlier, with none of Marcella O'Grady Boveri's advantages, Emily A. Nunn (1843–1927) tried to struggle her way to professionalism independently. Linda Tucker and Christiane Groeben describe her unfocused peripatetic study and investigations in both Europe and America. She had potential but clearly needed mentoring that was provided by none of the distinguished biologists she met along the way, presumably because of her unconventional and abrasive personality. Even after her marriage, at forty, when she expected to collaborate with her husband, Charles Otis Whitman (1842–1910), her potential was never realized. She turned to housewifery and motherhood. See Linda Tucker and Christiane Groeben, " 'My Life Is a Thing of the Past': Whitmans in Zoology and in Marriage," in *Creative Couples in the Sciences,* ed. Pycior *et al.* (cit. n. 3), pp. 196–206.

Where does Boveri fit among the diverse life patterns other researchers have discerned in their work on early American women scientists? As I noted earlier in this essay, her experience as a collaborator differs from that of other women, often in important ways. Except for the fact of her marriage, she may fit best where she started: with the single women who taught in the best women's colleges in the nineteenth century. The experiences of Boveri and Cornelia Clapp (1849–1934) include some surprising similarities, in spite of the difference in their ages.[71]

They were both innovators; they had some mentors in common; and their research interests were similar. Clapp introduced the laboratory method in zoology at Mount Holyoke as early as 1874 and designed a course in embryology. Boveri created a totally new curriculum in biology at Vassar and introduced experimental work in the laboratory. In the early 1880s Clapp worked briefly under Sedgwick's guidance at MIT (Boveri may have met her there) and with Wilson when he was at Williams. At Woods Hole in 1888 both women pursued research projects under Whitman's supervision; he was Clapp's Ph.D. advisor. Both of these women were gifted teachers and mentors, fostering the spirit of research in their students. Clapp taught at Mount Holyoke until her retirement in 1916. Boveri taught at Vassar for seven years in the 1890s and spent sixteen years at the newly formed Albertus Magnus College, where she initiated the science program.

Both women became well known internationally: Clapp through her association with the Woods Hole community, Boveri through the social access to other biologists brought about by her collaboration with a husband who was a leader in their field of research; such advantages were less available to single women. Clapp spent nearly every summer at Woods Hole from 1888 until her death, serving in a variety of capacities: as investigator, teacher of embryology, librarian, and longtime trustee. She also went to the Naples station and, I believe, was there when the Boveris were doing their famous dispermy experiments. Both Cornelia Clapp and Marcella Boveri, women with very different personalities and outwardly different lives, were influential figures.

At the start of her career Boveri challenged tradition to work toward professional acceptance in science. By the end of it she not only had won the respect of the international biological community but, through her teaching and mentoring, had stimulated many another young woman to meet the challenge of her own times. She was a quiet feminist: her wish was for women to be free to develop intellectually according to their abilities and interests, not barred from professional work in fields of learning traditionally dominated by men.

Why has Boveri's career been so hidden from view? My best guess is that because her life was sharply broken into three segments—she worked in three different contexts in three different time periods and places—few saw her career as a whole. This unique pattern hindered but has not entirely prevented her recognition.

---

[71] On Clapp see Maienschein, *Defining Biology* (cit. n. 10), pp. 44–45, 179; and *Mount Holyoke Alumnae Quarterly*, 1935, *19*(1):1–9.

Index